Philosophy and Medicine

Volume 152

The Philosophy and Medicine series is dedicated to publishing monographs and collections of essays that contribute importantly to scholarship in bioethics and the philosophy of medicine. The series addresses the full scope of issues in bioethics and philosophy of medicine, from euthanasia to justice and solidarity in health care, and from the concept of disease to the phenomenology of illness. The Philosophy and Medicine series places the scholarship of bioethics within studies of basic problems in the epistemology, ethics, and metaphysics of medicine. The series seeks to publish the best of philosophical work from around the world and from all philosophical traditions directed to health care and the biomedical sciences. Since its appearance in 1975, the series has created an intellectual and scholarly focal point that frames the field of the philosophy of medicine and bioethics. From its inception, the series has recognized the breadth of philosophical concerns made salient by the biomedical sciences and the health care professions. With over one hundred and twenty five volumes in print, no other series offers as substantial and significant a resource for philosophical scholarship regarding issues raised by medicine and the biomedical sciences.

Lucien Karhausen

Analytical Philosophy of Medicine

Scientific Philosophy and Philosophy of Medicine

Springer

Lucien Karhausen
Retired
European Commission
Brussels, Belgium

ISSN 0376-7418 ISSN 2215-0080 (electronic)
Philosophy and Medicine
ISBN 978-3-031-81256-9 ISBN 978-3-031-81257-6 (eBook)
https://doi.org/10.1007/978-3-031-81257-6

This Springer imprint is published by the registered company Springer Nature Switzerland AG
The registered company address is: Gewerbestrasse 11, 6330 Cham, Switzerland

TO
Micheline Iannuzzelli
1931–2019

Introduction

> Medicine is a curious discipline in some respects, because it is very nearly the sole professional specialty that claims the credentials of a science and renders its judgement chiefly in terms of prescriptive norms…
> Margolis[1]

Since ancient times, the aim of medical practice has been to reduce human suffering, whether physical or mental, due to biological causes. The aim has also been to prevent, cure or favourably modify the course of disease.

Medicine is one of the last human activities to emerge from ignorance, but it was very early, in Europe with Hippocrates, that it found its identity and defined its aims and intentions.

Yet, Lewis Thomas (1913–1993) wrote: "The reason why doctors have little interest in the history of medicine is that it is so incredibly disastrous, with its bloodletting, purging, and misunderstandings based on theories of the human body inherited from Antiquity." Nietzsche wrote: "Convictions are greater enemies of truth than lies." In order to progress, modern medicine has had to overturn the old convictions about illness and health, which were fortified by the support of systems of thought and the authority and power of institutions.

Medicine is a prudential activity, which was born out of a need to identify, classify, understand, explain, relieve, improve, cure or prevent physical or mental suffering, and over the course of its history, it has extended the concept of suffering to include interpersonal relationships. In the medical sense, suffering refers to all biological negativities, present or potential, such as pain, respiratory failure, severe pruritus, high fever, convulsions, vertigo, severe anxiety or obsessions, disability, etc., not forgetting processes that are still clinically silent or not but potentially harmful. It was often mistaken for an art, before becoming both a science and a technology. It is therefore important not only to gain a better understanding of the natural history of diseases but also to develop a methodology for measuring the effectiveness of therapeutic interventions.

[1] Margolis (1975, 96).

"Medicine," wrote James Le Fanu, "is one of the few spheres of human activity whose aims are unambiguously altruistic."

The present volume is *about philosophy of medicine, as a subset philosophy of science.* It integrates philosophy of medicine as a chapter of philosophy of science. It is neither about bioethics or applied ethics, nor about the history of medicine, but it is *comprehensive* and encompasses the whole field of clinical and community medicine including psychiatry.[2] Arthur Caplan claimed that philosophy of medicine as a subdiscipline of philosophy of science did not exist despite a great deal of literature, teaching and professional activity carried out explicitly in the name of "philosophy of medicine".[3] He gave the following stipulative definition: "The philosophy of medicine is the study of epistemological, metaphysical and methodological dimensions of medicine; therapeutic and experimental; diagnostic, therapeutic, and palliative." Broadbent, in his critique of what he calls the "natural turn" in the philosophical literature on health and disease, underscored the gap that separates philosophy of medicine and the grand tradition of analytic philosophy.[4] Philosophy of medicine, according to Jacob Stegenga, is a relatively recent field of study within philosophy of science.[5] "Philosophy of science," wrote Dr R. S. Downie,[6] "is a flourishing discipline and so also is moral philosophy. Medicine, which combines elements of both, has not had so much attention. This is partly a result of the attention to medical ethics which has obscured the need for a philosophical foundation for medicine."

According to the Oxford Dictionary of the English Language, medicine is the science or practice of diagnosing, treating and preventing disease, a definition similar to that in the Larousse dictionary. Similarly, most works or treatises on the philosophy of medicine and medical science focus on disease. Yet diseases are transitory conventions and represent only a fraction of the reasons and conditions that lead people to consult a doctor: what remains are what we call the iceberg of explained, unexplained or unexplainable symptoms, or MES (medically unexplained symptoms). A growing proportion of patients who visit a general practitioner complain of symptoms that are not medically significant, i.e., they are not illnesses. The world of medicine is like the starry sky: there are an unlimited number of medical problems or stars, and a finite number of constellations or diseases.

What distinguishes medicine from biology is *what is pathological*, i.e., all forms of physical or mental suffering. This notion of biological negativities is not scientifically definable, because it cannot be defined by its cognitive or descriptive nature. It comes under that way of describing the world of which Immanuel Kant spoke, that which concerns our interests, what is biologically good or bad, and which helps to guide our actions.

---

[2] Broadbent (2019, xix).

[3] Caplan (1992).

[4] Broadbent (2019, xvii).

[5] Stegenga (2018b, 1).

[6] Downie RS, "personal communication", 2007.

Furthermore, medicine is confronted with what is abnormal. And what is abnormal is usually gradual. It has to decide on the continuum of what is abnormal, what is pathological, a binary notion, because the medical discipline has to judge when to intervene.

David Hume's guillotine is that one can never derive an *ought* from an *is*. One can never derive what is abnormal or pathological from purely biological or physiological data. The distinction between normal and abnormal, between normal and pathological, is therefore not a scientific one. This distinction is the very foundation of medicine. And these foundations are not descriptive, because they are the expression of medical conventions and attitudes, even if medical language is inevitably bastardised since clinical practice makes it both descriptive and normative.

Philosophy of biology has won its spurs over the years ever since the biologist JH. Woodger and the philosopher Morton Beckner published major works on the philosophy of biology in the 1950s. However, philosophy of biology became a mainstream part of philosophy of science with the publication of David Hull book,[7] after which the field expanded, and it became one of the most exciting new areas in the field of philosophy of science. Philosophy of social sciences followed the same trend.

Contrariwise, philosophy of medicine has lagged behind. Even though in the history of philosophy, some of the really great thinkers had things of very great importance to say, an important part of medicine was neglected. All in all, philosophy of medicine is a very recent discipline that really began in the 1970s. Initially, it was mainly concerned with medical ethics, but it progressively got interested in the broad philosophical issues pertaining to medicine. Often, it also consists of literary reflection on medicine, what Pellegrino calls "medical philosophy", or in what Caplan calls "philosophy and medicine", a strategy of parcelled philosophical approach of medicine, that leads to various speculations as well as to a diversity of conflicting opinions.

In the past 10 or 15 years, philosophers moved to understand and work on the conceptual analysis of problems that are raised by medical science, medical practice and public health. In this way, philosophy of medicine progressively emerged as a new vigorous area, although it might be argued that what is lacking is a canon of unified consistent issues against the backdrop of philosophy of science.[8]

Although medical care is a profession that has for a long time been wedded to irrational and unjustifiable assumptions, it can now be taken for granted that medicine is a science, and as such, it has cognitive and instrumentalist dimensions, since it is to be regarded as a cognitive approach as well as an instrument for producing new techniques for controlling events.

"Im Amfang war die Tat" (*In the beginning was the deed/act.*) wrote Goethe (Faust I, Scene 3). In his *On Certainty* (On Certainty 1969: 402), Wittgenstein quotes Goethe: "Im Amfang war die Tat", which glows with a kind of primacy of

[7] Hull (1974).

[8] Marcum (2017), Solomon (2015), Gifford (2011), Schramme and Edwards (2017), Lee (2012).

practice, pragmatic nature of epistemic concepts, and perhaps with pragmatism. The deed, the activity, is primary in medical care. Goethe's line can indeed help us to understand this Wittgensteinian theme, by reminding us in particular that the "primacy of practice" which results from the gap between the knowledge and capabilities of trained physicians.

Medicine is being defined by what physicians do, and it includes clinical medicine, public health, epidemiology and biomedical research. It may also be defined by the diversity of topics included in standard medical textbooks.[9] Admittedly, medicine starts not as a philosophical issue, but at the clinical level, when health professionals attempt to help people who are suffering. It is probable that many disagreements in the literature of medical philosophy spring from the fact that some distinctions have been introduced and may seem legitimate for philosophers but could be bypassed if one looks at medical practice.[10] Medicine tells how physicians intuitively organise nature, not how nature is organised.

This volume is grounded on a first-order standpoint, namely the clinical gaze, and it strives to stay close to clinical medicine or epidemiology: it bestows an *epistemological bottom-up account* that arises from the clinical situation, the epidemiologic and the resulting public health account. It is not a review of the literature, and it is not intended to frame the debates, or to analyse and compare the various and substantial number of viewpoints. Some authors' accounts may form the basis of my presentation when they fit into the development of my philosophical view of medicine. Philosophic analysis should start from a set of initial platitudes and work up to some higher conceptual level.

The philosophical viewpoint of medicine should be *parsimonious* in that it should avoid any unnecessary complexity and be as simple as possible, avoiding venturing into far-fetched philosophical speculations. This is directly in line with what P.F. Strawson—the late English philosopher and professor of philosophy at Oxford University—termed "*descriptive metaphysics*",[11] which should bring into clear view the conceptual structures and interrelationships that constitute the central core of medicine. It falls a long way short of an inquiry into the ultimate nature of reality itself that is the target of orthodox metaphysics. Medical terms are tools, like a stethoscope, instruments to be used in the daily occasions of the health-care professions: our definition of the term "disease" should match the use which nurses, physicians and community physicians are making of it, which is also the way they have learned to use it. We should avoid, in medical philosophy, straying from actual linguistic practice. If our metaphysics does not fit it, so much the worse for it.

Philosophy of medicine is a set of *coherent beliefs* that fit together in a logical way so that everything in the set is consistent.[12]

[9] Porter et al. (2018), Goldman and Schafer (2019), Ralston (2018), Jameson et al. (2018).

[10] Karhausen (2023).

[11] Strawson (1959, 9).

[12] Bonjour (1985, 230).

The set of statements constituting this philosophical representation of medicine must be capable of being simultaneous true and *deductively closed:* this means that any statement, which is logically entailed by the theory belongs to the theory.

Finally, the interchange between medicine and philosophy leaves the medical topology unaltered, since philosophy makes nothing happen.

Medicine begins with its first question: "what is normal and what is abnormal?" This question is so specific to medicine that it might be considered as one of its criteria: *medicine is the human activity, which begins by a linguistic act that identifies the norms of health and ill-health.* These bivalent medical norms are not items in the basic inventory of the world. It follows that medicine is pervaded with vague dichotomous concepts such as semantic vs pragmatic meaning, descriptive vs normative discourse, function, and dysfunction, abnormal and pathological, needs and wants, causation and explanation, clinical vs. community-oriented care, physical vs. psychiatric diseases, mental illness vs. deviancy, organic diseases vs. functional disorders, biological faults vs. psychosocial distress, curing vs. healing and suchlike. These duple terms represent a complete range of features that are laid down in a spectrum conventionally categorised in binary order. Medical thinking has two dimensions intrinsically interweaved, namely a constant amalgam or admixture of biological and normative as well as of scientific and pragmatic aspects. This essential hybrid nature of the grammar of medicine explains the endless controversies wondering whether medicine should be naturalistic or normativist, biological or value-laden, realist or instrumental, reductionist or holistic, eliminativist or pragmatic, phenomenological or analytic.

Imprecision is common in medical language. Philosophy attempts through conceptual analysis, to define concepts and to use terms without ambiguity. Yet, these attempts might be confusing when imprecision in medically unavoidable, which leads to endless debates about the nature of diseases, of risk factors, of functions or of causality.

Medicine is thus understood as a theoretical science and an instrumental discipline, namely a body of knowledge and an activity concerned with a limited number of prudential interests, which pertain, minimally, to the preservation of life and the ability to use our bodies and minds as effective instruments. It is a blend of knowing that and knowing how, made of an agglomeration of loosely connected models, inquiries, and techniques, which coalesce through some common conceptual framework into an open-ended task: it is *forma mentis* rather than a system or some batch of matters of interest, labelled as belonging to various specialties or as an activity recognisable in virtue of some unique ontology or some specific method. Physicians are usually not interested in philosophical issues or in abstract generalities.[13] Physicians are doers and, at least in the typical case, have an extremely pragmatic attitude.

The question "What is medicine?" should be reformulated as "What is medical?" where "the medical" means a criterion for making some kinds of decision. Aesthetics

[13] Cassell (2004, 4–6).

finds such a criterion in the distinction between beauty and ugliness, and morality distinguishes between good and bad. The specific medical distinction to which medical motives or decisions can be reduced are between normal and pathological. One could paraphrase Ludwig Wittgenstein in saying that medicine is not a theory but an activity.[14]

This book is about occidental medicine, namely a discipline which is based on research, and which goes back to the tradition of the Enlightenment, but which was actually born in 1865.[15] This credo assumes that all medical questions can, in principle, be answered, that the answers can be discovered by careful look at the evidence, that they can be learnt, and taught to other persons and that they must be compatible with one another since nature for sure is a rational entity. It also relies, in the vision of progress, on our capacity to correct, develop and accumulate knowledge.

Meanwhile, medicine has specific priorities. Although constantly expanding, its demesne is narrow like that of engineering or chemistry, the reverse of that of literature or philosophy. Its representation is constrained by very specific and pragmatic imperatives. Physicians' way of knowing the world, the epistemic medical access, results from their training and mission, and is necessary for the completion of their clinical tasks. They dispose, within their own discipline, of a limited number of conceptual instruments that are sufficient for their own purposes. On this score, medical insight is exposed to a permanent temptation of broadening its scope to topics, which lie outside its remit and its language. But when applied beyond its purview, it forces the analysis into a straitjacket, which results in a loss of information as it delineates the limits of medicine. Small wonder, then, that as medical insight wants to save its soul, it endeavours not to wander too far from the centre of its compass card.

Leaving apart the ethical aspects, this book endeavours to uncover the implicit conceptual network, the chief implicit junctures of medicine, should they be found, and their articulation with clinical or community medicine.

I often disagree with Jacob Stegenga's *Medical Nihilism,*[16] but it is the most astoundedly, inexhaustibly stimulating book that questions our assumptions and habits of thinking about numerous medical issues. Alex Broadbent *Philosophy of Medicine*[17] is a very thorough common-sensical philosophical book that stays very near the clinical and epidemiological real world and avoids entering controversial philosophical speculations. Eric J. Cassell illuminating work *The Nature of Suffering*[18] is one of the most exhaustive, and the deepest books on philosophy of medicine and the closest to my account of the logic of medicine. Maël Lemoine's

[14] Margolis (1975, 98).

[15] Wooton (2006, 2008; 3).

[16] Stegenga (2018a).

[17] Broadbent (2019).

[18] Cassell (1982).

Introduction à la philosophie des sciences médicales[19] might well be the best introductory work representing a coherent viewpoint on philosophy of medicine. Mario Bunge's *Medical Philosophy*[20] is the best, the most readable, and the most exhaustive book on philosophy of medicine.

This book took its origins during my training in the Department of Internal Medicine, Memorial Hospital-Sloan-Kettering Institute, Cornell Medical School in New York, and at the Harvard School of Public Health.

I might never have written this book had I not acquired during my medical education a genuine vision of medicine and an intellectual heritage from three late outstanding teachers, Professor Henry Tagnon from Brussels Free University, Dr Mort Lipsett from the National Institute of Health, Bethesda, and Professor Brian MacMahon from the Harvard School of Public Health.

A special debt is owed to Jonathan Glover, professor of philosophy from King's College, London, for his motivation and sustained patient guidance, who provided detailed comments on an earlier draft of this book. I have gained much from my discussions with him, and I am overall grateful to him for his help and support.

I wish to record debts to Dr Frédéric Wittek and Dr RS. Downie who read and commented on some chapters on this book.

I am particularly pleased to acknowledge the encouragement and help I got years ago, from the late Chaim Perelman, Professor of Philosophy at the Free University of Brussels. I also take this opportunity to express my great appreciation for the opportunity to attend Jacques Bouveresse's lectures on Philosophy of Knowledge at the Collège de France.

The Argument

> The sickness of an age is healed through a change in the way of life of people, and the sickness of philosophical problems could only be healed through a changed way of thinking and living, not through a remedy discovered by an individual.
> Wittgenstein

Analytic philosophy covers a wide variety of philosophy associated with the English-speaking world; with its emphasis on language and logic, it is sometimes called "linguistic philosophy" or "linguistic analysis", following Frege, Russell and Wittgenstein, and often opposed to "continental" and "speculative" philosophy, coming from Hegel and Husserl.

Philosophy of medicine is a chapter of philosophy of science. It should not be displaced by sociology as a source of understanding. Medicine relies on scientific investigations as well as successful strategies. Like economic sciences or political sciences, it has a cognitive and an instrumentalist dimension.

[19] Lemoine (2017).

[20] Bunge (2004).

Reflection on the way, medical science influences philosophy shows that both of them are indispensable to understand each other; yet the question is still open whether and to what extent scientific philosophy is a prerequisite for understanding clinical medicine and public health.

A coherent integrated picture emerges from the chapters of this book that follow a logical sequence, starting from logical conceptual foundations, followed by an epistemological, methodological, causal and taxonomic analysis, by the identification of the goals of therapeutic and preventative medical activities, and ending with the moral and the existential dimensions of the therapeutic alliance.

Most philosophers of medicine claim that medicine pertains to health and disease, an assumption that has diverted their attention from a proper analysis of medical logic. People do not complain about diseases but about pain, suffering or unexplained somatic symptoms or signs, which are not necessarily disease manifestations. Moreover, diseases are a mere part of the conditions with which medicine is concerned.

Medicine begins with the need to help and to define physical or mental suffering. The medical convention that qualifies such biological negativities as abnormal or pathological provides a starting point for medicine and gives them logical priority. *Omnis determinatio est negatio*. This is the *first position*. Next, leaving these foundations, medicine, i.e. medical care, clinical medicine and public health, make a silent and implicit volte-face that leads to give epistemic priority to normalcy. Abnormal and pathological features are then being defined as departures from a normal theoretical background of physiology, biochemistry, or anatomy. This is the *second position*.

If, in the first *ontologic position*, abnormalcy is foundational and *logically* primary, the *second, naturalistic position* reverses priorities and affirms the clinical privileged status of normality; being abnormal, is then a counter-instance of being normal, not the reverse; and "normal" is defined (or redefined) as a positive standard of reference; hence, diseases are construed as privative in nature and medical conditions are deviations from some kind of ideal design. This second position ends up *as if* diseases or clinical features were factual in nature and value-free. With this pragmatic turn, physiology, which is essentially counterfactual, is true by default. The reversing of the first position raises endless debates and senseless controversies due to the need to define health, disease, illness, function, malfunction or normality. On this score, the *first position* selected factual physical and mental suffering as conventional value-laden norms, and the *second position* moved from a normative to a clinical naturalistic account.

Once identified and described, abnormal or pathological features are being classified and categorised in *disease entities* with a sizable *residue of unexplained physical symptoms,* as well as *somatoform disorders* and *functional disorders*. Nosology represents an advanced stage of development of medical science, since diseases are not foundational in medicine. Diseases are supposed to refer to discrete patterns of organic or mental, statistically recurring pathological processes or states, undergone by a creature and acknowledged by medical science. Their mechanism is internal to the body—except for mental disorders which may extend beyond the boundaries of the body. They have a temporally spread, a natural history that may be acute or chronic. It is assumed that if no treatment is available, it ought to and may, in principle, be found. Besides, diseases are not natural kinds and, actually, are open-ended, overlapping, shifting, conventional constructs, best defined by listing, namely by ICD-10.

Diseases embody an a priori prudential concern for prevention, remedy, redress or appropriate care. *Prudence* or practical wisdom is one of the basic virtues in medicine, and it canvasses the sequential pattern, which leads from a patient's ailment to a medical intervention. The need for help inherent to medical negativities and disease processes represent the starting point, which sets about the diagnostic and therapeutic process. Medical *needs* are extensional and forestall discounting the future. On the other hand, patients' *wants, wishes or preferences* are subjective and intensional, and provide no foundation for choices of therapeutic or preventative action.

Medical explanations rely either on the inductive-statistical model, or on inference to the best explanation. Medical explanations are contrastive since they point to anomalies contrary to normal expectations: genuine issues that call for medical explanation are deviations, perturbations, anomalies and irregularities. Pathophysiological research is methodologically reductionist, so that, should one forfeit the concept of reduction, clinical properties could perhaps be termed *emergent* or *supervenient*. A property of a complex system is *emergent* in case it is neither predictable nor reducible to the lower level· although it arises out of the properties characterising its simpler constituents. For two sets of properties, A the *super*venient set and B the *sub*venient, A supervenes on B just in case there can be no difference in A without a difference in B.

Diseases haven known or unknown *causes*. Causal relations give power to control events, through preventive or therapeutic interventions. Causality is probabilistic and relies on a tricky counterfactual logical structure. There are neither necessary nor sufficient causes in medicine, but tendencies towards necessity or tendencies towards sufficiency, which may be quantified by using epidemiologic methods. Causes in medicine are only partial, contributive causes. They are contrastive since they make a difference between circumstances in which they are present and normal conditions in which they are absent.

Medical interventions are a matter of need not of demand, and their goal is to prevent, control, alleviate or cure medical ailments, to reduce suffering and to avoid premature death. Most drugs have deleterious side effects and long-term undesirable effects, both known and unknown and they do not target just the intended condition. Yet biomedical research and medical care are increasingly engaged in targeted treatments of cancers or of congenital conditions such as haemophilia, retinal dystrophy or thalassemia. *Evidence-based medicine* and concepts of *effectiveness* and *efficacy* integrate clinical experience and patient values with the best available research information.

If evidence-based medicine is a necessary condition for good medical care, it is not a sufficient condition: *narrative-based medicine* helps health-care providers to understand the patient's narrative and its significance for their care. The clinical tête-à-tête of the clinical relationship is *intrinsic* to its exercise. However, medical care harbours a tension that is revealed by the medical precedence of *needs over wants* alongside the incommensurability between the patient's and the doctor's narratives. But the patient *expectancy* rather than his satisfaction may reach across the gap that divides these two narratives· Three models of the doctor-patient partnership extend from *paternalism to autonomy*: doctor's authority, guidance-cooperation and mutual partnership.

Medical care is confronted with *situations that are a threat to a meaningful life*, inasmuch as diseases and suffering are cries in the night, and death or the approaching death remove all meaning from life. There is a radical distinction between apprehending natural phenomena subject to the ordinary approach of the natural sciences, and the understanding of human beings' states of mind such as purposes, motives or desires. There are two sorts of illness narratives: the medical record and the patient's story. For one thing, the "detached" medical perspective requires a scientific replicable mould. The physician thinks in terms of diagnosis, prognosis and management, and he takes his patient's sickness and complaints as objective phenomena that are publicly observable.

For another, the patient apprehends his condition from his subjective standpoint, and his narrative is part of what Jonathan Glover calls "our inner story". The patient's attitude towards his disease is influenced by hope as much as by fear. For some patients, the why of the problem of suffering begs for a purpose. The experience of disease and suffering conveys existential anguish since it captures the intuition as to why life is worth living.

Furthermore, physicians have to meet various conflicting perspectives too, which may be of medical or of ethical nature. They have a double allegiance, to the constant inrush of science as well as to clinical empathic understanding. Clinical care, then, oscillates between the sacred and the profane. A placebo is a medicine or procedure, which has no therapeutic effect, and which may be used as a control in testing new drugs. *Placebo effect* applies to the outcome of inactive controls in clinical trials. Placebo effects are clinically important, being a component of any effective therapy as well as a major disturbing factor in clinical medicine and in the evaluation of health care: they may be positive or beneficial, absent or null, or else negative or detrimental. When it was not detrimental by its use of emetics, purging or bloodlettings, much standard medicine starting from Hippocrates until the prodigious intellectual achievement of the last hundred years, relied almost entirely on the placebo effect.

Summary

Chapter 1 introduces a kind of shuttling back and forth between a syntactic stance for which abnormal conditions are foundational so that physiology is a default, and a semantic stance, at the clinical level for which physiology is conceptually prior to pathophysiology, and diseases are altered states.

Chapter 2 introduces the primary features from which medical thinking is being logically deduced, namely the biological roots of medicine: suffering, discomfort and harm, which play a pivotal role in the normal/abnormal divide, that is, in the theoretical articulation of all medical concepts. The concept of "disease" does not belong to the starting point of medicine, but it represents a useful construct that helps to codify the brute clinical data. Medicine grew out of the fundamental call for the alleviation of suffering or potential suffering (be it physical or mental) and their consequential role in construing the normal-pathological divide.

Chapter 3 looks at the distinctions that split biological processes into two conventional parts, normal and pathological. Neither of them is a natural kind. Being abnormal is intrinsically bad and admits of degrees. Yet, abnormal, as it were, is real and empirical, while normal is factitious, so that being normal is counterfactual much the same as frictionless planes in physics. Being normal does not affirm anything positive, and it excludes suffering and pain: it is a default concept; but for clinical medicine, it represents a silent background.

Chapter 4 tackles the topic of explanation and its two dimensions: biomedical, i.e. vertical and hierarchical, and epidemiological, i.e. horizontal. Medical research applies a downward-reduction hierarchical biomedical and reductionist model, although it may well turn out to be a mere methodological contingent stance rather than a fundamentalist representation. Reductionism might have to be weakened and replaced either by a supervenience story or emergentism. Finally, Darwinian explanation has limited explanatory power in medicine.

Chapter 5 covers causality and aetiology. It analyses the various components of causation in medicine: manipulability, comparative form, causal roles vs causal capacities, plurality of causes, probabilistic causality and the formal logic of counterfactual dependence. There are neither necessary nor sufficient causes in medicine, but tendencies towards necessity or tendencies towards sufficiency may be quantified by using epidemiologic methods. What we call causes in medicine are only partial, contributive causes. Causation is *epistemic,* and not something in the world since causes are like *inference tickets.*

Chapter 6 analyses the concept of function and how to avoid teleological standpoints. Functions are not natural facts, but they depend on our set of values. Physiologists study functions just like physicists describe the ideal gas laws. Dysfunctions define functions counterfactually. The causal structure of physiological processes is intrinsic to the organism, but functions are never intrinsic but always relative to an observer.

Chapter 7 outlines medicine's prudential objectives as well as distinctions between needs and demand. Medical needs are prudentially minded, and medical care is essentially a prudential activity relying on need for intervention, namely prevention, care or cure. Personal demands or preferences usually do not provide a foundation for proper medical interventions.

Chapter 8 covers the epistemology of medical care, diagnosis and concepts of signs and symptoms.

Chapters 9 and 10 come to grips with diseases, injuries and impairments as well as with mental disorders. An attempt is made to define those conditions, how they are being construed and some of the features, which constitute them, though most of them are provisional conventions. Diseases are clusters of signs and symptoms. An important distinction is being made between manifestational or purely observational and causal disorders, i.e. disorders defined by their causes. Mental disorders differ from so-called physical diseases in the sense that their limits are broader than those of the affected body. Considering a given disease, each patient shares characteristics with many but not all the others. The categories of diseases are neither mutually exclusive nor jointly exhaustive, and there are neither necessary nor sufficient conditions for belonging to those classifications. What's more, what distinguishes physical from psychological medicine is not some ontological difference between body and mind, but that we grasp mental disorders in terms of reasons rather than causes through a dialogue with the patient.

Chapter 10 deals with mental disorders. This chapter attempts to show that mental disorders are illnesses and must be described as behavioural pathologies.

Chapter 11 deals with social deviant behaviour, namely social undesirable behaviour, which is distinct from, although often confounded with, mental disorders.

Chapter 12 covers unexplained symptoms and functional disorders. More than half of the patients observed in primary care or in population-based studies are medically unexplained, so persons manifesting some kind of symptoms, whether they seek treatment or not, may or may not suffer from some identifiable disease. The symptom iceberg represents the visible fraction observed by physicians. Actually, 26–33% of patients seen in a practice of internal medicine betray unexplained medical disorders. If one excludes respiratory infections and skin disorders, the incidence of unexplained conditions rises to more than 50–70%. If the concept of disease dominates our medical thinking, unexplained and functional disorders, by contrast, make up the lion's share of medical care as they make up the bulk of reported and unreported medical complaints and are responsible for a large amount of loss of absenteeism, productivity and medical expenses.

Chapter 13 outlines some of the conceptual and ideological confusions associated with the notion of health. What moves people are suffering, unmanageable disability and disease that is to say unhealth. Hence, health is contrafactual. When expectations are not met, a call for medical help goes up. When expectations are fulfilled, usually nothing is said, so that health consists of having the same disease as your neighbours.

Chapter 14 turns to medical interventions, that is, concepts of preventive, therapeutic and palliative care, effectiveness and efficacy, as well as to the complex question of placebos and nocebos, their role in clinical trials and in clinical medicine, evidence-based medicine, cancer screening and the paradox of health education.

Chapter 15 discusses the caring relationship, autonomy and paternalism, and the three models of acceptance and acceptability in the doctor-patient relationship: first, the traditional model of activity-passivity; next, the therapeutic alliance where the patient gives up certain freedoms in order to gain prudential benefit; and the third one, where decisions are a joint undertaking between physician and patient.

Chapter 16 asks what limits there are to the medical realm, epistemic, ethical, ontological or contextualist. Placebo, interactive kinds, non-natural kinds and comorbidity, merge randomly into one another with no clear border in between and are blurring the boundaries of medicine. In addition, do functional disorders and the question of medical enhancement lie beyond the edges of medicine?

Chapter 17, the last chapter, studies the core role of tragedy inherent to medical care, how suffering raises questions related to the meaning of life as well as to the physician's predicament.

The original version of the book has been revised. A correction to this book can be found at https://doi.org/10.1007/978-3-031-81257-6_18

Contents

Chapter 1
The Logical Roots of Medicine

Medicine is the only profession that labours incessantly to destroy the reason for its own existence.

Sir James Bruce (1838–1922)

Contents

Medicine relies on scientific investigations as well as on successful strategies. Like economic sciences or political sciences, it has a cognitive and an instrumentalist dimension.

For one thing, the physician is a pure spectator, characterized by descriptivist, empirical, and propositional knowledge.[1] The language of the medical scientist is merely fact-stating—semantic descriptivism. But for another, the norm-following stance is at the root of medicine: it does not state facts of nature but expresses a valuational attitude toward a range of situations such as suffering, unmanageable disabilities, diseases, handicaps, premature death, or some biological negativities that call for appropriate medical interventions. This is tantamount to saying that there is an abrupt alternation between two views of medical language, descriptive and normative.

The human sciences have two dimensions, either psychological or sociological, either that of the individual or that of the group. In the same way, medicine develops along two axes, one clinical, the other social or public health. The clinician deals with individual patients and their suffering, the hygienist with diseases in populations. The clinician works with anecdotes, the hygienist with statistics. The clinician lives in the day-to-day reality, the hygienist in conceptualisation. The clinician works within the practical reasoning of medical science, the epidemiologist is its conscience.

---

[1] Toulmin (1982, 237–248).

L. Karhausen, *Analytical Philosophy of Medicine*, Philosophy and Medicine 152,
https://doi.org/10.1007/978-3-031-81257-6_1

The *Arché* of Medicine and Its Logical Genealogy

Arché, according to Aristotle, is the first thing, the principle of which something consists of, and from which it comes to be. How does medicine originate, not historically and not culturally, but conceptually? Where and how does medicine split from biology? What are the premises of medical thinking? What are the conspicuous basic states or attitudes that undergird the medical gaze, that support the logical architecture of medicine and that give rise to medical science, medical practice, and public health? Understanding is rooted in atoms of intelligibility. There must be cognitive states that are more fundamental, and that support the rest of knowledge. Thoreau contended that '*there is a solid bottom everywhere*'[2]: we know we are in pain, not through inference but, immediately, through the pain itself.

Not Diseases

For most, if not all books, texts, or treatise of philosophy of medicine, medical science is about disease. "*The pivotal concept in clinical medicine is disease*".[3] The concept of disease is usually considered to be *central* to medicine.[4] Horacio Fabrega contends that "*Medicine, an institution of society, is defined in terms of its concern for disease*".[5] Fred Gifford adds: "*The concepts of health and diseases appear to be quite fundamental for medicine as we take medicine to have the goals of diagnosis, prevention and cure of diseases or the achievement of health.*"[6] Jeremy R. Simon in a long and detailed essay on Medical Ontology,[7] has tendered the suggestion that; "*In philosophy of medicine, this calls for an understanding of the nature of individual diseases.*" Maël Lemoine states that disease is the object of medicine.[8] For Jacob Stegenga, disease is a foundational question of medicine.[9] Corbellini at the end of his book '*History and theory of health and disease*' writes an appendix on the "*epistemological evolution of medicine*" that covers essentially the concept of disease.[10] From Hippocrates and Galen until the second half of the twentieth century, the term—if not the concept of— 'disease' was what medicine was all about. All in all, for most authors of philosophy of medicine, health and diseases are the first-order standpoint that is up for grabs: they assume that medicine is that which

[2] Thoreau HD, Chapter xviii, § 14.

[3] Thompson and Upshur (2018, 17).

[4] Gert et al. (1997, 175) and Ruse (1977, 137).

[5] Fabrega (1981, 423–522).

[6] Gifford (2011, 2).

[7] Simon (2011).

[8] Lemoine M., 6.

[9] Stegenga (2018b, 21).

[10] Corbellini (2014).

pertains to disease, and this assumption has diverted their attention from a proper analysis of medical logic.

However, is this contention pertinent, or is it a misconception?

People do not complain about diseases but about pain, suffering, unexplained somatic symptoms, or signs, which are not necessarily disease manifestations. Moreover, diseases are frequently transient conjectural constructs that are only a small part of the conditions with which medicine is concerned. Peter Schwarz, professor of medicine at Indiana University, claims that diseases are not interesting or not coherent enough theoretical entities, and that there is no general underlying concept of disease within the biomedical sciences.[11] Diseases do not lie in the core of medicine, any more than constellations lie in the core of astronomy.

In these circumstances, how can health, an elusive concept, and disease, a moving target that has evolved with the advance of medicine, be defining characteristics of medical science or of medical practice? But do we need to appeal to the concept of health to define 'disease'? We have, in one way, health, a vague and nebulous concept that is either defined negatively by excluding, among other things, diseases, or positively, by an infinite, innumerable list of identifying characteristics; in another way, diseases, a primal medical concept that pertains to a very clinically concrete finite list of described medico-biological processes.

On this score, health is not the absence of diseases, because diseases are a fraction, often a small fraction of medical conditions observed in clinical medicine. Several studies have shown that in 50–79% of all patients presenting to a family doctor, no evidence for a specific organic diagnosis could be found.[12] Pointless, unstructured conditions, unidentified or unidentifiable sickness processes, where the patient manifests the sickness role in the absence of diseases, make the largest portion of primary medical care. During the twentieth century patients became more and more prone to anxieties about their health, and readier to consult their doctors. What is called the symptom iceberg instead of the disease iceberg becomes a new and major socio-medical phenomenon.[13]

"*Doctors do not treat diseases, they treat patients*" writes Eric Cassell. Diseases and dysfunctions are highly elaborate and advanced explanatory concepts that cannot be primary, foundational questions in medicine. Although essential in medical thinking and practice, disease, and the medical sense of disease, are not the springboard of medical thinking, whether socially or medically constructed.

Thus, if the concept of 'disease' does not prima facie belong to the basis or the starting point of medicine, it represents a useful construct that helps to codify and clean up the brute clinical data.

What then distinguishes medicine from natural science? How do medically relevant events such as diseases, signs, or symptoms, differ from biological ones? How do we differentiate what pertains to medical science from what falls within the

---

[11] Schwarz (2007).

[12] Stewart et al. (1995; 33).

[13] Feldman and Ford (1994).

scope of biology? The border that separates biology from medicine is not part of the furniture of the earth.

Not surprisingly, the answer is that medicine, before anything else, needs to choose some normative system based on some set of rules and criteria that separate normal from pathological features. Eric Cassell concludes: "*Knowing diseases, in the old-fashioned sense, is not nearly as important as knowing pathophysiology.*"[14]

Then, beyond this, those norms are medical conventions, and are thus either correct or incorrect, being grounded on biological, empirically describable facts. But surely, once adopted, the norms form part of clinical medicine: being then stored in medical texts and of constant use in medical care, bestows them descriptive status, so they become true or false.

To sum it up then, what divides biology from medicine is that the first one is descriptive, and the second one both descriptive and normative.

First Position and Second Position

Medical thinking rests on *the sequence of two obverse basic tacit conventions*, which create the possibility of medical care.

For one thing, it divides biological features into medically good and bad ones, between harm, detriment and suffering and the absence thereof; it decides what is normal and what is pathological and lays out the medical topography along an asymmetrical spectrum in which pathological features have a *logical factual priority*.

For another, a *second* convention reverses the order of priority: standards of normalcy termed health, physiology or anatomy and the need to maintain or restore normalcy take *epistemic* and *pragmatic priority* over abnormalcy and pathological features.

We may now attempt to grasp, even though in metaphor, how these two starting conventions might cohere.

Firstly, the *ontological vantage point* is tacitly implied rather than wilfully stated: nobody ever knowingly construed or carried out the logical path I shall sketch. What follows is not a description of some premeditated and intentional rendering carried out by physicians or philosophers. But plainly, there is no path, but this is, I should conjecture, a mere way of picturing logical steps towards a coherent position that seems forced upon us.

The brute fact of suffering, incapacity, and harmful biological conditions directly affecting bodies or minds are legitimately allowed at the beginning of the story: they are the port of access to medicine. They are bad or aversive, have serious consequences for the agent and they call for help.[15] Having acknowledged the existence

[14] Cassell (2004, 9).

[15] Hofman (2017, 130–134).

of suffering, infirmity, complaints and disabilities, medicine comes into existence with a decision to divide biological features into normal and pathological. If normal and pathological features are factual, the line of demarcation that separates them is not: it is a tacit agreement, so that ascription of these two terms is conventional, normative, and prescriptive, and is thus neither true nor false; this convention is anchored to the intrinsically negative value-loaded side: abnormal features are the prior core of reference and normal traits represent the contrast case.

It follows that, with this syntactic perspective, such bivalent medical norms and the boundary that separates them, though grounded on descriptions, are not items in the basic inventory of the world but are prudential and deontic. It follows that there is no true way—although there might be a correct or incorrect way— to decide where the line of demarcation lays, even though whatever line we choose may have momentous consequences.

This demarcation is so specific to medicine that it might be considered as one of its criteria: medicine, the human activity, which is concerned with separating normal from pathological. So, if we took away or deducted every single abnormal feature or pathological processes from the biological reality, we would be left with a remain, namely a default position called '*normalcy*', usually christened "*health*". Physiology, anatomy, or biochemistry at this point are true by default and are grounded in the absence of pathological features. This is the starting point of the bottom-up ontology of medicine, what Broadbent calls the metaphysical stance.[16]

Call this the *first position.*

Secondly, this first position then dissolves in clinical medicine. The previous line of argument performs a sharp and unanticipated dizzying volte-face, moving from cognition to action, from ontology to pragmatics. The medical rationale is thus suddenly swinging round and, from then on, as though to forget it has done so, it takes those two *normative* categories *as if they were scientific descriptive epistemic truths*: the terms 'normal' and 'abnormal' are no more conventional, but are now taken as naturalistic, and so true or false.

If, in the first position, abnormalcy is *logically* primary, *the second position* reverses priorities and affirms the clinical privileged status of normalcy, which is then in the driving seat. The picture is now the mirror image of the starting situation. By way of metaphor, the negatives of the photographic film are now printed, and the print is now light where the negative is dark and conversely. In the initial asymmetrical splitting, abnormal or pathological situations were pivotal, but in the language of clinical medicine, priorities are being reversed and normality is now *epistemically* in control of the situation. For most clinicians, being normal, is a counter-instance of being abnormal and 'normal' is defined as a positive standard of reference; hence, diseases are construed as privative in nature and medical conditions are deviations from some kind of ideal design, as departures from the norm.[17] This is a case where the order of logical priority is not the same as the order of dis-

[16] Broadbent (2019, 183–192).

[17] Thagard (2003, 243).

covery. If 'normal' and 'abnormal' are equally meaningless for a biologist, *qua* biologist, normalcy is apprehended as factual by physicians.

This top-down account admits a background of some sort of abstract standard of normality against which abnormal, intrinsically negative features such as diseases, disabilities and dysfunctions will be constructed and defined. It follows that the wedge that has been driven in the clinical findings splits them in two *descriptive features* arising out of the surface of the world. The logical path laid out ends up *as if* they were factual in nature and value-free. By this pragmatic turn, physiology, or anatomy, which are essentially counterfactual, disregard their ontological and conventional origins and hence appear to be empirically true: they describe regularities of 'natural order'. Medical students learn anatomy and physiology before pathology and pathophysiology. It follows that, with this turnabout, normative language turns into indicative language that is true or false. Physiology and the fixed life cycles of living organisms are taken for granted: they are the neutral contrast state. Anything that breaks with this assumed regular course of events, deviations, anomalies, or irregularities are chunks of abnormal states of affairs, which, if harmful, i.e., if pathological, need to be reverted back to, or close to, the normal paradigmatic course of physiology. With this second position, medical thinking becomes naturalistic. This is what Broadbent calls the epistemic stance that leads to the practical stance.[18]

With the first convention of the first position, pathology is made of anxiety, biological negativities, the nearing of death, which are *additive intrinsic negativities*, so that one single affected organ is sufficient for the patient to be sick. Contrariwise, Lars Bergström has proposed a *principle of non-additivity of utility,*[19] that applies with the second convention, when physiological standards of normalcy take precedence[20]: a normal functioning heart does not make up for a failing kidney.

Although medical insight springs, in the first position, from inaugural conventions and value-concepts, in the second position, it draws a veil over those grounds and brackets them off. Ruth Chadwick in an interesting essay on '*Normality as Convention and as Scientific Fact*' underlines the importance of understanding what is normal since it is an important background of the notions of diseases and health.[21]

With these erasures, medical thinking comes out clean as it gives way to a seemingly robust naturalism—a case eloquently described by Boorse— thereby avoiding the cumbersomeness of asking ontological questions about itself. It does not matter whether this picture is fraught with philosophical hurdles. Medical practice is unconcerned with philosophy. Physicians disdain theorists and philosophers since they are concerned with "what works?".[22] And, however philosophically

[18] Broadbent (2019, 192–200).

[19] Bergström (1978, 47).

[20] Knutson (2018).

[21] Chadwick (2017, 23).

[22] Cassell (2004, 5).

meaningful medicine's contorted birth process might be, it is clinically invisible and ought to remain so.

Even then, very few physicians, openly or consciously at least, might subscribe to what precedes. The reasoning is implied rather than openly stated. Stephen Toulmin captures the same intuition in the common law: "*When first promulgated, the force of verdicts and rulings, decisions, and findings, is unambiguously prescriptive ... Yet, those same decisions subsequently become items in the judicial record, to be cited as historical' matters of fact'. As such, they are reported and criticized, quoted, and glossed, using a descriptive idiom*".[23]

In summary, the first *ontological* position consists in forcing, for prudential reasons, the empirical, natural descriptions of the biological world into medical conventional norms by bisecting the biological realm into normal and pathological. Although this first position is located among biological processes, its conventional divide may be *correct or incorrect* but has no objective residency in the biological world. Hence, pathology is foundational, and normality is being defined negatively as contradictory to pathology, i.e., counterfactually.

All the same, this first position is being reversed in clinical and preventive medicine, as normality surfaces in the positive role of a fundamental seemingly *natural* though revisable standard against which diseases and conditions of medical concern will be being defined and identified. With this move, the demarcation between normal and abnormal, or normal and pathological, is being put to cognitive use to describe clinical facts in descriptive sentences that may now be taken to be *true or false*: it bestows descriptive status to the language of pathology. What started as a mere normative convention ends up as a medical quasi-naturalistic representation.

The Trajectory from Suffering to Disease

Pathos is defined by Aristotle: "*The painful and destructive evils are death in its various forms, bodily injuries and afflictions, old age, diseases, starvation.*"[24] "*Suffering which we may define as action that involves destruction or pain, e.g., death*", seems to play in the structure of empirical knowledge, a role similar to that of axioms in deductive reasoning.

First, *suffering is the starting point of medicine* since one knows it directly and not through anything else.

Second, suffering some intrinsic, biological evil,[25] be it physical or mental, is unpleasant, disruptive and *disvaluable.* Identifying among abnormal biological features those that are harmful, i.e., those that are being termed "pathological" is the foundational convention of medical thinking.

---

[23] Toulmin (1972, 93).

[24] Aristotle, Rhetoric, 28138a 6–9.

[25] Cassell (2004, 29–45).

Third, the occurrences of suffering and biological intrinsic negativities, are ligated with the *need for interventions*. *«The obligation to treat suffering*, writes Eric Cassell, *stretches back into antiquity".*[26] Aristotle claimed that the main aspect of *prudence*, or practical wisdom, does not lay in the field of knowledge but rather in acting.

And therewith begins medicine.

Not surprisingly, the following chapters will often sound like a kind of shuttling back and forth between the first position and the second position, namely between the normative and the clinical vantage point.

Summary

1. Medicine does not originate, as usually admitted, with the notion of diseases. Concepts of disease, dysfunction, or health are evolved, sophisticated and advanced constructs.
2. Medicine originates—logically and historically—from actual or potential *suffering*, and from the necessity to relieve or prevent suffering, pain, unmanageable disability, or anxiety.
3. Medicine rests on the sequence of two obverse tacit conventions.
4. In the first ontological position pathological features have a logical priority.
5. The first position dissolves in clinical medicine into the second position which reverses the order and takes epistemic and pragmatic priority of normalcy over abnormalcy and pathological features.
6. Medicine proceeds from suffering to the distinction between normal and pathological, and to the need for intervention. Therewith begins medicine.

[26] Cassell (1982, 2004, 31).

Chapter 2
Intrinsic Negativities: The Ontological Roots of Medicine Suffering, Discomfort and Harm

> *Ask a man why he uses exercise, he will answer, because he desires to keep his health. If you then enquire why he desires health, he will readily reply because sickness is painful. If you push your enquiries further, and desire a reason why he hates pain, it is impossible that he can give any. This is an ultimate end and is never referred to any other object ... Something must be [un]desirable on its own account, and this because of its immediate accord or agreement with human sentiment and affection.*
>
> David Hume (1975, 293)

Contents

This chapter introduces the primary features from which medical thinking is being logically deduced. Medicine grew out of the fundamental call for the alleviation of suffering or of potential suffering, discomfort, and harm.

In one way, social conventions appeal to things people want, to people's desires, to things we want everyone to want. Notions of right and wrong are enmeshed in the complications of local or particular social contexts, and there are values that not all of us recognize and that can lead to genuine disagreements.

In another way, medical conventions are value-loaded, but they seem to be universal together with one of our central ways to align our responses to the world. The normal-pathological divide derives from those biological *negativities* that are intrinsically bad and are abhorred in and of themselves. Diseases are not the initial situation, or the departure point of medicine since they represent complex, often

L. Karhausen, *Analytical Philosophy of Medicine*, Philosophy and Medicine 152,
https://doi.org/10.1007/978-3-031-81257-6_2

transitory medical constructs made of years of research and elaborated through a process of trials and errors.

Eric Cassell writes: "*Suffering occurs when an impending destruction of the person is perceived; … although it often occurs in the presence of acute pain, shortness of breath, or other bodily symptoms, suffering extends beyond the physical*". Suffering is then "*a state of severe distress with events that threaten the intactness of the person*".[1]

Suffering, in an important sense, is a group of experiences, sensations or moods that have an objective disvalue. There is an array of feelings of unreflective aversion, avoidance behaviour or desires of avoidance as well as need for remedial intervention that is an inherent part of the concept of suffering and makes it paradigmatic of medical conditions.[2] Those *negativities* may be ordered along some dimensionalized continuum describing the magnitude or intensity of the aversion. Pain ranks in the top and irritability may rank last. Side by side with pain we find various kinds of discomfort such as anxiety, stress, acquired blindness, nausea, panic attacks, phobias, compulsions, severe obsessions, dyspnoea or suffering from incapacitation.[3] Further on the scale we have various degrees of discomfort and fatigue, dizziness, cough, dysuria, paraesthesia, or anything of that ilk. Some gradedly different symptoms may be located anywhere on the scale depending on their intensity such as oedema, insomnia, depression, itching, deafness, or somnolence.

Some of those experiences *hurt* (such as all forms of physical pain or mental suffering), others *offend,* that is to say that they do not hurt but are universally disliked (such as irritation, itching, vertigo, frustration, anhedonia, guilt). In both cases they may be *harmful* whenever their presence becomes incapacitating.

All in all, medical norms, rather than diseases, dysfunctions or any such complex theoretical concepts and their significant departures, have a certain stability and authority even if what is seen as abnormal or pathological may, within limits, depend on socio-cultural factors.[4] Those norms are broadly unassailable and have remained fairly stable throughout history.[5] They '*represent a statistically dominant pattern of determinate interests, as nearly cross-cultural as possible, that serves the greatest or at least a great variety of ulterior and overriding objectives*'.[6] And this legitimizes their foundational role. The human body or the cheetah's body and their various organs and systems have altered relatively little since there are human beings or cheetahs on the earth, least of all through the whole of the history of medicine. In view of their biological roots, the resulting norms of prudence stem from antecedent interests, namely some of the most enduring though not unchanging cross-cultural features.

[1] Cassell (2004, 32).

[2] Mayerfield (1999, 189).

[3] Hare (1964).

[4] Tallis (2009, 46–47).

[5] Munson (1994).

[6] Margolis (1975, 99).

Physical Pain

Pain is the intrinsic evil: it is one of the most common symptoms met in the clinical encounter, and it is the paradigm of the kind of things that have intrinsic negative value.

Pain cannot be described or defined any more than pleasure, writes Locke: we can only know them by experience.[7] Gilbert Ryle writes that pain *"is what anodynes and anaesthetics exist to relieve or prevent*".[8] Putnam writes: " *... there is no satisfactory way of answering the question, "What does 'pain' mean?" except by giving an exact synonym (e.g. "Schmerz"); but there are a million and one different ways of saying what pain is.*"[9] So, pain cannot be described since 'pain' is one of those terms we understand through having felt or experienced what it applies to: we know how to apply it, when to apply it and when not to apply it.

Is a painful sensory experience a *modality of somatic sensation*? To be sure pain has an affective aspect, an *aversive feeling*, and its negative valuing in virtue of which pain is seen as an evil. Pain, as it seems, is painful. *Praedicatum inest subjecto.*[10]

But perceiving in its various modalities can be the subject of attitudinal qualities (agreeable, horrible, unpleasant) each with a different modality. Having a pain or listening to Beethoven *Ninth* are different experiences and so can being in a certain emotional state. But the experience, being that of pain or of Beethoven *Ninth*, does not have an agreeable or an aversive character, but the person who experiences the experience does.[11] If so, then, pain is not painful. Furthermore, one does not, as often assumed, have pain 'in the brain' (other than headaches). Pain is not "in the head". This is not to deny that in the absence of a proper functioning brain, one would feel no pains.

David Hume wrote that it is impossible to give any reason why we hate pain: "*This is an ultimate end and is never referred to any other object.*"[12] Suffering drains meaning from everything that is valuable in life. "*To be in pain*, writes Raymond Tallis, is *to be in the brazier of unmeaning*." Being in pain is thus a mongrel concept: it is descriptive since it states a fact, an inexpungible fact; but it is also evaluative since one cannot be in pain without suffering and disliking it.

All those experiences are greatly influenced by the meaning the sufferer ascribes to his painful experience and by the context in which they are experienced. Epicureans praised ataraxia i.e., tranquillity of soul; they emphasized that, although pleasure consists in the absence of pain and pain is evil, suffering is part of nature.

---

[7] Melzack and Wall (1996, 44–45).

[8] Ryle (1971, 338).

[9] Putnam (1975, 328–329; 332; 334; 340).

[10] Pitcher (1970).

[11] Bennett and Hacker (2003, 274–277).

[12] Hume (1975), Appendix I, 293.

Life is not complete freedom from pain or illness and suffering is part of the furniture of the world.

The suffering level, then, does not necessarily correspond to the intensity of the experienced pain as it depends on contextual factors: what the attitude is of the individual towards the pain, whether he expects his pain will soon go away, whether it is interpreted as the sign of an incurable or permanently disabling condition, whether he is part of a close-knit family or of a circle of close friends. Freud in the last days of his terminal illness preferred being able to think clearly in pain than to being drugged for comfort. And people who have had a lobotomy although they feel severe pain may just shrug it off. Suffering of any kind is a symptom, and as such it is uniquely *accessible* to its bearer. I and I alone can experience my pain, and no other person can in principle determine whether I am in pain or not.

Physical pain is being pinpointed where it is felt, namely at the place where it hurts. Identifying the location of pain is one of the foremost tasks of physicians. Yet locating pain in the body may trap us into odd considerations. By way of illustration, people who had a limb amputated may feel pain as if in the limb that is not there. One can also experience a referred pain in which case its cause is situated elsewhere. Actually, a pain is located where it is felt it may thus be localized in the body or "*whatever is so recognized by the ego*",[13] namely the body image, in case of phantom limb pain. Pain is suffered *in* one's arm, leg, or head not *with* an organ.

Nevertheless, pain is usually a signal of physical harm, and it functions as such a signal because we spontaneously shun it.

Contra Putnam

If there is a structure of controversy regarding the nature of pain, it remains that suffering from pain is a foundation stone of medical thinking and that it is not species-specific, since it applies to all sentient beings.

Factual evidence warrants the claim that the higher mammals and conscious beings in general, experience pain sensations at least as acute as our own. Furthermore, their central nervous system is almost self-same to ours, and their responses to pain similar. The emotional component is evident, mainly in the form of fear and anger. Physiological reactions include immediate withdrawal reflex, screaming, moaning, weeping, and a rise in blood pressure, dilating pupils, gasping breathing, and tachycardia ending in a state of shock. These reactions are all observed in higher animals, particularly in mammals, and whereas we cannot stretch the point to the extent of crediting them with our own feelings and emotions, they feel pain, and it shows.

Hilary Putnam set out some audacious extrapolations, as functionalism licenses the view that 'feelings of pain' can be compared to logical states of a computer:

[13] Szasz (1975, 85).

"*machines can be equipped with pain signals.*" "*These "pain states" will normally be caused by damages to some part of the machine's body and will give rise to spontaneous inclinations to avoid whatever causes the pain in question.*"[14] It was the hope of Putnam that through those conjectures, talk of pain feelings would become autonomous and independent from neurophysiological descriptions of the brain.

The theory herewith becomes unwieldy. We now have two opposite extremes of mental activity and pain experiences representing the opposite ends of a spectrum. A medical, empirically rooted one and, poles apart, a philosophical one modelled on artificial intelligence. In between, Putnam contends that "*octopus, molluscs certainly feel pain*".[15] But where then lies the dividing line between the ability to experience pain and its absence? Could a cyborg have pain?

My sense is that the possibility of ascribing pain feelings to robots or Turing machines may seem crude and deliberately provocative. It seems there is something wrong in the characterization it suggests. Does this line of reasoning capture the intuition on which the notion of pain rests? Has the concept of pain been allowed to spill beyond the limits of its natural application? Has Putnam widened the idea of being in pain beyond any reasonable counter instance?

"*The question is,* said Alice, *whether you can make words mean so many different things.*" Meanings are given to words not built-in them. If so, words such as 'pain', 'feelings of pain', 'pain experiences' are usually learned within the circumstances in which they are properly used, and they secure meaning within that context.

Wittgenstein writes: "*You learned the concept of "pain" when you learned language*".[16] When is it legitimate to believe that someone else is in pain? It is to know what role does the word 'pain' play in our lives. It is to recognize when it makes sense to question whether someone else is in pain, and when this does not make sense. We have the notion of pain and we thus can believe that another person is in pain through participating in a form of life (*Lebensform*), in a way of living, a certain social and cultural background that involves a person who is in pain being cared for, being tended by others.[17] So, the whole meaning of the term 'pain' is rooted in that context, and it becomes vacuous outside that frame of reference.

Of course, we know what the word pain means and what robot means when taken separately but it does not follow that we understand their meaning when used together. Robert Musil mentioned the zoologist who would classify dogs, tables, chairs and fourth-degree equations among four-legged animals.

One could perhaps retort that ascribing pain to robots is metaphorical. Metaphors act by similarity. They represent extensions of the established usage of those words. But, again, one might then wonder whether the extended meaning has not become so thin that it turns out to be meaningless.

---

[14] Putnam (1975, 419).

[15] Putnam (1975, 436; 451).

[16] Wittgenstein (1953, 384).

[17] Wittgenstein (1981, 542, 548; 1953, 241).

Against this background, we all agree that we, human beings as well as *sentient beings* have pain experiences. Can we then believe, deny or doubt whether molluscs or insects have similar feelings? After all, this is a quest for adjudicating between conflicting claims. If it is a matter of ignorance, we should be able to check, namely, to shore up our belief or reject it by evidential overthrow. But it seems that it is not a matter of believing vs. doubting or knowledge vs. scepticism, or of truth vs. falsity, because the question might very well be untestable and pointless. To be able to answer a question, it must be intelligible. And the notion of being in pain becomes less and less intelligible as we try to apply it further and further into the biological world, farther away from our conscious experience. And when we come to robots, it may be hardly translatable into the language of our lives.

First Person Privileged Access

Pain differs from most of what takes place in the body. The complex biochemical processes and the activity of our organs occur without our being aware of their occurrence. But pain is different. I do not have to check with a gastroenterologist to know whether I have a stomach-ache.

Is pain an inner sensation? Bishop Berkeley (1685–1753) who held that unperceived objects do not actually exist, contended that we only have knowledge of our own sense experiences; there are not both trees and corresponding sensory representations of trees; we only perceive our experience of tree. He used pain to illustrate his argument in favour of *esse est percipi.* Knowledge here is unmistakable and this is why pain is a favourite in the epistemological quest for certainty. But is this correct?

Physiologists construe and measure pain as a sensation, which can be sparked off independently of the emotional state of the individual. As a matter of fact, the term 'pain' is a noun so that there is an irresistible urge to assume that, like most nouns, it stands for some object. Hence for Locke, pain is a mental object; the word 'pain' has meaning by standing for the 'idea' of pain. On this view, the language of pain reports a mental perception.

Are pains then substantive objects in the realm of conscience? Are there entities such as feelings identifiable as pains? What is the mode of existence of pains?

Pain experiences are first order standpoint that are up for grab. They are unmistakable and provide foundations of unimpeachable certainty. We usually say: '*I have a headache*' or 'I have a throbbing pain in the head', but we do not usually say: '*I know I have a headache*' except in some rare situation e.g., if someone else casts some doubt on my truthfulness."[18] Pain experiences are indexical inasmuch as I am directly aware of my pain experiences. I make no inference when I declare to be in pain: there is no logical space between being in pain and the pain itself. According

[18] Wittgenstein (1953, 82).

to Wittgenstein: "*If anyone said: "I do not know if what I have got is pain or something else", we should think something like, he does not know what the English word "pain" means*". He added that when we are in pain, we tend to think that there is "*something there*[19] *but not in the sense of the model of object and designation*".[20]

Within the first-person standpoint, pain, as it were, is incorrigible. One does not *know* one is in pain since it is impossible to doubt it: if there is no possible doubt, there is no knowledge! This is what makes suffering and more particularly pain, bedrock of the conceptual architecture of medicine. Whitehead observed that we should not ask: 'Where am I?' – I am here. Similarly, we never ask: "*There is a pain around here, but I do not clearly know who suffers from it?' Suffering shores up the individual, the gravitational character of persons and sentient beings*."

Pain is an exception to Aristotle's dictum[21] according to which knowledge and its objects are non-simultaneous correlative. Not knowing that one is in pain is impossible: if doubt is impossible, there is no knowledge! This is what makes suffering and more particularly pain, a bedrock of the conceptual architecture of medicine.

On this score, "I am in pain" is not assertable but *interjectable*.[22]

I do not possess any recipe either for talking about my pain experiences, even to myself, without using some accounts that belong to a shared public language. But plainly, no third-party can see or experience my feelings of pain in the direct and first-hand way that I do, but it can only collect indirect verbal or behavioural support about it. A physician has no access to my pain, neither can he feel nor find my pain in the transitive way that I do since it cannot be defined ostensibly.

This raises the *problem of other minds*: what ground (if any) does one have for thinking that anyone else has a mind, and is not, for example, just a body with external appearance and behaviour like his one? A person alone can find about the contents of his mind, but other people need to infer what is in his mind from his external behaviour. This epistemic primacy is the supposed fact that one's own mental states can be known directly only be himself.

This is to say that if I know that someone else is in pain, I do not know it by feeling it myself but I infer from his behaviour that he is in pain: we identify pain in other persons by some primitive natural expressions of pain (such as wincing, frequent grimacing and sighing, rubbing the affected part, screaming, flinching, sobbing, gritting the teeth, recoiling movements, sweating, moaning, and suchlike), as well as by their explicit avowals. This judgement, in this case, can be seen as a criterion, i.e., as a condition providing good evidence for it, rather than as a sign of pain because it is part of what we mean by pain, not a consequence of it.

Small wonder, then, that if there is such a concept of 'myself' as interior, based on awareness of my 'internal' experiences, there must also be a concept of 'myself' as 'exterior'; and these conjugate notions of an internal and an external purview of

---

[19] Wittgenstein (1981, 482).

[20] Wittgenstein (1969a, 296, 293).

[21] Aristotle (1971), Book Gamma, 21: 1010^b30.

[22] Danto (1968, 33–38).

oneself must also apply to others. Saying that there are first-person and third-person uses of terms like ‘pain or anxiety’ could put this across. Peter Strawson may thus contend that Giovanna’s pain or depression “*is something, one and the same thing, which is felt but not observed by her and observed but not felt by others*”. We may indeed speak of feeling depressed but also of behaving in a depressed way. The concept of depression or that one of pain thus spans both what is felt and what is observed. Feeling and behaviour are not causing each other but are rather part of the same concept.[23]

Considering the untrustworthiness of introspection, objective knowledge, and science itself need a publicly defined language for describing the facts of the world and among them, our mental states or occurrences and unpleasant feelings. But this implies that we apply to ourselves the same rules of language as those we use in talking about others. Words we use to mention mental states must necessarily be publicly understandable and be closely related to features that could in principle be observed by more than one person.

We define and explain the meaning of colour words by reference, pointing or comparing samples. But we have no samples of pain, except if we remember with which sensation we previously or originally associated this feeling of pain, but of course remembering presupposes the concept of pain.

In short, referring to pains as entities is a grammatical convenience that is utterly indifferent to the question of the existence of corresponding mental entities. And even if pain is a psychological state, it is not the mind that has the pain, neither is it the body. It is the person or the individual animal that has the pain.[24]

The Nature of Pain

Are mental states identical with brain states? Many if not most neurobiology-oriented discussions of the problem also seem to be based on some form of identity theory.[25] However, identical things have the same qualities, but they literally *are* the same thing.

In this theory, mental states or events are assumed to be, in some respect, identical to certain physical states or events, i.e., certain brain states or events.

We then have in terms of logic:

(For all things x, y) (if x = y, then (for all properties P) (x has F if y has P))

or $(x)(y)(x \equiv y) \supset (P)(Fx \equiv Py)$

But how is this possible?

[23] Strawson (1958b).

[24] Wittgenstein (1953, 286).

[25] Changeux (1983) and Crick (1994).

Don't conscious states have different qualities from physical states? Words describing mental events (I have an earache) do not have the same meaning as words describing a neurological event (My brain is in such and such condition). Furthermore, brain processes are in the brain, namely in physical space, but if consciousness *depends on* the brain, it is not *in* the brain: nobody found it there. Beyond this, a brain process is a *public* event, just like all physical events, and it can in principle be observed by other, from a *view from outside.* But conscious states or pain are not and can only be observed by the person who experiences them, from a *view from inside.*

From this, we have two types of endeavor, brain states are objects of scientific legitimate inquiry, whereas the subject of human consciousness is the person, not the brain.[26]

Dysfunctional Pain

Pain behaviour may itself be abnormal in the case of learned pain behaviour i.e., psychogenic, or non-organic pain when a patient describes more pain than appears warranted from any pathological process.

An 8-year follow-up study of the US Centre for Health Statistics found that 32.8% of the general population suffered from persistent or chronic pain syndrome.[27] Causalgia, tic douloureux or attacks of severe abdominal pain in acute intermittent porphyria are pain states in which damage or potential tissue damage is absent. Physicians when confronted with patients complaining of chronic so-called functional pain proceed to rule out all possible organic diseases and thereby unwillingly comfort their clients and concurrently substantiate their belief that they are sick. The patient is now consigned to a diagnosis where there is nothing to be diagnosed. Medical care may, in such cases, be inherently iatrogenic and do the exact opposite of what it is intended to do. Hence, pain states may themselves become entrenched, dysfunctional, and pathological whether or not they have a legitimate cause.

The Reversal of Values and The Symbolic Meaning of Pain

Feeling pain is not a purely intrinsic physiological feature but it has a symbolic meaning, which lies in its communicative aspect. There are aspects of pain that depend on the community and the social context to which the sufferer belongs. Pain expression may for example be "*a desperate cry for help to anyone who will*

[26] Bennett and Hacker (2022, 239–244).

[27] Magni et al. (1993) and Merskey et al. (1986).

listen",[28] a complaint about being unfairly treated or a sign that the person wishes to occupy the sick role. This game of *painmanship*[29] is central to the understanding of so-called *Somatic symptom disorders*.

Volenti no fit injuria. This old maxim of Roman law states that no one can ever be wronged if he gave his consent to being harmed. It is precisely because it is intrinsically aversive that sadomasochists pursue pain by attaching to it a value through a secondary move; they mediate between pain's primary aversive nature and its use as a means to an overriding pleasurable end; the alliance between pain and pleasure are then constitutive of their sexual gratification.[30] Furthermore, the Christian teachings played a major role in this effort to link pain and sexual pleasure, suffering and love. "*Pain* writes Thomas Szasz, "*is, among other things, a currency with which we repay damages done unto others*".[31] Suffering once possessed a mystical and redemptive power, although for most of us it ties us down to terrestrial trials. Pain was a central element in Caterina di Siena's devotional life; she was inverting our scale of values as she wrote that when truly religious people are suffering, they are happy and when they are not suffering, they are in pain. Thus, Gian Lorenzo Bernini's white marble masterpiece of the Ecstasy of Saint Teresa d'Avila in Rome illustrated the erotic suffering of mystical vision.

For all that, refusal to recognize the alienating essence of suffering prompts its own confusion. "*Disease magnifies both the sufferer and those who tend him,*" writes Dr Richard Selzer.[32] He finds beauty in illnesses and suffering, in body deformity and in surgical scars, in rotten limbs as well as in body cavities, which dissolve in his writings into some raving mysticism. But this sounds like some voyeuristic interest in the distressing character of pain and a lack of understanding of the tragedy of illness and suffering. Dr Raymond Tallis puts the matter vividly: "*Pain stands to pain-free consciousness not as silence to sound but as Smetana's tinnitus to his music. Consequently, it does not elevate but degrades, does not bring human beings together but isolates them.*"[33]

I leave aside, as too vast a subject, the question of the meaning of pain and suffering and whether or in what circumstances it may be wilfully accepted as a means of some self-fulfilling behaviour. Ballet dancers, rugby players and boxers measure their short-term pain against their long-term pleasure: they expose themselves voluntarily to pain as a price they are ready to pay for the achievement of some sort of project that structures and gives meaning to their life. Not surprisingly, expectations of future responses following administration of a placebo modulate the level of pain perception.

---

[28] Szasz (1975, 90).

[29] Szasz (1975, xlix).

[30] Largier (2007).

[31] Szasz (1975, 249).

[32] Selzer (1982).

[33] Tallis (1992).

Mental Suffering: Clinical Anxiety and Clinical Depression

By pain, I mean not only physical pain but also any unpleasant feeling of anxiety, sadness, and displeasure in their various expressions.

The medical concept of suffering is broader than mere physical pain as it includes pervasive negative feelings such as despair, grief, depression, anxiety, anhedonia, loss of freedom or opportunity, panic, terror, dread or self-doubt, distressing obsessions, delusions of persecution as well as auditory hallucinations, disorders of speech and language, of memory or of intellectual performance or anything of that ilk.[34] It follows that the first-person perspective has furnished the agenda for psychiatric practitioners, except for clinical neurologists for which the third-person perspectives are more important, because in their day-to-day practice, they deal with observable signs.[35]

Emotions are always positive or negative, but there are more negative emotions than positive ones.[36] Like pain, anxiety is an adaptive mood which is ubiquitous since our everyday life often entails some ingredient of danger and challenge: it brings about changes that enable to protect oneself from physical or social threats[37]; those threats are thus the objects as well as the causes of anxiety. However, in some circumstances, depression and anxiety become excessive, so that psychological suffering may be incapacitating and pathological. Whenever the intensity, duration or frequency of fear or anxiety exceed the usual response to stress and interferes with functioning, it is maladaptive and merits treatment. Anxiety and depressive disorders may then be diffuse and grow into feelings, which are not directed at any object.[38]

It should now be plain that the various forms of mental suffering, just like pain, divest life of its meaning, colour the whole world, strongly interfere with the well-being of the individual and serve no purpose. Pain, anxiety, or depression in themselves, not their causes, are primarily aversive and are reasons for acting. It is their unpleasantness coupled with their pointlessness, which makes them awkward and taxing. Anxiety can be as obsessive as pain since it can dominate everyday life to exclude all else and imply suffering as intense as extreme pain. To feel anguished or despondent, severely, or not, for long periods of life is abominable. But being stricken by short episodes of gloomy mood is a natural part of normal life.

---

[34] Culver and Gert (1982, 79–83).

[35] Hofmann (2017), Green and Palpant (2014).

[36] Glas (2013), Szasz (1975, 59), and Sims (2003, 299).

[37] Nesse (1999).

[38] Szasz (1975, 59–60).

Harm and Detriment

The value-laden concept of harm is anything that promotes physical or psychologic injury or damage, that impairs or affects a person's safety, and causes pain, discomfort, disabilities, diseases, or death. Medicine is essentially concerned with the avoidance of detriment and harm, the correction of their manifestations or their effects and the medical norms of health by reference to which harm is being defined.

Harm is not a privative term indicating the absence of a quality, but it refers to something intrinsically undesirable: no one wants it, everyone wants to avoid it; it must be reduced because it frustrates one's basic prudential interests[39]: what makes harm important is that harm creates *a priori* needs. We do not need to give an account of what constitutes human good and welfare to understand the concept of human harm, since harm is intrinsically negative.

According to the Oxford dictionary harm means *"evil (physical or otherwise) as done to or suffered by some person or thing; hurt, injury, damage, mischief"*. Harming means *"to do harm, to injure (physically or otherwise); to hurt, damage."* Being harmful thus also means *"evil (physical or otherwise) as done to some person or thing; hurt, injury, damage, mischief"*. Something that is harmful to an individual is thus harming him, it makes him worse off than he would have been in its absence: it thus causes harmed states.

Harm, when severe enough and persistent, is an intrinsic evil regardless of its causes since badness is constitutive of harm, as it deprives us of our bodily, physical, and psychological integrity. Suffering is then a common ingredient of harm. So are diseases and disabilities.

But harm is also contingently detrimental beyond the fact we merely dislike it; it also means an outward damage done, the further result of harm making or harm doing: a traumatism is a physical harm, a harmed condition. Physical or mental diseases are thus harmful in themselves (intrinsically harmful) but they may also have further harmful consequences (extrinsically harmful). *Being harmful hence purports either to the process of being harmed or that of causing harm.*

First, a disease is harmful because it is *intrinsically bad* and undesirable: it consists in either actual aversive condition, or potential as well as forthcoming bodily negativities which interfere with our capacity for growth or development and diminish our well-being such as absorbing physical pain, groundless anxiety or depression, injuries, sickness, disability, grotesque disfigurement, loss of integrity and normal functioning of one's body or forthcoming death.

Secondly, a disease may be harmful because it is *extrinsically bad* insofar as it deprives the individual of valuable aspects of living and disrupts the course of his life. In medicine, features defeating a person's ultimate goals and blocking his aspirations—such as achieving a successful career or writing a great piece of music or successfully raising a family—are usually seen as extrinsically harmful so that they are not part of the disease concept. Conversely, an unmet need, abstaining from

---

[39] Thomson (1987, 8).

enjoying good wine, may not be pronounced harmful in this second sense if, after balancing the risks, the resulting state of deprivation either promotes some future prudential good (the healing of an alcoholic hepatitis) or forestalls some future harm (the avoidance of liver cirrhosis, brain damage or cancer).

Thirdly, 'harmful', still according to the Oxford dictionary, means "fraught with harm or injury; injurious, hurtful": something or someone may thus be harmful, harming, detrimental, impairing, damaging or some harmed condition without resulting in harm. Being harmful is *dispositional* as it implies a greater risk of harm. One may be exposed to harmful air pollution without suffering any harm or to stressful life events without diagnosable mental illness. Cancer *in situ* is a harmful condition although it may never yield an invasive tumour. So are silent gallstones, which represent a risk factor for cancer of the gallbladder.

Fourth, secondary aging is a harmful condition, namely aging resulting from external influences originating throughout the lifespan and resulting from communicable diseases, environmental exposures, and social behaviour such as overnutrition or lack of physical activity.

To sum up, harmful features are either intrinsically bad or causally detrimental. But being harmed is one thing, disposition to be harmed is another. Thus, something may be harmful either by a disposition to some illness or by the illness it causes.

"*Medical harm*, Jonathan Glover observes, *is "directly" harmful: the harm must result from my mental or physical state itself, without the reactions of other people to my state being a necessary condition of the harm. Harm suffered hence consists in not meeting some of our basic primary prudential interests. Diseases are self-destructive natural evils: they are harmful for lack of which they would not be diseases.*" Glover adds: "*We should only regard someone as ill, and hence in need of treatment, where he is in a physical or mental condition that is harmful to him.*"[40]

In other words, a condition's intrinsic and outward harmful features are manifold. 'Harm', 'harmful' and 'harming' are terms that do not have a single precise use but a whole family of uses. They refer to concepts whose identity is fixed by a cluster of rules, none of which is properly thought of as the definition of the concept. Thus, harm has a constellation of connotations which involve the agent—embodied or not—that causes harm, harm itself and what kind of harm it ends with; 'pathological', on the other hand, is a predicate which applies, without explicit or implicit reference to some specific causes or to harm itself. That a disease harms a patient may be described pluralistically, namely by using different terms which are not cognitively equivalent in the sense that the listing of harmful features in one optional vocabulary does not necessarily correspond to the listing in the other: there is no unique true way of dividing the harming outcome of a disease process into token harming properties. Medical intrinsic evils can be carved up in different ways since there are many correct ways of describing and listing the damages, injuries, or harming manifestations observable in a case of illness.

[40] Glover (1970, 124, 151, 119).

The concepts we use to determine them reflect the clinical evidence just as much as they are determined by our ways of classifying and interpreting it. Hilary Putnam here uses the metaphor of the 'cookie cutter'. We have on the one hand, the dough, which represents what independent is of our conceptual choices and on the other, the shape of the cookie cutter, which is our conceptual contribution.[41]

Details and objections aside, should not the concept of harm be defined by contrasting it with what it is not? *"Being harmed"*, writes Harry Frankfurt[42] *"has to do with becoming worse off than one was …"* although the prior condition might itself have been a borderline or abnormal case. Therefore, there is an additional grammatical possibility pointed out by Jonathan Glover: a condition is harmful because it reduces someone's interest satisfaction by comparison with what it might be, and not by comparison with what it was. Yet there is even a more radical view that can be taken in construing our contrasting case: a condition is harmful when contrasted with what it ought to be. Thus, the line of argument, which began with an indicative account, was then expanded into a modal or counterfactual context, and ultimately ended up in normative status. Whatever our choice among these rival views, this last move blends the notion of harm with that of pathology and dissolves the logical priority of the latter over the first. By then, the concept of 'pathology' ceases to arise or be construed from that of 'harm'.

Disability is a condition that someone has a strong rational preference not to be in, and, in addition, one that is a harmed condition, i.e., that someone is badly off in a medically meaningful way. The important insight here is that harming claims are comparative in form since they express the lowering of a baseline by reference to some counterfactual norm of thriving.

When Harm or Detriment are Dilemmatic

The Babylonians, the Egyptians, Greeks, and Romans knew that sunlight promotes health. Hippocrates believed it was favourable for most ailments. In the beginning of the twentieth century, following observations that it kills bacteria, medical interest established that a deficiency of sunlight is associated with rickets and by the late 1920's sunlight was being promoted as a cure for pretty much every illness. When ultraviolet UV-B rays produced by sunlight irradiates the skin surface, they spur the synthesis of vitamin D3.

On the flip side, ultraviolet UV radiation from the sun is the number-one cause of skin cancer and melanoma is the most serious form of skin cancer. Epidemiological studies have confirmed the hypothesis that most of all melanoma cases are caused, at least in part, by excessive exposure to sunlight. In Europe, the survival of patients with a diagnosis of melanoma between 2000 and 2002 was 86%. The profile at 5

[41] Putnam (1987, 33).

[42] Frankfurt (1984, 6).

and 10 years is similar, even though somewhat smaller at 10 years. The survival diminishes with age from 88% for 15 years old until 74% for those more than 75 years old.

Yet, the perplexing finding is that people with large sun exposure have higher life expectancies than sun avoiders despite an increased risk of skin cancer and of melanomas. A large Swedish study conducted by Pelle Lindqvist and his colleagues at the Karolinska Institute into risks associated with melanomas and breast cancer among 30,000 women showed that, on average, women, who spent more time in the sun had a higher life expectancy than sun avoiders, even after adjusting for factors such as disposable income, cancer, education and exercise.[43] The reduced life expectancy of sun-avoiders was mostly due to a greater risk of cardiovascular disease and other non-cancer related illnesses, such as type 2 diabetes, autoimmune disease and chronic lung disease.

All this leaves health policy makers with a dilemma. Is sunlight harming or promoting health? This dilemma highlights why it is difficult to use a purely naturalistic explanation since it brings out the conventional nature of our concepts of harm and detriment, and their underlying normative connotation.

Conclusion

Medicine grew out of the fundamental call for the alleviation of suffering or of potential suffering, and their consequential role in construing the normal-pathological divide. Intrinsic medical negativities are the primary and first springboard of medicine, whereas concepts such those of disease, malfunction, cure, or health, reflect a highly sophisticated level of development associated with an advanced degree of medical care.

1. Actual or potential physical or mental suffering, incapacity, diseases, anxiety, depression, and harmful biological conditions directly affecting bodies or minds are aversive and legitimately allowed as the port of access to medicine. Physical pain is not properly an inner sensation, but it is an intrinsic evil: it is one of the most common symptoms met in the clinical encounter, and it is the paradigm of the kind of things that have intrinsic negative value as like mental suffering.
2. Unchecked and untreated suffering is unacceptable. The medical concepts of physical or mental suffering presuppose an implicit agreement in doing certain things, as well as avoiding to do certain things: suffering is a call for help, attention, and prudence.
3. Harm plays a pivotal role in construing the normal/pathological divide. Harm is a physical, mental, or emotional affliction, damage, hurt, injury, adverse effect of healthcare, impairment or detriment done to a patient.

[43] Lindqvist et al. (2016).

4. Harm has dual usage since it can be a verb or a noun, namely either a discrete action or an ongoing state. With harm as a verb, the emphasis on what led to pathology is the whole point of distinguishing harm from mere pathology. The notion of harm is prior to that of suffering so that it endorses a notion of change. It is a cause of badness. Harm as a noun means an outward damage done, the detriment or the result of harm making or harm doing:
5. The dilemmatic character of harm and detriment is brought out by the Karolinska Institute that show that sunlight both increases the mortality due to melanoma and increases life expectancy, a situation with significant philosophical implications.

Chapter 3
Normalcy or Abnormalcy

Ce n'est pas par l'analyse des lois de la nature que nous comprendrons pourquoi l'horloge est mal faite.

Descartes: VI Méditation

Contents

The starting claim of this book is that medicine is grounded on a medical convention, the conventional distinction between the notions of normal and pathological, whereas being normal is the default position. Yet, this claim demands grounds, and those grounds need to be warranted and backed within the frame of the argument.

Physicians are perfectly aware of what is normal and what is abnormal, but they would find themselves in a quandary if they were asked to be explicit about it. There are numerous examples in the history of medicine of severe disagreement among experts, particularly in psychiatry about non-existing conditions: are either masturbation, burnout or being attracted by people of the same sex, abnormal or pathological signs?

L. Karhausen, *Analytical Philosophy of Medicine*, Philosophy and Medicine 152,
https://doi.org/10.1007/978-3-031-81257-6_3

Be that as it may, there is a broad consensus in medical thinking, continually subject to revision, about what in the realm of abnormality is deemed to be pathological.

Norms

What do we mean by normality? Is it a meaningful question? "*Seeking for so-called normality,* writes Darrah, *is like looking for gold at the end of the rainbow.*"[1]

In Latin, the term *Norma* refers to a carpenter's or mason square, hence a pattern, a rule, and a standard of judgement. In medicine, 'normal' portends constituting, conforming to, not deviating, or differing from the common type or standard. It connotes the characteristics of what is usual (i.e., following the usual course of nature) or statistical (expressing some average: mean, median or mode) or else some sort of yardstick or referential value.

Normality could mean *frequent* in the sense of being the most common within a population, such as having brown eyes in Mediterranean countries or blue eyes in Nordic countries. It could mean *average* in the sense of a statistical mean, such as the average weight or height of a population, often represented with the familiar bell curve, or *typical*, as in representative of a group, population, or general type. Sometimes, normal means *adequate* in the sense of being free from defect, deficiency or disorder, and other times it means *optimal*—being physically fit or mentally sharp. Or it could refer to Darwinian more *adaptive* characteristics allowing to reproduce; it is a kind of a cold, heartless way of thinking: if one dies in his fourties. Finally, there is our basic everyday usage of the word, which often slip-slides among these different meanings and tropes, from the orthodox and standard to what is expected and good.

There are grammatical norms, rules of etiquette, moral as well as medical norms. Norms are conventions and represent an expectation perceived as legitimate. The definitions of social or moral norms[2] are *reportive;* they pertain to norms of cooperation, of reciprocity, which are socially beneficial, and that involve sanctions, both positive and negative: we all need to drive on the right or on the left.

In the 1800s, 'abnormal' meant either irregular, without rule or departing from the natural condition.[3] The term has shown a series of pseudo-etymological alterations: from the Greek ανωμαλοs (irregular, uneven, unequal) and Latin *anomalus*, stemming from alpha privative + ωμαλοs (equal, smooth, even); it was later adapted after *norma,* (a rule) due to a confusion with the medieval Latin term *anormalus* or *anormälis* or *abnormis* (differing or deviating from the ordinary rule or type) and

[1] Darrah (1939).

[2] Bicchieri (2014).

[3] The Compact Edition of the Oxford English Dictionary. Oxford University Press, 1971. Volume 1; p. 24 & p. 348.

stemming from alpha privative + *norma*. It later became *abnormalis* then *abnormous* and finally *abnormal*.

On this score, the term norm holds different and multiple meanings.

Normal as Natural

In medicine, being *normal* has a force such that it seems that human beings are bound to adjust their behaviour to realize such a value in their relevant projects. But can this term and its converse be used in neutral language? Can their meaning be captured in purely naturalistic or cognitive claims?

Naturalism has its robustness as well as its treacheries since the concept of nature functions in many ways and is often used ambiguously. As I am obliged to be synoptic, it seems sufficient to briefly discuss some principal meanings, which are relevant to our purposes.

For one thing, being natural is often taken to be the opposite of being artificial, resulting in a distinction between nature altered or manipulated for human purposes, and nature developing without deliberate human interventions. John Stuart Mill[4] observed that the term 'nature' refers to how things would be outside of any human influence. In the light of this, being natural tends to be seen, in the eyes of physicians and of the public, as a hallmark of acceptability. This yields to the contention that what is natural is commendable and should be blessed, and what interferes with it would have bad consequences and should be dispraised. These construals are at the root of some ideological crusades for the defence of the environment, which tends to depict human beings as alien to the natural world.[5]

For another, Aristotle[6] who was a physician's son, contended that is natural what is in accordance with the inner nature of a thing, i.e., its essence or its predispositions: nature is a source of processes intrinsic to a thing, namely the wellspring of its changing and staying unchanged. Is natural, whatever is allowed to develop without interference. "*The natural thing to expect is that,* writes Toulmin, *if left to themselves in their* usual *environment, things will follow out their normal courses of development.*" Nature cannot but act upon itself. What distinguishes something natural from an artefact is that nature is the source of its own changes, while in the case of an artefact changes are forced from without. So being normal, in a way, does not require an explanation.

Abnormalities, deviations, diseases, irregularities, disorders are departures from nature (φυσιs) and are features which we single out by contrasting them with the usual and natural pattern; hence interfering factors, complications and obstacles require to be exposed and delimitate the natural course of events. Galen wrote: "*So*

[4] Mill (1958, 6).

[5] Fuso (2016).

[6] Aristotle (1952, XXIX–XXX; 1970, Book 2, 1928–1932).

we have at least shown conclusively that a disease is one of the things that occur contrary to nature."[7]

On that account, if the natural is the given, the un-interfered with, the unnatural is what is interfered with, and something which we must dislike and disapprove of. This is a popular and probably old theory of pathology such that diseases come from outside interference which brings unnatural events to an organism.[8] Pathology is external, while physiology is internal and natural.

We have seen previously that following its turnabout, the *second position* vindicates a kind of counterfactual naturalism. Physicians apparently think in "naturalistic" terms, i.e., in some tacit ontological naturalism: they assume that all medical, initially normative concepts and processes can be defined or explained in descriptive language and in the indicative mode. This might be pragmatically or instrumentally legitimate, but it pertains to the mere surface of medical reality and clinical evidence.

Christopher Boorse's Naturalism

For an epistemic naturalist, our methods and explanatory resources are those of natural sciences and we must reject any principle or feature inaccessible to scientific investigation.

Shall we then accept the prospect that concepts of 'normal' and 'abnormal' as well as the boundary between them can be naturalized? Christopher Boorse avouches for it and seeks to provide an account whereby whether a condition is abnormal depends solely on value-free biological facts. He believes that "*the normal is the natural. The state of an organism is theoretically healthy, i.e., free of disease, insofar as its mode of functioning conforms to the natural or species design of that kind of organism*". On this score, mammals' teeth as well as human teeth wear out, eventually leading to bone loss and loss of teeth, which is normal since it conforms to the natural species design.

Boorse considers the human body being made up of numerous subsets, i.e., organs, each of which performs one or more function in a healthy human being. The function of each subsets contributes to accomplish the goal of a goal-directed system. Whenever a sub-system malfunctions, there is a disease.

Being abnormal, then, means being unnatural either by being atypical or being the prey of a hostile environment resulting in being below one's species' standards. It results from any condition that hinders proper functioning and interferes with the causal continuity of natural processes. Diseases may thus be seen as an alteration, a departure from some natural physiological pattern of the human life cycle due to some external agent acting on a passive prey. We are back with Aristotle who held

---

7 Galen (1991, 58).

8 Plato (1971b, 333, 345).

that all changes are 'sufferings' and depend on the external intervention of some sort of agent.[9] The normal and abnormal status as well as the boundary between them are thus purely descriptive concepts.

But are the terms 'normal' and 'abnormal' simple and unanalysable into more elementary terms? By bringing species design into play, Boorse merely procrastinates he tacks on an additional unknown but does not solve the riddle. For Boorse the theoretical though assumedly empirical notion of species design is equivalent to its 'nature'; it refers to the alleged "*internal functional organization typical of species members, which (as regards somatic medicine) forms the subject matter of physiology*".[10] Boorse offers the analogy of a car that is in perfect working order whenever it fulfils its designer's specifications. Since an organism has no designer, Boorse brings out a neo-Darwinian stance: *"a function in the biologist's sense is nothing but a standard causal contribution to a goal actually pursued by the organism*".[11]

Against this background it is not difficult to appreciate that, considering the theory of evolution and the omnipresence of variation, distinctions between species are surely indeterminate. The term *'species* has no agreed upon definition,[12] and many philosophers of biology recognize that it is hard to find traits that are shared by all and only members of a species.[13] What criteria should we use: morphological, phylogenetic, ecological, molecular, or the possibility of interbreeding? Immanuel Kant would have described the term 'species' as a *usurpatory concept* since there are no such entities in the natural world.[14] What's more, about two thirds of human diseases are zoonotic, meaning they can jump between animals, and are not 'species-specific'.[15] Gout has even been observed in dinosaurs. What's more, the fact that mice, dogs, pigs and monkeys are commonly used in laboratory experiments for modelling human diseases indicates that most of them are not species-specific.

Boorse assumes that these kinds consist on all and only those things that share some essential property, although it is recognized that no such property can be found to demarcate species.[16] Biological species are not natural kinds.[17] Boorse believes that if the world is divided in all sorts of categories, some of these divisions correspond to the way nature is really divided; we must aim at cutting nature "at the joints", but Boorse sees joints only where he carves.

---

[9] Boorse (1975).

[10] Boorse (1975, 57).

[11] Boorse (1997, 7).

[12] Ruse (2008).

[13] Sober (1994) and Hull (1965).

[14] Waterlow (1982a, 136–137).

[15] Hull (1976), Dupré (1981), Rosenberg (1985, 33, 201), Sober (1993), Dupré (1993, 234), and Kant (1965, I, A84, B117).

[16] Dupré (1993, 26–59).

[17] Sulmazy (2005) and Kitcher (2001, 48–49).

To avoid these objections, Boorse, more recently mentioned "a reference class of organism of uniform functional design". A normal function of a part or process within members of the reference class is a statistically typical contribution by it to their individual survival or reproduction. Health in a member of the reference class is normal functional ability. Elodie Giroux observed, that the comparison groups are the philosopher's stone of epidemiologists.[18] Some basis for comparison is required to select a reference class and the development of methods to identify a class of organisms of uniform functional design is an overriding challenge.[19]

Yet, Boorse assumes his conclusion as a premiss, as he defines a normal function or health in terms of the *norm* of uniform functional design. He claims that the distinction between health and disease is determined by empirical facts alone, even though as shown by Elselijn Kingma[20] his reasoning is circular since it assumes what it attempts to define: the terms to be defined or a version of them, occur in the definition. In other words, his definition of disease is tautological.

Simply put, there is no real population predicated on a uniform functional design and this is a hypothetical premiss based on some speculative norm. Right in the beginning of his definition, Boorse needs a value-laden concept, a universalised theoretical construct based on a norm of "uniform functional design".[21]

Biostatistical Theory. Methodological Naturalism

There is an alternative strategy, which offers a biostatistical account of normality initiated in the beginning of the twentieth century in Germany[22] and more recently promoted by Christopher Boorse.[23]

'Normal' may mean usual or any value situated between the upper and lower percentile points on a frequency distribution. The use of the Gaussian bell-shaped curve visualizes normalcy as the middle range, and both extremes as pathology and illness. The adoption of this perspective is legitimized by the variability, which is inherent to any kind of measurement in medicine or biology. Normalcy is understood either as what is seen most frequently, namely an average—the mean, the median or the mode or as a fraction of a distribution defined in statistics as the 'normal curve'—or else, as whatever is included in the normal distribution. In the latter case, a normal value may thus be far away from the mean, unlikely but still normal: its probability may be low, but it is not zero. Williams writes: "*if the term*

[18] Giroux (2009).

[19] MacMahon and Trichopoulos (1999, 187–190, 236–240).

[20] Kingma (2007, 4).

[21] Ereshefsky (2009, 222–223).

[22] Zola (1966, 616), Institute of Medicine (2001), Green et al. (2001), Biering-Sorensen (1929), Devereaux (2003), Mannix (2001), and Evans (2003).

[23] Boorse (2014).

normal is used at all in reference to the normal frequency distribution curve, it must include the entire range from the lowest to the highest value, the entire population must be considered as normal."[24]

But there is something wrong in the characterization that it suggests. The statistical normal set does not define what medical language calls normal or abnormal. It merely allows us to predict what is the probability for a given measurement that it is part of the normal distribution; or else what is the range of possible values given a certain probability level. The upshot is that there is no sense in excluding the 5%, or the 1% in trying to separate normal from abnormal, particularly when abnormal values belong to a different population than the normal one: 5% of normal people will have abnormal tests results because of the way normal is defined.

Next, different laboratories have different normal reference ranges: "normal" haemoglobin have a reference range of 10.0–18.0; if one repeats a test, one gets a slightly different value each time.[25]

Beyond this, the more tests performed on a healthy subject, the greater the likelihood of obtaining statistically abnormal results. If the normal range is limited to 95% of subjects, one person in twenty is expected to be abnormal in the absence of any disease. If two independent tests are performed, an abnormal result will be found in one among ten (a product of probabilities: $0.95 \times 0.95 = 0.90$); and if fifteen tests are performed, more than half of the population will have at least one abnormal test (0.95 to the fifteenth power); in the case of fifty tests, more than nine among ten subjects will display some abnormal result (0.95 to the 50th power). If one multiplies the number of independent tests, almost everyone will be classified as 'abnormal',[26] so that it has been said that a normal control is a man who has never been properly investigated.[27]

Be that as it may, the deflationist idiom of biostatistical theory results into confusing or misleading consequences.

First and foremost, as much as 90% of apparently healthy population samples have some physical aberration or clinical disorder and 46% have one or more chronic illnesses. Tension headaches affect more than 90% of adults and severe periodontal disease is present in more than 50% of adults. Furthermore, very few young adults possess 32 well-aligned permanent teeth with normal occlusion and without caries and receding gingivae, despite the picture displayed by texts of human anatomy. Uterine leiomyomas affect 70–89% of women, which thus makes it a normal condition.[28] Worldwide, back pain is the leading cause of disability, with as much as 80% of the population experiencing this condition at some point of their lives. This suggests that the presence of some illness or of clinically serious symptoms is a statistically normal feature of our lives. And Roger Williams contends,

[24] Williams (1956, 3).

[25] Hall (2019, 25).

[26] Boorse (1997, 8).

[27] Sackett et al. (1991) and Gräsberg (1984).

[28] Heart Protection Study Collaborative Group (2002).

"*Practically every human being is a deviate in some respects*". This is in sharp contrast with the case of artefacts, cars, clocks, or computers, which are expected to reveal no fault, or defect that makes them not work correctly.

Second, some conditions are *statistically normal but clinically pathological.* Hence, the desirable clinical range may differ from the population normal statistical range. The accepted 'normal' value of blood LDL cholesterol is about 125–130 mg.%. But there is now good evidence suggesting that the best cholesterol level might be the lowest one: the statistically normal cholesterol level found in Western populations is 'abnormal'.[29]

Third, the biostatistical model of abnormity assumes that malfunctions are always limited to hyper- or hypo-functioning.[30] But diseases often result from new, different, emerging harmful functioning, such as genetic disorders malignant tumours or so-called positive symptoms of insanity (delusions, hallucinations, abnormal emotional states and so forth); immunological disorders are neither hyper-, nor hypo-functioning.

Fourth, statistical standards of normalcy are (a) either determined on the general population in which case they include abnormal and sick individuals: so, pathological features are spilling over and tarnishing our estimates; or else (b) our standards are determined on a small number of selected subjects, eliminating abnormal cases by using some sort of criteria for normalcy, a strategy that begs the question. Determining a statistical normal distribution cannot be used to decide biological normalcy because it is circular since it surmises what it is supposed to define.

Fifth, the Gaussian model and its confidence intervals, or mere percentiles for the normal curve reveal themselves to be illusory. The biostatistical model *assumes that variates are normally distributed;* but many anatomical, clinical or laboratory test values are non-Gaussian in distribution: body weight, arterial blood pressure, mean red-blood-cell-diameter, haematocrit, blood cholesterol, potassium or bilirubin have a skewed distribution. Furthermore, using a bell-shaped spectrum distribution may occasionally yield negative values, which is nonsensical, e.g., for blood cell count.[31]

Sixth, 1.2% of the population are exclusively homosexual,[32] so that according to Boorse, it represents a disease.

Finally, if 100% of the persons in a given population were afflicted with some sort of infectious disease, we would hesitate to call them healthy: yet autopsies showed that in the nineteenth century, almost hundred per cent of urban populations of industrial cities had at some point of their lives developed tuberculosis.[33]

---

[29] McMichael in JD Swales (1985, 114).

[30] Kiple (1993, 1063–1964).

[31] Canguilhem (1966, ch. 8), Wachbrot (1994), and Murphy (1997).

[32] Epstein et al. (2012).

[33] Claridge (1987).

The Swiss psychiatrist Eugene Bleuler (1893–1988) was saying: "*Never give anyone a certificate of normality. I wouldn't even give one to my wife.*"[34]

Summing up, statistical distributions are descriptive but not normative and they are inadequate for separating health from disease. The normal range reflects how things are; it should be distinguished from how things should be. This statistical fallacy, which is a garden-variety of the naturalistic fallacy, occurs when there is confusion *between the normal range and the medically normal state*. It is a wild goose chase to attempt to define the normal in statistical terms although statistics have some bearings on the concept.[35]

The Homeostatic View of Normalcy

A different strand of thought invokes the dynamic process by which an organism maintains a constant internal milieu in spite of external changes: whenever a departure from some specified preferential state occurs, a process sets in motion which tends to bring the situation back to the baseline original state, called the *normal* state, which tends to be maintained despite environmental fluctuations. Organisms are regulative, self-maintaining and self-repairing, and they also develop.[36]

Hippocrates thought that diseases are cured by natural powers, a *vis medicatrix naturae*. Claude Bernard fostered this idea when he turned his attention on the maintenance of the internal environment: "*It is the fixity of the milieu intérieur which is the condition of free and independent life ... And all the vital mechanisms, however varied they may be, have only one object, that of preserving constant the conditions of life in the internal environment.*"

Homeostatic equilibrium consists in some invariant self-regulatory tendencies, which achieve a dynamic stability. It depends on circular causality or negative feedback, namely causal loops such that failure to reach an effect, e.g., maintaining a dropping arterial blood pressure produces a correction that brings the blood pressure back to proper levels.[37] Organisms which reach homeostatic states of growth, functioning, and maintenance are said to be 'normal'. Claude Bernard thus contrasted the maintenance of the *milieu intérieur* with its failure in disease.

But the point to be driven home is that the homeostatic mechanism may itself be disturbed, and what's more, the body may function homeostatically even when it is sick and when it is dying. A decerebrated patient may remain alive and his body may function in a homeostatic way for years. Malignant tumours are admirable homeostatic systems. What's more, a large realm of pathology results not from the breakdown of regulatory processes, but from maintaining a stable but different

[34] Bleuler E.

[35] Jamison (1993).

[36] Boorse (2011, 24).

[37] Smith (1947, 371, 373).

homeostatic norm, e.g., in hyperthyroidism, fever, bipolar disorder, or malignant growth. It follows that homeostasis is of no help for a theory of normalcy.

Medical Statements Are Both Prescriptive and Descriptive

But the question now before us is that medical claims differ from empirical or scientific claims insofar as they are normative and express attitudes. Biologically, pathological features or diseases are neutral-descriptive concepts like 'herbs', but for nurses or medical men they are value-laden and prescriptive like 'weed'. Disease is a concept sensitive to human concerns, like vermin, pests, or weeds; it is not a natural kind, but it is made up of natural processes (pain, tumours, physical disability) and can be object of scientific investigation.

There is a contrast between descriptive meaning, whereby language is used to describe facts — for instance, the language of biology or the diseases stored away in medical texts — and the prescriptive meaning that is characteristic of value judgements of a performative language using terms such as 'good', 'bad', 'right', 'normal' or 'pathological'. Value judgements commend or censure: they entail singular imperatives and are primarily used for guiding action.[38] Things are good or bad based on whether they contribute to the good or harm human beings *in general.*[39]

For all that, medical language may be descriptive or prescriptive. A patient's anamnesis or physical examination is written down in the medical record in descriptive terms, while his treatment requires a prescriptive language. Even so, the behaviour of a clinician, his thoughts, his gestures, his diagnosis, his treatment decisions are implicitly value laden. A patient takes his disease and its manifestations as an objective event, an unpleasant one, in his life, just like a car accident or a heat wave. But the same signs and symptoms perceived by the physician are double-faced. They are clinically objective facts that often can be quantified. But what gives them medical meaning is their relation to pathology and the underlying web of primeval reasons and constraining norms that prompt the physician to take charge of the patient's care. We are here faced with an apparent incompatibilism holding between two alternative levels of discourse having the same grammatical structure, one of them being antecedently normative, and the second being an object of knowledge. Even for Boorse, natural design is evaluative, since it carries in its meaning, a commitment to health care: he acknowledges that an illness is at once undesirable for its bearer and a title to special treatment.[40]

Furthermore, we have an ingrained tendency to cross the treacherous barrier that separates the descriptive from the prescriptive, a tendency to move from *what is* to *what ought to be.* This is known as "Hume's guillotine:" "*Systems of morality',*

[38] Hare (1970).

[39] Foot (1967, preface: Moral Beliefs).

[40] Boorse (1975).

argued David Hume, "*usually begin with reasoning about matter of fact … Then suddenly they switch, and instead of saying "is" and "is not" they begin to say "ought" and 'ought not'. But premises about 'is' do not lead to conclusions about 'ought*'".

Hence, since medical facts are descriptive and prescriptive, they are susceptible to two interpretations, nomological or normative, pragmatic, or theoretical, scientific, or instrumental, epistemologically naturalistic or expressing valuating judgements.[41]

But surely, even though clinical observation, mechanistic biomedical or epidemiological research are epistemically or methodologically naturalistic, underlying normative capacities are implicitly operative.

It follows that medical science looks at the world in terms of hybrid concepts that have two dimensions: they have 1 foot in the world and a second one in values. Medicine is a science, but medical theory and medical language are also to be regarded as instrumental, for controlling diseases or harmful conditions: they have instrumental value as they have value as a mean to some other end. Truth and falsity are then collapsed into positive or negative conventional norms. Pragmatism is sympathetic to instrumentalism, since it assumes that all belief is mere acceptance into the system deemed most useful.

Factual or scientific beliefs are not action guiding, they are neutral. It is the combination of cognitive knowledge and intentions, which commits an architect to draw and implement his work project. Medical concepts are crossbred of factual and intentional marks in which the implied 'ought'-sentence—the need for health care intervention—is inherent to the actual or potential deleterious quality of pathological features.

Even so, medical statements are not properly twofold, i.e., made out, on the one hand of factual utterances about signs and symptoms, and, on the other hand, of valuational claims about needs: both of them are intrinsically related to one another and merge together. On balance, medical terms are what Sir Bernard Williams termed *thick terms,* thanks to which the thoughts and judgments expressed by these utterances on the one hand are candidates for truth and falsity ("world-guided") and, on the other, supply defeasible reasons for action ("action-guiding").[42]

Taking Stock: Defining the Term "Normal"

The previous discussion indicates that although the medical term "normal" is cognitively meaningful, it is an indefinable word, standing for a property, which is not analysable into other properties.

---

[41] Whitbeck (1978, 210) and Margolis (1981, 562ff).

[42] Williams (1985).

We must keep in mind the distinction between defining and accompanying characteristics. We want to distinguish two kinds of statements:

1. Statements about the meaning of the noun 'normal' or of 'normal biological processes', just as one would define a 'father' as a 'male parent'.
2. Statements of the predicate applied to the biological processes, such as for example, what properties always accompany normalcy, and can thus be used as a criterium of its presence.

If we define 'normal' as the homeostatic equilibrium, the term "normal" could be omitted in the sentence, and we can substitute the phrase "contributes to the homeostatic equilibrium". More probably, it is a statement *about* normal features: it means to say that every biological process that has one characteristic or property contributing to homeostatic equilibrium, also possesses *another* characteristic, being normal, thus making the two classes of features co-extensive.

Or else, we can say that one characteristic or property that contributes to the homeostatic equilibrium also possesses another characteristic, namely, being normal. The presence of one characteristic would be a criterion, or a universally accompanying characteristic of the presence of the other.

A definition must be adequate to the *possible* as well as the *actual* cases. We want to know what the characteristics are, all the characteristics, the presence of which would entitle some biological process to be called abnormal, and the absence of which would keep it from being called normal. Any list of characteristics defining normalcy is bound to be insufficient, since a list of criteria that might define what is medically normal would be endlessly long to specify truly sufficient conditions.

The upshot is that there is an unavoidably long set of necessary conditions on the *simultaneous occurrence* of which a clock works; anything less than the complete set is not sufficient for the clock to work. But there are numerous conditions, which *alone* are sufficient for its breakdown.

Likewise, to give an adequate description of a normal kidney or a normal heart is impossible, since this would necessitate an infinite denumerable, exhaustive list of characteristics, anatomical, functional, and numerous clinical tests for assessing kidney or cardiac anatomy and function.

On the flip side, the presence of one single pathological characteristic may indeed have dramatic medical consequences. For instance, a small diminution of the glomerular filtration rate, a long ECG *QT* interval, are sufficient to define an abnormal kidney or an abnormal heart. The functional abilities of an organism are no stronger than their weakest link. Then, beyond this, a well-functioning heart does not offset a failing kidney: there are no effective trade-offs between healthful and harmful processes.

In one of Schulz cartoons, Linus is building a rock wall. A little girl asks him: "*Does it keep things in or does it keep things out?'* and Linus answers: *'It hasn't' t decided yet.*"

In brief, the medical term "normal" should be defined counterfactually.

The Nature of Medical Norms

The term "norm" has two different meanings in medicine. The first one refers to a conventional standard. The second one is injunctive and refers to what must be done.

Medicine works in two steps.

In the first position, it selects a set of rules and criteria by which normal and pathological processes are separated, a conventional distinction that is neither true nor false.

In a second position, medical statements become true or false. Clinicians thereafter perform their medical duties through observation of purely descriptive data, namely signs and symptoms.

For one thing, contrary to social norms that are *reportive*, the definition of "pathological" is *stipulative*. By becoming medical, biological facts split into a binary feature, and separate into normal and pathological. They are *correct or incorrect.* Nonetheless, once adopted, medical norms fall within the context of medical doctrines, are stored away in textbooks of medicine, and may then be *clinically true or false.*

For another, medical norms represent the prudential behaviour of the health care system towards diseases or suffering, namely medical interventions. Medical norms are prescriptive: they are not observable features like weight, size, or speed, but we feel a strong obligation to obey them.

Medical norms are *conventions.*[43] They are rules of behaviour that may be submitted to a social mediation—as they can change—or governed by specific rules, but they are essentially *medical, not sociological.* Right or wrong, medical norms do not imply praise or blame, but success or failure. There's no denying that medical norms are neither true nor false *of* the persons or the features to which they are ascribed; they may be correct or incorrect, but they are neither factual, nor non-factual.

Hence a blood sugar is not normal or abnormal in and of itself, but on account of its predictive meaning and adverse consequences, which separate biological facts from medical ones. Miscarriages are no less normal than births. There is nothing inherent to a piano that is prophylactic of false notes. False notes are not *less natural* than right ones but what differentiates them from one another are some systems of norms alien to natural laws. Right and wrong notes are relative to some written score—like symptoms to our texts of anatomy and physiology.

---

[43] Bicchieri (2014, 208–229).

Canguilhem and "Natural Norms"

In his historically important book *Le normal et le pathologique*, Canguilhem introduced the questionable idea of "*natural norms*" and maintained that each one of us creates and changes his own norms. He also attributes "normative power" to living beings. Since norms are, by definition, conventions, social or medical conventions, this is a contradiction in terms. Norms are parts of a practice that is institutionalized in some way. The long history of incompatibility between the natural and the normative dates back at least to Immanuel Kant. Furthermore, Wittgenstein insists time and again on the impossibility of following a rule privately. The set of correctness criteria for each normative practice must be potentially shared and sanctioned by a community, medical in the present situation.

Medical norms are, by definition, being adopted under a *third person perspective,* so that they are analogous to social conventions, except for their negative sense. Canguilhem's use of the notions of norms became extremely confusing inasmuch as he attributed norms to a motley collection of situations so that one wonders to what features norms don't apply, and what are Canguilhem's criteria for the application or the non-application of norms? Beyond this, Canghuilhem's book is filled with medical errors and misunderstandings on which he is basing his arguments.[44] Finally, Canguilhem is spreading false information, e.g. that vaccinations should be avoided and arterial hypertension does not need treatment! This author is highly unreliable.

The Semantic of the Normal/Abnormal Divide

Everything being considered, medical norms—normal and pathological—are the noncognitivist, intentional foundations of medicine without which the whole point of medicine would be unintelligible. We could echo Immanuel Kant in saying that being normal or abnormal is not a predicate: these are not properties like being red or being painful, neither are they facts or any other element in our world.

One can be a complete master of the concept of some biological process without knowing whether it should be seen as normal or abnormal. Even more, whether or not some process is normal or abnormal does not in the least affect the description of that process. On this score, the terms 'normal' or 'pathological are germane to other predicates such as 'existing' or 'non-existing', 'true' or false' or even 'good' and 'bad' since they are not properties: the ascription of one of those terms to some biological process does not affix any additional descriptive feature to their meaning.

Being pathological is not the result of the breakdown of some natural order lying underneath a world of complexity, multiplicity, difficulty, and suffering. Because there is no such order that can eventually be discovered, no formula, symmetry or

[44] Karhausen (2017).

rightness that virtually rules the biological universe. Disorder, as it were, is some different kind of order. The distinction between normal and abnormal reflects our need to understand and control, as well as to smooth and regulate nature's complexity through rational thought and logic. The concept of a pathological process is not transparent to the underlying biological structure of reality since it rather pertains to the surface of the world.

It sounds like an instrumental rather than a natural feature. Instead of pointing to the organism's inner structure, it brings some natural processes into the sphere of action. On that account, a pathological status, i.e., a medically abnormal status, does not constitute a natural kind: a condition is not pathological by its nature but because we value its consequences for the organism or the type of creatures to which it is attributed.

Since there are no norms in nature, a blood pressure measurement is neither normal nor abnormal *per se,* but only in reference to the clinical setting or to its adverse consequences if any. Hence, being normal or abnormal is not predicated of the biological process itself but it depends on the criteria of its ascription, and then merely contingently as a value the concept bears in its interlocking relations with medical prudential judgements. The meaning of 'normal' and 'abnormal' consists in their ascription and the ensuing consequences thereof: the procedure of ascription exhausts and achieves its normative meaning.

So, in a way, abnormality does not decide whether a condition is medically relevant. This is because we decide what conditions are medically relevant or are diseases, not on the basis of their underlying nature, but on the basis of what sort of consequences they have, namely if they have harmful consequences. What counts as harmful is a value judgment, and not something we discover by employing scientific methods.

Distinguishing Abnormal from Pathological

In clinical language, the terms normal, abnormal, and pathological are adjectival. They can be ascribed to various features which may be developmental, structural (subcellular, histologic, or anatomical) functional (biochemical, biophysical, physiological) or behavioural.

The categories of normal and pathological are the first step taken by medical science in the prudential ordering of life. Medicine must determine the threshold for medical intervention. Although the terms 'abnormal' and 'pathological' are often used synonymously, they differ in their connotative meaning.

To say of some biological or abnormal state or process that it is *pathological* implies that it is *intrinsically disvaluable.*

Our intuitions about what is pathological are derived wholly from the clinical level. Laboratory and post-mortem room results may be deemed abnormal, but being pathological implies being a causal factor of some sort of harm to the patient. It also means that qualifying some adverse anatomical or functional feature as pathological signifies that it is harmful to the welfare interests of the individual concerned, namely to his body or his health *as compared with what his condition would have been had such feature been absent.*

On that account, whether some feature is pathological is understood as *whether it does or may cause or increase harm or suffering* which, within certain limits, are verifiable matters; in some cases, being pathological refers to some relative proportion of harm (that is intrinsically related to the pathological process) and benefit. The '*may*' is sometimes understood in terms of risk or probability. Pathological features are tendencies or dispositions to deteriorate or decay; they are potential causes of, or pathways to sickness. The statement that some features or processes are pathological is a modal claim, which means that they *are statistically harmful.*[45] This means an increase in the cumulative incidence of the disease of interest, or of the number of new cases of the relevant outcome as a proportion of the total population, during a certain time period.

In contrary manner, '*abnormal*' is a seamless robe, an open-textured, interval-valued predicate, which comes in degrees: it represents a continuum of increasing predictive adverse consequences and a domain of medical uncertainty. If being abnormal is scalar, being pathological is non-scalar and stipulative. This distinction is essential since it separates a quantifiable phenomenon, from a predicate that, contrary to being abnormal, has *deontic* status: it calls for reparative intervention.

Ventricular ectopic beats are abnormal: they always represent a premature depolarization of ventricles, an abnormal electro-physiologic process; although present in more than half of apparently normal people, they may be pathological, namely evidence of serious disturbance of the rhythm.

Claude Bernard has shown that abnormality may not be a discrete but a scalar term, in which case there is a continuum of severity from health to very severe conditions. A kidney cyst is abnormal, but not pathological. Shortness of stature is a disadvantage in some particular social settings, which value body height. It may decrease the ability to play sports, to choose some professions but it becomes abnormal or pathological when, on the request of the family, the doctor decides that the child needs growth hormone. In a study of 101 healthy newborns, 46 had subdural haemorrhages which disappeared by 3 months. So, an individual may be clinically abnormal in a purely statistical sense or even a medical sense, in the absence of any significant disease. Adults with autism spectrum disorders are clinically abnormal, but some of them with limited intellectual disability are not seen as mentally ill and do not necessitate psychiatric or social help.

Sub-threshold, abnormal but not pathological forms of cognitive and affective deviations are encountered among relatives of schizophrenic patients. Abnormal, borderline or 'spectrum' forms related to affective psychosis such as the recognition of cyclothymic personality disorder are of limited therapeutic interest, since most

[45] Goosens (1980).

cases do not require pharmacological treatment, and there is little evidence that treatment might be effective.[46]

Everything that is pathological is abnormal, but the reverse is not true. The conceptual gap between what is abnormal and what is pathological describes a zone of uncertainty of penumbral cases subject to constant re-examination. Mistaking abnormal for pathological manifestations is a common source of unnecessary diagnoses and faulty treatments.

This perhaps merits a digression. Opposite terms may be *contradictory* (such as white and not white) if both cannot be true and both cannot be false, so that one must be true and the other false: they have opposite truth-value. On that account, 'normal' is contradictory to 'abnormal': whatever is abnormal is not normal. In the case of two *contrary* terms (such as being white and one of the ways of being non-white e.g., red), at least one of them is false and both cannot be true, but both may be false. Contrary means "*at least one is false*". Normal and pathological are contraries: they cannot both be true, but some feature may be abnormal yet not pathological e.g., in the case of minor dermatological disorders such as lipomas, skin tags or angiomas. On that account, normal and pathological are contraries.

The upshot is that the meaning of the term '*pathological*' is a matter not of nature but of convention. Pathological features are not in the world *qua* pathological but are presupposed in our medical theoretical doctrines and cognitive procedures. They betray the deforming lenses through which medicine looks at biological facts. For various clinical and diagnostic features, it is as if taking them as abnormal was *extensional* since it refers to some scale along which individuals are being truthfully graded and categorized. But being pathological comes into play, as it were, in an *intensional* context as it is a mere convention, not a fact.

The Logical Priority of the Pathological

Being pathological has the primary meaning as it takes some logical precedence over being normal. *Magna est vis instantiae negativae.*

"*In the case of organs,* writes G.H. von Wright (1916–2003), the Finnish philosopher, pupil of Wittgenstein, *badness (poorness) appears to be logically primary to goodness.*"[47] *And Elisha Bartlett (1804–1835) added: "Pathology is not founded upon physiology … Our knowledge of the morbid processes and susceptibilities of the several organs and tissues of the body cannot be inferred or deduced from our knowledge of their healthy processes".*[48]

[46] Jablensky (2012, 82–83).

[47] Von Wright (1963a, 55).

[48] Bartlett (1844, Pt II, Ch. 6).

Contrary to the received view,[49] norms, in medical logic, are negative. Several pairs of words, which are the cornerstone of medical systems, such as 'normal' and 'abnormal', 'ill' and 'cured', 'anxiety' and 'relief', 'health' and 'disease' or 'life' and 'death' give evidence of a significant asymmetry.[50] "*We feel pain, not painlessness,*" writes Schopenhauer. The negative members of the above pairs are irreducible as well as more determinate, explicit and basic than the positive ones, which are ill-defined and indicate derivative, secondary ways in the theoretical articulation of medical concepts. The factors and processes which make people suffer or be disabled are better known and far more firmly grounded and classified than those which make them healthy. Death is a definable fact and there are listings of causes of death, but we rarely wonder what the reasons are to be alive. Having something wrong: pain, disability or death are basic concepts that clarify the logic of the concept of "pathological'.[51]

And J.L. Austin suggests that "*so often, the abnormal will throw light on the normal*" and that "*it will not do to assume that the "positive" word must be around to wear the trousers*".[52]

In matters of health and illness, pathology has the first word since it is conceptually prior to physiology. A normally functioning heart is not described as normal because of some benefit it promotes but merely by not causing harm; defining some function as normal, implies privation or absence of some resultant painful or harmful features. *Omnis determinatio est negatio.* The oldest Greek biological books presented "*the evolution of physiology and pathology and they tended to begin with the sick man (with pathology) and from that have learned about men's normal constitution*".[53]

Even then, physiology at this point is true by default and is grounded in the absence of abnormalities or diseases. In medical language, physiology is true by deliberate omission, so that it is false: it describes non-existing processes to which actual processes may be compared, graded, and lined up in terms of similarity. Physiology echoes Newton's seventeenth century ideals of natural order such as his principle of inertia, i.e., his first law of motion, which was an abstract paradigm of natural motion, which is never encountered in the real world. "*He is providing us,* writes Toulmin,[54] *rather, with a criterion for telling in what respect a body's motion calls for explanation ... Only if a body ever were left completely to itself, would it move steadily along a straight line, and no real body ever actually is placed in this extreme position.*" And this concept of the 'natural course of events' is given in negative terms: "*positive complications produce positive effects and are invoked to account for deviations from the natural ideal, rather than conformity to it. This*

[49] Reznek (1987, 154–171) and Boorse (1977).

[50] Tranoy (1967).

[51] Culver and Gert (1982, 27, 64–85).

[52] Austin (1970, 180, 192).

[53] Hall (1969, 66–68).

[54] Toulmin (1961, 56; 79).

being so, the appearance of words like "inert" and "inertia" in our theories is perhaps significant. For these are essentially negative terms, indicating how things will behave of themselves if nothing is done to them from outside". If left to themselves, things will follow their "natural" course. Physiology is true by default.

Medicine does not maintain that one *ought to* be normal: what is desirable is not the condition of carrying normal features but rather that of not displaying pathological ones. This implies, it seems to me, that *normalcy is privative*, that it is constructed counterfactually as it corresponds to the absence of abnormal features. In contrary manner, being harmful or detrimental means to affect someone's health unfavourably, which is not the same as not affecting it favourably: it is logically primitive.

Likewise, pathological features also have a psychological precedence. Plato writes: "*Well, you must have heard people say, when they are ill, that nothing is pleasanter than to be well, though they never knew it until they were ill.*"[55] Being pathological is undesirable: it results in pain or unpleasant experiences and limits the liberty of action, of flourishing and growth.

Medicine is basically concerned about suffering and those bodily or mental states that may be consequent upon thwarting prudential needs. The relief of suffering exerts a stronger claim than the promotion of positive outcomes such as happiness or health. Public health itself has been governed much more by avoiding threats than by seeking goals.[56] One does not resort to medical help in the hope it will offer some receipt for happiness or well-being: von Wright contends that if happiness is positive, health is negative.[57]

The primacy of what is pathological contrasts with the machine analogy. In cases of instrumental or technical concern—for instance if your watch is stopped— the positive, namely the forms of goodness, are logically primary to the privative: good cars are *conceptually prior* to defective cars. We do not need to observe defective cars to establish what we mean by a good car.

The Clinical Epistemic Priority of Normality

The previous *first position* is now being reversed in clinical medicine with the *second position*, while normality surfaces as a natural standard setting that converts the *normative convention* of being normal or pathological into its contrary. Physicians indeed learn physiology before pathophysiology and biochemistry before pathological biochemistry. Once physiology has been adopted as a default position as shown previously, sentences which describe and classify some features as

[55] Plato (1945, 309).

[56] Vickers (1958).

[57] von Wright (1963b).

abnormal, unhealthy, or pathological, are henceforth being determined by reference to this standard. Counterfeit banknotes presuppose genuine banknotes.

From now on, clinical language's cognitive approach endorses some form of epistemic naturalism, namely describing patients, analysing their complaints, and trying to categorise and to explain their conditions.

With this move, what reason or justification do we have to call some feature pathological? A reason or justification is something that is or can be given in reply to a question, and it depends on the kind of question that can be asked in the circumstances: this method which was illustrated by Socrates is called *elenchus:* it consists in asking questions in a comparative form; when we raise a question, we often have a *contrast* in mind which hides some implicit subordinate clause: 'Why do cigarette smokers (*compared to non-smokers*) develop lung cancer?' What we attempt to explain is not simply: 'Why this?' but rather 'Why this rather than that?' We compare a fact with a foil and a fact may have several foils, various contrastive alternatives depending on our interests.

Since normalcy is now a theoretical contrast case, the epidemiological investigator often requires several sets of controls going in for allegedly normal populations. This reflects the difficulty if not the impossibility of identifying a counterfactual population, which might be considered as normal. Surrogate groups may be selected in medical and epidemiological research from the general population or from assumedly healthy people. Depending on the question, the contrast group may be matched for sex, age and other characteristics known to influence the distribution of the relevant pathological variate.

Stuart Hampshire writes: "*Anyone who applies concepts necessarily applies also the distinction between a standard or normal case of something falling under a concept and an abnormal or imperfect case. He cannot avoid making this comparison*" ... "*The comparison and ordering of specimens as more or less imperfect specimens of a kind is as unavoidable as the comparison and ordering of statements as more or less certain*".[58]

The notions of 'normal' and 'abnormal', health and disease, physiology and pathology are comparative in form and 'normal' serves the role of a conventional reference or neutral value, a kind of zero level like the tonic in a musical scale.

To make a long story short, clinical medicine, with the *second position*, reverses the original ontological sequence in the way that the logical priority of abnormality turns into its converse, namely the epistemic priority of normality that is, by a medical convention, treated as *a given*. This epistemic priority depends not on how things are, but rather on how we can tell what things are.[59]

If so, then, *once adopted by clinicians*, statements about medical, biochemical, or physiological norms may be true or false, or even spurious. Thus, medicine proceeds in two steps. Gilbert Ryle insisted about the momentous difference that separates two dimensions in epistemology: he held that epistemologists very frequently

[58] Hampshire (1965, 225).

[59] Hurley (1989, 10).

describe the path-making as if it were a piece of path-using.[60] Medicine starts with a *path-making move*, by choosing a normative system that divides physiological from pathological features. Once it has been decided to accept and to conform to this first normative system and its specific extensions—the medical insight becomes openly propositional, through a *path-using move*, and finds itself on the 'fact' side. Medical statements then become true or false, since they result from premises that medicine considers as true.

In other words, normative and non-normative features turn naturalistic, factual, and descriptive within the *second position,* in clinical medicine and in epidemiology, and as soon as a text of physiology is being written and agreed with. From then on, diseases represent some negativity, defectiveness or an error marring the wholeness of our living, which needs to be corrected.

In the light of this, we may now come back to the question of epistemic naturalism. Are the terms 'normal' and 'pathological' naturalistic or not? By now, the question may sound a bit awkward: it is one thing to say that these normative terms are not epistemically naturalistic since they are not descriptive as they designate some sets of rules and standards, and departures from them. But it is quite another that the clinical gaze about diagnostic procedures and nosological categories is a purely descriptive exercise, even if it is implicitly subsumed under some tacit, underlying, theoretical articulation implying normative facts or properties, in virtue of which clinical statements may be true or false. In other words, a text of physiology proposes facts and properties that are theoretically normative but become naturalistic when used empirically in clinical or public health practice. And clinical facts are true on the condition that basic medical concepts have a relevant normative credit.

Pathological processes are part of nature but only *qua* processes. But the boundary we are obliged to countenance between normal and pathological processes is not part of the furniture of the world. Despite their anti-naturalist framework, even though they occupy no natural or non-natural space, normative practices are compatible with epistemic naturalism, because they express the commitments and evaluations that physicians construct when they cope with other natural entities, such as stethoscopes, laboratory tests and imaging.

There are thus good reasons to probe the grip that naturalism holds onto the medical territory. A physician does not need to entertain some explicit, recognizable regular instruction for acting in a normative way: following a norm, deciding that such or such a clinical manifestation might be pathological or not, does not implicate repeating to oneself general instructions that work as maxims. There is no algorithm for normativity. Clinical medicine is often a product of inevitable "muddling through" Yet, the ascription of the normal/pathological divide is being made according to criteria which are external to the architecture of biology but which medicine endeavours to internalize. The correctness criteria must be potentially shared and sanctioned by the medical community.

[60] Ryle (1963, 272–273).

Specification of Pathological Boundary: The Case of Hypertension

"*It has been relatively easy*, writes Homer Smith, *to slip into the error of dividing everything into the normal and abnormal*".[61] But the question before us is whether the yarn of biological facts shows some crack fault, which might divide them into two different sets, those that are normal & those that are pathological. But surely, some diseases, symptoms and signs or clinical variates are discrete while others are continuously distributed. Plainly, some states of affairs have a clear-cut line of demarcation: to the question whether a categorical or *non-scalar adjective* is ascribable to them, the answer is 'yes'; it admits no borderline cases. Such discrete variates clearly separate pathological sets of characteristics from normal ones, such as phenylketonuria, chromosomal anomalies, inguinal hernia, malignant tumour, tuberculosis, or leukaemia. So, it seems, to say the least.

However, medicine in the second half of the last century became progressively more concerned with quantitative or naturally continuous variates. There is a greyish penumbra around the demarcation between normal and abnormal traits. The range of features to which the adjective 'pathological' applies is not sharply defined but fuzzy at the edges.

It follows that abnormal features usually shade progressively into normalcy so that a clear distinction cannot always be drawn between pathological and normal and between indisputable, doubtful, and non-eligible cases.

For all that, medicine, being driven by the need for intervention, needs single cut-off points; to dispel vagueness and at the risk of being too inflexible, it resorts to stipulative definitions or a decision to divide the spectrum of traits into two (or more) conventional categories, each with a range of variability. The need to choose some boundary that separates normal from pathological, and how medicine proceeds to set up such boundaries seem to refute Boorse's approach better than any long discussion.

The cut-off points for what is pathological may be *clinical*, namely '*the level of a variable above which symptoms and complications become more frequent*' or *prognostic*, i.e., defining the level beyond which intervention will improve symptoms or course of the disease.[62] This conventional fiat pigeonholes individuals into sick and non-sick, those needing care and those who do not, those who are normal and those who are pathological. In this way, natural continuous distributions are being transformed into discrete categories. Of people with an intraocular pressure (IOP) above 21 mmHg, 1–2% per year develop glaucoma. Additionally, about one third of patients with glaucoma do not have an IOP > 21 mm Hg.

Such boundaries are always subject to revision and have occasionally been progressively shifting in the last century, for example in the case of sexual 'deviations', alcohol abuse, blood cholesterol, blood pressure, infertility, schizophrenia. Changing

[61] Smith (1947, 374).

[62] Rose and Barker (1978).

environmental conditions may abruptly alter what is usually accepted as normal. The expansion of the iodide pool from dietary sources resulted in a dramatic decrease of the 24-hour thyroid radioiodine uptake obtained from clinically normal patients.[63] Moreover, emphasizing treatment tends to minimize normalcy in the effort to avoid false positives but giving priority to prevention and screening tends to lift it up, as it will include false positives by trying to avoid missing false negatives.

The historical controversy between Sir Robert Platt and Sir George Pickering[64] revolved about the true character of arterial hypertension and more fundamentally about the distinction between normal and abnormal, physiological, and pathological as well as normotension and hypertension. Are hypertensive patients a discrete group or is there no natural dividing line between normal and high blood pressure? Platt defended the first and Pickering the second view. Platt claimed that the distribution curve of blood pressure was bi- or tri-modal, while Pickering argued that it was unimodal and Gaussian but slightly skewed. Platt used hospital propositi with their inherent selection biases while Pickering preferred to study populations of unselected individuals. Pickering's view has prevailed, and it is now generally acknowledged that there is no line dividing normal pressure from hypertension: we have to think of hypertension in quantitative not in qualitative terms. Since there is no natural dividing line between normal blood pressure and hypertension, hypertension does not constitute a natural kind.

Considering that the population distribution of blood pressure is continuous, an arbitrary cut-off point is needed to define hypertension.[65] A person with a systolic blood pressure of 160 mm Hg is hypertensive and someone who has a systolic pressure of 110 mm Hg is clearly normotensive; but between these two extremes there are penumbral cases in which one cannot say whether the blood pressure is normal or not. Since the probability of future complications shows an exponential relation to blood pressure over the whole range and with no threshold, it is important to define the level above which intervention is necessary. This shows how physiological norms, though conventional and remaining conventional, may change under the pressure of new evidence.

If the guidelines had lowered the systolic to 110 mm Hg, very few persons with putative hypertension would be missed, but many normal individuals would be wrongly classified as hypertensive; conversely if the limit was set high at 140 mm Hg. systolic, a smaller proportion of putative hypertensives would be misclassified as normal, but a greater fraction of normal individuals would be classified as hypertensive. A low threshold leads to false positives, and a higher threshold leads to false negatives. The common territory shared by the *overlapping normal and abnormal distributions* represents the *fringe*, a twilight area of doubtful application, which includes the borderline cases, an intermediate so-called 'pre-hypertensive level'.

---

[63] Greenspan (1970).

[64] Swales (1985).

[65] Pickering (1999) and Murray and James (1999).

Thus, translating quantitative into qualitative variates by setting up a dividing line implies trade-offs: to avoid misclassifying normal individuals, one must accept failure to include all abnormal ones; conversely to make sure of identifying all abnormal individuals, we must accept misclassifying as abnormal and increased fraction of normal ones. Opting for the best boundary value for hypertension, becomes quite tricky when the normal and the abnormal population distributions overlap. In like manner, several other measurements such as glucose tolerance and diabetes, or intra-ocular pressure and glaucoma, or osteopenia and osteoporosis can be plotted against the number of cases in which the relevant disease is either present or absent, yielding two partially overlapping probabilistic curves.[66]

In short, in the case of a discrete variate, we usually *know* whether it is abnormal; but, in the opposite case, we make allowance for a wide set of clinical, epidemiological, and statistical information, then *infer* that to a given level of a continuously distributed variate corresponds a given probability of being abnormal, and finally *decide* what will be our operational threshold of pathology.

Jacob Stegenga does not seem to understand the rationale of the trade-off's predicament, which he attributes to the *disease creep*.[67]

Vagueness is inherent to qualitative predicates; they are *scalar adjectives* and correspond to a gamut of properties which come by degrees; they involve comparative properties such as some clinical manifestations (height, blood pressure), as well as the majority of characteristics singled out by diagnostic procedures including signs and symptoms (e.g. urea nitrogen blood level, the length of *P-Q* segments in ECG), functional capacities and most behavioural characteristics. In addition, some pathological features such as cataract, diabetes mellitus or depressive manifestations are scalar concepts. We usually can classify them by comparing them with a conventional standard of normalcy.

On this score, hypertension does not come in ontological chunks: there are seamless gradedly different levels of hypertension with corresponding continuum of potential harm. The New Guidelines if the AHA and the American Heart Association 2017 Scientific Sessions[68] recommends new targets for treatment:

- Normal: systolic blood pressure less than 120 mm Hg, and diastolic blood pressure lower than 80 mm Hg.
- Stage 1 hypertension: average systolic blood pressure 130–139 mm Hg, and diastolic blood pressure 80–89 mm Hg (previously christened *prehypertension).*
- Stage 2 hypertension: average systolic blood pressure equal or higher than 140 mm Hg, and diastolic blood pressure 90 mm Hg or above 90 mm Hg.
- Older patients: older patients have the same treatment target as younger adults, although several health authorities (American College of Physicians and Academy of Family Physicians) recommended a target systolic blood pressure of less than 150 mm Hg for the elderly; it should be 100 plus the patient's age.

[66] Bhopal (2016, 202–204).

[67] Stegenga (2018b, 86).

[68] Welton et al. (2018).

Such interval valued classifications and the quantification of clinical judgement are helpful for prognosis, clinical decision-making, and treatments.

The problem of defining hypertension illustrates the poverty of the naturalist standpoint that, for sure, provokes interesting philosophical top-bottom debates, which too often blossom far away and in exile from clinical medicine and epidemiology, their pitfalls, and dilemmas.

Harmful and Pathological

Harm, be it done to or suffered by an organism, is a defining or essential characteristic of being pathological. It is logically impossible that biological 'harm' or 'harm doing', would apply to some feature to which the term 'pathological' does not apply. And it is also logically impossible for the term pathological to apply to some feature to which the term harm would not apply. Being pathological is a convention that refers to a deleterious biological manifestation or condition, namely a harmful one.

Yet, being harmful differs from being pathological in that a harmed condition, process or situation is itself harmful, while being pathological is not pathological. Something harmful, such as a disease, is aversive for two reasons: it produces harm and it is itself a bad condition, namely the condition of someone that has been harmed insofar as it results from causes that are harmful.

However, 'harm' is a term of common use, while 'pathological' is a theoretical term; the first one refers to a process or the result of a process, the second to a norm. Harm comes by degrees but being pathological is bivalent. 'Pathological' contrasts with 'normal' but harm contrasts with 'well-being', 'benefit' or 'satisfaction of needs and 'presumptive interests' or some range of 'agreed-upon interests'. The concept of harm seems more fundamental conceptually than that of being pathological since it is independent of our representations. Being pathologic ascribes a convention while being harmful, as it were, is a finding.

One consequence follows from this analysis. Being pathological is attributed in the presence of an aggregate or of the emerging sum of disparate harmful processes or situations or states of affairs. "*Our 'empirical propositions' do not form a homogeneous mass*" writes Wittgenstein.[69]

Yet, for the sake of translating the utilitarian language of harm into the conventional language of pathology, harmful features must be neutralized since they need to be made additive: each one of them is being treated as if it was homogeneous in quality, divested of descriptive traits and merely differing from one another in terms of intensity or duration. Their qualitative differences are taken away, leaving a remain of neutral aversive features.

[69] Wittgenstein (1974, 213).

When Being Abnormal Is Advantageous

Beyond this, there is a further consideration. If some descriptiveness of 'normal' or 'abnormal' were forced on us, this move should not lead to contradictory statements. The truth is that some conditions can be at once beneficial as well as abnormal. This stems from a conflict of obligations since the force of the claim advanced, on the one hand, by the adverse effects has to be weighed against consideration of its healthful impact.

By way of illustration, sickle cell disease is a haemolytic anaemia characterized by sickle-shaped red blood cells and a serious and often lethal disease in individuals who are homozygous for the gene for sickle cell haemoglobin S instead of haemoglobin A. Nevertheless, red cells of heterozygotes (haemoglobin AS) are more resistant to the malarial parasite *Plasmodium falciparum* than are those of homozygotes (haemoglobin AA). There is an odd arithmetic acting here. Briefly, malaria is a serious disease, and the sickle cell gene is deleterious; both are negativities, but the last provides an advantage against the first; the harmful gene makes the heterozygote fitter to its environment. The sickle cell gene is at once beneficial and deleterious, and the 'normal' gene makes one susceptible to malaria. The applicability of our normative terms is now stamped by ambiguity.

So, in a way, a condition may have positively and negatively valued effects, beneficial effects as well as pathological traits, features or consequences that are conflicting with one another. Schizophrenics are better than normal in some cognitive tasks as well as at discriminating between genuine and sham emotions in other people.[70] Aristotle already linked melancholy and depression with creativity; similarly, it is now well known that depression may enhance literary and artistic creativity and some of its aspects may be seen as normal adaptive responses to stress.[71]

Summary

1. Medicine springs from the conventional distinctions, which *split biological processes into two conventional parts, normal and pathological.* The concepts of being abnormal or being pathological are the primeval points of departure of the logic of medicine, after which the medical picture of the world depends on blended prudential and descriptive gauges.
2. Abnormal or pathological features are *not natural kinds.* Being normal or being abnormal have no descriptive meaning and they are neither factual, nor non-factual. They are not secondary qualities either for they are not qualities at all. They do not consist in a mere breakdown or shortcoming of nature, but they are

[70] Nesse (2000).

[71] Lewy (1968).

construed as *caused alterations,* and not mere deviations or approximations, from some conventional counterfactual physiological norms.

3. The philosophical literature generally manifests a heroic effort to define the concept of normality—a non-existing property of living beings—with the mediation of theoretical concepts such as the biostatistical theory, the homeostatic view, or Christopher Boorse's naturalistic theory of health, which all run aground when confronted with the clinical reality.
4. Medical norms—abnormal, pathological, dysfunctional— are medical conventions pertaining to whether some feature *does or may cause harm or suffering*. Medical norms are neither true nor false, though they may be correct or incorrect. They are constructs cut out of the biological reality. It follows that medical science looks at the world in terms of hybrid concepts that have two dimensions: they have 1 foot in the world and a second one in values.
5. Being *abnormal is dimensional and admits of degrees*: one moves seamlessly from normalcy to abnormalcy. Medicine being an applied science, needs fence-sitting boundaries between physiology and pathology so that *being pathological should be a precise all-or-nothing matter.* Those boundaries are not resident in the world; they are neither true nor false; they are conventions that result from some medical decision based on predicted.
6. Being 'pathological' is intrinsically bad and ontologically prior to 'normal' so that the latter is construed counterfactually: normal is a default position, a mere conventional yardstick. However, clinical medicine reverses the situation and gives epistemic priority to normality and converts the convention of being normal into a 'natural' position.
7. Medical language is at once descriptive and prescriptive, empirical, and normative, assertive, and performative. They manipulate kinds of concepts that both have a significant degree of descriptive content and are evaluatively loaded.

Chapter 4
Explanation

There are four types of things contrary to nature: the activity that is damaged, the disposition responsible for it, the cause which produces the disposition, and the symptoms that follow upon it.

Galen (1991, 151)

Contents

Natural sciences explain occurring phenomena, which means, by and large, that they make them intelligible or rational by showing what their structures are, and how they may have evolved out of their immediate and direct conditioning

L. Karhausen, *Analytical Philosophy of Medicine*, Philosophy and Medicine 152,
https://doi.org/10.1007/978-3-031-81257-6_4

antecedents. One might then characterize explanation as the imposition of intelligible order into what was previously unintelligible chaos.[1]

Explanation or Causation?

The question now before us is: how do we differentiate causation from explanation?

James Lovelock writes: "*As Newton found long ago, logical thinking does not work with dynamic systems, things can change over the course of time. Quite simply, you cannot explain the working of something alive by cause-and-effect logic.*"[2]

In ordinary speech and even in philosophical thinking, causation is often not differentiated from explanation and the epidemiological literature also tends to obliterate the distinction. Causation comes first in medical thinking. That *Helicobacter pylori* is the major determinant of gastric ulcer was known more than 20 years before it became accepted by the medical community.[3] This causal relation was rejected because it seemed impossible that some bacteria could attack the stomach lining in its harsh acidic environment. Critiques expressed at the level of explanation delayed acceptance of a specific causal factor.

In most analytical epidemiological studies, causal relations are up for grab, not explanatory theories: we knew that cigarette smoking increases the probability of lung cancer before knowing what kind of carcinogenic substances were contained in tobacco smoke. So, the evidence of a cause-effect relationship between smoking and lung cancer has been unequivocal for several decades, even though the *why* and *how* of this relationship has long remained incomplete. In other words, causal statements tell us *that p* gives rise to *q* (causation), whereas scientific explanatory theories and laws tell us *why p* leads to *q*. Investigating a causal effect in medicine does not necessarily require knowing its mechanism.

On the one hand, to cause something is to *produce* something, to *bring about* something. On the other, scientific knowledge enables us to *explain why* things occur as they do.

Diseases do not cause their signs and symptoms since they are not the effects of the disease process, but part of it; they are internal to the disease process whereas the effects of a disease are external to the disease process.

For Donald Davidson, causation is an extensional relation that holds between coarse events, while explanation is an intensional relation that holds between the coarse events under a description.[4] Causal connections obtaining in nature behave extensionally since a cause is external to its effect, and since its effect is its

[1] Humphreys (1989) and Schaffner (1983, 1993).

[2] Lovelock (2019, 18).

[3] Thagard (1999, 40–70).

[4] Davidson (1967).

extension, namely the object to which it refers. By contrast, explanatory relations, and contents of explanations, being linguistic, behave intensionally.

Causation describes some natural propensity, liability or disposition that ligates one type of occurrences to another type. According to Strawson[5] it is often seen as a relation, which holds in the natural world between things to which we can assign places and times. But explanation is a rational relation, which holds between facts: the objects related are not found in nature and the relation is between objects, "in our minds". Explanation clarifies or produces understanding of events heretofore not understood. In medicine one talks metaphorically about the *mechanism* of a disease when referring to its pathogenesis and explanatory theories. Causes contribute to the existence of their effects whereas explanation, i.e., pathogenesis or mechanism, refer to how a given causal relation operates and requires complexes of processes.

An *explanandum*, namely something to be explained, for instance a disease, describes a change. Causes and effects are temporal endpoints of changes. An explanation thus consists in filling the gaps between them and in spelling out the intervening steps or the course of events that lie in between the temporal endpoints of a change. It follows that the gaps are like blanks in an already prepared framework of conditional expectation, Strawson writes.[6]

Meanwhile, knowing this range of fillings is not essential for resolving causal issues. But knowledge of the latter may allow us to start a prevention programme despite our ignorance of the first. Lack of vitamin B_{12} is the *cause* of pernicious anaemia, and the disorder is being treated successfully by injections of vitamin B_{12}. Yet, pernicious anaemia results from vitamin B_{12} deficiency due to an autoimmune atrophic gastritis with loss of intrinsic factor.

Two Strategies: Biomedical or Epidemiological

Bertrand Russell pointed out that a system must be isolated for any explanation to be appropriate.[7] So that medical research, as well as scientific research in general, endeavour to neutralize the causal factors that are irrelevant for the purpose of the ongoing investigation, i.e., the *ceteris paribus* modifier. *Ceteris paribus* are generalisations applying only when other things are being equal. In the light of this, there are perhaps two important considerations.

Firstly, mechanistic biomedical investigation or experimental medicine, may be minimally characterized as operating in a world of "*closed systems*", of systems insulated from their surrounding biological or clinical background. If so, then, all changes go on within the system. Nothing passes in or out. The objects of analysis

[5] Strawson (1985).

[6] Strawson (1985, 126).

[7] Russell (1959, 197–198).

are then cut off from interfering causal *ceteris paribus* factors by working on experimental models in the close world of a laboratory. "*Fortunately,* writes Claude Bernard, *it is enough for us completely to isolate the one phenomenon on which our studies are brought to bear, separating it by means of comparative experimentation from all surrounding complications.*"[8]

Granted that all there is in medicine are regularities rather than laws, this model lends itself to experimental medicine within a closed set up, where all interfering, contextual factors are carefully controlled.[9] What's more, it assumes determinism (at least for methodological reasons) so that it supposes that every event has prior events as its sufficient or necessary cause. It thus captures an important intuitive feature of mechanistic biomedical explanations by giving information about how some phenomena fit in the causal structure of the world.[10]

Even so, Woodger warns us: "*The process of isolation is absolutely indispensable in many cases in physiological research ... But from the point of view of interpretation of those results it is extremely important to bear in mind one obvious methodological point: it is never safe to assume uncritically or assert dogmatically that an isolated relatum ... exhibits the same properties in isolation as it does in its place in the level–even when it is furnished with an artificial organic environment which is as 'normal' as possible.*"[11]

Secondly, epidemiological research operates in populations, i.e., in *open systems*, and its strategy is probabilistic. Animals, plants, and human society are open systems as they are really interrelated sets of processes and causal networks. The explanations provided by mechanistic biomedical investigations on the sheer basis of underlying intrinsic physical-biological properties do not precisely hold in the clinical or the public health context where allowances must be made of a *ceteris paribus* nature—that is, that they only apply, all other things being equal.

To patch up this difficulty, the strategy is this: an open system may selectively be closed to causal influences from outside through an alternative *comparative strategy:* epidemiology or clinical trials, hold other causal factors constant by using a comparative design which blindly and randomly assigns the individuals either to the experimental or the control group. If we observe some regular sequence between two types of events, the epidemiological strategy consists in that, time after time, producing the first event is followed by the occurrence of the second event. And conversely, whenever the first event is missing, the other event usually does not occur. The concatenation of counterfactual conditions is such that things would have been different from what they otherwise would be, had nothing interfered with their 'natural' course.

Acquired aplastic anaemia is a condition characterized by a low blood-cell count and fatty bone marrow. For one thing, it has several variegated causes or

[8] Bernard (1961, 128).

[9] Von Wright (1971, 60–64, 186–187).

[10] Salmon (1984, 19).

[11] Woodger (1967, 313).

epidemiological determinants, such as ionizing radiation, chemotherapeutic drugs, exposure to toxic agents, use of certain drugs and some antibiotics, parvovirus infections, particular HLA backgrounds or anything of that ilk. But for another, instead of a mosaic of causal factors, *mechanistic biomedical* explanation involves an immune attack on the hematopoietic stem cell: in vitro studies and clinical investigations suggest an immune pathophysiology with the presence of autoantibodies expressed on all haematopoietic cell lineages; furthermore, an anomalous population of cytotoxic T-cells has been noted in affected individuals; these activated T-cells release interferon gamma and tumour necrosis factor alpha, which inhibit haematopoiesis in vitro and are present in blood and bone marrow of patients with aplastic anaemia.

What clearly emerges from this is that mechanistic biomedical and epidemiological research are complementary. Causal/mechanical explanations are construed either by bracketing off certain factors (in experimental medicine), or by controlling them (in epidemiological studies) to isolate specific factors responsible for some observable effect.

Scientific Explanation: The Hypothetico-Deductive Model

The so-called hypothetico-deductive or law-covering model of Carl Hempel is a way of getting sound scientific beliefs by the generation of hypotheses from which predictions are deduced. If the predictions are true then we have evidence for the hypothesis and if they are false, we have evidence against it.[12] The form of such an explanatory story must, then, be deductive and "deterministic".

Such mechanisms and causal stories yield what has been called by Marsden Blois[13] a *vertical* explanation that pertains to causal chains and a hierarchical model of knowledge, which proceeds from the general to the particular. But are there scientific laws in medicine? Must we allow for law-like[14] statements relating clinical observations with physicochemical processes? Is the hypothetico-deductive model relevant to medical research?

Firstly, there is little concern in the medical literature with discovering laws of nature and universal laws because, except for trivial statements, there are very few general statements in medicine or epidemiology which one would call laws. Medical research is actually a set of case studies oriented projects.[15] Most of medical research is made of minute patient and diversified activities: studying the multistep process of carcinogenesis, identifying the single gene mutation responsible for b-thalassaemia, describing the distribution of a disease in a population, establishing

[12] Hempel (1965, 231–243).

[13] Blois (1988).

[14] Law-like, since they may be false.

[15] Smart (1963, 52) and Rosenberg (1985).

a causal relation, identifying some syndrome, evaluating a diagnostic test, applying decision analysis to some specific medical condition or anything of that ilk.

Secondly, due to the number and diversification of initial conditions as well the genetic variations, stochastic lines of arguments are pervasive in medicine. To this we must now turn.

Scientific Explanation: The Inductive-Statistical Model

In contrast to the stringent covering law explanation based on deductive inferences, epidemiological reasoning turns out to be inductive rather than deductive. From true premises that have some empirical content, it supports some conclusion not by showing that it is true but merely that it is reasonably probable. Epidemiology is the study of the distribution and determinants of disease frequency in the community or in the clinical setting, in the complex clinical or social reality and not merely in some closed experimental system.

Epidemiology is apt to be '*horizontal*'. It makes great play of contextual determinants such as time, place, and person, including age, sex, socio-economic level and factors like tobacco smoking, alcohol consumption, lifestyle, albeit also molecular body components or processes. In contrast to the first approach, this *contextualist approach* is less concerned with absolute guarantees. Its method strives to arrive at the most probable or the most plausible explanation and it does not acknowledge any statement as basic. It relies on probable hypotheses supported by inductive modes of reasoning and appeals to evidence that may not resolve all possible doubt. Epidemiology hence gives way to probabilistic-statistical reasoning and pertains to Hempel's *second model:* the *inductive-statistical explanation* (I-S explanation),[16] which is of stochastic rather than deterministic nature.

It should now be plain that causal factors, which are basic to preventive medicine, are neither necessary nor sufficient but are *ceteris paribus* probabilistic causal factors that indicate an increased or decreased liability of falling ill. Causal factors are not entities but predicates, which become medically meaningful only when related to some disease entity or health problem. In epidemiology, risk factor does not necessarily imply that the characteristic has a causal effect. Furthermore, some risk factors may have more than a mere statistical association with disease. When a causal relationship is agreed, the terms '*causal factor*' or simply '*cause*' or '*determinant*' are used. Smoking is a cause of coronary heart disease (CHD). Yet, epidemiologists rarely claim that most CHD risk factors have a causal role, for instance: hyperhomocystinaemia, low levels of high-density lipoproteins (HDL), high C-reactive proteins, stress etc.).[17] There is a great imprecision in contemporary epidemiology in the use and interpretation of such terms linked either to the error or to

[16] Hempel (1965, 377).

[17] Bhopal (2002, 222).

the avoidance of the error of attributing cause where it is merely association and conversely.

For all this, there should be felt a certain looseness in causal explanations in medicine. Tobacco smoking increases up to twenty times the incidence of lung cancer in males who are heavy smokers. Even then, many people who smoke do not develop lung cancer before dying of some other causes and some people develop lung cancer although they never smoked.

Summing up, medical explanation, ultimately, is methodologically two-dimensional and consists in two ways of thinking and two methods of epistemic access to reality that are perpendicular and complementary, though often blended—mechanisms and determinants of disease processes, namely what is usually termed pathogenesis and aetiology. For one thing, mechanistic biomedical research intervenes in closed, deterministic experimental set-ups to clarify the mechanism connecting cause to effect; for another, epidemiological research evolves in an open, probabilistic context; it tries to identify causal factors or determinant of diseases and uses statistical surrogates for experimental controls.

Epistemic Risk: Inference to the Best Explanation IBE

Hypothetico-deductive and inductive-statistical models embody a low-risk strategy inasmuch as they avoid the danger of immediate false beliefs. Yet, there is a price to pay for playing it so safe. A strategy, which accepts a greater risk that some of the results may sometimes be wrong would have chance of providing useful truths. The payoff for accepting epistemic risk, namely the possibility of some false beliefs, was to be a greater chance of getting true useful beliefs. The general idea proposed by Peter Lipton, is that a hypothesis gains inductive support if, when added to our stock of previously accepted beliefs, it enables us to explain something that we observe or believe, and no competing explanation works nearly as well. A hypothesis provides the best explanation when it is more explanatory, broad, modest, powerful, falsifiable, and generally more conservative than any competing hypothesis. Such strategy is the *inference to the best explanation* (IBE).[18]

One of the most common forms of inductive argument is a concept formulated by Gilbert Harman,[19] but it is sometimes also known as "*abduction*", a term introduced by Charles Sanders Pierce (1839–1914). Gilbert Harman says: "*The inference to the best explanation" corresponds approximately to what others have called "abduction," the method of hypothesis," "hypothetic inference," "the method of elimination," "eliminative induction," and "theoretical inference.*"

In problems of everyday life, we commonly adopt the solution that best explains it. In the hypothetico-deductive method, we try to make sense of a hypothesis by

[18] Lipton (2004) and Sinnott-Armstrong and Fogelin (2015, 195–204).

[19] Harman (1965).

deriving it deductively from premises that are well established. With inference to the best explanation, we reason in the opposite direction, since one derives the explanation from the observation instead of deriving an observation from its explanation.

IBE means we should favour hypotheses that are more modest, simple, natural, falsifiable, powerful, broad, and conservative, and that cohere better with other scientific theories, than any competing hypothesis. Such hypotheses that are both good and better than any competitor, will provide strong inductive support.

The logic of abduction, namely inference to the best explanation, is a major concern in science. Most piece of reasoning in science and in medicine are reached by arguing to the best explanation. The treatment of mental illness, which has been dominated by psychoanalysis has been transformed by inductive explanation with the development of pharmacological research such as chlorpromazine that dramatically improved the condition of schizophrenic patients, electroconvulsive therapy, and cognitive psychotherapy. Of course, whether an inference is the best explanation is strong enough depends, among other things of the context.

Yet, medical scientists may continue to use a theory in the face of counterevidence, and this sometimes perpetuates errors. Until the beginning of the twentieth century, doctors used to reduce high fever by inducing diarrhea with purgative combined with bloodletting, which amounted to deplete patients of fluids and valuable electrolytes. In 1927, a discussion presented by W.G. Spencer on bloodletting, its past and present was held at the Royal Society of Medicine,[20] one century after the groundbreaking publication of Pierre-Charles-Alexandre-Louis (1787–1872) that showed not only that bloodletting was ineffective, but also that it increased mortality.

Beyond this, in the twentieth century, scientists and statisticians have developed methods for handling epistemic risks. and quantifying degrees of uncertainty and evidential support focusing primarily on the use of p values together with a threshold of statistical significance of $p < 0.05$ and avoidance of Type I error.

Jerzy Neyman and Egon Pearson provided a method for choosing between hypotheses. They described two types of errors: Type I error, or false positive and Type II error, or false negative. In Type I error, the experimenter rejects the null hypothesis of no treatment differences, when it is in fact true. In Type II error, the experimenter fails to reject the null hypothesis when it false.

Reduction in Type I errors can only be had by paying a price in increasing Type II errors. This means that reducing the number of false positives over the long run will increase the number of false negatives.

Beyond this, in the patho-physiologic studies and in epidemiological investigations, one must deal with complex multifactorial processes involving modest or week associations and the more variables are included in the analysis, the less plausible the null hypothesis.

In other words, the increasing sophistication of our methodological and statistical tools, allow medical scientists to detect more subtle phenomena, thereby

---

[20] Spencer (1927).

increasing the chance of erroneous findings. Excessive emphasis on *p* value, may distract attention from other potential weaknesses in study design, such as potential confounders.

Finally, there is to determine what degree of epistemic risk is acceptable since one needs to consider external consequences, i.e., non-scientific constraints when setting up evidential standard. Mark Parascandola in his landmark report on Epistemic Risk showed how, in the twentieth century, scientists and philosophers developed methods in public health and medicine for managing epistemic risk and quantifying degrees of uncertainty and evidential support.[21]

There Can Be No Laws in Biology

Paul Thomson and Ross Upshur in their original texts on philosophy of medicine aver that "for *most of science, a commitment to materialism is also a commitment to determinism*".[22]

These authors present a syntactic account of the structure of models and theories, formulated in first-order predicate logic, with a set a deductive system consequence of a set of axioms, which they compare to the foundations of physics by Newton. However, they subscribe to a semantic account that, according to them, mirrors a better actual theorizing in biology which yields to a conceptual modelling of Gregor Mendel's genetics, which they develop in great details. This may come as a surprise after reading Alexander Rosenberg's essay[23] on *Reductionism (and Antireductionism) in Biology*, and Lindley Darden's careful analysis[24] of *Mechanisms and Models*. Rosenberg writes, "*Mendel's laws are not laws: they began to be riddled with exceptions almost from the moment they were first discovered in the early 1900: cross-over, linkage, mitotic drive, autosomal genes, and so forth.*" ... "*And when protected from exceptions by ceteris paribus clauses, they particularly defy derivation from more fundamental principles.*" Rosenberg adds that "*there can be no laws in biology, nor anything that could satisfy the criterion of connectability between an item functionally and structurally characterized.*" Darden countenances that "... *anomalies guide the generation of hypotheses about alternative, variant mechanisms that do not fit the hypothesized schema*".

Meanwhile, there is an alternate possibility. In the same **Companion**, Roberta Milstein and Robert Skipper contend that "*Biologists call models that predict one specific value* ***deterministic*** *models; this should not be confused with the Laplacean or philosophical sense of determinism, which generally refers to a property of a*

[21] Parascandola (2010).

[22] Thomson and Upshur (2017, 43).

[23] Rosenberg (2007).

[24] Darden (2007).

model. Models that provide a probability distribution for a range of results are called ***stochastic*** *models*".[25]

This approach would give way to a pragmatic approach in which the syntactic account of theories would not be formulated in truth-table first-order logic but in modal logic that displays what is expressed by *must* and what is expressed by *possible.*

The Pragmatics of Explanation

Hilary Putnam[26] has defended the view that explanation is an interest-relative notion: *why questions* and consequently explanations presuppose a range of interests and always take place relative to a background space of alternatives. If so, then, an adequate explanation depends on the constraints the investigator *nolens volens* places upon the terms of reference of the exercise. Explanation, Bas van Fraassen contends, is not part of science proper, but reflects various pragmatic interests. It is an application of science, which is brought into play to satisfy some of our desires for information; and the evaluation of how far those desires are satisfied varies from context to context. Explanation involves why questions; instead of an inferential procedure, i.e., a conclusion-deriving, van Fraassen prefers a question-answering approach. So, he starts from the logic of questions.[27] But which questions? What are the best available answers in view of the information at our disposal? How do we avoid getting bogged down in the web of causation?

The traditional view holds that an explanation is an attempt to explain an event. But we never conclusively explain an event, a phenomenon, or a disease. We are only capable of explaining them *in reference to* or *under a certain description.* Thus, for van Fraassen, there is no such thing as explaining an event, separately from the investigator's particular interest. In epidemiology, our question, and the answer we get depends on which set of controls we use. If so then, any kind of explanation is bound to be incomplete as it leaves many things unexplained. What is explanatory in one context may not be so in some other context. Rarely are we able to provide a complete account of all the factors that affected a given outcome.

For instance, an epidemiological prospective study suggests that 12.5% of cardiovascular deaths could be avoided in the UK if the population increased their adherence to the Mediterranean diet.[28] Yet, numerous other factors are associated with cardiovascular mortality, but single epidemiological investigations can only explain a fraction of them.

---

[25] Millstein and Skipper (2007, 28–29).

[26] Putnam (1987, 38–39).

[27] van Fraassen (1980, 134, 156).

[28] Wise (2016).

As medical explanation is interest-relative, the interest determines the appropriate contexts. It differs from biological explanation much as maps differ from itineraries. In other words, medical explanation is concerned with this or that particular kind of medical event that we are interested in preventing, counteracting, or producing. But if medical explanation effectuates a partial retreat in instrumental science, could it find itself in the grip of pragmatic success—which might be no road to truth?

Science has been dominated during four centuries by Bacon's *Novum Organum* and guided by data and the practice of formulating models. Scepticism has been growing during the last 10 years about the importance of formulating hypotheses, of setting up adequate models and of understanding what the most important questions are; the idea has gained ground that science is associated with the accumulation of data, of constructing and implementing regressive algorithms and of identifying statistical correlations, shifting the focus away from causal networks. The success and the amazing efficacy of analysing *big data* is a revolution in science and in epidemiology: the intelligent use of an enormous amount of data made available by computer technology. This trend is being illustrated in medicine by the Framingham Study, the National Health and Nutrition Survey (NHANES-1) and its follow-up (NHEFS), the Follow up Nurses' Health Study, the European Prospective Investigation of Cancer (EPIC), or anything of that ilk.

For all that, "*The scientific method has now become a tremendously vague thing, but this we expected to happen, anyway…*" in view of the difficulties inherent to the formal results of inductive logic: scientific methods often manage to muddle through, provided it presupposes prior notions of rationality.[29]

Nancy Cartwright argued that there is often more than one compatible theoretical account that could be applied to the same phenomena. Scientists often take a pragmatic approach to selecting which theoretical treatment to use depending on which is most appropriate to any particular case. Appropriateness depends in ease of application and of accuracy, but if the theories used in explanation had to be true, no such pragmatic approach would be possible. In other words, explanation and truth are incompatible goals.[30]

The Contrastive Counterfactual Nature of Medical Explanations

But there is a further consideration of immediate interest. Stephen Toulmin writes: "*I drop a glass on a concrete floor and it breaks. So what? That is so unsurprising an event that it raises no issue for science. If it had not broken, that might have prompted a scientific inquiry: "Why not?" Had there been some substance in the glass just before it was dropped that protected it from breaking? Was there*

[29] Putnam (1981, 174–200).

[30] Cartwright (1983).

something about the exact angle at which it felt? Or how else are we to account for its failure to break as expected?"[31]

The need for explanation is now of the form: "Why shouldn't it have happened?" This parable points to the importance of anomalies, events that run counter our reasonable expectations. What are the kinds of medical facts that need to be explained? Not all facts or events demand an explanation but merely some aspects of them. Certainly, not everything that can be observed during an epidemiological investigation or during medical care needs to be recorded and explained. A physician may wonder why his patient's heartbeat is irregular, not why it is regular, even though he is supposed to write it down in the medical record. Medical investigation pertains to those observable features that are either pathological, or that raise a legitimate question for research, or else that are a matter of concern for medical insight or a challenge to some background of beliefs and knowledge. Presumably, genuine issues that call for medical explanation are deviations, perturbations, anomalies, and irregularities.[32]

Should an anomalous state of affairs be defined as anomalous with respect to accepted normative knowledge, relative to normal species functioning? To be sure, in medical language, we call "normal" what makes the backbone of anatomy, physiology or biochemistry. Newton's first law states that objects *with no external forces on them* will continue to move in the same direction at the same speed. Physiology serves the style of explanation according to which, if left to themselves, things will follow their natural course. It marks off those phenomena that need to be explained by contrasting them with the normal course of events, which does not. It is a negative factive situation.

For one thing, this definition of natural order is given in negative terms: the absence of harm, suffering, unmanageable anxiety, or disability. For another, anomalies, deviations, or irregularities are clinically positive complications that produce positive effects. If we take for granted this natural order, we thereby acknowledge the proper relation between physiology and pathology.[33]

It follows that anomalies are not refutations directed against the relevant context. Epileptic fits, for example, are something anomalous. But for the neuroscientist, the problem is to investigate all the neurophysiological processes going on in the brain equally, be they functional or malfunctional alike. So, the neurophysiologist, contrary to the clinical neurologist, treats normal and pathological processes on an equal footing. Even so, a certain contrast is needed for a fruitful *medical* explanation: a physician might ask the neurobiologist why his patient displays epileptic seizures *rather than* none. In other words, explanations are comparative in form. And it is the choice of the contrast case that partly determines what counts as a successful explanation. Abnormal phenomena may thus be explained by comparing

[31] Toulmin et al. (1979, 235).

[32] Ducassé (1925), Toulmin (1961, 43), and Humphreys (1989, 61–94).

[33] Toulmin (1979, 79).

them with other self-explanatory events of the same kind, which are thought to be intrinsically more normal or more natural.[34]

Methodological Reductionism

Traditional philosophy of science ventured the conjecture that, *ontologically*, everything in the universe was made of the same substance (quarks, strings, branes etc.) and could be reduced to fundamental physics at the end. Furthermore, complex things were held to be made of simpler things: sociology should be reducible to psychology, which in turn is reducible to biology, and the latter in physical chemistry and so on.

The alternate possibility is that scientific theories are not objects or things that are part of the furniture of the world: they are constructs of the human mind. This is why theoretical reduction is not a logical consequence of ontological reduction. Theoretical reduction was the philosopher's stone of the Vienna Circle and of the theory of unified science.

Reducing one science to another raises several difficulties, not least because the different sciences use different classifications, each of which is specific to the discipline from which it derives. Add to this the heterogeneity of the categories used by biologists, for example, and reduction becomes almost impossible.

If we forfeit the concept of reduction, different perspectives, and points of view, which are equally legitimate, can and must necessarily coexist in the sciences. And this is what justifies their multiplicity.

Nancy Cartwright (1944–), who succeeded Karl Popper (1902–1994) as chair at the London School of Economics, believes that the project of the unity of science is excessively optimistic. She thinks that the evidence suggests that nature is disunited and that there are pockets of phenomena that are totally distinct from each other, but that behave, each of them, in a sufficiently regular way that we can manipulate them. Science is a structure made up of patches and pieces with tangled combinations. "*Our knowledge of nature is highly compartmentalized*," she writes. Confronted with a problem, scientists deploy a set of empirical strategies and methods that owe nothing to a general theory, but which have proved useful in that they provide the right answer to the problems raised.[35]

Scientific theories are human creations and scientific reduction does not logically follow from ontological reduction, even if it presupposes that complex systems are potentially reducible to properties of simple parts, a proficiency never achieved. John Dupré showed that reductionism does not work in biology.[36]

---

[34] Toulmin (1979, 63) and Glymour (2003).

[35] Galison and Stump (1996).

[36] Dupré (1993).

Meanwhile, there is a reduction of the basic inventory of reduction in scientific research, albeit in the medical sciences where it has a privileged status. Medical reductionism is a system of explanation in accordance with which biological processes, including diseases, are of the form *A's are nothing but B's.* Crudely, clinical entities are nothing but aggregates of physicochemical entities. Ontologically, medical reductionism assumes that complex wholes, diseases, or pathological processes ought to be fully explainable in terms of (and sometimes replaceable by) features involving parts of which those wholes are composed. This is termed micro-reduction. Diseases and clinical facts considered 'macroscopically' are to be explained in reference to biological or physicochemical processes considered 'microscopically'. According to this view, the global properties that characterize some disease process allegedly arise as a simple aggregate of independent biochemical processes plus eventual interaction effects. Reductionism's privileged status sounds appealing to medical researchers. Is not science based on the premise that everything should, at least in principle, be fully explained? It is *foundational* since it holds that most of the causal power is centred on physicochemical processes which are the ultimate constituents in the study of diseases. It follows that, if indeed there are such things as clinical facts, they should ultimately be explainable by factoring them with reference only to rock-bottom physicochemical processes, and not in reference to other clinical facts. Thermodynamics has allegedly been reduced to mechanics in that propositions about heat can be translated, as it were, without remain into statements about molecular motion. Without remain means without presupposing another additional sentence.

Medical research supports *methodological reductionism* or at least fleshes it out. What is the point at issue? Medical reduction is *not a theory of meaning*. It does not contend that every clinical statement is really a statement about biological, immunological, or biochemical features. We should not look for biconditionals expressing some semantic equivalence between clinical facts and patho-physiologic or molecular facts. It is *not an analytical theory*, in accordance with which statements about clinical facts are held to be, at least potentially, translatable without loss of meaning into statements thoroughly about some basic sciences. It is *not an ontological theory* either, since advocating that diseases ought to be reducible to molecular events does not suppose that there are no clinical facts. It assumes that the best strategy is always to attempt explanation in terms of more minute entities without necessarily identifying them.

To be sure, there are descriptive levels and patho-physiologic levels to explain the 'upper' more complex by the 'lower' and simpler. These levels are related in some causal fashion. Yet, each of these descriptive levels is autonomous, has its own vocabulary and, as it happens, its own explanations and causes. Clinical medicine is a discipline, which attempts to achieve its own ends, and pathology is another discipline that is being played with its own rules. Yet, the dream of medical reduction cannot be fulfilled. Karl Jaspers contends that: "*There are no examples in natural sciences or in biology where this reductive model has been fully achieved although progress may have been made in that direction.*"[37]

Medical reduction requires that, between two relevant classes of statements—those from clinical science and those from some more basic sciences—there be some kind of determined relation that is a locus of medical research. This is tantamount to saying that medical research should be striving, among other things, to explain clinical facts in terms of some more basic science. If so, reducibility merely implies the *pro tempore* possible fruitfulness of medical research.

Hierarchical Systems

Explanation by mechanistic biomedical research, namely the *vertical* style of explanation, assumes that there are different levels of phenomenal reality. Living organisms are hierarchically organized in terms of levels of description. They require a multilevel causal/mechanical explanation which purports to explain complex medical phenomena in terms of simpler, more fundamental features. This implies a search for regularities at different levels of explanation, even though these generalizations often cross-levels, with effects at one level being related to causal factors from a different level. Our organisms are made from cells, our cells out of proteins and biochemical molecules, and our molecules out of nanoparticles and atoms and so forth. The metaphorical multilevel model of explanation admits that there are 'higher' and 'lower' levels of organization in biology which must all be ultimately explained in terms of physicochemical mechanisms.[38]

In this view, it is expected that higher-level generalizations will hopefully be restated in terms of lower-level ones. A geneticist would tell us that the genetic susceptibility to type 1 diabetes mellitus is probably polygenic, including genes within some major histopathology complex (MHC), in concert with genes outside the MHC, which regulate insulin production. On the other hand, an immunologist might, on his turn, point out the role of cell-mediated immune mechanisms and their cytotoxicity to pancreatic insulin-secreting ß-cells as a promising area for therapeutic research. Finally, several viral infections such coxsackies viruses, Epstein-Barr virus, rubella, and cytomegalovirus have been linked with the onset of juvenile diabetes.

Thus, clinical medicine is of a higher hierarchical rank than pathology or pathophysiology, which themselves are of a higher level than biochemistry and immunohistochemistry which themselves rest on biophysical and molecular processes and so on. In this process of explaining and understanding, medicine sees living bodies or plants as complex mechanisms, so that disease processes can be understood by breaking them down to their smallest constituent parts. Medical thinking is

[37] Jaspers (1962, 789).

[38] Braithwaite (1953, 345–347).

methodologically reductionist insofar as it considers that (or proceeds as if) the objects or processes at each level contain as their parts the objects or processes of some lower and less complex level and can be analysed (at least in principle) with or without remainders, being micro-reduced, into such more fundamental systems. This view also assumes that any lower level constitutes a concealed determination, a complete underlying causal *explanans* of the higher level. The belief is that explanation flows downward, and truth flows upward into the system from level to level.

Prima facie, one might then say that we have two distinct strata in medical discourse, namely a descriptive-clinical order and a multi-layered explicative account.

The clinical gaze has a distinct descriptive voice: it includes narrative anamnesis, physical findings, imaging studies and laboratory tests, signs, symptoms, and laboratory evidence.

Even so, pathogenesis, namely the processes by which morbid conditions originate and develop, pertains to reductive reasoning. More specifically, it expresses itself in narrative explanations of structural, functional, genomic, or physicochemical features or whatever is causing the disease at some more basic level, and it ends up with the disease, not with the patient.

For instance, questions about heart failure can be answered at the mechanical level of the heart pump but also in terms of ion channels or neuro-humoral responses.

Even before addressing those questions, let us pause here, and take stock of the situation. In the last sections, we have broached some of the major problems of the methodological ladder-like system of explanation.

Ladders are for going up and going down. Properties of a given level realize properties of the above level. Properties of a given kind that are explained in reference to properties of a different underlying favoured kind are those that are reducible; and emergent properties are those that are not properly reducible.

Let us try to sort out the various philosophical assumptions, which seem to underlie the explanatory hierarchical system. We get something like this:

(1) Sentences about types of clinical facts are logically independent of sentences about respective underlying pathogenetic processes. The theory thus does not hold that every statement about clinical facts is ultimately a statement about physicochemical and structural events. Explanation does not portend that the meaning of predicates ranging over clinical facts will be rendered by predicates, which range over structural or functional anomalies. The relation between the macro- (hemiplegia) and micro-levels (cerebral thrombosis) is not semantic but causal.
(2) Kinds of clinical events are epistemologically distinct from the respective underlying kinds of pathogenetic processes. In the analysis of the two levels, we are dealing with distinct kinds of phenomena.
(3) A one-one relation between the clinical and the respective explanatory level is often assumed if only for pragmatic reasons and research methodology. The same, single, specific, simple, or complex, known, or unknown, relevant and necessary pathogenetic process, for instance some anatomical abnormalities or

some genetic defect, gives rise to the same disease, e.g., diabetes mellitus or Down syndrome.

The Pivotal Clinical Level

The above considerations are some of the ways in which the problem of explanation in the biological, as well as medical sciences has been articulated. They require similar and predictable lines of argument.

But medical research has some distinct and idiosyncratic features in addition to the preceding ones. The lines of medical reasoning are circling round clinical issues. And clinical facts are pivotal because the normal/pathological conventional split hinges on them and grants a kind of epistemic priority to the clinical situation, which permits a good deal of ground shifting. All medical negativities, abnormal and pathological events or processes stem exclusively from the clinical level.

Small wonder, then, that in biology, physical and chemical events are the ultimate constituents of the biological world, which is an ontological thesis, whereas human (or sentient) beings' sufferings, in a way, are the ultimate constituents of medicine. The latter belongs to a different representation albeit conditional upon the epistemic priority of the clinical picture. If human (or sentient) beings are to some measure, the central determining agents of medical explanation, it does not follow that they are the only elements of medical explanation, but only that what makes whatever else a piece of medical explanation is determined by its relationship with clinical processes. This is not because individuals are the only causal agents but rather because the normal/pathological split and the consequent clinical conditions are the chief junctures of medicine and the tacit premises of medical explanation. So, in a way, medicine is rooted in the gravitational character of individuals, persons, or sentient beings, at the level of their clinical plights.

But in what ways is the quest for patho-physiologic explanations expressly related to the clinical situation?

The layered model explains the properties of any upper-level pathological process in reference to the properties of its constituents. For one thing, its constituents are sort of embedded in some higher-order clinical or phenomenological events. But since disease processes result from a causal web, the lower-order events could take place indifferently as to whether they were embedded in medically relevant higher-order events: an individual may interiorize acid-fast bacilli and never develop clinical tuberculosis. But for another, this line of argument is paradigmatic of explanation in the natural sciences where one investigates the causal nexus of processes which gives rise to some phenomenon, e.g., the interaction between the shape of two proteins, actin and myosin, and muscular contraction. But since the explorative medical inquiry originates with a clinical issue, it looks "upstream" within the causal web in the attempt to identify causes, determinants, or some causal chains, which might be manipulated for therapeutic or preventive reasons. To this we must now turn.

Pathogenesis and Mechanistic Vertical Explanation

What precedes licences the view that we have two parallel hierarchical models of explanation: a scientific model and a medical model.

For one thing we have a *bio-physiological* (or rather a non-medical biological) "natural" model which is bidirectional, i.e. top-down and bottom-up; but for another, we have a *patho-physiological* "conventional" model which is also bidirectional but with an additional constraint: the top-down and bottom-up are not logically symmetrical: they are not the reverse of one another since the medical model depends on the salient starting point, some *clinical pathological feature,* which fixes what requires description and explanation; since clinical harm, sufferance and disease are paradigmatically first introduced, *the medical line of reasoning begins from the clinical level and proceeds with a top-down scheme.*

Both biological and medical research, then, climb down the step-ladder-like systems explaining some facts about a given level with reference to other facts about some more fundamental level.

For one thing, in biology, the strategy of explanation of a physiological process is that one descends or ascends the ladder, step by step, reducing some high-level properties to some subordinated and more basic ones (physiological, biochemical, and biophysical) until one reaches some sort of seemingly ultimate process. This layered model examined above, assumes that an organism is a combination of hierarchically subordinated structures that are reducible or supervenient to lower levels constituents.

For another, in explaining clinical events, an appropriate form of explanation in each disease instance, sunders them under a series of sub-events until some set of ultimate "basic events" are being reached. But this move is more selective, being compelled to follow those explanatory itineraries, which foreordain the clinical features, which need to be explained. The path it follows when going down the medical ladder is constrained by the epistemic priority of the pathologic: the clinical investigator tries to explain a given disease, some sort of pathological processor, a single malfunction, or some of their aspects. It is the latter which defines the appropriate boundaries of this line of argument. In other words, the lower-level explanations are selected in terms of the extraneous clinical higher-level. Lower-level explanations are thus formulated for the specific purpose of explaining clinical pathological features, and not merely some kind of biological process.

Ultimately, if medical explanations are to be justified, their payoff can only be the richness of successful explanations and predictions they provide and the possibilities of preventive or therapeutic interventions.[39]

Against this background, it is not difficult to appreciate the significance of what Jaegwon Kim has termed the '*causal inheritance principle*'.[40] The principle holds that the causal powers of a property realized on a given occasion by one of its

[39] Thagard (2003).

[40] Kim (2003).

realizers are identical with those inherited from this realizer. So, an abnormal clinical condition with its pathological properties cannot have causal powers beyond those it inherits from its realizers. Causal analysis brings the abnormality down the scale countercausally. The causal inheritance principle serves as the thread that guides us in this second construal—i.e., in the causal transmission of negativities—when stepping down the ladder.

Hence, medical research follows a line of reasoning that starts from the clinical level and goes downwards the hierarchical system from realized to realizers, that means opposite the directionality of interlevel causation. What is real at the top, namely what makes the clinical manifestations abnormal or pathological, goes all the way down to the level where something abnormal splinters off.

The strategy is this: tracking down this thread broaches a sequence of successively lower necessary mechanical/causal determinants; thus, analysing smaller and smaller units, there comes a point when we face the essential building blocks at the level where normal mechanisms split from abnormal ones, namely the initial deviant entities or the inaugural, or one of the inaugural departures from physiological processes; this ultimate level of the exercise reaches the grounds of medical explanation. This bottom level where the explanatory process departs from normal biological processes and sets in motion a pathological causal sequence, represents one of the boundaries of the medical realm. Such basic processes are those that initiate the explanatory chain that ends in the emergence of clinical facts.

Beyond this, in climbing down the explanatory scale beyond this bottom level, we meet processes that are no more pathological but essentially biological. Claude Bernard writes that "*the explanation of pathological phenomena must be deduced from the same laws which govern all normal phenomenon of life*". It is in this sense that Claude Bernard was right in his claim that the processes that underlie normal and abnormal phenomena are fundamentally the same.[41]

It should now be plain that a medical explanatory account ascends this specific hierarchy of realizations springing from the surmised initial or first anomalous grounds and ending in the clinical situation. Less metaphorically, the initial deviant change (i.e., *Helicobacter pylori* infection, at the bottom) resulted in a sequence of anomalous changes paralleled by a sequence of causes concluding with the final changes at the clinical level (gastric and duodenal ulcers). The mechanisms are usually complex and the explanatory pathways often arborescent. Gastric ulcers result from the interactions with *H. pylori* of environmental factors such as stressful events, tobacco smoking, as well as genetic predispositions, which may increase gastric acid secretion and accelerate gastric emptying.

In explaining a disease, whether the principles thus emerging would be justified is a disputable issue. But if the characterization suggested by this layered model is correct, it captures the intuition on which the notion of patho-physiology rests.

Take in short, mechanistic biomedical research proceeds as follows: assuming some pathological process (disease, malfunction, or any of their aspects) observed

[41] Bernard (1947).

at the clinical level, the explanation consists in going down the successive explanatory levels of the hierarchical ladder until some level is reached beyond which the biological processes are those of "normal" physiology or biochemistry; this last level reveals the initial explanatory anomaly.

Critique of the Hierarchical Model

The previous discussion assumes a multilevel system, and a causal hierarchy in which relations between properties at a given level are to be explained by relations between properties at a lower and somewhat more 'ultimate' level, like a building erected upon carefully prepared foundations. It also tells apart the intrinsic nature of pathological processes from the external position of what causes them. It might at this point be appropriate to ask to what degree this version of explanation, which assumes a one-to-one correspondence between a specific mechanism, and a given clinical condition, or different mechanisms in case of multiple realization, can deal with more complex situations.

This bottom-up approach is tainted with the assumption of a linear causal chain from molecular variants to cellular, from cellular to physiological and from physiological to clinical levels. Small wonder, then, that the clinical phenomena may not be predictable insofar as the causal nexus is linear or branched and involves complex components that interact in non-linear manner inside the organism as well as between the organism and its environment (e.g., gene-gene and gene-environment interactions).

To begin with, this hierarchical view is primarily an epistemological thesis. It relies on an oversimplified and a questionable, although widespread metaphysical assumption in accordance with which there is in biology a complex interdependence of entities at many different levels.[42] But these levels of analysis might very well be mere levels of discourse devised to make our instrumental knowledge intelligible. Philip Kitcher[43] has cast a critical eye on the claim that classical Mendelian genetics can be reduced to molecular genetics. This supposition results, he believes, from the assumption that because a system is composed of smaller constituents, an explanation at the level of the constituents must be better than an explanation at the level of the system.

Clinical statements get their meaning partly by their connections within an interdependent system of statements and concepts, i.e., from a web of belief. The entire edifice of medical thinking thoroughly penetrates every statement at every level of the system. Medical explanation never judges hypotheses on their observational consequences taken in isolation.

[42] Dupré (1993).

[43] Kitcher (1999).

But there is an additional consideration. The hierarchical view admits one-way-only causal interactions from the lower level upon the upper-level properties, not vice versa. But any clinician is aware that clinical events, the top-level, are not causally inert. At each level of analysis, clinical events may have downward causal efficacy: there are genuine causal entities at different levels of biological organization. A fever of 106 °F is a clinical sign but it may well give rise to seizure disorders in children in the absence of intracranial infections. And social factors such as a chronic exposure to stressful situations may cause a decrease of immunological surveillance.

Hence, the question before us is now: can properties of things really be ordered into related levels some of them being more basic than others and the latter resulting from the former? The putative levels of organization, physical, chemical, biological, psychological, and social, might very well be autonomous springboards at many levels. Should we then rather opt for some ontological pluralism and reject prioritizing any particular level? Didn't Von Neurath claim that scientific knowledge rather than being like a skyscraper, sounds more like a much-weathered ship that cannot put into port and which must constantly be repaired at sea?[44]

Simply put, reductionism and its weakened versions should be allowed to be, in large measure, the reflection of a methodological commitment, a pragmatic working hypothesis, rather than an ontological claim. But it is probably false.

The Limits of Reductionism

Reductionism in this ladder-like system has all the hallmarks of an ontological claim about the nature of pathological processes but it might well turn out to be a mere epistemological one, or even only a methodological one, owing to the epistemic limitations of human understanding. It might very well be that there are more levels of organization, or fewer levels of organization than those needed to conduct our studies; or that there are no real levels but only a chaotic causal nexus of interacting systems with properties that cannot be reduced to the sum of individual components. Even though there are stochastic components in the explanation of some pathological processes (e.g., atrial fibrillation), reductionism is perhaps freighted with a deterministic metaphysics, a somewhat outmoded view of scientific theory, and linked with the hypothetico-deductive explanatory model. One cannot help wondering whether medical or biological explanation can be completely reduced without residue or should be fully subordinate to physicochemical explanation. Biological phenomena are so complex that a full account, a complete unique explanation of them sounds like a mirage that passes beyond our cognitive capacities. Furthermore, adequate explanations in medicine apply to some mere fragment of the biological world: they are always partial and finite in length. They are often

[44] Neurath (1983, 92).

embroiled in conflicting explicative models, psychological vs. neurological or individual vs. sociological. Some topics are selected not because of any intrinsic interest but because of mere available evidence.

So, reductionism may be deeply embedded in scientific belief but is more and more being stigmatized as a dogma. On this score, medicine seems to be faced with a set of questions. Should reductionism be jettisoned? What shall we make of the undeniable though partial methodological success of reductive analysis in medical research? Or else should we accept some sort of weaker version of explicativity?

Emergence

Emergence is a concept introduced by Lloyd Morgan (1852–1936) in his book *Emergent Evolution* published in 1923. Some qualities of things are said to be emergent if one could not have predicted them from even a complete knowledge of what went before. Water is composed of hydrogen and oxygen, which are gaseous at ordinary temperatures. The two together are forming water that is not gaseous but liquid at ordinary temperatures. The more complex properties are emergent when we cannot deduce them from the properties of their constituents: they are then different altogether and cannot be predicted from or reduced to lower-level properties. They are novel over and above the properties of the underlying levels. In some way, they transcend the properties of their constituent parts in the lower levels. On this score, when their constituents are associated together, the resulting complex may exhibit new properties, which cannot be expressed by the constituents in isolation.

A melody is not a mere succession of tones of different pitch. Nothing needs to be added to the sequence of sounds to generate a melody. If terms like tune, melodic contour or motive were absolutely and totally defined in terms of acoustics, there really would be no difference between them and terms which describe sequences of sound—terms like frequency, overtones, or beats. But these are two different vocabularies that do not mix because a melody is an emergent property generated by some acoustic realizer.

Diseases and pathological processes are usually not predictable based on their physiological, biochemical, or molecular component parts, or combinations thereof. Claude Bernard held that "*In a word, when we unite physiological elements, properties appear which were imperceptible in the separate elements . . . All this proves that these elements, though distinct and self-dependent, do not therefore play the part of simple associates; their union expresses more than addition of their separate properties.*"[45]

Clinical manifestations that characterize a disease are emergent characteristics: they are properties that are held by diseases considered but not by their constituents and that cannot be deduced from the properties of their constituents. It follows that

[45] Davidson (1980).

the record of clinical findings is something over and above its biochemical and structural correlates in that it requires new concepts to relate to it and new terms to describe it. In the reductive multilevel model, each entity at a given level is presumably being wholly realized from entities belonging to lower levels. But matching this hierarchy, there is a parallel hierarchy of properties of those entities. If so, then, each level has its own specific set of new, unforeseeable properties that first emerge at that level, like pathological properties at the anatomical or histologic level; trisomy, spindle formation and segregation failure at the cytological level; genetic code and protein synthesis at the molecular level and suchlike.

The aching, deep, substernal pain of angina with radiation to the left arm does not belong to the vocabulary of pathological physiology but to the clinical language. Angina pectoris is an emergent property of reduced coronary blood flow: the angina is a property that is over and above the myocardial ischaemia; one can observe the angina without observing the acute thrombus, the plaque obstruction or the vascular spasm that are causing it.

Realization and Supervenience

What indeed clearly emerges from the preceding discussion is that, should we forfeit the concept of reduction, we would still need to capture the intuition that clinical facts and phenomena arise out of biological and biophysical processes. This is tantamount to asking whether the biological and biochemical levels of the hierarchical system can realize, fix, or determine the clinical level without allegiance to a reductive strategy. For all that, clinical properties could perhaps be termed resultant, consequential or *supervenient* in that they depend on some other more fundamental properties. Supervenience, a term introduced in 1952 by Richard M. Hare, a British moral philosopher, satisfies the need for a non-reductive dependence relationship. What is it about?

Clinical facts, although distinct from and eventually irreducible to physical-biological properties, are nevertheless realized by them. If so, then, realization turns out to be a relation between properties, which indicates how properties at different levels are related to each other. The multilevel model is thus a hierarchy of properties, with properties at some level realizing those at the lower level. This supposes that both the realized and the realizer are distinct from one another and ontologically homogeneous.

But then, may a disease have multiple realizers? Can different underlying abnormal configurations lead to a single clinical condition? This position is usually termed *variable or multiple realizations.*

Down syndrome can be realized either by a trisomy of chromosome 21 or by a chromosomal translocation. These two conditions are descriptively similar but have different preceding causal sequences. If so, then, do we have one or two diseases? The answer is that a multiple realizable disease such as the Down syndrome is held to pertain to two distinct types of entities or syndromes, two mutually exclusive subclasses.

Diabetes mellitus can be realized either in type 1, by absence of insulin, usually by autoimmune destruction of the islets of Langerhans, or, in type 2, when insulin secretion can no longer compensate for insulin resistance. 75–85% of people with diabetes have the more common type2.

Be this as it may, a question remains before us: what is the nature of this interlevel relationship between the realizer and the realized?

We need some sort of dependence relationship that is not reductive, i.e., a necessary dependent variation or covariation that strives toward being either causal, or logical, or emergent, or probabilistic. Such a property, which depends on some other property, is termed *supervenient* in the philosophical literature.

Things of one kind A supervene upon those of another kind B, when things are A in virtue of being B (the base property), namely, when the presence or absence of things of kind A is completely determined by the presence or absence of things of kind B. Elias Savellos and Ümit Yalçin write: "*The concept of supervenience is supposed to denote this dependence relationship that appears to be weaker than reducibility.*"

Medical research, as it were, assumes that emergent properties met in the clinical encounter are supervenient properties. This is perhaps a solution to the problem of interlevel dependence: kinds of clinical events merely occur in virtue of or are instantiated by respective kinds of pathogenic processes. To say that clinical facts (*A*-properties) supervene on patho-physiologic facts (*B*-properties) is to say that clinical facts are irreducible to but consequential upon those underlying properties. If so, *B*-properties determine *A*-properties, but *A*-properties cannot be reduced to *B*-properties. If this intuition is correct, it follows for one thing, that there are, or can be, no differences in the clinical level without there being differences in the underlying levels, and for another, that any patient's base-properties will determine all its clinical manifestations.

A causally supervenes on B, is to say that differences of sort A at time *tj* are completely but contingently explainable by differences of sort B at time *ti*, where $i \leq j$. If so, *B*-states determine *A*-states. Supervenience is different from reduction for which to be something of sort A is nothing but something of sort B. It follows that, contrary to reduction, the subvenient relation that binds A to B is asymmetrical. By way of example, symptoms of mental disorders are supervenient relative to neurophysiological processes but cannot be defined by them.

Causal Versus Non-causal Supervenience

Two consequences follow from this analysis. To begin with, in causal relations, the *explanans* precedes the explanandum: for instance, hypertensive vascular disease precedes the eventual development of cerebral haemorrhage. Such supervenience is a causal process between two types of ontologically and epistemically distinct events. It comes down to saying that the disease is being realized or *caused* by a preceding underlying malfunctional process.

But there is a second consideration regarding non-causal supervenience, namely a situation in which the explorative realizer occurs at the same time as the explanandum so that their relation is not causal but *constitutive.* A non-causally supervenes on B, if and only if, necessarily, differences of sort A at time t must be explainable by differences of sort B at time t (although there may be *A*-differences without *B*-differences).

This merits an explanation. In the first case, that of *causal supervenience*, the macroproperties are required to be of a different kind from the microproperties. But surely, an explanation to be acceptable to non-causal realization, must assume, among its premises, that the clinical disease is not ontologically different from its underlying mechanism.

Generally, we individuate most disease entities by means of aggregates of empirical signs and symptoms. And once individuated, what is being sought for an explanation is a set of causes. In contrasting manner, some conditions, such as infectious diseases, are defined by their unique necessary cause, when available. If so, aetiological factors which, in the first case, are essentially contingent realizers and externally related to the disease they determine, may eventually be turned, as in the case of infectious diseases, into intrinsic properties implying a stronger conceptual relation. It follows that an *a posteriori* necessary cause then becomes constitutive of the condition it determines. Meanwhile, defining a disease by its necessary causal structure is not a pure semantic move but the pursuit of what makes things of a given kind things of that kind. It is an ontological claim.

What clearly comes to light from this is that an infectious disease is perhaps identical with, or nothing but the underlying aggregate of necessary intrinsic properties and realizing processes which define it, up to the clinical manifestations. Just like something's mean molecular kinetic energy does not cause but constitutes its temperature. Any change in the clinical picture necessitates a change in the underlying pathological process and a same pathological process will generate a similar clinical picture. So, such non-causal realization absorbs the causal chain. It merely skirts the issue of causal supervenience by relegating it to some outpost out of the boundaries of medical discourse. The pure biological standpoint of *causal supervenience* is being coaxed out of discussion as it turns into a *constitutive supervenience*. The perspective afforded by this model amounts to a conceptual reducibility.

Biologically, HIV virus causes and explains the process of AIDS, and, in one sense, this relation is contingent: it explains the relation from a pure biological standpoint. But medically, a virus stipulatively defines AIDS. In the light of this, AIDS is not consequential anymore, i.e., it is not resultant or derivable from the presence of the virus. This tantamount to say that the relation between an infectious disease and its agent, though empirical is necessarily true, if true, instead.

But there is a further consideration. The preceding line of argument seems to license the view that signs and symptoms are part, but not effects of the disease process inasmuch as they occur within and are not external to the disease process.[46] So that, at least in the typical case, shortness of breath—i.e., the observable clinical fact of dyspnoea—displayed when a patient climbs the stairs or during physical exertion may be caused by heart failure. But dyspnoea is not merely indicative of but also constitutive of what we mean by a diagnosis of heart failure.

The final account leaves us either with a non-reductive causal dependence or with a position in which diseases and their causal powers are of a piece or are closely identified with one another. The *explanans* either causes or constitutes the *explanandum*.

The Twofold Context of Explanation in Psychiatry

Back in 1923, Karl Jaspers introduced an important distinction between causal and meaningful explanation and implied that psychiatry needs both: meanings and causes.[47] At this critical juncture, a gap opens, which is meant to set the objective strategy of physical medicine and natural-scientific psychopathology off from 'internal understanding', that is, the patient's moods, wishes and fears or the content of delusion and dream.

In natural sciences and in medicine, we try to grasp causal connections and we attempt to identify some sort of regularities, such as the rules of Mendelian hereditary transmission. This is plain scientific knowledge. The concrete reality of causal connections is obtained inductively through repetition of experience, something we do in our everyday life.

Yet, it is rare to find any such regularities in psychopathology where we merely observe particular causal connections. By way of illustration, we may consider the next statement: "Lucy is depressed because of a neurotransmission disturbance in her brain": Karl Jaspers contrasts the objective nature of such string of words, he calls explaining *(Erklären)* with the subjective nature of what he calls understanding *(Verstehen)*, giving reasons, not causes: "Lucy is depressed because of her husband's death."[48] Contrary to the strategy of physical medicine, this form of explanation appropriate to mental disorders is essentially bifocal; for one thing, it is similar to those used in the natural sciences, namely *extensional*, by gazing out at a patient's behavioural, clinical and biomedical manifestations; but for another, it is a special method embodied in a jump across the boundaries of the self into the world of the other. It follows that the evidence of such meaningful connections is acquired through *intensional* understanding (not through repetitive experiments of inductive

[46] Whitbeck (1977b, 632–634).

[47] Jaspers (1923, 1963, 30–336; 451–462).

[48] Jaspers (1923, 1963, 27, 302–303).

reasoning), a way to engage directly with another individual, be it a human being or a pet.[49]

If this act of understanding, which Jaspers calls *empathic understanding*, is a sympathetic identification with the other, is it a correct ascription of mental predicates? This empathic intuition is a special form of reasoning by analogy; yet, it is not a form of knowledge, since knowledge entails the truth of what is being known, neither is it a wherewithal to knowledge, but a kind of internal understanding of unique and never duplicated episodes in which the patient is engaged, and which leads directly to the psychic connection itself. It implies nothing about the truth or falsity of what is understood.

Meanwhile, a psychiatrist appeals to his patients' beliefs as a means of understanding their conduct. He can only speak of his patients' worlds in terms of their beliefs because their beliefs define their worlds. It follows that, for a psychiatrist, his patients' worlds are their worlds. In his own case, his own world is the world, and he regards his own beliefs as true. He has a medical and scientific education, so that his own beliefs are, in an uncanny way, perfectly obvious to him, since it is the world, rather than his beliefs about it, that he thinks of. It follows that for a psychiatrist, to understand the world of someone else or of a patient, is not to understand the world, unless those beliefs are also his beliefs, at which point they become understandable. If so, then, intensional understanding depends on the psychiatrist's own self-understanding, that is, on the self-evidence of his own representations. The empathic leaps of the therapeutic relationship are an attempt to throw oneself into a reflection in the mirror of someone else's world, "*a literal, vicarious occupation, as it were, of the interior of Other Minds*".[50]

"*There is no limit*, writes Jaspers[51], *to the discovery of causes and with every psychic event we always look for cause and effect. But with understanding, there are limits everywhere.*" To the degree that the life of others really is being different from ours, empathic understanding is fallible and unreliable and comes out underexposed and in blurred shape. And it is this process by which we discover that their beliefs and their world differs or are similar to ours, which is at the core of psychiatric understanding.

Darwinian Medicine and the Panglossian Fallacy

Evolutionary medicine is a sub-field of medical research, which grew rapidly in the last two or three decades, and which, until now, has been irrelevant to health care and has created a quandary for medical explanation. This field that applies the principles of evolutionary biology to health and disease, is a dawning, but growing

---

[49] Jaspers (1923, 196Z, 316, 347); Taylor C, vol 2, 22.

[50] Danto (1985, 338).

[51] Jaspers (1962, 305).

response to this demand of integrating evolution with medicine. It has grown exponentially since the early 1990s, contributing to a greater understanding of topics paramount to human health including aging, reproductive health, immune function, infectious disease, cancer, behavioral disorders and mental health, microbiomes, veterinary medicine, inflammation, and diet. An evolutionary perspective is a powerful and broad lens for medical advancement but can also be used as a comprehensive scaffold for organizing medical knowledge that otherwise remains unconnected.[52]

DNA from the bones of people who died 37.000 years ago, identified diseases such as malaria, hepatitis B, leprosy, and leptospirosis, all of which are still present, despite natural selection.

On this score, to support his naturalistic theory of disease, Boorse draws on a notion of fitness and adaptation maintained by natural selection even though "*the notion of adaptation cannot be "Darwinian fitness," or pure reproductive success.*"[53] With L. Wright, functions are also those which were favored by natural selection.

The dominant naturalistic metaphor for the body has been a machine, and disease has been viewed as a defect arising in an otherwise perfect mechanism. However, there is a structure of controversy regarding the idea that every organ is adjusted to the living conditions of the organism and owes its existence to natural selection; and that the fitness advantage of any biological or behavioural trait or structure has some positive influence on the survival of the organism. Details and objections aside, the human body is badly designed.[54]

This common fallacy, which is pervasive in the biological and in some of the recent medical literature, sets medicine's goals to maximize Darwinian fitness and to further the understanding of the evolutionary function of various physical or emotional processes. Some unpleasant physical or emotional features may, it is assumed, serve some yet unidentified purpose since Mother Nature, under the Darwinian guise, allegedly never proceeds in vain. Such medical explanations seem to be governed by norms determined by anthropological hypotheses or research about human past rather than by actual causes as determined by current clinical or epidemiological investigations.

It is almost breath-taking to observe the widespread trend to assume that an adaptationist explanation can and should be given to all characteristics of an organism. For one thing, features or traits of an organism are assumed a priori to be present because they provided greater fitness to its ancestors, so that they were selected because they were useful. But for another, any feature of a creature that attracts our attention must be useful or must have some fitness advantage since it has been selected. This viciously circular reasoning implies that traits have been selected because they were adapted, and conversely, they are adapted since they have been selected. Features and traits can and should be given some adaptationist explanation

---

52 Alvergne et al. (2016).

53 Boorse (1977, 557, 548).

54 Hafer (2015).

since whatever exists, exists because it is for the best. Voltaire's Dr Pangloss was seeing a purpose for whatever exists: "*everything is made for the best purpose*".

Medical evolutionists are forced to re-evaluate the human body, not as an optimal designed machine, but rather as a series of compromises that indeed has left us vulnerable to a variety of conditions, particularly as we age.

This adaptationist programme is begging the question and has often been denounced in the literature, and among others, by Stephen Gould and Richard Lewontin.[55] It is not a disfigurement of Darwinism to reject this untestable metaphysical scheme ascribing causal and explanatory sufficiency to natural selection and claiming that all features of organisms are optimal features or are specifically produced by the Caudine forks of ultra-Darwinian selection. So, why did natural election leave our bodies with traits that make us accessible to diseases? Why do people get frequent infections due to the suppression of the immune system? Why do our gums recede so that teeth loosen and fall out? Why do our bodies lack vitamin C that has to account for the millions of cases of scurvy, whereas cats can produce vitamin C, but humans can't?

This Darwinization of humanity has been christened *Darwinitis* by Raymond Tallis.[56] Darwin himself warned that "*natural selection has been the main, but not the exclusive means of modification*".[57]

Gould illustrates this abuse of natural selection as an explanatory tool by quoting Robert Wright for whom our fondness for sweetness was designed for an environment in which fruit existed, but candy didn't. Gould comments: "*These ranks as pure guesswork in the cocktail party mode; Wright presents no neurological evidence of a brain module for sweetness, and no palaeontological data about ancestral feeding. This "just-so-story" therefore cannot stand as a "classic example of an adaptation" in a sense deserving the name of science.*" Life and the human body are a whole lot messier and makeshift than adaptationists would incline us to believe. "*The standards of human bodies are just too badly put together to stand up even reasonable design specifications, much less infallible ones.*"[58]

Stephen Jay Gould and Richard Lewontin's article "*The spandrel of San Marco and the Panglossian paradigm*" is a highly influential paper that indicates how some non-adaptive features result unavoidably from the process of natural selection. The evolutionary process imposes a set of attendant sequelae upon any adaptive change. These sequelae—*spandrels* in the terminology of the paper (a term borrowed from architectural design refer to the pendentives of San Marco in Venice)—designate the class of forms and spaces that arise as necessary by-products of another decision in design, and not as an adaptation for direct utility in themselves.

Evolutionary biology needs such an explicit idea for features arising as side effects, rather than adaptations, whatever their subsequent *exaptive utility*, namely a

---

[55] Gould and Lewontin (1979), Eldredge (1995), Fodor (1998, 172–180), and Gould (1997).

[56] Tallis (2011, 5).

[57] Darwin (1859, 6).

[58] Hafer (2015, 16).

function that was not acquired through natural selection and for which it was not originally adapted. Darwin himself provides examples of highly useful functions that have not been selected so that there is evidence of a *non*selectionist hypothesis.[59]

For one thing, strict Darwinism often narrows the process of adapting to all aspects of a feature's evolution. The primary mechanism of natural selection is then viewed as a direct causal basis for the entire sequence.

For another, the spandrels arise nonadaptively—as architect byproducts— but may regulate or dominate the late history of a lineage because of their capacity of coaptation to subsequent utility.

This is not to deny Darwinian principles, but to overthrow the centrality of adaptation in evolutionary theory by a proper appreciation of the interaction between structural channeling (including the nonadaptive origin of spandrels) and functional adaptation, conventionally analyzed in studies of natural selection.

Raymond Tallis is complaining about "*the mistaken belief that Darwinism requires us to accept Darwinitis which purports to explain everything about people in terms of biological evolution*".[60]

This Panglossian fallacy is being the source of a systematic application of natural selection to the problems of medicine, covering sex, menstruation, pregnancy, allergy, cancer, senescence, mental disorders, infectious diseases, and anything of that ilk.[61] It helps to explain many aspects of our concept of a disease as side effects of evolved benefits. The everlasting conflict between humans and their enemies, viruses, bacteria, parasites, and the physical and chemical environment, is seen as a Darwinian arms race. Our bodies are an assortment of trade-offs, moulded by natural selection to optimize reproductive success. The question is to find the adaptation, not to ask whether it exists at all: but plainly, it is a caricature of Darwin's vision.[62] Agreeing with Darwinism or Neo-Darwinism as an accepted scientific doctrine is one thing, but tempering faith in the power of natural selection and its reliance on mere plausibility for setting up speculative narratives, is another. Shouldn't we question this unwillingness to consider alternatives to adaptive stories? Shouldn't we take a more pluralistic approach into account, including genetic drift, developmental constraints and correlation with selected features, or the separability of current utility of a characteristic and the evolutionary reasons for that characteristic's existence?

Be that as it may, the point to be made here is that Darwinian explanation in medicine is a misconception or has limited explanatory power. The philosophical analysis of function is based on the hyperadaptationist model. It should proceed with prudent choice of paradigmatic examples rather than with a systematic blind use of natural selection.

---

[59] Darwin (1859, 197).

[60] Tallis (2011, 153).

[61] Stearns (1999) and Nesse and Williams (1995).

[62] Lewontin (1984).

Conclusion

1. A principal objective of biomedical research is to provide understanding of medical issues as it considers the field of ill health as intelligible.
2. The covering law model, namely the hypothetico-deductive model, which attempts to identify general laws, does not apply to medicine, since there are no proper laws of nature in biology. More fruitful are the sort of explanations offered by the inductive-statistical model. Inference to the best explanation proceeds in the opposite direction, since one derives the explanation from the observation instead of deriving an observation from its explanation.
3. Explanation is contrastive to a conventional-normative, counterfactual background. All diseases reveal some derangement of some counterfactual normal processes and explanation in medicine is contrastive.
4. Synoptically, medical explanation evolves along two axes, which are complementary and mesh with each other, but which are, in metaphorical language, perpendicular to one another: the mechanistic biomedical vertical one, which springs from a hierarchical viewpoint, and the epidemiological horizontal approach, which deals with the causal skyline and its stochastic uncertainties.
5. Explanation by mechanistic biomedical research, namely the vertical style of explanation, assumes that there is a hierarchy of different levels of phenomenal reality. Methodological reductionism claims that the best explanatory strategy is to attempt explanation in terms of entities at lower levels. Mechanistic biomedical research goes down the hierarchical ladder, and the explanation ends up when it reaches the level of normal states or processes. Although highly successful in mechanistic biomedical research, reductionism has been highly controversial, for it is essentially deterministic and associated with the hypothetico-deductive method.
6. Patho-physiological research is grounded in a hierarchical explanatory model, namely a *vertical* style of explanation: it assumes a multilevel causal/mechanical explanation which purports to explain complex medical phenomena in terms of simpler, more fundamental features. The logic of this model rests on the premise that clinical facts are pivotal because all medical negativities, abnormal and pathological events, or processes stem exclusively from the clinical level.
7. For medical reductionist projects, complex wholes, diseases, or pathological processes, are dispensable and ought to be fully explainable in terms of (and sometimes replaceable by) features involving parts of which those wholes are composed. Subsequent recognition of the holism of meaning and the apparent failure of these reductionist programmes switched attention to other ways of obtaining the benefit of reduction.
8. Diseases are usually not predictable based on their physiological, biochemical, or molecular component parts, or combinations thereof. One may have recourse to *supervenient* properties. For two sets of properties, A (the *super*venient set) and B (the *sub*venient set or supervenience *base*), A supervenes on B just in case there can be no difference in A without a difference in B. We need some kind of

dependence relationship that is not reductive, namely, a necessary dependent variation or covariation that strives toward being either causal, or logical, or emergent, or probabilistic. Such a property, which depends on some other property, is termed *supervenient* in the philosophical literature. A clear example of the supervenience of the clinical on the physicochemical: patients have a diabetes (A) in virtue of physiochemical abnormalities (B), and there can be no diabetes (A) without such biochemical abnormalities (B). Similarly, the mental supervenes on the physical.

9. Finally, the common Panglossian fallacy consists in the widespread and misleading trend to assume that an adaptationist explanation can and should be given to all medical or non-medical characteristics of an organism.

Chapter 5
Causation and Aetiology

Felix qui potuit rerum cognoscere causas

Virgil

Contents

"*It has frequently been remarked,* Israël Scheffler wrote *that the notions of cause and effect recede progressively into the background with the advance of science, though they remain of relatively constant importance in everyday practical affairs and in applied science.*"[1]

[1] Scheffler (1963, 53–54).

L. Karhausen, *Analytical Philosophy of Medicine*, Philosophy and Medicine 152,
https://doi.org/10.1007/978-3-031-81257-6_5

In medical reasoning and in health care, we are constantly using knowledge of causal relations. A characteristic feature of the notion of being abnormal consists in having a causal history namely an aetiology. Causes intrude at the chief junctures of medicine such as:

- It contributes to explanation and to identify factors that promote a disease (smoking and lung cancer) or that worsen its course (pneumococcal infection supervening on a viral bronchitis).
- It has predictive usefulness.
- It gives power to control events, through preventive or therapeutic interventions. "*Causal laws cannot be done away with, for they are needed to ground the distinction between effective strategies and ineffective ones*," writes Cartwright.[2]

Caroline Whitbeck countenanced that one must distinguish among laws in general, those that express causal relations in the sense that they express the situation to be produced or prevented as a function of the situation that produces or prevents it. To decide whether a particular law does express a causal relation, we must fall back on our causal judgements. The context determines whether the disease we are concerned about controlling is one we wish to produce a desired outcome, in which case we are interested in sufficient conditions: with treatment, we focus on causal related sufficient conditions. In contrast, if we wish to prevent the occurrence of a disease, we are then interested in causally related necessary conditions.[3]

Furthermore, we do not need to understand the mechanism leading to a disease, we do not need to explain the pathogenesis of a disease on its entirety to effect preventive or therapeutic measures.

Sufficient and Necessary Causes

Suppose some event *C* is causally connected with some other event *E*: this observation invites two questions: (1) Is *C* a sufficient condition for the truth of *E*? (2) Is *C* necessary for the truth of *E*?

In his *System of Logic,* John Stuart Mill proposed that a cause is a mere element in a set of conditions that unitedly necessitate its effect. J.S. Mill invokes "*the conditions of a phenomenon, or, in other words, the set of antecedents which determine it, and but for which it would not have happened. The real cause is the whole of these antecedents, and we have, philosophically speaking, no right to give the name of cause to one of them exclusively of the others ...* ".[4] The assessment of causal conditions takes place against a contextual background and the interplay of various components, some of which are standing features of the context, some of which are

[2] Cartwright (1983, 22).

[3] Whitbeck (1977b, 628).

[4] Mill (1970, III, V, 3).

changes which occur prior to the effect, and all of which constitute the causal constellation we call 'the sufficient cause' or simply the 'cause'. On this score, to say that a cause is sufficient for its effect is to say that if the purported cause occurs then the purported effect occurs too. In this case the cause necessitates its effect.

On the other hand, to say that a cause is *necessary* for its effect is to say that if the purported cause does not occur, the purported effect does not occur either. No-one develops tuberculosis without prior internalization of *Mycobacterium tuberculosis,* a necessary cause of the disease. Thus, a necessary cause is an invariant part of a causal process.

Meanwhile, this micro-organism is not sufficient, since not everyone exposed to this agent will develop the disease. Few people do. A cause is seldom sufficient to give rise to an effect without the participation of other causes. Innate defences often prevent its initial invasion. The probability of infection depends on age, ethnic group, sex, socio-economic status, lifestyle, and hygienic conditions. If so, then, a *sufficient* cause is a set of minimal occurrences that inevitably produce disease. In what sense should this speculative entity be understood is not clear since there are no sufficient causes in medicine. But surely, sufficient causes, and in some cases necessary causes, presuppose a deterministic universe.

Even so, to deal with this, and since he took determinism for granted, J.L. Mackie introduced what he terms INUS conditions.[5] Each singe factor (INUS factor) is an insufficient but necessary part of a condition which is itself unnecessary but sufficient condition for the effect *E*. If *C* is necessary for *E*, it is a *posteriori* necessary component of a broader condition. Still, *C* is not sufficient for *E* either, but the broader condition is sufficient for *E*. It follows that the cause *C* is an insufficient but necessary part of a combination, which is (unnecessary but) sufficient for *E*. It is hard to evaluate the practical value of Mackie's model proposed by Rothman in medicine or in epidemiology,[6] a surprising claim since the tools of epidemiology are essentially statistical and do neither need nor support the vague, speculative, and unjustified hypothesis that every event is necessitated by its previous causes.

By way of illustration, HIV is insufficient, although necessary for AIDS. But there is a broader combination of factors (sexual transmission or else blood contamination, genetic susceptibility, social factors, age, sex, and suchlike), which are individually unnecessary but collectively *a posteriori* sufficient.

In science and in medicine, most causes are neither sufficient nor necessary. Effects have, typically, a plurality of causes since a certain effect can be brought about by several distinct clusters of factors. Thus, instead of determinism, Karl Popper supported the concept of probability propensity.

[5] Mackie (1986, 62–63).

[6] Rothman (1976) and MacMahon and Trichopoulos (1999, 29).

Aetiology and Deviation from the Norm

At different periods of history, in various cultures and in variegated guises, diseases have been construed as a departure from health due to the action of some evil will, noxious agents, miasmas or microbes or various kinds of agents thought to be exogenous.

Physiology represents a background of expectations and successions, which require no explanations. But causal explanations are needed for anomalous or pathological course of events that contrast with the regular, physiological, or intelligible case.[7] In like manner, we do not ask for the cause of gravitation in Newton's system, but we may ask why a bridge collapsed. A cause, in medicine, is first and foremost a cause of harm. The attribution of an aetiology is constitutive of the notion of disease. A failing heart asks for some sort of causal explanation, but a normal heart does not.

According to Aristotle,[8] for every process of change, any *kinesis*, there is some causal agent uniquely related to the change and there are conditions whose presence could have prevented the process. Genuine change is caused change and consists in the subject being exposed to an undesirable or opposite factor, i.e., "*receiving a contrary*"[9] which means receiving a παθos[10] (παθειν, to suffer, to be hurt by, to experience joy, pain, surprise &c) or "*suffering an affection*"[11]; a process of kinesis passes from one contrary to another. It thus depends on a causal agent (which is active) and takes place in a subject (a 'patient', who is passive). For Galen, « *we call any change that has already occurred an affection* ».[12] For Aristotle, "*A cause is a condition which departs from the ordinary or regular course of events*".[13] Aristotle uses the terms 'affection', 'patient' and 'παθos' in a broad sense, that is much wider than their ordinary associations in these days; the subject of change might be inanimate bodies as well as living things.

Von Wright puts it simply as follows: "*The cause brings about a disturbance of a state of equilibrium and is thus responsible for some evil or wrong in nature.*"[14]

Hart and Honoré countenance that "*In ordinary life ... the particular causal question is most often inspired by the wish for an explanation of ... a departure from the normal, ordinary or reasonably expected course of events.*"[15] Causes are thus abnormal conditions, i.e., conditions that lead to deviations, catastrophes, or diseases. "*What is abnormal ... makes the difference*" between deviations and things

[7] Toulmin (1961, 56–57, 44–46).

[8] Emmet (1992, 2).

[9] Aristotle (1969, 5, 4a10–b19).

[10] παθειν, to suffer, to be hurt by, to experience joy, pain, surprise &c.

[11] Aristotle (1971, D.21; 1969, 4b4–18).

[12] Galen (1991, 163–164).

[13] Hart and Honoré (1985, 466).

[14] von Wright (1971, 64–65).

[15] Hart and Honoré (1985, 34–35, 39).

going on as usual. Why did this patient contract this disease while it would normally not occur? What causal factor explains this contrast?

Admittedly, *the privative concept of health* implies that when a person suffers from some illness there must be a cause which has befallen him, and which is responsible for the disturbance of a normal equilibrium.

The Greek word αιτια means cause but also guilt, i.e., liability for harm. To be sure, some causes i.e., aetiological factors, are harmful, detrimental, damaging or injuring, noxious or toxic; yet others are protective or incidence-reducing. Whereof, causes in medicine either split biological processes into normal and pathological ones, or else, in case of preventative factors, they seal the gap, which divides them.

What's more, new disease definitions must displace old ones if they have greater predictive, prognostic, or therapeutic potential, i.e., if they sort patients into categories, which permit better, precise, and useful inferences relevant to our practical interests. Clearly, the most useful method of classification for medical purposes, namely preventive, therapeutic, and explanatory needs are those that regroup ill people around a necessary etiologic agent. This bestowal often redefines the class of patients, which are referred to: some patients may be excluded, and new ones included. Parts of our classificatory system of diseases may be subjected to explanatory shifts, so that new disease entities based upon newly discovered necessary causes may crop up.

Are Causes in Medicine External or Internal to Their Effects?

Are causes extrinsically conferred, or are they constitutive of their pathological effects? How to adjudicate amongst those alternatives depends on whether we canvass diseases as events or as processes. If we have independent definitions of a disease and its etiologic factors, they are then both taken as autonomous events and their causal relationship is medically informative. But if the etiologic factors are part of the definition of the disease, the relation that binds them is analytic: it is necessarily true since the aetiology is contained in the concept of the disease, which is now construed as a process.

Events have external and processes internal causes. For one thing, events are general or token types of changes, i.e., happenings, occurrences, or episodes. Epidemiology and the language of medical science take events, explicitly or implicitly, as concrete particulars that are capable of being described by purely referential designators. Medical events are thus completely describable by observation alone without reference to anything outside their present scope of examination: they have natural boundaries and a causal history; and their causes are neither part of them nor part of their concepts. So that if we canvass individual instances of diseases or medical conditions as complex events, thereby untying them from their causes or their proximal causal conditions, the latter become exterior to the concept of the disease. Causal attribution may be true or false.

For another, the concept of disease, disease-type, is not an event, but a process, and it is intrinsically, semantically, related to a causal attribution. The causal story is then internal and is being absorbed into the disease process. Stegenga refers to *the causal basis of disease,* namely the pathophysiological causes of disease that is about the causal constitution of a disease, and not the causal aetiology that is often external to one's body.[16]

Even so, medicine needs to identify effective strategies, and within its conceptual apparatus and methods of investigation, it splits causes from their effects and keeps them parted, thereby echoing the basic divide between normal and pathological. Causes, as it were, are thus semantically internal to the disease process, yet epistemically exterior to the sequence of events that constitute the clinical cases of disease.

Contributing Causes

We are never able in medicine, as well as in science in general, to provide a complete account of causation, that is to say a complete set of conditions or an exhaustive list of all the factors that affect some outcome. The specification of causes and determinants of diseases is unavoidably incomplete. "*Experience* wrote Claude Bernard, *teaches us ... that we cannot go beyond the how that is, beyond the proximate cause or the conditions of existence of phenomena*". We only know 'partial' causes, never 'whole' causes, if such exists. This is the reason why causal selection is so important, which feature of the causal history is the most relevant for our purposes. And Lewis writes: "*We may select the abnormal or extraordinary causes, or those under human control, or those we deem good or bad, or just those we want to talk about*".[17]

It follows that in medical research and in clinical practice, we do not need to include the whole of the antecedent conditions, as suggested by J.S. Mill. We are concerned with relevant factors that have a causal role in the circumstances[18] in a background of what Mill termed 'permanent conditions' which are almost never explicitly mentioned. All these causally relevant factors (termed *ceteris paribus*, all things being equal, or if everything else has remained the same) are held fixed prior to the one in question. Mackie's INUS conditions contend that *C* is sufficient for *E*, given certain other conditions, that is: in the circumstances.

Epidemiology consists in a progressive localization of contributive causes by means of inductive methods. Brian MacMahon held that: "*The special contributions of epidemiology are its search for concordance between the known or suspected causes of the disease and the known patterns of the distribution of the disease, or*

[16] Stegenga (2018a, 26).

[17] Lewis (1986b, 162).

[18] Mackie (1986, 31–37).

the use of these patterns to postulate elements of the environment that should be investigated for possible causal roles."[19]

At this point, it seems advisable to introduce two concrete examples. Suppose someone dies from the injuries resulting from a motor vehicle accident. For the policeman, the death is caused by the car crash. For the pathologist, it is due to multiple traumatisms. For the internist, it resulted from a hypovolaemic shock. The forensic physician must decide to what extent the accident contributed to the death. And the family of the deceased will sue the driver of the car.

In 1981, physicians in the US observed clusters of cases of immuno-depression accompanied by rare conditions such as Kaposi's sarcoma, *Pneumocystis Carinii* pneumonia occurring in young men. They termed it AIDS, acute immunodeficiency syndrome. It was then observed that all patients initially showed large numbers of HIV particles in the blood. Subsequently, scientists became convinced that a group of retroviruses are a necessary cause of AIDS. They were necessary under all circumstances because they were intrinsically capable of producing AIDS. HIV viruses being inherent to AIDS, they became a criterion for its diagnosis and part of what we mean by AIDS. But once this necessary factor was being absorbed into the concept of the relevant disease, once it became constitutive of its notion, the causal claim became tautological in the manner of 'insomnia causes sleeplessness'.

Yet, causal analysis of AIDS does not stop there: again, there are further causal contributing factors related to homosexual or heterosexual practices, intravenous drug use, and genetic susceptibility.

Incidentally, emotions cannot influence cancer risk or recovery as opposed to certain gratuitous assertions which are not based on any evidence whatsoever.[20]

Causal Role and Causal Capacity

Counterfactuals or contrary to the fact conditionals also allow us to differentiate causal capacities from causal roles. *Causal capacities* are one of the main research targets of epidemiologists and are usually measured by probability differences (rate difference) or else by probability quotients (rate ratio, odds ratio, or their logarithms); causal capacities remain counterfactually meaningful even in populations where the relevant causes are absent.

If causal capacities correspond to causal support for the effect, *causal roles* correspond to the degree to which a cause caused the effect in some specific situation i.e., the causal contribution to the causation of E provided by C; causal roles are causal facts among particular populations and are estimated by population attributable proportions.

---

[19] MacMahon and Trichopoulos (1999).

[20] Harrington (2008, 98).

Contrary to causal capacities that are two-place relations between cause and effect, causal roles are three-place relations between cause, effect and a given population. How many people die of bladder cancer attributable to tobacco smoking in the United Kingdom in a given year is a matter of causal role. Causal roles suppose that the relevant causal factor is present in the base population under study. If asbestos was eliminated from the United Kingdom, asbestos would lose its causal role in the induction of mesotheliomas in the British population, but its causal capacity would remain unchanged.

In brief, causal capacities support counterfactuals. They do not refer to any token base population, but they hold consistently across a range of different populations and causal background situations, that they hold in hypothetical versions of the real population. On the other hand, claims about causal roles are made in reference to actual (present or past) token base populations. Public-health decision makers tend to be more interested in causal roles observed in base populations, and epidemiologists and medical scientists in causal capacities and aetiological claims.

Causal capacities reflect what is termed *internal validity,* and causal role reflect *external validity.*

The Concept of Causality

There are several ingredients that are central to the concept of causality.

Manipulability

Hume's critique of natural necessity is that of a passive observer who observes the natural course and the regular successions of clinical facts. But this standpoint leaves out the principal stratagem we use for making causal inferences, that is experimental, therapeutic, quasi-experimental, or statistical manipulation.

Reality runs its own course and is independent of our interests, but our interests determine what aspects of reality we decide to select for attention since we wish to bend nature to our purposes. Scientific understanding attempts to explain the world's appearances, but medical concepts, including causality, are unavoidably intentional, at least under certain descriptions and intentions commit one to act. Knowledge of causes is necessary to fill our need for power and control. It follows that medicine is gazing at the world in terms of concepts with which we act upon, i.e., concepts of intentional objects, such as those belonging to ill health, disorder, infection, suffering, harm, pathology, or causality. The concept of aetiology pertains to human agency since it springs from the need for interventions in preventing or correcting abnormal processes.

Meanwhile, causes are not resident in or constituents of the world. They are parts and sometimes grounds for explanation and intervention. This is why Gasking[21] has tendered the suggestion that causes in medicine and engineering are broadly framed generalizations which are not mere inference licences but are like recipes for producing or preventing certain effects. Collingwood wrote, *"The cause of an event in nature is the handle so to speak, by which we can manipulate it."* Causality in medicine must be understood in terms of producing, preventing, or correcting some pathological process by manipulating some state of affairs. What is being taken as a cause and what is not, depends on the framework of our medical interests. Proper proportion of oxygen in the air we breathe is a necessary condition for survival; yet it is not part of our therapeutic strategy in a case of breast cancer to prescribe proper air breathing because this is not relevant to our workaday medical instrumental interests. But it may be so, in cases of heart or respiratory failure.

The manipulative notion of cause[22] is the fundamental and primitive one: we assert causal relationships between events when we are able—every so often *per impossibile*—to obtain, through manipulative techniques, changes which do not usually occur without our intervention. Causation can thus be defined in reference to the potential (counterfactual) interference of agents: if we were to influence (bring about or prevent) *C*, we would influence (bring about or prevent) *E*, the asymmetry between causes and effects thus stems not only from the temporal priority of causes, but also from the asymmetry between means and ends. Causes lay on a blueprint for hypothetical control.

Clinicians and public health officers are primarily interested in sufficiency, in causes that necessitate, i.e., in promoting an effect, in investigations involving explanatory hypotheses and therapeutic efficacy. But they are more attentive in necessary causes when they wish to prevent some effect. What we have, then, is perhaps this. Causes in medicine are like recipes or manipulative techniques to produce (in the case of sufficiency) or eliminate (in the case of necessity) pathological processes; they are intrusions in what we consider the normal course of events: they may be facts, events, processes, phenomena, relations, conditions, states, or deeds which are abnormal when compared with the normal contrastive background which accounts for the difference in the outcome, i.e., the occurrence of a disease. Causal explanation thus involves identifying salient factors and their salience depends on pragmatic concepts of relevance, contrasting explanation and need to intervene. Knowing that dovetails into knowing how.

A longstanding ill-founded debate in counterfactual theory relates to whether sex, race, or biological characteristics such as obesity can be treated as causes, since they do not correspond to specified interventions. M.A. Hernan claims that causal questions are well-defined if interventions are well specified, a claim that has been the subject of an ongoing controversy.[23]

---

[21] Gasking (1955).

[22] von Wright GH (1971, 71).

[23] Hernan (2004), Broadbent (2015, 75), and Glymour and Spiegelman (2017).

Meanwhile, manipulation can hardly take attributes into account, such as age, gender, race, that are essential from an epidemiological standpoint. On this score, manipulability is not an essential condition for causality, but it is one of the possible ways to test for causal relations.[24] Presumed cause-effect relationship in the area of age, gender, race and ethnic group, or marital status are not manipulable: nevertheless, the belief that the presumed effect *would* change if the cause changed is the defining concept of the causal relationship. The term '*determinant*' proposed in 1970 by Brian MacMahon[25] is now used in epidemiology to cover both manipulatable and non-manipulatable factors; the latter such as age, sex, profession, and socio-economic level are being seen as proxy variables for manipulatable causes. Determinants thus have a wider scope than causal factors, as they are parts of medical explanation while the latter are closely linked to preventive or therapeutic intervention. In medical philosophy, conceptual debates often turn out to be merely lexical.

To sum it up, causal statements attempt to explain why a disease occurs and include both proper causes and what John Stuart Mill called mere conditions, both of which are equally necessary for the disease to follow; but, contrary to conditions, proper causes are alterable causes of diseases, namely those by the manipulation of which we can produce or prevent its abnormal or pathological effect. Manipulative causation presumes that we know what will happen if we do not interfere with it. To this we shall now turn.

David Hume and Causal Regularities

What do we mean by causes necessitating their effects?

For David Hume (1711–1776), the necessary connection we suppose exists between cause and effect is mistaken and unsupported by perception; the causal order of events is a projection of or own mental confidences in the way they follow from one another. Hume defines a cause as follows: "*An object, followed by another, and where all objects similar to the first are followed by objects similar to the second.*"[26] All we know is that causes have regularly be followed by their effect in the past. We get our idea of cause from a habit of mind founded on the assumption that the future will be like the past. Our knowledge of causes does not result from demonstrative reasoning but from what merely happens to be the case. This he called *constant conjunction*. So, if one has observed a vast number of instances of *C* followed by *E*, one would pronounce them causally connected. Hume does not concern himself about whether there is some element of necessity beyond this close

[24] Russo et al. (2010, 8–9).

[25] MacMahon and Pugh (1970).

[26] Hume (1975, Sect. VII, Pt II, 76).

connection but about what we can know about it rather. His main interest is epistemological.

If so, Hume's idea of causation combines three elements:

(1) A causal relation presupposes that cause and effect are *distinct events*. They should have, in Hume's terms, distinct existence. Galen formerly held that causes are necessarily distinct from the diseases that result from them.[27]
(2) Causal relations require *regular succession,* and this constancy is all there is in them.
(3) Causal relations, as far as we know them, include an *idea of necessary connection*, which we infer from observed regularities rather than a causal law at work. "*Perhaps 'twill appear in the end that the necessary connection depends on the inference, instead of the inference's depending on the necessary connection*".[28]

A major difficulty is to avoid confusing causal with non-causal association. These errors, *biases* in medical terminology, result usually from the association of both categories with a third category. On this score, if C is causing E, and E and F are associated, this association is non-causal: manipulating E will not change F, and conversely, F is an example of what is called a *confounding factor.*

For example, drinking coffee and cigarette smoking are associated, non-causally, both of which are being affected by some common cause. But tobacco is a cause of lung cancer, and coffee is a confounding factor. Sir Bradford Hill put it bluntly: "*The fact that in areas of Scandinavia the level of the birth rate varies directly with the prevalence of storks is not likely to mislead us.*"[29]

But what we are looking for is a causal proposition of the form, "q because of p". This assertion seems to entail more than the joint affirmation of p and q. What do we then need in addition to Hume's "if p, then q"? What is the nature of this natural necessity, of this "*law that we know not*", which is neither observable nor empirical nor logical? What distinguishes correlations from causal relations, mere invariable succession from causal string, and bare *post hoc propter hoc* sequences from causal explanation? What is this unobservable extra bit, which will provide the idea that a cause in some way necessitates its effect?

David Hume and Counterfactual Dependence

In seeking to define causality as constant conjunction, Hume added a second and different one: "*or in other words: if the first object had not been, the second never existed.*" In like manner, for Galileo, "*cause is that which put, the effect follows; and removed, the effect is removed*". This second definition that emerges from Galileo

[27] Galen (1991, 40).
[28] Hume (1973, I, I, III, I 6).
[29] Hill (1967, 276).

and Hume goes beyond constant conjunction and corresponds to what logicians call *contrary-to-fact conditionals, unfulfilled hypotheticals counterfactual conditionals, contrafactuals, subjunctive conditionals, subjunctive hypothetical,* or *sine qua non* causes. Some of these terms reflect distinctions, which does not need to detain us.

Daniel Hausman remarks that: "*Counterfactual reasoning should permit one to work out the implications of counterfactual dispositions, so that to be prepared in case one supposes what actually happens*".[30]

Counterfactuals provide a general framework for designing and analyzing epidemiologic studies.[31]

Conditionals are sentences of the form "if *p* then *q*", where *p* is called the antecedent, and *q* the consequent.

- "*If it is raining, then Ariane is inside.*"

First, this conditional is expressed in indicative mood: it is called an *indicative conditional.* Now, this conditional is a very complex matter, since the core of that meaning is modelled in logic by what is called the *material conditional* and is represented as $p \supset q$. It is defined as a non-causal logical connective, which is read as "*if p, then q*".

Yet, the material conditional is a rough match of the English conditional: '$p \supset q$' means that if the antecedent *p* is false, $p \supset q$ *is true,* and it is *false* only when *p* is true, and *q* is false. This and several odd results of modeling the English conditional by the material conditional are known as the paradoxes of material implication. The material conditional does not properly model the English language.

Propositional or sentential logic is an artificial language or a *formal* language since it has no *content*: it merely describes how certain arbitrary signs behave together. In propositional logic, such *material conditionals* are represented by truth-functional conditionals: '$p \supset q$', in which *p* implies *q* is read as either 'if *p,* then *q*' or as '*p* only if *q*'. The above sentence is *truth-functional* since its truth or falsity depends only on the truth or falsity of the component sentences (*p* and *q*): it is true whenever both components are true, false otherwise. There is no necessitation, *no causal link* or whatever is required between the antecedent (*p*) and the consequent (*q*), but *a logical link:* only their respective truth matters. Propositional logic allows us to express in a concise and formal way, inferences that people would make in the *indicative* mood. Indicative sentences describe what happens under circumstances that are *actual* or *existential.*

- "*If it were raining, then Ariane would be inside.*"

Secondly, the material conditional does not model the above *subjunctive conditional.* This subjunctive conditional is sometimes called *counterfactual conditional,* on the ground that it implies the falsity of the antecedent, i.e., *that it was not raining.*

[30] Cartwright (2007b, 237).

[31] Maldonado and Greenland (2002).

A counterfactual assertion states:

- how things *would have been* (q) if the antecedent (p) had obtained.
- but it did not obtain.

"If kangaroos had no tail, they would topple over."

Hence, counterfactual conditionals are conditional statements whose antecedent is false: they express a situation contrary to facts.

In many contexts, counterfactual statements presuppose p, or leave the question open. "If p than q" is then being *taken nonexistentially* with a vacuous antecedent, and it asserts a causal role by way of a local entailment. The usual answer is that different counterfactuals correspond to different causal relations, namely like 'causes', 'prevents', 'inhibits' or 'triggers'.[32]

However, counterfactuals are non-truth-functional conditionals, since their antecedent is subjunctive and hence modal, so that they involve the language of modal logic.

Conditionals '$p \supset q$' are false only if p is true and q is false, so that they are guaranteed to be true, whenever p is false, whatever q says. But, in formal logic, counterfactuals seem to be true, vacuously true, since if p happens then q happens, but *p does not happen*.

But how to formalize counterfactuals encounters difficulties.

Since causal relations are neither logical nor empirical, they cannot be expressed in terms of indicative conditionals. On this score, the counterfactual outcome model has become routine for causal inference in epidemiologic studies.[33]

Deterministic causal statements, 'p causes q', are then equivalent to: (1) p is true (2) q is true and (3) if$p$ were false, then q would be false, if all else remains the same. This reading of causation, and particularly statistical counterfactuals, fits the medical need for interventions and for model building and testing: if p were to change, q would accordingly change. Causal effect of the exposure on the outcome is the difference between the counterfactual probability of the outcome had everybody in the base population been exposed and the counterfactual probability of the outcome had everybody in the population been unexposed. This definition of the counterfactual outcome can be applied in deterministic as well in non-deterministic counterfactual outcomes.[34]

Lewis added additional structure on modal systems, and counterfactuals are then represented in modal logic as follows:

'$p \Box \rightarrow q$, which means that if p were the case, q would be the case (where $\Box \rightarrow p$is called "a necessitation" and means "locally entails"). The antecedent *does not imply* the consequent since it is contingently possible that the antecedent be true and the consequent false; the connective '$\Box\rightarrow$' should be read, "*locally entails*".[35] This means that counterfactual conditionals represent *local entailments* with the result that causal statements are not laws of nature.

[32] Lewis (1973) and Hall and Paul (2003).

[33] Höffler (2005).

[34] Hernan (2004).

[35] McCawley (1981, 311–316).

There is one objection to the counterfactual model of causality that has not been considered. Indicative conditionals satisfy a law of transitivity, but counterfactuals do not.

One of the most popular treatment of counterfactuals is that of David Lewis's *theory of possible worlds semantics*; the counterfactuals are then true or false according to whether *q* is true in the most relevant similar possible world to ours in which *p* is true. In order to avoid the ontological extravagance[36] of possible worlds, it is possible to treat them as collections of propositions through a linguistic turn.[37] After Lewis, philosophical research went astray and departed from purely counterfactual analysis of causation and highlighted the intuitive idea that causation is an intrinsic relation between events, namely a local link depending exclusively on the inherent properties of the events as well what takes place between them. On this account, causation is a structural relation that underlies causal dependences.

Incidentally Alex Broadbent wrote a devastating critique of the use made of counterfactuals by epidemiologists. The main difficulty is the fact that effects do not always counterfactually depend on their causes. Less well-known difficulties include the fact that effects typically depend on many more events than we are usually willing to count as causes.[38]

The Probabilistic Nature of Causality

> While the individual man is an insoluble puzzle, in the aggregate he becomes a mathematical certainty. You can, for example, never foretell what any one man will do, but you can say with precision what an average number will be up to.[39]

Causality does not require determinism. The medical and epidemiological literature often assumes tacitly that we live in a deterministic universe in which medical events come and go according to immutable causal regularities. Rothman's pie model[40] accepts blindly and without any reservation, the notion of determinism. "*For most of science* write Thomson and Upshur in their text on philosophy of medicine, *a commitment to materialism is also a commitment to determinism.*"[41]

Clinicians, epidemiologists, and medical scientists are methodologically probabilistic though often keeping a largely deterministic approach.

Be that as it may, causation should be analysed in probabilistic terms as causes make their effects more likely and revolve around the transmission of probabilities

[36] Scruton (1994, 171).
[37] Illari and Russo (2014, 91).
[38] Broadbent (2013, 44–49).
[39] Conan Doyle (2014).
[40] Rothman et al. (2008).
[41] Thomson and Upshur (2017, 43).

within ordered pairs of events: causality is a transmission of probability distributions.

We say that *C* causes *E* in a given situation if and only if, in that situation, the chance of *E* conditional on *C* is greater than the chance of *E* conditional on not-*C*.[42] A factor *C* is a positive (or negative) causal factor for *E* (distinct from *C*) if and only if *C* precedes *E* and *C* raises (or lowers in the case of preventive factors or counteractive causes) the probability of *E* in the circumstances (i.e. *ceteris paribus)*, that is in each way of holding other causal factors for *E* (or of not-*E*) or other factors which interact with *E*, constant.

If *Cj* represents any alternative relevant factor, we can define a cause C of a condition *E* as follows: *C* is a positive cause of *E* if and only if: $\Pr(E|C \,\&\, Cj) > \Pr(E \mid \text{not-}C \,\&\, Cj)$ for all j.

A negative causal factor (e.g., a preventive factor) is a positive causal factor for not-*E,* or else it will be represented by the same expression where the sign $<$ is substituted for the sign$>$. For an irrelevant causal factor, the sign $=$ will be substituted for the sign $>$.

Smoking does not produce lung cancer but increases the probability of lung cancer. Such probabilities may be *epistemic*, i.e., stemming from our present or future limited cognitive abilities and our subjective rational expectations, or, if we formally reject determinism, *ontic*, i.e., supposedly referring to the nature of the world and its individual propensities. Yet, even though both choices are theoretically legitimate, it is surprising that, although most epidemiologists and clinicians are constantly confronted with randomness and are handling statistical instruments, few of them are inclined to think that probability laws could be ultimate, and that chance might not be a mere sign of our ignorance.

Everything being considered, my sense is that we should forfeit the concept of determinism, so that the final account leaves us with a complex and multifaceted concept of probabilistic causation, to encompass Hans Reichenbach, Patrick Suppes, Wesley Salmon, Nancy Cartwright, and Kenneth Schaffner. The lack of predictability is inherent to our situation. The laws of the universe are merely saying that certain events are probable, given other events.

Small wonder then, that Karl Popper[43] has refuted all kinds of determinism. In a universe in which predictions would be theoretically possible, it would be impossible to predict, with a desired degree of prediction, the future state of a system based on information coming from a predictive element operating in the system. This is because knowledge of the future development of a system, guaranteed by the predictive capacity of a theory, is an integral part of the universe being described, and thus the influence of one prediction can never successfully predict everything that covers the system to which it belongs: the self-referential nature of the prediction means that it can never predict the future development of the system to which it belongs, since, to do this, it would have to wait until the end of its evolution.

---

[42] Humphreys (1989).

[43] Popper (1982).

Analytical Epidemiology

It was a poet, one of the greatest Italian poets and humanists, Francesco Petrarca (1304–1374), who, as early as the fourteenth century, sketched out the future of medical science: "*I believe and affirm the following: if a hundred or a thousand men of the same age and lifestyle were all simultaneously afflicted with the same disease, and half of them followed the advice of doctors such as those we have today, and the other half, on the other hand, without any doctors, governed themselves according to natural instinct and their own judgement, I have no doubt that more will die in the first group, and more will be saved in the second.*"

A principal objective of epidemiology is to identify alterable causes of disease.

Case-control studies are a type of epidemiological study in which cases of disease are identified and enrolled and a sample of the source population that gave rise to the cases is also identified and enrolled, i.e., the controls. Exposure histories of the two groups are collected and compared. Epidemiologists use a case-control study when little is known about the disease because this type of study allows to examine several exposures in relation to a disease.

Cohort studies (previously called *prospective studies*) are a type of epidemiological study in which subjects are defined according to their exposure level and followed for disease occurrence, also called follow-up, incidence, or longitudinal study. Epidemiologists use a cohort when little is known about an exposure, because this type of study allows to examine several outcomes or health effects in relation to an exposure.

The principal limitation of case-control studies is that information on the supposed cause must be retained (in the memory of persons, in biomarkers or in documents) until the person can be identified as having the disease, which raises issues concerning the reliability or lack of it, of most indicators.

Bias is a systematic error that results in an incorrect estimate of the measures of association There are two main types of bias: selection and information bias. When evaluating a study for the presence of bias, it is necessary to identify the source, assess the strength and determine its direction.

Confounding is a mixing of effects between an exposure, an outcome and a third extraneous variable, the *confounder*. Epidemiologists consider confounding an annoyance as it is one of the main reasons for completely distorting attributability by either exaggerating or minimizing a true association. A confounding variable is associated with the exposure; it is an independent cause of the disease, and it is not an intermediate step in the causal path to the outcome between exposure and disease. There are three ways to control for confounding, randomisation, matching whereby study subjects are selected so that potential confounders are distributed in an identical manner, and restriction whereby admissibility criteria for subjects are limited, which also limits the generalizability of study results.

Cohort studies have some major methodological advantages since it takes the logically proper sequence from cause to effect, and it is less susceptible to selection and information bias. To be sure, in observational studies showing a consistently, strong, unconfounded association, residual confounding and bias can never be totally excluded. Results of every study should be judged in the context of the overall empirical evidence and their medical plausibility and supporting biological information.

Causal Tendencies

The question before us now is that since we dispense with sufficient and necessary causes, sufficiency and necessity are becoming limiting concepts, by substituting tendency toward sufficiency or tendency towards necessity.

Causal tendencies, according to John Stuart Mill, do not indicate what effect occurs when the cause is present but rather what the cause contributes to an effect in more realistic circumstances where other causally relevant factors have not been eliminated. Cartwright defends the view that causal claims are really about capacity claims or tendency claims, namely about local truth-makers.[44]

On this score, we need to weaken causality to causal tendency.[45] There are *grades of causal sufficiency* and *grades of causal necessity*.[46]

If *C* (cause) tends to cause *E* (effect), then:

(1) The probability of sufficiency is, in a way, prospective: it is the probability that *C* would produce *E* in a situation where there are no other causes of *E* present and where *C* and *E* are absent.
(2) The probability of necessity is retrospective: it is the probability that *E* would not have occurred in the absence of *C*, given that *C* and *E* did in fact occur.

We now turn to David Hume's two questions concerning causality. First, is *C* regularly followed by *E*?

Cohort studies raise that same question with a modifier: *to what extent* is *C* regularly followed by *E*? Then, *C* is a positive cause for *E* means that *E* is more probable given *C*, than given Not-*C*. *C* is held to be positively statistically relevant to E. If so, in the exposed hypothetical population, where all individuals would be exposed to *C*, there would be a certain percentage of individuals exhibiting the effect *E* which reflects the *probability* Pr $(E|C)$. The higher this probability, the closer *C* turns into a sufficient condition. This probability of a sufficiency of *C* is a random variable with a numerical value in the closed interval [1, 0] with a value of 1 when *E* is present and 0 when it is absent. Hence, in the counterfactual unexposed population, there will be a certain percentage of individuals exhibiting the effect *E*. So, the outcomes usually lie on a continuum between the two extremes, sufficiency, and necessity.

---

[44] Cartwright (1989).

[45] Good (199–200).

[46] Skyrms (1988, 172–176) and Good (1988).

In the case of sufficiency, at one end, Pr $(E|C) = 1$ in the hypothetical exposed group. And at the other end of the spectrum, we would have: Pr $(E|\text{Not-}C) = 0$ in the counterfactual control group. That means that the causal contribution of C increases if *C* tends towards sufficiency. The lower the percentage of non-exposed subjects in the contrast group, the more *C* nears necessity. If there are no non-exposed subjects, *C* is necessary.

In the light of this, *C* has a *tendency towards sufficiency* if and only if the conditional probability of *E* given *C* is greater than the conditional probability of *E* given Not-*C*; at the end of the spectrum, in the case of *sufficiency*, *E* and *C* are absent. *C*'s tendency towards sufficiency for E (attributable risk) can thus be measured by the difference between two conditional probabilities: Pr $(E|C) - \Pr(E|\text{ Not} - C)$.

Its value ranges from 0 to 1, that is from causal irrelevance to causal quasi-determinism. The higher the difference, the more *C* proves to be sufficient.

Case-control studies raise Hume's second question, "*if the first object had not been, the second never existed ...* " again with the same modifier: to what extent can we predict the absence of *E* from the absence of *C*? Or else, to what extent can the presence of *C* be predicted from the presence of *E*?

Case-control studies look at the *probability* of finding a putative contributing cause given that the disease has occurred. This probability Pr$(C|E)$ also has a full range of degrees. The higher the percentage of exposed subjects among the hypothetical study group, the nearer *C* closes in on necessity and conversely: it thus measures the probability of necessity. On the other hand, the lower the percentage of exposed subjects among the contrast group, the more *C* nears sufficiency. If *C* is absent in the control population, *C* is sufficient for *E*.

So, *C* has a *tendency towards necessity* for *E* if and only if the conditional probability of C given E is greater than that of *C* given Not-*E* if the disease and exposure did actually take place. The following difference measures the tendency towards necessity of *C* for *E*:

$$\Pr(C|E) - \Pr(C|\text{Not} - E)$$

If it is not statistically significant, *C* is causally irrelevant to *E*.

What precedes licences the view, details, and objections aside, that cohort studies are of the form if *C* is a contributive cause for *E, C* as an alternative to Not-*C* would have a tendency towards being a sufficient condition for *E*. And case-control studies specify the alternative: if *E* rather than Not-*E* had happened, *C* (the putative cause) would tend towards being a necessary condition.

In brief, Cohort studies are measuring causal tendencies towards sufficiency, and Case-control studies are measuring tendencies towards necessity. Be this as it may, if we forfeit determinism, sufficiency and necessity become limiting concepts.

Risks and Measures of Effects

Three factors should always be considered when measuring how commonly any outcome of interest (disease, death, hospital admission) occurs in a population: the number of affected individuals, the size of the population from which the case arise and the amount of time that this population is followed.[47]

Incidence rate is defined as the occurrence of new cases of disease arising over a specified time period. It is sometimes called *incidence density* or *hazard.* A rate equals the average population *during an interval.*

A *risk* relates the probable number of events *at the beginning of the interval.* The word *risk* if often used synonymously with rate, but this should not be: it is the accumulated effect of a rate operating during some specified period of time. Hence, contrary to rate, risk increases proportionately as length of observation increases.

Risk can be expressed as a fraction. The numerator is the number of people who experience the outcome, and the denominator us the number of people capable of the outcome of interest, often called the population at risk.

Risk is often used to describe the *cumulative incidence* of a disease during a transient epidemic, often called an *attack rate,* although this is not a rate. It is the proportion of a candidate population that become diseased over a period of time.[48]

Risk factors may also be distal factors, a few steps upstream from the proximate cause of diseases. For example, well-known risk factors for breast cancer are high socio-economic status, never married and being a Jew Ashkenazi.

The term *risk* is vague. For example, a hospital may express its mortality rate as the number of maternal deaths per thousand deliveries. This is a proportion although the use of the term *rate* is common. The women constitute a group within which the deaths occurred, but they do not form a geographic population, even though such a group is often referred as the *population at risk.*

When there is an association with an increased probability of disease in those exposed to some environmental, individual, or social factors, those characteristics are called *risk factors.* However, the phrase *risk factor* does not necessarily imply the characteristic has a causal effect, since association is not causation. When a causal relationship is agreed between disease and the risk factor, the phrase causal factor, or simply cause, is used.

Yet, epidemiologists often use the term *risk factor* instead of cause, which reflects epidemiologist's caution about making causal inferences.

$$\textbf{Cumulative Incidence} = \frac{\text{Number of new cases of disease}}{\text{Number in candidate population}} \text{ over a time period}$$

Risks differ from rates in two respects. First, risks have no time dimension as a fundamental component (except descriptively, in indicating the time period or an

[47] Long (1991).

[48] MacMahon and Trichpoulos (1999, 46–47).

age limitation). Second, the population is the unexposed population at the beginning of the period of observation, as distinct from the average population during the observation period.

While incidence measures the occurrence of new cases, *prevalence* measures the frequency of existing cases.

There are numerous methods of estimating risk and the number of diseases, that must be put in perspective. Measures of disease frequency are contrasted in either absolute or relative terms depending on the goals of the epidemiologist.

(1) To measure the public health impact, one can measure the *population attributable rate* or *risk difference* (incidence rate, cumulative incidence rate, or prevalence) between the exposed and unexposed group:

$$\textbf{Incidence rate difference}: \mathrm{IRD} = \mathrm{IR}_{\mathrm{exposed}} - \mathrm{IR}_{\mathrm{unexposed}}$$

Causal interpretation using probability raising applies to rates or incidence.[49]

(2) On the other hand, relative measures of comparison, the *attributable proportion* among the exposed, are based on the ratio of two measures of disease frequency, and they describe the strength of the relationship between exposure and disease:

Attributable Proportion among the exposed:

$$\mathrm{AP}_{\mathrm{exposed}} = \left(\frac{\left(\mathrm{R}_{\mathrm{exposed}} - \mathrm{R}_{\mathrm{unexposed}}\right)}{\mathrm{R}_{\mathrm{exposed}}} \right)$$

(3) A *relative measure of comparison* is based on the ratio of two measures of disease frequency:

$$\textbf{Incidence Rate Ratio or Risk Ratio}: \mathrm{RR} = \frac{\mathrm{R}_{\mathrm{exposed}}}{\mathrm{R}_{\mathrm{unexposed}}}$$

Where R is incidence rate, cumulative incidence, or prevalence. The size of the rate ratio is a better index than is the rate difference of the likelihood that a causal relationship exists between exposure and the disease involved. The Doll and Hill study indicates that the mortality rate ratio from heavy cigarette smoking is far greater for lung cancer and chronic bronchitis than from other causes, but the rate difference (attributable mortality) is greater for cardiovascular diseases than for lung cancer mortality.

Furthermore, the likelihood that an association between two variables results from an association of both with a third variable would appear to decrease as the rate ratio increases because the higher the rate ratio the stronger, and therefore,

[49] Long (1991).

presumably more obvious, must be the association between each of the variables and the third variable. Thus, one would be inclined to accept the associations with cigarette smoking (RR = 32.4) as evidence of a causal relationship more readily for lung cancer and for chronic bronchitis than for cardiovascular disease (RR = 2.61).

Yet, if it is accepted that the observed association is a causal one, then the rate difference (attributable rate) gives a better idea than does the rate ratio of the impact that a successful preventive programme might have. Because the associations of cigarette smoking with lung cancer (IRD = 2.20) and with cardiovascular disease (IRD = 2.61) are both causal in nature, then elimination of cigarette smoking would prevent even more deaths from cardiovascular disease than from cancer.

(4) In instances when the investigators do not know the size of the total population, they calculate a number called an **odd**, which functions as a rate or risk, for instance in case-control studies. The odd of an event is the probability that it will occur divided by the probability that it will not occur. In an odd, the number of persons with a characteristic (an exposure or a disease) is expressed relative to the number without the characteristic.

The **Odds Ratio** can be computed as:

$$OR = \frac{x_e}{N_e - x_e} \Big/ \frac{x_0}{N_0 - x_0}$$

where N_e and N_0 are respectively the number of exposed and unexposed at the start of the observations, and x_e and x_0 are the number of cases observed among them. Why compute an odds ratio when the risk ratio is already available? Because the odds ration can be calculated in a case-control and in a cohort study.

In case-control study, we usually cannot obtain the actual rate or risk in the exposed and unexposed groups so that it is not possible to obtain an absolute comparison such as the risk and rate difference. However, it is possible to obtain a measure of the public health impact of the exposure: the **attributable proportion of the exposure AP_e**:

$$AP_{exposed} = \frac{OR - 1}{OR} \times 100$$

This represents the attributable proportion of disease among the exposed population.

Relative Risk and Odds Ratio (Njolstad et al. 1996)

Disease.

Exposure		+	–
	+	a	b
	–	c	d

Myocardial infarction

Smoking		+	–
	+	355	3140
	–	140	2507

$$\mathbf{RR} = (a/(a+b))\text{divided by}(c/(c+d))$$
$$= (355/(355+3140))\text{divided by}(140/(140+2507)) = 1.92$$

$$\mathbf{OR} = (a/c)\text{divided by}(b/d)$$
$$= (355/140)\text{divided by}(3140/2507) = 20.2$$

OR approximates RR fairly well when the disease is rare.

It goes without saying that those expressions of causal relation or of causal strength should be judged in the context of the overall empirical evidence and their biomedical plausibility.[50] Previous guidelines of the past decade were more number based. If a cholesterol is high, then maybe it should be treated. If it is not, it is probably all right. But now we are not saying "What is the cholesterol?" but instead "What is the overall risk?". This risk-based approach is evidence-based even though we are not as able yet at determining risk as we would like to be.

Contrast Case

Contributing causes are comparative in form. They indicate the contribution the cause makes compared with its absence or with some standard situation. In the process of causal selection, the choice of the contrast case, that is what epidemiologists call "control groups", entails some matching. This strategy is an application of J.S. Mill's inductive method of difference; it is the process of choosing a contrast case, a neutral state that is the same as the study group except for the specific characteristics, which are to be studied. Hence, a variable used for matching cannot be investigated for its causal role since cases and controls are alike with respect to that characteristic.

"*Contrastive statements,* writes Dretske*, embody a dominant contrast, a contrastive focus, a featured exclusion of certain possibilities. Something similar to a figure-ground distinction is at work in these statements.*"[51]

A contrastive phenomenon consists in a fact and a foil, and the same fact may have several foils. What do we mean by that? Some feature C is a cause, and some other feature E is an effect only in some relevant sense. Since events are enmeshed in a causal web, explanation draws attention to, and selects salient features, which constitute the cause(s) of why an event or a type of event happened: why did E happen, as against such and such alternatives? The meaning of a why-question thus depends on the context of our (instrumental) interests within which it is raised: a causal explanation in one context—may be no explanation in another context. The

[50] Broadbent (2013, 142–144).

[51] Dretske (1972n 411–412).

impossibility of providing a complete causal account is not a sign of our ignorance or of the weakness of our models, but an inherent part of what we mean by causality.

The contrastive approach serves several purposes: excluding changes due to other causes, so-called confounding factors, ruling out effects caused by no intervention such as placebo effect, and artefactual changes in methods of measurement or ascertainment. In brief, it serves to select and analyses causal conditions. The contrast case is then construed as the antecedent clause of a counterfactual, the if-clause that is inconsistent with the context of the individuals under study.

According to this stance, we explain event E rather than event E', or E rather than not-E: why does Susanna now have cancer of the lung rather than not? We never explain the occurrence of an event simpliciter, but always in contrast to some set or sets of occurrences, which are to be neutralized, and which constitute a counterfactual contrast class e.g., the set of relevantly normal individuals.

Associations between putative causal factors and their effects are observed in actual samples or base populations, whereas causal relationships cross the divide between indicative and subjunctive modes and are defined between hypothetical versions of actual populations. The claim that cigarette smoking is a cause of lung cancer is not equivalent to the mere observation that there are more cases of lung cancers among smokers than among non-smokers (with some *ceteris paribus* conditions), but to the claim that there would be more cases of lung cancer if everyone smoked than if no one smoked (with some ceteris paribus conditions).

The Causa Vera Fallacy

"A police surgeon investigating … a case of murder, found a bullet in the heart. This, he decided … was the causa vera … of this case of murder." He then went to war, and since he thought that war was murder on a large scale, … "*he investigated the appearance of many bodies, again finding bullets, he declared that bullets are the cause of war, as of murder.*" ... Furthermore, the occasional absence of pellets in some bodies "*disconcerted him until he realized that he had once found gas poisoning, the causa vera in a case of murder, and he therefore came to the conclusion that several wars here existed, side by side, each one sui generis, and boasting a different causa vera.*"[52]

There is a tiresome cliché in the medical and philosophical literature, which affirms that there is or should be a uniquely correct grouping of ailments into natural kinds since every disease must have its own particular cause, that is to say that all diseases entities should, as it were, be modelled on those resulting from host-parasite interaction.[53] Thus, the *causa vera fallacy* assumes first that diseases are entities, i.e., things existing by themselves, stemming from singular specific causes,

[52] Crookshank (1920, 1778–179).

[53] Cassell (2004, 7–8, 102–103).

and second, that the classification of diseases must be based strictly upon their aetiology. The individuation of a disease entity, e.g. Parkinson's disease, is allegedly tainted with the view that there must be a common cause; if we reject this belief, we cannot develop a uniform conception of *paralysis agitans* since the various symptoms would then have no status beyond their own nature.[54] Much of the past and current medical research is a quest after single, proximal, biochemical, and immunological, physiological, structural or genetic causes of diseases. Ehrlich claimed that once we know the singular cause of a disease, we could develop a curative treatment that is a 'magic bullet'.

We can readily identify the theoretical, though implicit, presuppositions of such a view. The human body supposedly runs like clockwork in which parts exquisitely harmonize with one another with manifest concert and regularity and without redundant or conflicting elements. Clockworks are goal-directed systems in which every element is necessary to achieve the preordained purposes for which they were built; within a clock, parts and pieces are each necessary and sufficient for the clockwork[55]; it is usually possible in case of breakdown, to identify a single change from which the breakdown stems. Thus, there is a body of literature, which attempts a drastic reformulation of causality, and holds that it is self-evident that agency is constitutive of causality and that the 'real' causes, most of them unknown yet, of diseases are always necessary.

But the clockwork analogy is misleading. It reduces complexity to simplicity; it wishes to extend agent-host-environment triads and Koch's postulates from their original domain of communicable diseases to the whole field of medicine. Even worse, this lofty pursuit that obfuscates the distinction between necessary factors, and the *causa vera* which usually goes along with it, that philosophers attribute erroneously to medical thinking[56] has been denounced by several epidemiologists as a dogmatic slant of the nature of scientific activity.[57] Such a programme subscribes to the view that medicine must allegedly aim at ultimate categories such that to each disease would correspond a single explanation. The usual doctrine of the multiplicity of causes is thus attributed to our present inability to analyse the effects with the same thoroughness as the causes.

[54] Nordenfelt (1995, 169).

[55] Boorse (1977).

[56] Stegenga (2018a).

[57] Evans et al. (2008), Rothmann (1988), Skrabanek (1994), and Stehbens (1985, 1992).

The Web of Causation: Plurality of Causes and Causal Diversity

Nancy Cartwright summarized the concept of causation at its core: "*About causation I argue ... there is a great variety of different kinds of causes and that even causes of the same kind can operate in different ways ... Recognizing this should make us more cautious about investing in the quest for universal methods for causal inference*".[58]

Cartwright does not contend that we have several competing theories but rather a host of theories each appropriate to one of a large plurality of different kinds of causal systems. Causation is not one thing with different methods, but there are different types of causation with different characterizing features. "*Causation is a highly varied thing*," concludes Nancy Cartwright.

The occurrence of everyday phenomena as well as of natural events has been known ever since Aristotle to be the result of multiple causal influences: diseases are usually the result of the sum total of positive and negative influences taken together, a causal nexus, i.e., the net effect of external (e.g. infectious diseases, traumatism) and internal (e.g. genetic factors, metabolic anomalies) factors operating within a causal network that MacMahon has christened *the causal web*.[59]

Meanwhile, medical research has no need to unravel this causal nexus. It attempts to identify and select some relevant causal strands the correction of which may potentially increase the opportunity for preventive or therapeutic intervention. What clinicians and epidemiologists usually call the cause is quite different from the *causa vera;* it refers to the factor to which they wish to call attention or the most obvious one, bracketing off temporarily the other contributing factors or standing conditions. In causal analysis, selection is not only desirable but also inescapable.

Suzan Haack raised a challenging issue pertaining to the assessment of causation, and whether and if so, when, and why, can a compilation of pieces of evidence, none of which is sufficient by itself to warrant a conclusion to the legal degree of proof, better than any of its elements? There is no mechanical procedure, algorithm, or statistical method, which spell out when combined evidence is stronger than its components and why; and this necessitates articulation of the elements that establish whether, and to what extent, evidence warrants a conclusion, and how these elements may be enhanced when evidence of a distinct kind is combined.[60]

With the *web* of causation, we have a network of causes that interact with each other, and that may or not reinforce each other or inhibit each other. So-called 'manifestational entities" are not multicausal but can better be represented by a spider-web model.

---

[58] Cartwright (2001).

[59] MacMahon and Trichopoulos (1999, 26).

[60] Haack (2008).

Back to Sufficiency and Necessity

Philosophers of science have succeeded to avoid the shortcomings of Hume's theory.

But when epidemiologists analyse the concept of causation, it usually seems like sort of a rear-guard action that raises more questions than they can answer.

Do we know a single example of sufficient cause or of sufficient causal constellation? If we do not, and it seems that we do not, does this reveal some sort of abysmal failure in our explanatory activities, or does it mean that the prevailing view should be abandoned or severely revised? Is there no way to salvage it? Or do we need to decide whether the world is deterministic or not?

First and foremost: *Sufficiency*. Rothman assumes, in the absence of evidence, that a disease is a whole pie, and that the "slices" of the pie are the component causes; when all the components of the sufficient cause are present, the disease occurs; a disease is supposed to have a finite number of constellations of sufficient causes, but the component causes may be present in several sufficient causes; if a component is a member of all sufficient pies, it would be a necessary cause.[61] This is a pure figment of the imagination.

Next: *Necessity*. The internalization of the AIDS virus is not necessarily followed by a clinical disease. Monogenic diseases are caused by one gene, such as thalassaemia, Huntington's disease, or Duchenne muscular dystrophy. Yet there may be gene variants, incomplete penetrance, or variable expressivity, and "modifier genes" so that genetic diseases do not work in "all or nothing" way.

So-called *post factum* necessary causal factors when available or identifiable are a natural bounty in medicine. Medical history thus reveals a constant pursuit for *sine qua non* causes since they provide effective ways of intervening. Necessary causes are one of medicine's philosophical stones and are the object of a constant and tireless quest. This justifies persevering with the genome project as well as the call for a proteome project and the interest in oncogenes and in RNA interference, in the hope of developing specific treatments for some of our ills. We also canvass our disease classification whenever possible around such causes.

Even then, necessary or sine qua non-causes, at least in the typical cases, and when specific, are being incorporated into the definition of such single criterion entities. They become internal to the diseases brought about by them and part of their definition. Consider, for example infectious diseases, genetic diseases such as phenylketonuria, post-traumatic stress disorder, nutritional deficiencies, or diseases due to and defined by external agents, such as beriberi, pellagra, or scurvy. Such conditions are often called 'causal entities', in opposition to diseases defined by their signs and symptoms, i.e., 'manifestational entities' that are seen as being externally produced. This corresponds to the medieval distinction between *causa transiens* (transitive) and *causa immanens* (immanent). An immanent cause is a cause whose effects lies in it, as opposed to causation from external factors. In the first case, the view defended by David Hume, the cause was said to lose nothing of its

[61] Rothman et al. (2008).

own nature in producing the effect, whereas in the second case, that of agent causation, the cause was exhausted (or partially exhausted) in its effect.

When an epidemiologist looks for the cause of a disease, he is looking for an extensional relation. But Donald Davidson showed that once a causal relation has been established between events under a description such as relating the AIDS virus to the condition of immunodeficiency, the causal factor becomes explanatory and hence intensional and immanent inasmuch as it holds between events under a description.[62]

Necessary causality springs from a conceptual confusion. There are two types of necessity.

For example, to declare that oxygen is necessary for combustion is an empirically necessary condition: the world is such that you cannot get combustion in the absence of oxygen; yet we wouldn't call it a cause.

But when we talk about causality, *we* introduce the necessity by our stipulative definition.

On this score, it was a *matter of fact* that a disease entity had been identified early in 1837 on purely clinical grounds. However, it became a *matter of meaning* since 1910, when Howard Taylor described the micro-organism (*Rickettsia rickettsii*) from the blood of patients in Mexico: the internalization of the micro-organism became a necessary part of the definition of the disease.[63] A stipulative definition proposes necessary and sufficient conditions and is like a suggestion: "Let us use the word *Rocky Mountain spotted fever* to mean…". What clearly emerges from this is that the necessity is not in the causation process but stipulated in our language: such a definition and the necessity that it introduces are not discovered, cut off or carved from epidemiologic data, but intentionally defined.

Reverse Causality

Reverse causality means that *C* and *E* are associated, but instead of *C* causing E, it is the other way around: *E* causes changes in *C*.

On this score, an unmeasured or imprecisely measured confounder may prevent causal inference being drawn between any two factors, and reverse causality.

There are numerous examples of risk factors or health behaviour that can be influenced by reversed causality. For example, when lifelong smokers are told they have lung cancer, many then quit smoking, so that ex-smokers being diagnosed after the disease develops, may then seem more likely to die of lung cancer than current smokers.[64]

---

[62] Davidson (1967).

[63] MacMahon and Trichopoulos (1999, 29).

[64] Trompet et al. (2009) and Armitage et al. (2016).

Low systolic and diastolic blood pressure are associated with greater mortality in the elderly. Yet, observational studies show that systolic blood pressure diminishes in people destined to die compared with those who survive. Moreover, randomized trial showed that lowering systolic blood pressure to less than 120 mm Hg in individuals over 75 years of age reduced all-cause mortality by 33%.

Low blood cholesterol is associated with higher cancer risk. Even so, serial date shows that blood cholesterol declines in advance of cancer diagnosis. Besides, randomized control trials of statins have shown no increase in cancer risk. On top of that, genetic epidemiology did not show higher cancer incidence in people with lower cholesterol.

Observational epidemiologic studies are not conclusive regarding cause and effect, even with exhaustive listing of all risk factors and concomitant illnesses, especially when the topic of reverse causality may influence multiple key exposures, and therefore exaggerate the strengths of associations, partly because of the presence of subclinical illness.

The Structural Equation Framework

In his landmark book Judea Pearl[65] rejects causal pluralism and, contrary to Nancy Cartwright defends a substantial theory of causality, defined, analysed, and formulated in his Structural Causal Models (SCM). Presenting probability calculus as a set of logical propositions, he brings together with a unique impact a probabilistic and a counterfactual approach to causality as well as the effects of potential interventions. On this score, he understands causality on the model of a causal diagram. To evaluate sentences in this logic system, he uses structures that are causal and modelled with Bayes nets, with sets of arrows expressing some hypothesis about what affects what. The resulting toolkit visually summarizes the theory with diagrams, using directed acyclical graphs' (DAGs) which, when drawn, are connecting (causing) together nodes (variables). DAGs are single-headed arrows such that no variable can be cause of itself. The absence of arrows between two variables indicates it is assumed there is no causal relationship between them. Pearl's logical language allows for indicative as well as for subjunctive conditionals. If so, then, the causal graphs show when one factor would raise the liability of another variable. It shows not only what occurs, but it also shows what would happen if something else happened.

Even so, Nancy Cartwright[66] wonders whether there is any method of causal inference that is applicable on all cases. Under the influence of Hume and Kant, causation is often seen as a monolithic concept. There is, she thinks, a variety of causal relations that operate in a variety of different ways and a variety of different

---

[65] Pearl (2000, 2011).

[66] Cartwright (2008).

questions can be asked. She also casts doubt on one of Pearl's assumptions that it is possible to alter causal relationships in one part of the system while leaving causal links elsewhere unchanged. She adds that Bayes-nets methods, she says, do not apply for instance when positive and negative effects of a single factor cancel, when factors can follow the same time trend without being causally linked or when populations with different causal structures or different probability measures are mixed. N. Krieger and G. Davey Smith[67] hold that DAGs can of course be useful, but that they can also restrict the meaning of causality, and lead to erroneous causal influence, which requires among other things to being pluralist.

For all that, epidemiology should be grounded in the search to find solutions to problems, rather than to adhere to one school of thought about causality. Causal inference remains a judgement based on all the available types of evidence from diverse branches of science and needs to integrate diverse strategies and diverse types of knowledge.[68]

The Unnatural Nature of Causality

Causal processes connecting cause and effect are a treacherous topic. Hilary Putnam writes: "*The materialist metaphysician often takes* **causal relation** *as an example of built-in structure.*"[69]

For one thing, if one assumes that the description of the world in science in general, as well as in medicine, is mind-independent, it could be given in purely physicalist terms. Yet, in that case, one cannot locate intentionality within the physical realm. In this so-called epistemic naturalism, events have causes and objects have causal powers, and causes are in the world itself, they are built into the world itself. Causal relations are physical relations. Aristotle thought that causes are part of reality.

But for another, Putnam claims that "the cause" is roughly "*that part of the total cause that may reasonably, given the interests appropriate to the context, be regarded as the bringer about as opposed to a background condition, and hence that concept "the cause" involves something* ***intentional***".[70] Putnam adds that the fact that something, like a causal relation, is interesting relative to something intentional does not mean it is non-objective. It follows that the search for causes in medicine or in epidemiology "*smuggles in intentional/semantic notions*". On this score, causation as well as truth, are an intentional, context-driven notion that depends on the specific interest and saliencies of the inquirers, and not "a view of nowhere".

---

[67] Krieger and Davey Smith (2016).

[68] Vandenbroucke et al. (2016).

[69] Putnam (1983, 211).

[70] Putnam (2016, 71).

Kant saw causality as a feature of experience, but he was reluctant to admit whether there were causes in the world, apart from our experience of it. Causes should not be understood as actually existing properties like mass, weight, or other natural properties. Mackie contended that causation is *epistemic,* and not something in the world. They are like *inference tickets,* to quote Gilbert Ryle, that allow us to go from one statement to another, which is not the same as contending that there is some special, occult force or hidden property that is responsible for that change. Danto concludes that whether the world is causally ordered apart from the way we think about it is almost a meaningless question. Maybe the Causal Principle is not a statement about the world, not something we learn from the world, but something we bring to the world.

Against this background, it is not difficult to appreciate how Nancy Cartwright gets us out of this quandary with her benchmark book *Hunting Causes and Using Them.*[71] Causality, according to her, is actually a label we use to classify various events and processes such as 'causes', 'allows', 'helps', 'enables', 'forces', 'requires', 'permits', 'helps', 'inhibits', 'prevents', or anything of that ilk. It is useless to wonder whether a certain factor in a given situation is really a cause or what causation really is. Cause is a plural concept. Different causal accounts seem to be at odds with one another only because the same word means different things in different contexts.

"*Causal laws cannot be done away with, for they are needed to ground the distinction between effective strategies and ineffective ones,*" writes Cartwright.[72] For an epidemiologist or a physician, causal claims are nothing beyond their evidence or the gathering of that evidence: the meaning of a causal claim is identified with its verification writes Arthur Danto[73]: *"the concept of causation dissolves into our evidence for it"*. That is to say that the methods of epidemiology provide an implicit definition of causality in epidemiology: the best definition of causality in medicine resides in the methods of analytical epidemiology. The progressive methodological sophistication of the last 40 years is in perfect alignment with a gradual implicit overhaul of our concept of causation. Any definition of causation in medicine will therefore inevitably dissolve into the use made of epidemiological methods and the process of methodological revisions moves us to implicit or explicit, partial, or full-fledged re-examinations of our concept of causality. Thus, specifying how the term 'cause' is used, leaves nothing further to say what it means.

---

[71] Cartwright (2007b).

[72] Cartwright (1983, 22).

[73] Danto (1989, 258).

Conclusion

1. A main objective in epidemiology is to identify alterable causes of disease. It is therefore necessary to understand what is meant by *a cause* in medicine. *Prima facie,* a causal relation is an association of events or characteristics in which an alteration in the frequency or quality of one category is followed by a change in the other category. Causes reveal what one would have to do to bring about specific kinds of outcomes. They are context dependent and interest dependent. They are manipulative since they are intrinsically connected with goals and effective strategies. Causation is not one thing with different methods, but there are different types of causation with different characterizing features.
2. Causes are abstract and not real entities. They are not natural kinds. They do not cut nature at the joints, but they pertain to the mode of description of the world. Causal role differs from causal capacity.
3. The nature of causality in medicine, just like that of pathology, is thus not merely descriptive but also intentional. Cause, in medicine is a 'credit' for being good or bad: it is that on which the responsibility for a given medical condition can be laid.
4. The counterfactual outcome model that stems from David Hume, has become routine for causal inference in epidemiologic studies, despite its tricky logical structure.
5. There are neither necessary nor sufficient causes in medicine, but tendencies toward necessity or tendencies toward sufficiency may be quantified by using epidemiologic methods. What we call causes in medicine are only partial, contributive causes. They are contrastive since they make a difference between circumstances in which they are present and normal conditions in which they are absent.
6. Population measures of disease frequency are of two types: prevalence and incidence. Disease frequencies can be compared between different populations or between subgroups within a population. One calculates absolute and relative measures of comparison, including rate/risk difference, population rate/risk difference, attributable proportion among the exposed and the total population, rate/risk ratio, and odds ratio.
7. The two principal types of observational studies in analytical epidemiology are cohort and case-control study.
8. Causality is probabilistic: a prior event causes a subsequent event if the probability distribution of the subsequent event changes conditionally on the values of the prior event. Yet, probability raising needs to be supplemented in epidemiology by some risk evaluation that measures the strength of association.
9. Causes in science or in ordinary language are external to their effects. But in the case of diseases, causes are often intrinsic—i.e., immanent— to their effect, namely the disease under consideration.

10. The *causa vera fallacy* and the magic-bullet fallacy assume that diseases are entities, i.e., things existing by themselves, stemming from singular specific causes, that the classification of diseases must be based strictly upon their aetiology, and that medical interventions are based on the model of 'magic bullet.'
11. The meaning of causation and of causal lawlike generalizations is coextensive with the use we make of epidemiological methods.

Chapter 6
Function and Medicine's Hybrid Concepts

Admire, mon fils, la sagesse divine qui a fait passer les fleuves juste au milieu des villes.

Henry Monnier

Contents

'Function' stems from *fungi*: to do. It is the role of an organ, part or physiological activity or mental modules in an organism. The best definition of function in medicine would be a listing of those functions described in physiology texts. Even so, the function of a component part or process of an organism consists in the role it plays or is expected to play, and its contribution to the whole bodily economy.

The Received View

Prima facie, we may define the function of an organ or of some system *O* as follows:

(a) *O* (the heart) causes *F* (blood circulation)
(b) *F is* a capacity to reach some sort of goal or purpose of *O*

L. Karhausen, *Analytical Philosophy of Medicine*, Philosophy and Medicine 152,
https://doi.org/10.1007/978-3-031-81257-6_6

(c) O is logically prior to F

The third claim (c) serves mainly to rule out treating F as the initial cause hence avoiding recourse to final causes.

Functional ascriptions are thus end-state explanations that presuppose success or achievement (or a specified level of statistical success in reaching its target). If so, then, talking about some function presupposes firstly that it is apt for truth:

(1) The function of hearts is to pump blood.

Secondly, Aristotle[1] writes: *"Just as for a flute player, a sculptor, or any artist, and in general, for all things that have a function or activity, the good and the well is thought to reside in the function ..."*. If so, then, the function of the heart is not the mere effect of its contraction; it also contributes to some necessary and useful outcome. It is adaptive.

Thus: (2) "The function of hearts is to pump blood properly".

Functions are plastic, in the sense that the goal can be reached by alternative roads. Such plasticity *"is the property of the organism with respect to a certain goal, namely that the organism can attain the same goal under different circumstances by alternative forms of activity making use frequently of different causal chains."*[2] By way of example, acid-base balance is being maintained by three main mechanisms, by chemical buffering, by renal and by pulmonary elimination, so that blood H^+ concentration remains within a narrow range; in case of impaired renal function, a resulting acidosis may be regulated by hyperpnoea, i.e., by long deep breaths at a normal rate.

The term 'function' is thus factual, adaptive, and plastic; it bestows descriptive, i.e., causal, and evaluative status, as it embodies its usefulness.

Biological Functions and the Machine Analogy

The received view defines as normal anything that functions according to its design. Machines are kinds of things that have been built by human beings for their own purposes. The functions of the gear and cogwheels or of any part of a machine are clearly defined and limited. They are not discovered: we do not have to set up some scientific research project to uncover them. Constructing a clock *means* that the train of wheels will fulfil some function. Machines are intentional devices : they have intentional content; they are *purposive* and reflect the intention of the maker. They are *constitutively teleological.* Clockworks are systems in which every element is necessary to achieve the preordained purposes for which they were built.[3] A

[1] Aristotle (1969, 1976, 1106a).

[2] Braithwaite (1953, 329).

[3] Hobbes T, I, Ch. 1, 2.

good instrument fulfils its promises; a defective one is lacking in excellence and good quality: what makes it good takes here logical precedence on what makes it bad.

In 1802, the English clergyman Archdeacon William Paley wrote: "… *there is precisely the same proof that the eye was made for vision, as there is that the telescope was made for assisting it*". Even though it is now common knowledge that life forms are the result of an evolutionary process, William Paley's contention raised the following question: does the human body run like a clockwork?

Even so, Descartes held that animals are machines, but he rejected final causes and teleology.[4] The *mechanistic* view implies that all the behaviour of a living organism or its organs are to be explained without remain in terms of material causes, such as those used to explain the behaviour of inorganic bodies in terms of preceding events. A 'mechanist', then, rejects the tenet that concepts of purpose have a genuine and explicative role in science and philosophy. Immanuel Kant argued in the *Critique of Pure Reason,* that only mechanical concepts can give knowledge of nature and that the teleological argument cannot extend our knowledge of the world in whatever way.

Meanwhile, biology does not propose a mechanical metaphor but an organismic scheme in which all processes are interrelated to the whole organism. Kant wrote: "*its parts, both as to their existence and forms, are only possible by their relation to the whole*".[5]

Beyond this, one defines the breakdown of a machine in syntactic terms, and the breakdown of an organ in semantic terms. In the first case, the breakdown represents a failure of the logic underlying the machine: it is internal to its mechanism. But in the second case, the breakdown is contingent to the explanatory model the physiologist assigns to the relevant organ: it is external to our conventional representation and to the counterfactual nature of physiology.

One main kind of positive answer is that of Christopher Boorse: his naturalistic theory of functions in biology and on statistical normality is perfectly compatible when applied to machines, which suggests that his analysis of functions is mechanistical.[6]

Indeed, naturalistic normality consists in the fact that the quality of functions of the parts and processes make a statistical contribution to string and durable products. A machine of good quality should have no or few defects.

Teleological Explanation Versus Naturalistic Understanding

What does a function's usefulness consist in?

---

[4] Descartes (1978, 127–143).

[5] Kant (1952, §65).

[6] Boorse (1977).

Aristotle wrote that *"Nature does nothing purposeless or in vain"*.[7] He differentiated efficient from final causes. In one way, the efficient causes of something bring about their existence, while the final causes are its purposes or aims. In another way, teleological explanations—i.e., explanations based on ends and final causes—account for the existence of some attribute of a system by showing the attribute's positive contribution to reaching some preferred state of the system.[8] Dr Pangloss in Voltaire's play *Candide* explains the shape of the nose of human beings in terms of its function in holding a pair of glasses on the face.

The classical puzzle of functions is how the consequences of a trait could explain why the trait is in fact present.

When biologists and physicians use the concept of 'function', they describe organs and features of organisms in terms of their proper or characteristic activity: the activity of the lungs and of the kidneys exist *for the sake of* keeping blood pH constant, and the heart contracts *to* pump blood, which comes roughly to presupposing purposes that may help us to understand the natural world.

For one thing, artefacts are objects made by some designer, so that teleological features are the prerogatives of human or rational beings; they could not spring from organisms devoid of intentions and of a capacity for reasoning; they could not flow from unintended and unplanned interactions between mere material bodies.

But for another, in the absence of intentions as in the case of physical events, teleological explanations imply that the present is determined by the future; the heartbeat is explained with regard to blood circulation: this would suggest backward causation i.e., philosophical, and scientific anathema. *"This doctrine of final causes,* writes Spinoza, *turns Nature completely upside down, for it regards as an effect that which is in fact a cause, and vice versa"*.[9] Spinoza reduced explanations by purpose to causal explanations.[10]

On this score, the advocates of the teleological argument countenance that the harmonious functioning of organisms induces us to attribute intentional characteristics to biological organisms; however, purposiveness in nature presupposes either some sort of intelligent or some God-given design. Thomas Aquinas wrote: *"Nothing that lacks awareness tends to a goal, except under the direction of someone with awareness and with understanding; the arrow, for instance, requires an archer."*[11] But plainly, such goal-directed explanations are fraught with difficulties as they raise more questions than they can answer. The scientific enterprise dispenses with theological explanations, which are unverifiable and cognitively meaningless.

---

[7] Aristotle (1952, 1.8.12).

[8] Aristotle, (1970, II, 194^{a}).

[9] Spinoza (1982, 59).

[10] Spinoza (1998, IV. Preface).

[11] Aquinas T, Part I, question 2.

Boorse, with his naturalistic approach, defines function as the causal contribution of something to a goal within a teleological system.[12] Notions of physical and mental health appeal to habits and virtues, which, if we were to use Boorse's terminology, could probably be expressed as kinds of "goal-directed functions." Function is a non-normative concept, itself part of a non-normative concept of disease and health. However, Boorses's account is not concerned with malfunctions. His views raise difficulties to establish a clear border between health (normal) and unhealth (abnormal) levels of functioning without additional speculative considerations.

It is clear from what precedes, that it looks as if we got ourselves into a quandary. If it turns out that anterior design drops out of sight, are we left with the philosophical difficulty of forward-looking teleological explanations whose outcome is yet to come? But what if it does not? With the missing goal, we enter the realm of pathology, which is also part of nature. What content could our concept of function have if we reject teleology based on Darwin's natural selection and take account of the existence of pathological features?

The physiologist Homer Smith observed long ago that to define normal as the relationship between design and function is not very useful: "*If it is normal for an intact wheel to roll in accordance with its design, is it not equally normal for a broken one to fail to roll, in accordance with its design?*" "*Is it not normal for a diabetic to exhibit glycosuria, for a cretin to be just what a cretin is, for a malignant tumour to do just what a malignant tumour does, for a schizophrenic to behave like a schizophrenic?*"[13]

Function as an Activity, a Biological Role, or a Historical Concept

How can biological functions and purposefulness be accounted for without appealing to teleological or quasi-teleological notions? Philosophical analyses of the notion of biological functions are all attempting to legitimate and explain design for a purpose.

With Boorse, functions contribute to needs, purposes and goals[14] and with L. Wright, functions were favored by natural selection.[15,16]

First, Larry Wright[17,18] suggests the following analysis of functions:

The Function of *O* is *F means*:

[12] Boorse (1977).

[13] Rothschild et al. (1997).

[14] Boorse (1976).

[15] Neander (1991).

[16] Millikan (1989).

[17] Wright (1973).

[18] Achinstein (1983, 263–290).

(a) *F* is a consequence (or the result of) *O*'s being there,
(b) *O* is there because it does *F.*

According to Wright, this definition applies both to biology and to artefacts; it offers a criterion to distinguish a function from a mere effect; and it captures the normativity of functional ascriptions so that malfunction is a possibility. To say that the function of hearts is to pump blood means that the circulation of the blood is a consequence of the heartbeat *and* that, on the ground of the operation of natural selection, hearts are present in the body because they pump blood.

Three consequences follow from this analysis.

Larry Wright's analysis is *essentially causal*: it is through its constitutive selective success that a function strives to persist, *in suo esse perseverare conatu*r.[19] And this is a reverse explanation of temporal succession: the effect explains the cause. The obvious, although hypothetical explanation, substitutes a causal instead of a teleological analysis of functional explanation. It is now proposed that a trait has a function because it has been favoured by natural selection. Functions, for Wright, are thus natural kinds, if natural kinds are the ones that support certain modal implications needed in biological science.

But there is another consideration. Ernest Nagel presses the following objection. Wright's analysis requires that *F* is a function of a feature *i*, if and only if the feature *i* has been selected in some way to be present in the organism just because *F* is an effect of *i'* presence. If so, then, *F* can be *asserted* to be a function of *i* if and only if it is *known* both that *F* is an effect of *i* and that *i* had been selected to be present in the organism just because *F* is an effect of the item *i.* The scope of evolutionary explanation has thus been broadened from causation to definition, as if it could capture the intrinsic nature of our concept of function. Selection becomes definitional, analytical, or constitutive of functions. If so, then, functions are defined by the role they play, which itself explains their evolutionary success.

Secondly, Robert Cummins endeavors to avoid having to explain the presence of the pumping heart by what it *does*—its function—because this is to 'explain' its presence by appeal to factors that, according to him, are causally irrelevant. Cummins is fully agnostic about purpose and takes a distinctive scientific strategy based on biological role or capacities: for him, neither Darwinian evolutionary, nor contemporary purposes or goals play a role in the analysis of function.

To ascribe a function to something is to ascribe a capacity to it, which he intends to analyze by appealing to the capacities of the system's component parts, those that are assumed to contribute to the higher-level capacity being analyzed. Cummins assigns functions only to those capacities of components which are invoked in a functional explanation. Yet, some of those capacities are utilized, some are not, so that he stealthily introduces goals and purposes by choosing for analysis only traits which are already known to be purposive.[20] With his emphasis on causal capacities

[19] Spinoza B, The Ethics, III proposition 6–7.
[20] Rosenberg (1985).

of components and the absence of essential reference to overall systemic goals, Cummins might be throwing the baby with the bathwater.

Thirdly, functional processes are allegedly constituted by their Darwinian selective success. Functions are fixed by their history, that a function is anything or everything that it is naturally selected to do. A function, natural or artefactual, needs to have some useful end or "appearance of end" as Conrad Hal Waddington called it; and in addition, it is often assumed that there must be some way of *explaining its existence by reference to its usefulness.* On this score, Immanuel Kant proposed a 'provisional' definition of natural purpose: "a thing exists as a natural purpose *if it is (though in a double sense) both cause and effect of itself*".

The way living organisms work has to do with their causal history since they are the outcome of evolution which leads to increments that shape living organisms into complex systems. The line of argument is that what appears to be purposeful in living organisms is purported to be the result of random as well as self-organising evolutionary processes based on genetic mutations and recombination, natural selection, genetic drift, and suchlike. We merely see the end-result and believe that the organism was either designed or evolved for the purpose of those appropriate adjustments while it merely manifests some successful tasks.

But how do we then take account of the pervasive prevalence of autoimmune processes, i.e., when B-cells of an individual's immune system attack their own body components, a process present in arthritis, multiple sclerosis, Alzheimer disease, and which accompanies most disease processes? Despite natural selection, the function of B-cells, that, supposedly, is of protecting the organism, turns against the individual's own tissues, which they are expected to protect. To claim that this is a case of malfunction seems to be circular. How can we define functions as the result of natural selection as long as they seem useful, whereas when they become harmful, we call them malfunctions and exclude them from functions: are malfunctions persistent biological features that also result from some evolutionary process, i.e., from natural selection? If functions as well as malfunctions are either capacities or dispositions, respectively positive and negative, and since both results from natural selection, natural selection cannot be used as a defining criterion of functions since it does not differentiate function from malfunction. And if the criterion is the distinction between the positivities and the negativities of function between functions and malfunctions, then the recourse to natural selection becomes irrelevant.

Is this analysis plausible? Is it possible to account for functional statements by reducing them to the mere end stage of some causal process? And when this *historical view* becomes a *definition,* it meets severe difficulties.

Darwin held that: "*To suppose that the eye, with all its inimitable contrivances for adjusting the focus to different distances, for admitting different amounts of light, and for the correction of spherical and chromatic aberration, could have been formed by natural selection, seems, I freely confess, absurd in the highest possible degree.*"

The function of the heartbeat was discovered long before Darwin, and so was the word '*function*'.[21] Yet, this Darwinian interpretation proposes an empirical scientific theory about the evolutionary origins of functions, as well as a definition. But how could we discuss this interpretation if we did not know beforehand what 'functions' mean? Doesn't this view commit a form of *genetic fallacy* confusing the origin of something with its nature, or looking for support in the origin of a thing to determine its character? Understanding the origin of a function does not automatically tell us its end or present purpose: the purpose does not necessarily follow from its origins. It seems that in this case, the cause of a thing piggybacks the definition of its effect to protect its limits. It fails to discriminate between process and product, explanation, and definition.

Functions are what they are, regardless of what causes them. At one point, Charlie Dunbar Broad suggests: "*Now we are all extremely liable to confuse a history of the becoming of a thing with an analysis of the thing as it has become. Because C arose out of B, and B out of A, people are inclined to think that C is nothing but A in a disguised form. To analyse anything, you must analyse and reflect upon it; and the most elaborate account of what preceded it in the course of history is no substitute for this.*"[22]

"*This program,* write Stephen Jay Gould and Richard Lewontin, *regards natural selection as so powerful and the constraints upon it so few that direct production of adaptation through its operation becomes the primary cause of nearly all organic form, function, and behaviour.*"[23] Natural selection is a mere filter, and "*filters cannot be the sole cause of the coffee that comes out of them*", writes Mary Midgley.[24]

Function is not a historical concept, i.e., a concept which presupposes that the system to which it applies had actually had such and such a history, in the manner of notions such as 'hybrid', 'polyploid', 'genetic disorder' or infectious diseases. Although it may well be the case that the genealogy of functions can be partly or fully accounted for in Darwinian or neo-Darwinian terms, and if this is so, this cannot stand in for a definition or a conceptual analysis of what we mean by 'function'. Moreover, it is questionable whether all functions can be identified with survival enhancing propensities: endorphins lack these properties but actually have the function of shutting off pain.

It is probably more economical to define with Robert Cummins, the purpose of a component of organic systems as the function it presently fulfils, independently of the latter's derivation, whether or not it is the effect of some selective process or convergent fortuitous factors.[25] Textbooks of physiology explain functional regularities occurring in an organism through current causes. Plainly, a function is "*some feature of the organism–morphological, physiological or behavioural–which*

[21] Woodfield (1976, 118–119).

[22] Broad (1923, 12–13).

[23] Gould and Lewontin (1979).

[24] Midgley (2014, 16).

[25] Cummins (1975).

serves some proximate end that the observer believes he can discern fully by direct observation and without reference to the history of the organism".[26]

With functions, there are questions of how they fit here and now into the natural order of things, and how we should go about studying them, regardless of how far upstream they are to be found. Arno Wouters summarises clearly the philosophical debates, arguments, and wrangles on the notion of biological function and on functional explanations.[27]

Teleology: The Naming-Explaining Fallacy

Consider again biology. Is there a different type of teleological explanation, which does not credit the heart with intentional purposive behaviour and excludes the negative connotation of goal-seeking processes?

Prima facie, physiology follows Aristotle and assumes that organisms are like machines.[28] By sheer analogy we talk of the heart *pump* or the *lens* of the eye helps to see. Metaphorical machine analogies are pervasive in biological language to identify adaptive and functional processes. Inasmuch as machines act as if they were guided by a built-in intelligence to accomplish some inherent purpose, they offer a good model in understanding organic functional processes. But biology knows of no watchmaker.

So, the language of teleology has been tentatively banned-in-principle from natural sciences as unwarranted metaphysical fiction. Functions are directive, not directed. Colin Pittendrigh[176] suggested that the term *teleological* (i.e., intentionally purposive, or goal-intended) should be reserved for intelligent design and he proposed to use the term *teleonomic* for biological complexes of means-for ends,

Hence, teleological explanations are not completely ruled out, but now more innocuously hinge on empirical observation and causal accounts. Behaviour is purposive in the sense of being intentional and aiming at ends, whereas teleonomy is purposeful in the sense of being needed for the performance of biological function.[29] If artefactual functions are intentional and teleological, natural functions are seemingly intentional and are said to be quasi-teleological *or*, following Julian Huxley, *telic*: their functions are reckoned to be both discovered and assigned. Normal functions are those, which, independently of human or animal intention, contribute to the teleonomic goals, which are assigned to them by the medical model of rationality; and this allows scientists to detach themselves from theological considerations while bringing the debate back into a naturalistic view. The end of a function is a guide to its study but not part of the explanation of its operation. In

[26] Pittendrigh (1958).
[27] Wouters (2002).
[28] Aristotle (1986).
[29] von Wright (1971, 60).

linguistics, this manoeuvre is termed a *logodaedalist* neologism, namely, manipulating words with great cunning.

The Alternate Possibility: The Semantic View of John Searle

How is the term "function" used in actual biological and medical practice? What accounts for functions? Can the ascription of functions be made based on mere empirical inspection: can functions be discovered?

Immanuel Kant contends that: "*Medical physiology extends its very limited empirical knowledge of the ends served by the articulation of an organic body, by resorting to a principle for which pure reason has alone been responsible; and it carries this principle so far as to assume confidently, and with general approval, that everything in an animal has its use, and subserves some good purpose*".[30] The teleology that we see in the natural world is only apparent. Kant countenanced that purposes are not observable and are not principles constitutive of reality: they are merely regulative guides in our judgements and our investigations: "*For, strictly speaking, we do not observe the ends in nature as designed. We only read this conception into the facts as a guide to judgement in its reflection upon the products of nature.*"[31]

For Nietzsche, the whole idea of final causes is nothing less than a projection of human practical life onto the whole of nature. There are no purposes in nature. Purpose, he claims, was a human not a natural category. It follows that a claim of knowledge about any natural purpose is not an epistemic claim but is reducible to a hidden desire of self-exertion over the natural world. Final causes are imputed by humans to make sense of the world.[32]

In like manner "Curt John Ducassé contended: "*If we label it an 'end' or 'end result', rather than an 'effect', it is only because we then import into it our own interest in it and our desire that it occurs, but not because we find a purpose objectively and intrinsically present in it as a necessary part of its description.*"[33] Similarly, Ernst Mayr thought that teleological notions attribute actions of the mind, where there is none.[34]

In the light of this, our concept of 'function' is an inescapable convention that is indispensable for the contrastive system of medical beliefs. For all that, the construal of functions depends on our *cognitive interests*.[35] There could have been no understanding of the nature of blood circulation until a pump was invented. A

[30] Kant (2015, 35.2.1).

[31] Kant (1952, II, §14).

[32] Nietsche (1974; number 125).

[33] Ducasse (1968, 38).

[34] Mayr (1988).

[35] Grene (1974, 211).

physiologist who gives a functional explanation has some sort of expectation at the back of his mind. As important as it may be in clinical science, the goodness of functions expresses our own choices and biases, not some natural fact. They are projections of our own desires into nature. It follows that although the concept of function is not provided by experience, we need to bring it in to structure our understanding of processes given by experience in biology and in medicine.

The question is now: granted that organisms and their parts differ from artefacts, could or should their quasi-teleological purposeful functions or faculties nevertheless be analysed into radically non-teleological language? Is it possible to avoid the classical puzzle of functions noted above, about the fact that the consequences of a trait should explain why the trait is in fact present?

What clearly emerges from this is that taking cognisance of functions presupposes the identification of the *prudential interests* and saliencies of human beings and the discernment of aversive patterns.

For the anthropologist A. Radcliff-Brown, the concept of function in social sciences is the contribution, which a partial activity makes to the total activity of which it is a part. The concept of function involves the notion of a *structure* consisting in a *set of relations* amongst *unit entities*, the *continuity* of the structure being maintained by a *life-process* made up of the *activities* of the independent units.[36]

When we discover the function of the heart, we discover, contends John Searle—one of to-day's leading analytic philosopher—how certain causes operate to serve certain purposes, but the notion of purpose does not reflect laws of nature: they do *not belong to mind-independent nature,* but they are *relative to our set of values.*[37] Functions are always observer-relative, since they depend on interests, values and purposes that we attach to things, on properties or states of affairs by using them as a particular manner or by looking at them in a certain way. Functions are never intrinsic to any phenomenon.

For one thing, Searle defends the thesis of *ontological dependence* of functions on values. Since values are subjective, it follows that functions are subjective. We talk about a better or not so good heart, but never about better or worse stones, unless we attribute them some function, for instance a paperweight. Functions may be assigned in what occurs in nature in terms of our human interests and activities: *"we anthropomorphise the heart or the kidney just as we do with the sun or the rain.*[38] "*When we speak of functions,* writes Searle, *we are talking about those of its causal relations to which we attach some normative importance.*"[39]

For another, he holds that functions are not natural facts, even though they imply causal statements such as "*contractions of the heart muscle cause blood circulation*" that are perfectly objective and natural. When an experimental biologist describes how the heart is pumping blood, he is stating a fact potentially true or

[36] Radcliff-Brown (1935).

[37] Searle (2010; 58–59).

[38] Collingwood (1940, 1998; 335–337). Margolis (1975; 104).

[39] Searle (1992, 237–239).

false. In contrary manner, when a clinician is concerned with the function of the heart, he is expressing an evaluation that may be correct or not.

We say that the heart's function is to pump blood just like we declare that rivers serve to irrigate the fields that aromatic plants serve to improve our cooking, that the sun serves to illuminate the surface of the earth. Everything, which surrounds human beings and is of some use to them, gains a function in the web of our goals and prudential and instrumental interests. It results from a tacit expectation, which tenders prudential functional models and certain paradigmatic conventional regularities against which ill persons or disease entities are assessed.

Except for those fragments of the world that are endowed with conscience, there are no intentions, no functions, no teleology in the physical or biological world. We may also make objects to serve some function such as chairs, road, umbrellas etc. These functions are never intrinsic properties of the physical nature of these phenomena: they are assigned by external conscious observers, so that they are relative to the observer, and therefore external.

Since it is impossible to call upon values, we never talk about the function in the case of malignant tumours or of osteoporosis. We discover functions in nature if this is taking place within the context of a prior assignment of values, which includes goals or some kind of teleology. If we assume that, for an organism and for a species, survival and reproduction have value, we can declare that the function of the heart is to pump blood. On the other hand, if one doesn't consider it for granted that survival and reproduction are values, if what one appreciates above all is death and extinction, then one can claim that the purpose of cancer is to accelerate death, that of aging is to hasten death, and that of natural selection would be extinction.

To sum up, when we think of functions, we are referring to those relations to which we attach some *normative* importance. The causal structure of physiological processes is intrinsic to the organism, but functions are never intrinsic but always relative to an observer. It is intrinsic to nature that the heart pumps blood and that it causes the circulation of the blood. But when we say: "the function of the heart is to pump blood", we are not merely mentioning this intrinsic physiological process, but the purposes come from intentional human values. It is in this sense that functions are related to intentions and therefore *mind dependent.*

John Searle did not solve the problem of the origins of functions: he dissolved it.

Functions and Their Effects

Functional ascriptions are not predictive since the end object may perfectly well be missing. Goal failure, missing goal, objects or hindering factors are part of reality and partly coextensive with the domain of pathology.

Plainly, functional statements describe propensities directed at some end with no guarantee of reaching it. A function is not a mere occurrence or a sequence of occurrences, as it is not occurrent but dispositional. It represents a causal power or potential, an ability, a capacity, a propensity, or suchlike.

Nancy Cartwright, professor of philosophy at the London of Hygiene and at Durham University, writes: *"the logic that uses what happens in ideal circumstances to explain what happens in real ones is the logic of tendencies or capacities"*.[40]

In the Framingham study, the prevalence of myopia was 60% in people 23 to 34 years of age. This observation does not defeat the contention that the function of the eyes is that of seeing well. But it means that the function of the eyes is seeing well *with the implicit proviso that pathological conditions are being ruled out or bracketed off.* The function of an organ is related not to what it does, since it may never succeed to achieve it, but to it is supposed to do. Propensities are permanent in physiology although they may not be continuously displayed. It follows that functional ascriptions are dispositional statements that refer to earmarked capacities from which negativities and disturbing features have been subtracted.

If so, how do we construe our concept of function of some organ or system? For John Stuart Mill: "*We might, indeed, guard our expression ... by saying that the body moves in that manner unless prevented, or except in so far as prevented by some counteracting cause. But the body does not only move in that manner unless counteracted; it tends to move in that manner even when counteracted*". He adds: "*These facts are correctly indicated by the expression tendency. All laws of causation, in consequence of their liability to be counteracted, require to be stated in words affirmative of tendencies only, and not of actual results.*"[41]

This is tantamount to saying that the assertability of functional statements does not depend on their actual occurrence, or of the occurrence of their effect.

Since functional processes are intermittently present on account of occurring malfunctions, functions cannot be identified with their own token manifestations. Ascribing functions does not hinge on dated particular occurrences but on universal, repeated event forms, which may or may not be realized. Even if pumping blood efficiently is a criterion of a well-functioning heart, it is by no means obvious that it is an absolute property of the heart. Physicians and veterinarians who are observing hearts are aware that a sizeable fraction of them happens to be functionally defective.

Dispositions are conditional or hypothetical statements, that is, statements containing an actual or implicit 'if': "If p then q". The antecedent clause p represents *a priori* necessary normal conditions in the absence of which the function is not achieved: this includes *standing* conditions (the physiological background displayed by the whole organism), *internal activating* conditions (e.g., electrical impulses from the sinoatrial node, the heart's own pacemaker) or the absence of *surrounding or internal hindering* conditions (such as a heart block). The function could not have been achieved if the factors necessary for its completion had been absent or if those, which hinder it, had been present.

Let $\Sigma Y^i = (y^1 + y^2, + y^3 \ldots + y^n)$ being the set—assuming that it is denumerable—of enabling and activating factors that are necessary and sufficient for the proper

[40] Cartwright (1989, 190).

[41] Mill (1950, 248, 1971, Ch 7, section 14).

fulfilment of a function: these occurrent conditions *must be jointly present for the function to be truly ascribed.* The Löwenheim-Skolem theorem in mathematical logic shows that any denumerable set of sentences that has a model has denumerable infinite model. In other words, formulas in predicate logic are satisfiable only in countable domains.[42]

Yet, this definition of function takes the no-obstacle clause for granted. Suppose, then, that we could also make an exhaustive listing of all potential hindering factors, disturbing causes, or pathological conditions individually liable to cause goal failure: call it X^i; members of the set of hindering factors $X^i = (x^1 \text{ or } x^2 \text{ or } x^3 \ldots \text{ or } x^n)$, if present, are *individually sufficient for goal failure*.

It follows that if one of the necessary factors is absent or one hindering factor is present, the ascription of a function capacity is subject to cancellation, i.e., the heart does not actually fulfil its function. We then have:

(1) *If the causal factors ΣY^i are all present and none of the obstacles X^i are present, then the heart performs its function F.*

But propensities are counterfactual properties. To apply to malfunctions (1) must be extended to counterfactuals. Considering a case of heart failure, we have:

(2) *If the causal factors ΣY^i were all present and none of the obstacles X^i were present, then the heart would perform its function F.*

To sum up, the function of an organ, as it were, is not merely the outcome it is bringing about, but the outcome it is expected to achieve: in a way, *it still has the same function if it fails to achieve it.*

Functional Role and Functional Capacity

Causal explanation of the process of circulation of the blood is a contingent relation made in terms of "*how does it work*", not "*what does it do?*" and it specifies a mechanism not a goal. Hume held that causes and effects must be logically distinct events since causation connects wholly distinct things, so that the heartbeat and its effect are two logically independent occurrences. It must be possible that the first event should obtain while the other does not. Under this description, it is *a contingent truth* that the heartbeat causes the circulation of the blood.

But plainly, a function necessarily harbours its goal since the outcome of a function is not the mere effect of a causal work. *Claiming that the function of the heart is to pump blood, incorporates the heartbeat and its effect in a single concept.* Even though the heartbeat causes the blood to circulate (an empirical, contingent fact) the goal attainment becomes now *constitutive of the function of the heart,* so that the functional ascription is now necessarily, not contingently true. If so, what were

[42] McCawley (1993, 182–183).

initially two distinct logically independent events, are, under this second description, two logically connected features.

Goal failure being excluded, we then have:

(3) *A function can be truly ascribed to any system (organ, tissue, activity, and suchlike) if and only if it is a priori necessary that the system realizes its goal if reasonably normal conditions obtain.*

But this procedure is wrong in the characterization it suggests. Seeing remains the function of the eye in a person who is losing his eyesight. And in cases of heart failure or of cardiac standstill, the function of the heart remains that of circulating blood even when this disposition is not achieved, or its function is decreased. To patch up this difficulty, what more that is needed has to do with a distinction between functional role and functional capacity.

Functional roles are ideal functional abilities stored away in textbooks of physiology (akin to the law of ideal gases or Newton's universal law of gravitation), and which remain true even in cases of abnormal functioning: thus, functional roles support counterfactuals. If so then, physiology describes functional roles, which prove to be adaptive. The corresponding counterfactual conditionals make them the contrast case and capture the permanency, that is, part of physiological talk.

Functional capacities, by contrast, are empirical and factual: they are measured and evaluated in a clinical set up on individuals or in an epidemiological survey on some selected population.

So, the final account leaves us with this:

(4) *A functional role could truly be ascribed to the system, if and only if it is a priori necessary that the system would realize its goal if normal conditions did obtain and if no hindering factor was present.*

Functional roles are *de re,* while functional capacities are *de dicto.* Functional roles ignore malfunctions, whereas malfunctions represent decreased functional capacities: in such cases, the goal may be missing but the functional description is still appropriate. Hence, a failing heart has a decreased functional capacity; and medical care attempts to bring it nearer to its functional role, to its ideal normal physiological function, that is, to the counterfactual case.

The British philosopher Gilbert Ryle (1900–1976), mainly associated with the British Ordinary Language Philosophy movement, broached the subject of episodic words, which he labelled "*achievement words*" or "*success words*", or "*got it*" words with their antithesis the "*failure words*" or "*missed it*" words. There are episodic words such as "function" for which it is proper to say that they score a goal, or that they miss it. There is an important distinction between *task verbs* and *achievement* or *success verbs*, between the heart pulsating and the heart of *De Motu Cordis.* For example, "to see": as soon as it is correct in saying that someone sees something, it is also correct in saying that he has seen it. Competing in a race is a task verb, but winning it, is an achievement verb, and the same with striving and achieving.[43]

Physiology Is False

Marcel Proust claimed that "*Physiologists … describe a functioning of normal organs such that one never really encounters them*".[44] Since not all hearts are pumping blood and some of them are pumping poorly, claiming that hearts are pumping blood is contingently true, but claiming that it is a law of nature that the role of hearts is to pump blood is false.

We demarcate functions from malfunctions, but the gap that separates them from one another does not segregate natural kinds, i.e., ultimate constituents of the biological world. Biology, biochemistry, and physiology are indifferent to that distinction since the discriminating criteria cannot be read from the surface of biological reality but are *brought in by us.*

In view of this, once singled out and entered with the *second position* into the clinical realm, functions are printed in textbooks of physiology, and are then removed from their normative, conventional, unnatural features. By a *Principle of Ontological Commitment*, we then find ourselves committed to the existence of quasi-naturalistic functions.

Immanuel Kant writes: *It is common knowledge that scientists who dissect plants and animals, seeking to investigate their structure and to see into reasons why and the end for which they are provided with such and such parts, why the parts have such and such a position and interconnection, and why the internal form is precisely what it is, …say that nothing in such forms of life is in vain, …and that nothing happens by chance. They are, in fact, quite unable to free themselves from this teleological principle."*

Dana Copeland writes in a similar vein: "*The pure scientist can afford to disdain purpose but medicine which views biology from the human angle, is a prejudiced science. Accepting as desirable only that which benefits human beings, it invents concepts to both promote and justify its aggressive action against other organisms, concepts which are only useful in that they serve its purpose.*"[45]

This brings us back to Nancy Cartwright's argument that simple physical laws such as the law of gravitation are literally false.[46] The laws of physics lie and so do the functions described in physiology. Functions only represent facts in closed experimental systems or standard circumstances. They are identified as if they occurred isolated, i.e., cut off from *ceteris paribus* conditions, contrary to the real

[43] Ryle (1949, 143–147).

[44] Proust (2011; 41).

[45] Copeland (1977).

[46] Cartwright (1983).

world where functions act together and interact in a living organism. Contrary to the case of a car or a vacuum cleaner, if the science of physiology was exhaustively completed and thoroughly written up, and no functions of an organism remained unknown to us, we might still be unable to explain and predict the actual functional behaviour of an organism with their composite interwoven interactions.

It follows that physiology, and its functions are literally false. Their descriptions avoid deviant processes. Physiology is a default position that separates the wheat from the chaff and, according to Immanuel Kant, assumes as an axiom *"that no organ will be found for an end which is not the fittest and best adapted to that purpose"*.[47]

Function models as given in texts of physiology are brought to light counterfactually. They are convenient fictions, cognate with rules and guidelines, yet they reflect normal expectations. This privileged status gives them very great explanatory power. They are theoretical paradigms and resemble physically impossible construals such as ideal gases, perfectly rigid bodies, frictionless planes, instantaneous velocity, and bodies moving in a medium totally devoid of resistance. We know perfectly well that there are no ideal gases or perfectly rigid bodies. It is a standing joke that physicists are modelling spherical cows to make the equations easy to handle. These are limiting construals, or theories since they only approximate what is actually observed. In this sense, what physiology teaches are lies.

Even so, if biology is written in descriptive langue, physiology is formulated in prescriptive or potentially prescriptive medical norms, which are contrastive, counterfactual paradigms that physicians mean to pursue.

Functions Are Hybrid-Concepts: Empirical and Normative, Descriptive and Prescriptive

Ascriptions of functions do not come to us with a flag; goal-seeking processes are not, as it were, *natural* regularities and are not part of the furniture of the earth; they are not intrinsic to nature, but we are bringing them into our description of hearts, eyes, or kidneys as indispensable tools to goad us in correcting malfunctions.

In one way, functional ascriptions make *factual claims*.[48] They are in the indicative, listed in textbooks of physiology, true or false, in quite straightforward ways. Incidentally, Boorse argues that functions are objectively discoverable. Ascribing a functional process consists in isolating a fragment of some organisms, namely some process or some causal tasks, and getting to know its causal role. Experimental medicine as initiated by Claude Bernard, isolates from some outside interferences, and thereby magnifies the role played in the body by some specifiable parts or processes in the performance of some earmarked activity, which tends to maintain the

[47] Kandel (1972; 60–61).

[48] Grene (1974).

organism through mutually supportive independent parts. The physiologist selects and emphasises, in this way, those surface features, which seem to him relevant and disregards those he deems irrelevant. In conferring a function to some item such as the heart, he assumes it is "normally" successful, he mentally deletes some of its inconvenient properties, and leaves out altogether some of its effects which are irrelevant, those which result from disturbing causes or those which might be harmful for the organism; he vests what remains after the subtraction of those hindering or irrelevant features of the activity of the heart, with a 'power' he calls its function.

In another way, if statements describing functional processes qua processes are factual, functions *qua* functions are not given as a state of nature, as a natural kind: we do not discover functions since they are not observable features; when we finish describing the various physiological processes in which kidneys are involved, it makes no sense to say that in addition, the function of the kidneys is to eliminate water and waste products. Purposes are neither objectively present in nature nor intrinsic features of biological reality. It follows that physiological depictions delineate functional models that are normative conventions and commendable yardsticks.

Philosophers of science are very reluctant to explain natural events in terms of purpose, but physicians have no reason to be embarrassed because their heuristic merit calls for no underlying metaphysics: *it merely results from the pragmatic concerns of medical care.*[49]

When we define words such as 'function', 'normal' or 'pathological' we may be making two things. For one thing, giving a true report of how those words are used by physicians. This descriptive account is a *reporting* statement, which may be true or false depending on whether it correctly or incorrectly describes how it is applied. But for another, those words may be used to *commend* which is neither true nor false. With this move, the term function expresses a recognised standard, which implies a quasi-prescriptive attitude. Functions are clinically prescriptive medical norms that may call for intervention, which only make sense relative to human or animal agents or sensitive beings that value life and survival and fear suffering.

On this score, a teleological approach is a significant advance in the attempt to explain malfunction, since it is determined in terms of how a system is *supposed* to function. Pathology is already between the lines of physiology: to describe the function of an organ is a preparation for discussing its breakdown. The intention was put there right in the beginning. A function is a contrary to facts medical norm in the sense that it is a biological standard to which organisms ought to conform in default of which they tend to turn sick. Functions, then, serve as paradigms of normalcy governed by justified standards, that it to say that to accept them means to subscribe to them.[50]

Contrary to biology, medicine is *Janus-faced.*

For one thing, it follows that our intuitions about *what is pathological are derived wholly from the clinical level.* For another, qualifying some adverse anatomical or

---

[49] Cassell (2004, 5).

[50] Woodfield (1976; 122; 130–134), von Wright (1963a, b), Craver (2001; 72).

functional feature as pathological signifies that it is harmful to the welfare interests of the individual concerned, namely to his body or his health *as compared with what his condition would have been had such feature been absent.* Hence, *prima facie,* medical facts are susceptible to two interpretations, descriptive and epistemologically naturalistic, or prescriptive, normative, and conventional, expressing valuating judgements.[51]

But surely, normative capacities are already operative with the *second position*, in clinical observation, in patho-physiological and epidemiological research themselves. Medical statements are not twofold i.e., made out on the one hand of factual utterances about signs and symptoms, and on the other hand of valuational claims about needs: both are internally related to one another and merge together. Medical statements are at once descriptive and prescriptive. Medical science is reporting, but it is normatively constrained so that its valuating capacities are at work as soon as medical care intervenes.

It follows that medical science looks at the world in terms of hybrid concepts with which it acts on it, i.e., concepts of intentional objects, such as those belonging to ill health, disorder, infection, fatigue, suffering, anxiety, depression, harm, pathology, function and malfunction, frustration of needs and wants as well as those of freedom, responsibility, autonomy, reasons for action, justice. Medical statements have two dimensions, empirical and normative, descriptive, and pragmatic: they have one foot in the world and a second one in values. Medical judgements convey assertoric as well as imperatival force.

Are Diseases Malfunctions?

According to Boorse, a disease is a malfunction of a sub-system of the body. "Sub-system of the body" is used in the broadest sense imaginable, referring to organs, systems in the body such as the nervous system, and sub-systems of the mind, for example those devoted to memory or language comprehension. The overall aims of the organism are to survive and reproduce and the different sub-systems function to contribute to the attainment of these goals. Boorse considers that diseases are an internal state that impairs or limits function ability. "*Normal functioning is defined in terms of a reference class which is a natural class of organisms of uniform functional design (i.e., within a specific age group and sex), when a process or a part (such as an organ) function in a normal way, it makes a contribution that is statistically typical to the survival and reproduction of the individual whose body contains that process or part.*" Diseases are determined by empirical, biological facts alone, and are value-free statistical states with respect to species design, but Boorse's reference class cannot be established in the absence of normative judgment. Some departures from those natural functions may be indifferent or beneficial, but others

[51] Whitbeck (1978; 210), Margolis (1981; 562ff).

are not. The latter are diseases. Boorse claims that the determination of a body malfunction and the determination whether a body malfunction is detrimental are objective matter to be determined by science.[52] However, no justification can be provided for a value-free justification of Boorse's assumption of normal, healthy reference classes.[53]

Ferdinand Schoeman defined disease state as a condition of *impaired functioning*[54]; PH. Schwartz mentions a malfunction-requiring approach[55] and for Smart, diseases are impairments to natural functions.[56] For L. Nordenfelt, "*The disease is identical with the subnormal functioning of the organ or other part.*"[57]

If two things are identical, they must have the same properties. Are such definitions tautological?

Malfunction does not designate the sum of the characteristics that a disease must have to be called a disease. If malfunction was a defining feature, something wouldn't be a disease if it lacked malfunction. Malfunction is the failure of a function, a failure to work or operate correctly, namely, an abnormal function.[58] But an abnormal function is not necessarily a pathological function, so that it can be neither manifestational nor synonymous of a disease process.

Reznek denies that malfunctions are a necessary condition for disease. An injury, a bone fracture is not a process, but an event and it does not repudiate physiological principles. Ventricular extrasystoles are malfunctions but most of them do not reflect a disease. Furthermore, somatic symptoms disorders and factitious disorders, and several so-called functional disorders are not malfunctional. Incidentally, mental disorders are symptoms or extremes of symptoms, but not malfunctions or dysregulated conditions. Whatever relates diseases and malfunctions is a matter of discovering, not of definition.[59]

On this score, half of patients with *heart failure have a normal function*, namely a preserved left ventricular ejection fraction (HFpEf), called diastolic failure. Morbidity and mortality in HFpeF are like values observed in patients with heart failure and reduced ejection fraction.[60]

In any case, the weakness of this assumption is brought about by Jerome Wakefield, for whom mental disorders are characterised by "harmful malfunction".[61] He believes that a condition's being a malfunction should determine whether or not it is a disorder. It follows that he relabels many conditions considered as disorders

[52] Boorse (1977).
[53] Kingma (2007).
[54] Schoeman (1994; 186).
[55] Schwartz (2017).
[56] Smart (2016, 39).
[57] Nordenfelt (2013, 24).
[58] Hoendorf et al. (2010).
[59] Reznek (1987).
[60] Konstam (2003).
[61] Wakefield (2007).

as non-disorders. Even more, he places many conditions out of the realm of disorder altogether, relying on a shoestring undesirable account of health.[62]

But the Dutch philosopher E. Kingma showed that Wakefield's evolutionary account of disorder is misguided since not all our mental traits have been selected for effects that they themselves perform. In some cases, the presence of a trait is explained by the effects of a different trait. For example, the presence of blue eyes is not explained by an effect of blue eyes, but by the increased ability of lighter skin to absorb UV B radiation (which helps with vitamin D production). This can happen because the trait "blue eyes" is *linked* to the trait light skin. Such linkages can happen in various ways; traits can be genetically linked when they appear close together on chromosomes; they can also be pleiotropically linked, when they result from genes that give rise to or are involved in the development of multiple traits; or they can be developmentally linked, when physical and other constraints on human development are such that the development or evolution of one trait cannot happen without giving rise to or changing another.

For instance, Wakefield's example of dyslexia could be caused by a selected effect, but it is just as plausible that it was the by-product of the selection of a different, linked trait, and thus devoid of function; or that the normal ability to learn to read and dyslexia are on a spectrum of normal variation in non-selected effects brought about by a functioning underlying mechanism; or that such dyslexia itself is explained by its linkage to an adaptive trait, such as a superior visual-spatial ability, so that dyslexia is then indicative not of a dysfunction but of a superior function.

What precedes is not really about dyslexia, but it is about whether Wakefield is able to provide a successful account for *all* mental disorders: we have an abundance of traits that fulfil an important role in our culture, particularly in the mental realm, but whose effects may not be what drove their selection.

A Naturalistic View: What Functions Are Not About

George Engel writes that "*biomedical dogma requires that all diseases, including 'mental' disease', be conceptualized in terms of derangement of underlying physical mechanisms*". The language of science must be public.

Consider two accounts of the naturalistic thesis.

Boorse argues that functions are objectively discoverable and contingently true. He defines dysfunction as a deficiency: it means a "*deficiency of function*", namely "*less function, less contribution to the goals, than average, "a reduction of one or more functional abilities below typical efficiency, or a limitation on functional ability caused by environmental agents*".

This is an arithmetical not an evaluative concept. It means a lowered degree of achievement of the relevant physiological goals. Hyperthyroidism vs.

[62] Kingma (2013, 372–380).

hypothyroidism, high-output vs. low-output cardiac failure, myopia vs. hyperopia, manic vs. depressive episodes is various kinds of excessive vs. insufficient functional activity. Aristotle mentioned that "*As the physician say, there are two modes ... These are opposites; in general, they are excess and deficiency or failure*".

However, as shown by the following examples, Boorse misrepresents dysfunctions since he delineates dysfunctions as if they were a mere reduction of functional abilities below typical efficiency.

On this score, schizophrenic patients present what is being called *positive symptoms,* which are not present in a normal population, such as hallucinations and delusions. In autoimmune disorders (diabetes mellitus type I, rheumatoid arthritis, Crohn's disease and so forth), the immune system produces harmful antibodies to an endogenous antigen: their *functional ability is not diminished but altered.* Chondrocalcinosis consists in an intraarticular abnormal —*neither excessive nor diminished*—crystal deposition.

Dysfunctioning

Spinoza[63] wonders how do we take account of *"disasters, such as storms, earthquakes, diseases"? ... "Nature has no fixed goal and final causes are figments of the human imagination" ... "Things are not more or less perfect, according as they delight, or offend human sense, or according as they are serviceable or repugnant to mankind". "It follows that in construing the relation which binds an item to its function we must account for the possibility of goal failure."*

Medicine navigates the straits between the conventions of physiology and those of pathology. It is a tale about the relation and interplay between function and dysfunction. It is concerned with departures or *alterations* from those singled out ideal and fictitious norms of functioning in situations in which they break down or do not apply.

Felix Mainx indicates how functional ascription results from a selective process: *"The rules which assert something about single processes occurring in a system then hold conditionally, with reservations, since in such cases the mutual relations of the processes considered to all other processes are neglected or deliberately simplified."*[64] It is within an intentional space that we unfold the function of the heart and the understanding of its breakdown.

In severe arrhythmia or ventricular standstill, the heart is unable to keep up its workload, yet its function remains that of sending blood to the lungs and the rest of the body.

If so, then we have:

---

[63] Spinoza B, Ethics, part I, Appendix.

[64] Mainx (1950, 24).

(5) *If all the causal factors Y^i were present and if none of the obstacles X^i were present, the function F would come about.*

Thus, for a correct analysis of goal failure, we must have recourse to the additional following subjunctive conditional, so that the final account leaves us with:

(6) *If some of the necessary factors* $\Sigma Y^i = (y^1 + y^2 + y^3 \ldots + y^n)$ *had been absent or if any single obstacles* $X^i = (x^1 \text{ or } x^2 \text{ or } x^3 \ldots \text{ or } x^n)$ *had been present, then the function F would not come about.*

Functions may not be in regular working order, either because they are working excessively or insufficiently (hyper- and hypothyroidism), or else because their outcome is sharply divergent from physiological norms and detrimental to the body (autoimmune diseases, hallucinations). A function may also be prevented from functioning because of resistance to the peripheral signals (resistance to insulin in diabetes type II, resistance to leptin, which is common in the case of obesity, resistance to blood circulation in case of hypertensive vascular disease, resistance to insulin in diabetes type II).

Goal failure supposes that some goal may be attained physiologic circumstances.[65] But, since the goal is initially given in the concept of a function, the prediction of goal-attainment concerns its contingent functional role rather than its capacity. If so, then, the truth of a functional explanation can be upheld even if the expected goal attainment fails to occur. The functional capacity of a blind eye remains that of seeing. But the degree to which this goal is attained or the failing expectation, measure an organ's functional role. Functions and malfunctions have no ontological residency in the biological world, and merely need to be adopted for heuristic and pragmatic reasons to begin medical enquiry.[66]

Henrik Von Wright points out that *"The relation between the functioning organ and its effects on the body is a causal and thus extrinsic relation. But the relation between the badness of the effects and the badness of the organ, whose functioning is responsible for those effects, is a logical and thus intrinsic relation."*[67] If so, then, what makes a malfunction pathological is embedded in the functional relation and constitutes what we mean by a malfunction. By way of illustration, diminished cardiac output may cause cyanosis, dyspnoea, peripheral oedema, and other pathological manifestations of congestive heart failure. This implies, I believe, that the relationship between the malfunction of the heart and its manifestations is a causal attribution.

On the other hand, when a cardiologist, because of signs of dyspnoea and peripheral oedema, diagnoses a case of heart failure, this relationship between malfunction, *qua* heart disorder, and its negative consequences, signs and symptoms is no more causal but semantic. This semantic relationship is constitutive of the clinical disorder insofar as it holds on to account of the meaning of the terms "abnormal",

[65] Scheffler (1963, 114–115), Ruse (1977, 118–121).

[66] Schwartz (2007).

[67] von Wright (1963a, 56).

"pathological", "disorder" "malfunction" and suchlike. The relation which binds a malfunctioning organ to its effects is thus both causal, that is, extrinsic, and semantic, that is, intrinsic. This reflects once more, the primacy of negativities in the structure of medical representations.

Physiology texts separate the inseparable, i.e., functions from malfunctions and side against the latter. Much of what does not fit or that which disfigures some clear purposes or definite notions of function is being conveniently discharged into the category of pathology. Empirically, functions are seamlessly blended in with malfunctions, but they are also being defined counterfactually against the background of malfunctions, and conceptually separated from them. Functions, Austin would have said, are "*constitutionally iffy*".

To cite still another example, angiotensin is a substance that is part of a class of proteins involved in regulating blood pressure. What is its function? It stimulates the release from the adrenal cortex of the hormone aldosterone, which causes an increase of blood pressure. In addition, angiotensin causes vasoconstriction or narrowing of the small blood vessels resulting in an increase of blood pressure. In case of severely bleeding injury, its role is to reduce the loss of too much blood and fluid by constricting the vessels and encouraging blood clotting.

Yet, this is not the whole story. Angiotensin can also cause considerable damage in the long term: while raising blood pressure, it also fosters the development of atheromatous plaques in the arteries, helps provoke the rupture of those plaques, and increases the degree to which the heart muscle enlarges after a myocardial infarction; here we have a body's own protein with hormonal function that plays a significant role in promoting chronic disease. In understanding its mode of action, we thus split the effects of angiotensin into the useful ones we christen its functions, and the harmful ones, which we relegate to pathology.

It follows that the very same notion of function, as it were, supposes the prior notion of malfunction and that malfunction delimitates the boundaries of function. In a contrary manner, to say that an organ is 'functioning well', primarily means there is nothing wrong with it. 'Function' just like 'normal', 'fit', or 'healthy', is a privative term. It does not float on the surface, but it hinges on the hypothetical subjunctive mood. Again, *Omnis determinatio est negatio*.

And the concept of 'function' springs from the delusive idea that behind imperfection lays perfection. "Normal" is the default position and means nothing more than "not abnormal". This contrasts sharply with the case of artefacts: being defective for a machine is privative so that normalcy is here logically prior. A good watch has logically precedence over a defective watch.

In the light of this, it seems that abnormal function is foundational; it has the first word and is logically prior to normal function. We seek causal explanations for malfunctions, not for functions. When we believe that some process is hindering another process, the second one is a function: the hindrance delimits the function. Granted that sight depends on the eye, the contrast class of defective vision defines the function of the eye. It follows that organ functions are not judged to be normal by virtue of some good effects but by it causing no harm. Normal functions have a mere permissive role: they are *a posteriori* necessary but not sufficient conditions for health and well-being. The silences of physiology about clinical reality tell us much more about what we should avoid than about what we seek. Plainly, only rules that hold, can be infringed. There is no hinder where there is no function. We need having goals to be apt to fail of them. Hindering factors are thus constitutive of the concept of 'function'.

So, the final account leaves us with this: only abnormal functions or malfunctions, their degrees and their shades lie at the surface of the earth, whereas functions are hypothetical and paradigmatic medical norms, just like ideal gas laws, point masses, frictionless planes, and perfectly rigid rods. Small wonder, then, that physiological research, allegedly the study of 'normal' functions, proceeds since Claude Bernard by studying highly abnormal conditions and that functional attributions have often been tested in experimental medicine by adding and removing parts, or by setting up extreme pathological situations, such as animals deprived of thyroid or patients missing a large portion of their brain.

An abnormal function thus is not pathological *per se,* and its presence may not be necessary for there to be a disease; it becomes malfunctional whenever it contributes to some disease.[68]

Conclusion

The concept of functions is related to the treacherous question of natural purposes. Functions are both factual and normative, processes and counterfactual conventions.

There are four accounts of function: a causal explanation, a biological role, namely a mechanistic explanation, a historical role, namely an evolutionary explanation, and finally a semantic explanation that puts the whole issue in jeopardy and escapes from the philosophical controversies.

1. The classical puzzle of function and purpose in biology is responsible for the emergence and proliferation of philosophical and speculative debates, which attempt to explain how the desirable consequences of a trait could explain why the trait is in fact present. Several solutions are being put forward to explain final causality.[69]

   Chistopher Boorse's naturalistic approach analyses teleological language as primarily descriptive and not explanatory, even though human beings are not like machines that perform intended action. Larry Wright's so-called etiological analysis holds that the function of *X* is that particular consequence of its being where it is, which explains why it is there; Cummins holds that the proper target of explaining biological functions is not the presence of the trait, but the capacities of biological organs or organisms; finally, functional naturalized explana-

[68] Nordenfelt (1995, 29–31).

[69] Woodger (1929, 246).

tions turn to Darwin 's theory of natural selection: for Paul S. Davies, a trait's function causally explains the existence or maintenance of that trait via the mechanism of natural selection. Voltaire Panglossian fallacy evokes the abuse of natural selection is an explanatory tool.

2. In any case, there is a heroic irrelevance of the series of those philosophical quandaries. John Searle showed that we anthropomorphise the heart or the kidneys to which we attach normative importance. It is in this sense that functions are related to intentions and therefore *mind dependent.* Final causes are not laws of nature, but always relative to an observer; and functions are no natural facts although they imply causal statements such as "contractions of the heart muscle cause blood circulation".
3. It follows that the relation between the heart and blood circulation is a causal one; but the relation between the circulatory failure (not providing tissues with adequate blood for metabolic needs) and the ventricular malfunction of the heart, whose functioning is responsible for the decompensation, is a logical one and thus an intrinsic relation, but functions are never intrinsic.
4. Besides, functional models, functions, like most medical statements, are hybrid concepts: functions are empirical and normative, descriptive, and prescriptive. Functions are seamlessly blended in with malfunctions, but they are also being defined counterfactually against the clinical background of malfunctions, and conceptually separated from them.
5. Next, one should distinguish functional capacities and functional roles: *Functional roles* are *de dicto, i.e.,* those *ideal* functional abilities stored away in textbooks of physiology. *Functional capacities,* by contrast, are *de re* and are empirical and factual: they are measured and evaluated in a clinical set up on individuals or in an epidemiological survey on some selected population, so that failure of goal attainment or malfunctions have a functional role different from their functional capacities. A failing heart has a normal functional capacity, but a poor functional role.
6. The language of purpose and functions is unavoidable in physiology and medicine since it acknowledges how and within what limits our bodily and psychological processes and abilities satisfy our prudential and instrumental interests: Textbooks of physiology separate the inseparable, i.e., functions from malfunctions and side against the latter. Much of what does not fit or that which disfigures some clear purposes or definite notions of function is being conveniently discharged into the category of pathology.
7. The concept of 'function' springs from the delusive idea that behind imperfection lays perfection. They are theoretical paradigms and, as such, they resemble physically impossible construals, such as ideal gas laws, point masses, frictionless planes, perfectly rigid bodies, instantaneous velocity etc. Functions are counterfactual and physiology is false. Abnormal function is foundational. This contrasts sharply with the case of artefacts: being defective for a machine is privative so that normalcy is here logically prior. A good watch has logically precedence over a defective watch.
8. Finally malfunctions are not diseases, neither are diseases mere malfunctions

Chapter 7
Prudential Objectives Medical Need and Demand

What is desired is life, ability to procreate, physical capacity, strength, little fatigability, absence of pain, and a lasting state in which the body, apart from pleasurable feelings of its existence, is disregarded as much as possible? This is so obviously desired by everyone that the concept of physical illness attains a face-reaching consistency.

Karl Jaspers (Jaspers K, 1962, 789)

Contents

Prudence or practical wisdom is one of the basic virtues in medicine. It is a concern about one's own future well-being. Aristotle defined prudence "*to reflect upon and determine good ends*". It is an attitude that is pointed to practice in the range of goods and linked to correct planning. It does not lay in the field of knowledge but rather in acting. On the other hand, wisdom is the quality of being wise and knowing.

Why should a person care about his health? And if a person to-day is part of the same person to-morrow, shouldn't she or he care about his future body?

Prudence requires being able to conceive and empathize with one' own future concerns. The answer to these questions is that since medicine is concerned with any present or future state or change of states of one's own body or mind and of other people's body or mind, that are averse or that one wants to avoid, medicine

L. Karhausen, *Analytical Philosophy of Medicine*, Philosophy and Medicine 152,
https://doi.org/10.1007/978-3-031-81257-6_7

then issues its prudential advices in the form of imperatives or hortatory statements addressed to the patients, such as "Do not smoke", "Get sufficient exercise", or "Stay in bed" or anything of that ilk.

Even so, all welfare needs are not *medical needs*. And it is altogether striking that if medical practice is confronted with self-centred valuing activities, i.e., *patients' demands*, desires and preferences, clinical and public health decisions spring from their sheer needs.

What creates medical needs, be they diagnostic, or therapeutic? What marks off those medical needs? What must needs be like to be medically relevant? These questions are at the root of medical thought and their answers must be brought to bear upon the queries of medical care.

Prudence and the Authority of Medical Judgment

The term 'prudence' stems from *prudentia*, the quality of someone who is *prudens* which itself springs from *providens* (seeing ahead into the future) provident, foreseeing. Prudence (φρονησις), writes Aristotle,[1] is the capacity *"to be able to deliberate rightly about what is good and advantageous for oneself; not in particular respects, e.g., what is good for health or physical strength, but what is conducive to the good life generally"*; prudence is *"imperative (since its end is what one should or should not do)"*. Kant calls it *"private wisdom"*.[2] Thomas Nagel defines 'prudence' as a disposition to respond appropriately to *"the possibility of avertable future harm and accessible future benefits"*.[3]

Prudential considerations demand a capacity to project and be motivated by the future consequences of present actions or decisions. A prudent being pursues and protects—not necessarily exclusively—his own rational autonomous self-interest, namely those things which one ought either to avoid or to do out of regard for oneself and not on account of others. In short, prudence is enlightened self-interest. One does not decide to be or not to be prudent because this deliberation is already in itself an act of prudence. Prudence in medical care purports to various ways of looking after oneself such as treading carefully or else taking calculated risks and trading off some high stakes (such as a risky surgical intervention) for some probable gains. Cost-benefit and cost-effectiveness analysis are prudential instruments of medical rationality.

Prudential reasoning can be valid or invalid, successful, or unsuccessful. It is valid if it is justified, i.e., if and only if it is logically valid and well suited to finding means to the anticipated end. A valid prudential reasoning is not necessarily successful and an invalid one may occasionally be successful, but in medicine, success

[1] Aristotle (1976, 1140, a24-b30; 1143, a8–9).

[2] Kant (1972, 79, 82).

[3] Nagel (1970-1971, 37).

takes precedence over validity. On the other hand, a successful prudential reasoning involves choosing an action that has the predicted consequences and that is causally effective.

Furthermore, prudence consists in taking a temporally neutral viewpoint that is to stay outside the present and independent of immediate desires. Life is not a mere staccato of 'nows'. A prudent person cares for his future and does not trade in some greater future good for a lesser present one: he *"discounts for uncertainty,* holds Derek Parfit, *but not for mere remoteness*".[4] *"A smaller present good,* writes Sidgwick[5] *is not to be preferred to a greater future good."*

What tightens the concept of health care needs is that medical needs are universal while our wants are contextual and depend on our forms of life and the cultural context within which they arise. This is why prudential ordering in the sphere of health has been rather stable over human history since it reflects broad cross-cultural prevalent patterns of interests, which satisfy some of the main ulterior goals, and which allow each one of us to use his body as an effective instrument. Unlike social institutions, the human body has hardly changed over millennia, so that, contrary to the norms of law, the prudential values, which determine the functional norms of medicine, are comparatively conservative. Admittedly, prudential concerns imply choices and preferences but are logically prior to them: this is why medical decisions deriving from prudential reasoning constantly need to be evaluated and brought up to date.

Except for mental health, our prudential interests usually mirror the cultural background, the social expectations, or the characters of the social fabric. That "hallucinations are a frequent manifestation of schizophrenia" is empirically true: it is a fact; but hallucinations are pathological by reference to a warranted psychological standard that is contravened. But hallucinations are not regarded as abnormal features in all cultural settings: at one time, the priestess of the oracle of Apollo at Delphi or the seer of vision received special honour and their seeing of visions was accepted as some sort of special endowment.

It follows that medical needs, i.e., prudential medical norms of maintaining the body's and mind's capacities at a certain level, are the least contentious among prudential valuing activities; they result from the endorsement of a rather narrow, stringent, selected set of basic minimal prudential motives that are much more binding than general prudential reasons; such present and future interests include avoidance of death, minimizing pain, bodily harm, disability, disease, as well as maintaining physical strength and bodily or mental abilities.[6]

---

[4] Parfit (1981, 162).

[5] Sidgwick (1981, 381).

[6] Alexandrovna (2014, 12).

Time Preference

An inscription on the *Allegory of Prudence* (1550–1560) painted by Titian and exhibited in the London National Gallery reads: *"To the past, the present should turn so as not to put the future at risk"*. Hence, there is a further consideration, which tightens the role of needs in medicine. Wants usually imply some time-discounting, needs do not. *"In the case of an individual,* writes John Rawls,[7] *the avoidance of pure time preference is a feature of being rational."*

Prima facie, and by choice, the line of medical reasoning is committed to an impartial concern for all parts of people's lives. If so, it ignores discounting the future albeit other considerations may weaken this standpoint within the doctor-patient relationship. It would be irrational to prefer the present to the future and hence to ignore long-term or lifetime consequences that may be some decades away, on the mere grounds that they are in the future. Health needs, as it were, are timeless reasons for medical intervention and they should be allocated evenly over time.

Contrarywise, considered, or unconsidered preferences, choices, attitudes, desires may vary over a lifetime. The Gospel of Matthew says: *"Take no thoughts for the morrow for the morrow shall take thought for the things of itself. Sufficient unto the day is the evil thereof."*[8] So, at least in the typical case, patients' wants (in contrast to needs) are rooted in the present and reflect their desire to do away with present suffering.

The Structure of Medical Need

Spinoza[9] made an important distinction between 'privation' and 'negation'. *Privation* is a state of need resulting from lack or loss of some significant feature. *"We say, for example that a blind man is deprived of sight because we readily imagine him as seeing, or else because we compare him with others who can see or compare his present condition with his past when he could see."*

But *negation* is *"denying of a thing something, which we do not think belongs to its nature"*. Human beings, contrary to dogs do not hear ultrasound although they are not deprived of hearing them.

Spinoza's distinction is important for three reasons: firstly, privation contrary to negation delimits what we might call a state of need; secondly, privation is brought about by some causal process; thirdly, privation "is not the act of depriving" but "a mode of thought framed in comparing one thing with another", so that privation (like harm) is a contrastive concept.

---

[7] Rawls (1971, 293).

[8] The Gospel according to Matthew, 6, 34.

[9] Spinoza (1955, letter XXXIV, 339).

We may perhaps canvass the differences between two contrasting uses of the term 'need' from our medical vantage point: 'need' can be either descriptive and instrumental, or normative.

To begin with, the notion of necessity may be merely *descriptive*. Medical needs are factual, turn on observational evidence, and result from natural law-like causal sequences. They are intrinsic to our nature. They are not intentional, planned, or self-conscious. This is like the statement that for water to be boiling it needs to be heated or that roses need fertilizer to bloom or that we need water to survive.

So, medical needs convey information, and they appeal to facts alone. We need adrenal glands to live. We need to eat and to sleep. They are comparable to technical norms concerned with the means to be used for the sake of attaining some ends. From the standpoint of their external meaning, statements describing needs thus express tacit biological and psychological goal-directive capacities.

This is an *instrumental need* as it depends on some person's or some conscious being's aim. It refers to valuing activities of human beings, i.e., attitudes or activities that connect persons with some of their goals.

But in the second place, medical needs may be looked at differently since they are *prescriptive*, that if they are not satisfied, serious harm will result. The avoidance of harm is not some adventitious, or some contextual factor that is associated or causes the need, but it is one of its defining or essential characteristics. The resulting duty to meet a need is grounded in the avoidance of harm: the logical antecedent of a need is the avoidance of harm.[10]

Granted, then, there is a hint of circularity in such definition. Harm is the result of frustration of needs and needs are explained in terms of harm and, in the bargain, harm avoidance is part of the concept of 'need'. But harm and the avoidance of harm are the empirical touchstone in relation to which the concepts of basic needs and their satisfaction are being delimited.

Medical needs for treatment or preventive interventions are expressed as conditional sentences.

If Natalie has a mediastinal Hodgkin's disease Ann Arbor stage I or II, *then* she will need an extended field of radiotherapy on the mediastinum.

Should the patient accept it? Why?

The answer will be that she must accept it "*Or else the disease is going to spread, and the patient will become incurable and die*." But this answer drives home two points. One concerns the empirical nature of medical needs and the other, their normative valuing.

David Wiggins[11] proposes the following line of argument:

> *I need [absolutely] to have radiation therapy:*
> if and only if*: I need [instrumentally] to have radiation therapy if I am to avoid suffering harm [incurable – lymphoma].*

[10] Kurz (1965, 144), Thomson (1987, 35–54), Doyal et al. (1991).

[11] Wiggins (1987, 10).

And, if and only if*: it is necessary, things being what they actually are, that I avoid being harmed.*

Then I shall have radiation therapy

In this analysis, medical needs are instrumental (my purpose is to avoid harm), normative (harm is intrinsically bad) and prescriptive (they recommend or require help). Interpenetration of fact and valuing is ingredient in medical thinking.

Three Types of Needs

We may begin by distinguishing three types of needs:

1. *Felt need* that is an individual's perception of his own need for care. It usually expresses felt or occurrent wants for care. If one has a headache, he is usually aware that he needs paracetamol. Or else, people suffering from mental disorders view their own health status as worse than those who have no mental disorder, and this self-assessment does not depend on their actual state of physical health.[12]
2. *Expressed need*, which is the need for care that some individual mentions to physicians or health professionals. It depends on the degree to which he is aware of his medical need.
3. *Need as defined by health professionals,* which is the need of an individual regardless of whether or not he recognizes this need. The evaluation of health care needs is of practical relevance to the provision and the organization of health care. It does not depend upon tastes, attitudes, or interests although such needs are partly conventional just like concepts of malfunction, ailments, or harm. Needs may be given fresh interpretations when the medical doctrine changes.

The Instrumental Necessity of Health Needs

When a physician endeavours to resolve complaints from his patient, the latter's statements are *first-person reports*, that is to say, claims someone makes about himself, and which he is uniquely suited to make with a privileged authority; the doctor obtains answers expressing preferences, wants, desires, wishes, or demands.

But surely the imputation of medical needs results from a *professional judgement*. The decision about what is to be taken as pathological or as to what conditions constitute a disease does not depend on the opinion of the afflicted person. In this sense, medical needs are *second* or *third-person ascriptions*. So that, at least in the typical case, it is rather dictated by the formation and acceptance of a current

---

[12] Clarke (1985, 45).

medical doctrine. If so then, *unsatisfied, and unsatisfiable needs are still needs* since they respond to an *a priori* necessity. Patients suffering from amyotrophic lateral sclerosis will merely receive palliative care since no treatment is yet available; still, they badly need care, palliative as well as curative care.

Then beyond this, needs are about very abstract properties; they are neither about objects, nor physiological events, states or processes occurring in the organism: they are not like colours or shapes as they answer to nothing objective in the world. They are neither inside nor outside the world. They merely link descriptions with injunctions about what we ought to do. There is a conceptual interpenetration of harm and needs and a conceptual interpenetration of need and therapeutic obligation.[13] There is nothing in common to look for, between a panic attack, an acute appendicitis, a broken leg, or a chronic rheumatoid arthritis except for a need for remedial intervention.

It follows that medical needs do not have to be justified. Necessity is here *a priori practical necessity*, that means needs are in-the-circumstances-inescapable, unnegotiable, or indispensable states of dependence (relative to being harmed) and hence express strong reasons for intervention.[14] We can neither opt in, nor opt out of the web of health care needs: we do not choose them, they choose us.[15] But insofar as they are there, promoting our health interests may require an explanation but no justification. Medical needs are neither socially relative nor conventional, but they are self-justifying and self-evident.

Statements of health care needs impose moral responsibilities on the health-care professions or collective responsibility on the public health system: those needs have a moral force and are more compelling than, as well as overriding, our desires or preferences.

Needs Are Objective and Demands Subjective

The relation between needs and preferences is contingent and not necessary. The logic of wants is indeed very different from that of interests.[16] Patients' medical needs refer to patients private or public interests connected with what is best for them, and what is best for them depends more on what they need than on what they want. What is best for their interest may not be what they are interested in. Reasons for acting, writes Grice, are logically independent of desires.[17]

---

[13] von Wright (1963a, 183).

[14] Thomson (1987, 4), Wiggins (1998, 40).

[15] Griffin (1986, 54).

[16] Nordenfelt (1995).

[17] Grice (1978).

But surely, needs may occasionally overlap with wants, although wants are not a subclass of desires,[18] and desires are not a reliable touchstone of need. Needs and wants are conceptually distinct and logically independent from one another. Yet, they may be contingently dependent from each other: a need may depend upon a person's will. Some medical interventions depend on the patient's blessing.

Medical needs are *objective,*[19] even though objectivity may come by degrees. Medical needs and the effectiveness of meeting needs are empirical features which are discovered and can be evaluated quantitatively in human populations—just like needs for treatment in agronomical settings—and arise from assessing the balance of benefits and harms.[20] Statements of medical needs are true or false (within the limits of a given body of knowledge). They are objects of knowledge: questioning needs implies correcting knowledge. It is legitimate to ask: How do I know I need radiation therapy? But odd to ask: How do I know I want radiation therapy? Needs are thus related to the state of the world, not to states of mind.

Against this background, wants, preferences and wishes are psychological features: they are *subjective*. Contrary to needs, wants and preferences are patient-centred, reflect individual or personal valuing activities, and stem from individuals who seek medical care whether or not they need it. They are expressed in first-person claims, while needs and interests pertain to third-person account. Wanting is primarily expressive of an attitude. It would be sheer nonsense to declare: "I want to maintain my physical abilities, but I believe this is not true". Wants are objects of belief and are neither true nor false: questioning somebody's wants implies questioning his sincerity. They have motivational content; they are psychological dispositions and are a logical part of emotions.

Wants are *immune to doubt:* one cannot dispute about whether a patient wants a tranquilizer, but we can question his need for it. It is one thing to question his good faith and quite another to improve his understanding as shown by expressions like 'I am thirsty' or 'I am in pain, please relieve me': those expressions are incorrigible and cannot be false provided the speaker understands the words and is trustworthy. And if one may be unaware of one's needs, one cannot be ignorant of one's desires although one may be oblivious of them. One may indeed be mistaken about a need but not about a want.

Furthermore, wants may be legitimate, absurd, self-deceiving, illogical, contradictory, or naïve, but needs cannot. They may be grounded on inadequate or uninformed beliefs, but they may also be unjustifiable in case of drug addiction. One can therefore be blamed for one's wants not for one's needs.

We do not decide about our medical needs, but we may choose our wants. Wants are conceptually related to free will, needs are constraints that limit our free will. When patients have to share part of the cost of health care, this decreases demand

[18] Griffin (1986, 41).

[19] Copp (1995, 172).

[20] Doyal et al. (1991, 172–179), Rice (2000), Baines (1991).

for health services, but it may also disrupt the fulfilment of the sick poor's need.[21] Contrary to our needs, the limits of our wants are the limits of our awareness: we cannot want something about which we have no idea. One can get a sense of this by thinking that only conscious beings such as mammals or birds can be said to have wants or desires although any living being can have needs.[22]

Extensional and Intensional Context

The *extension* of a term or of a predicate is all the objects that it describes, to which it refers. The extension of 'red' is all the objects that are red. The *intension* is the condition a thing must satisfy to be truly described by the predicate.

Descriptive terms such as 'toothache' have extension and intension, that is, 'standing for' and 'meaning', or alternatively, denotation and connotation. The 'extension' of a term, i.e., its 'denotation' or 'reference', is the set of things to which it applies and that it names, i.e., what the word refers to.

The *'intension'* of a term or an expression is the concept, which the term relates to, or the idea it connotes, that is to say all the properties that are part of its definition. So, *extention* is a relation between some linguistic expression and a real thing, about something being said that might be true or false. And intension is a predicate describing properties of terms in virtue of which the predicate is true of them. It refers to a representation or the fragment of a representation held by the author of the utterance. There is a slippery risk of confusion between the terms 'inten**t**ional' or 'intensional'. '*Intensionality*' (with a **s**) refers to linguistic properties. '*Intentionality*' (with a **t**), that is to say 'aboutness', is applied to mental events and the projective nature of thoughts. Yet they are germane to each other insofar as mental states and linguistic expressions are both characterized by their topic-directedness. In the case of psychological ascriptions, the intensional stance regarding the embedded sentence 'that *p' which* describes some states of affairs is obvious. But it has equal application in linguistic meaning as in mental content, insofar as sentences and narratives are *about* existing or non-existing facts.

Co-referential expressions are not interchangeable within an intensional context *salva veritate*. On this score, two terms may have the same extension but different intension. Down's syndrome and 'mongolism' refer to the same syndrome resulting from trisomy 21, translocation or mosaicism. A lady may believe that her child suffers from Down's syndrome; yet it does not follow from the fact that trisomy 21 is identical with Down's syndrome, that she *believes* (an intensional term) her child suffers from a chromosomal anomaly. Expressions that have the same extensions with different intensions cannot generally be substituted for one another in an

[21] Steinbrook (2004, 2539).

[22] White (1975, 1197).

intensional context (such as mental states, beliefs, or representations), while preserving truth.

Expressions referring to needs are *extensional,*[23] so that their truth depends on the way the world is. What one needs, one needs it under no matter what description. So, needs are essentially public since they draw upon a common appreciation of what are the best ways and means to avoid harm. We can in any statement substitute for the term that refers to the object of a need, an expression referring to the same object without changing the truth of the original sentence: a patient's need may indifferently be termed a thyroidectomy or a thyroid exeresis. Needs are said to be *referentially transparent.*

On this score, a relation is transitive if and only if when any item *a,* bears that relation to any item *b*, and *b* bears that relationship to any item *c,* it follows that *a* bears that relation to *c.* If *a* is taller than *b,* and *b* is taller than *c,* then *a* is taller than *c.* Hence needs are transitive, wants are not.

Conversely, demand, preferences, desires and wants, are *intensional*; they are intensional by being the way things are intended to be understood or are being perceived by individual persons or patients.

They are subjective since there is no guarantee that there is anything in the world corresponding to their intensional content. Being intensional, wants are *referentially opaque,* because their truth depends on the way the person describes their object. In other words, if there are two terms referring to the same object of wants, they cannot be substituted *salva veritate:* one may desire something under one description but not under a different one. For instance, a patient suffering from chronic headaches who is on paracetamol may well, trying to ease the pain, refuse a tablet presented under some different proprietary name, which, known or unknown to him, is made of the same medication; yet it would satisfy his need.

Hence, the satisfaction of demands depends on the subject's ability to form representations, namely how he looks onto the world. Wants are vectorial beliefs. They give expressions to some seemingly inner occurrent pointing that is directed toward an object, concrete or fictitious, real, or ideal, though primarily accessible to their owner.

A woman who *wants* her breast cancer to be treated, may refuse surgery, chemotherapy, and radiation therapy. However, if she *needs* to be treated, she needs surgery, chemotherapy or whatever treatment is appropriate for her. Moreover, one may want something under one description (treatment for breast cancer) and refuse it under another description (surgery); demands are intensional, while needs, being extensional, remain needs under whatever description.

[23] Griffin (1986, 41), Doyal et al. (1991, 42).

The Trade-Offs Between Need and Demand: Acceptance and Acceptability

Physicians' authority is based on the contention that needs pre-empt demands and that whenever there is a competition between need and demand for different therapeutic choices, the need starts with an imperative edge. This is the underlying premise of medical care. We must here introduce an important already mentioned distinction between acceptance and acceptability, which is cognate to the corresponding difference between needs and demand. Acceptability refers to what patients ought to accept according to the best available practice, while acceptance refers to the fact of what patients actually accept. Acceptance is simply acquiescence and, as an expression of preference, the counterpart of demand.

To begin with, public opinion and patients in general often find themselves being guided by their desires for health care, by health care demand. By way of illustration, the misuse and abuse of tranquilizers, sleeping pills, antibiotics, hormone replacement therapy, food supplements or homoeopathic preparations result from giving greater weight to preferences over needs.

On the other hand, physicians have always been essentially concerned with the needs of their patients. So, needs take priority and precedence over preferences in clinical as well as in community medicine. Harry Frankfurt calls this *the Principle of Precedence*[24]: "*It is because making things better is, from a moral point of view, less important (measure for measure) than keeping them from getting worse.*"

Allegations of needs thus are internal to the medical discourse whereas desires are external. This is tantamount to saying that medical needs are embedded in the disease concept even when they cannot be fulfilled. The medical obligation results from this intrinsic relation between the need for intervention and the notion of disease we ascribe it to some sentient being. The upshot is that, unless people's or patients' desires coincide with their prudential interests, no medically significant harm will probably result if their demands do not obtain since the latter poorly reflects their basic interest.

Surely, demand and preferences could not be neglected in medical or public health practice because it is an essential element for good care. However, clinical and community medicine would remain purely academic exercises if needs were not at a certain point supported by wants. One of the most fundamental philosophical and practical issues in medicine consists in translating needs into wants, convincing healthy or sick people to adjust their wants and the motives of their wants, to their needs, and converting scientific acceptability into public acceptance.

There is a line of controversy regarding the distinction admittedly fundamental between need and demand. This is doubtless why doctors and patients as well as health care authorities and the public often talk at cross-purposes.

All in all, reasons for medical interventions are independent of the patient's motives or desires just as reasons for public health decisions are independent of

[24] Frankfurt (1988, 110), Baybroke (1987, 314).

their public acceptability even though their acceptance is needed for their implementation.

Patients sometimes cannot separate wants from needs and are unable to properly evaluate the quality of therapeutic care they receive.[25] There may thus be a fundamental tension between what the patient wants and what the doctor believes he needs, that is to say between an opinion and an epistemic virtue. When patients' judgements and preferences conflict with medically informed judgements, it is the responsibility of physicians (or public health policymakers) to bridge the gap between acceptability and acceptance, or between acting on the patient's behalf and acting at his behests. Medical care attempts, with various successes, through health education and medical dialogue, to bring needs and wants, as well as acceptability and acceptance of medical choices and decisions, to coincide.

The handling of communication in order to narrow or close the perception gap between need and demand is more than just factual, since it also depends on emotional and instinctive components of the patient's response. Presenting the facts alone may occasionally deny the patient's feelings, implying that if his perception does not conform to the facts, it suggests he is wrong, with the implicit message that he is dumb. It is presumed that cognitive messages alone may be counterproductive.

In cases of irreducible conflict between needs and wants, the doctor ought to make his best to override the patient's opinion and convince him—or in some cases, compel him—to do what is in his best medical prudential interests, at the risk of encroaching upon his autonomy and self-respect. Patients suffering from anorexia nervosa may occasionally be considered for involuntary treatment when there is a risk for their health or lives. It seems reasonable to suggest that such medical decision should not be taken unless imperative health care needs overpower the patient's desires.

Such a divorce between needs and preferences, when it occurs, pertains to the technical complexity of the issues not likely to be understandable to the patient and reflecting a certain incommensurability between the patient's story of illness and the medical narrative.[26] A patient may be poorly informed of his condition; he may refuse treatment for lack of discounting of future adverse effects or for self-destructive or self-deceptive reasons; he may express wants which conform to no need, or which are not properly medical, and which would be more fittingly dealt with by some social service or home help; finally, an irrational person, due to anxiety, conflict of wants, or religious prejudices (e.g. the case of blood transfusions for Jehovah's witnesses), may in some circumstances disregard his own health care needs, or those of his children, and refuse some medical intervention even though he knows perfectly well it is necessary.

Beyond this, if the source of wants is subjective, the truth of wants grows as they become closer to needs. Usually, acknowledged needs are wants: my needs can be

[25] LaCombe (1995), Weingarten et al. (1995), Pickering (1996).

[26] Hunter (1991, 123–147).

identified with my wants as soon as I recognize my own needs for care. Informed consent is supposed to express some of the patient wants in terms of his needs. If so, then, the quality of care will depend on how successful demand will become congruent with needs. However, patients' wants or preferences often prevail over medical needs, particularly in chronic diseases, so that prudential objectives may be overridden by other interests. The economic crisis of health care in occidental countries and the paradoxical success of irregular medicine stem from the conflict between wants and needs and from the epistemic and logical gap between them, which become more and more difficult to bridge. When a practitioner uses some alternative 'treatment' and does not use it as an intentional placebo, it usually answers the patient's want but it may be confusing demand with need.

Besides, there is an increased demand for medical care that often results from cultural fashions induced by consumerism that generate very high implicit health care governmental liabilities in European countries.

One of the fundamental trends of the history of medicine, has been the necessity of giving the patients what they want. However, medicine has ceased giving patients what they want, a tendency backed by health insurances. On this score, medical care is being plagued by a growing dissatisfaction, because physicians find or tend to believe that it is no longer necessary, or that it may be increasingly difficult to enlist the psychological comfort of the doctor-patient relationship in supporting a patient through an illness.

The Genealogy of Our Prudential Concepts and Practices

What kind of rationality is operating between first-person singular present-indicative, self-regarding discourse of wishes, wants and preferences, and third-person discourse crediting medical prescriptions? How are we led from personal suffering, malfunction or diseases to bodily obligations and medical needs?

Clinical medicine pivots on those minimal prudential medical norms relating to individual persons, while public health places special emphasis upon political prudence i.e., what Dante called a kingly prudence *(regal prudenza)*.[27]

We need to move from a personal to a neutral vantage point, from a first to a third-person account; or from a presumably self-protective to a utilitarian reckoning; hence we must turn personal prudential interests into normative goals and personal wants into neutral needs to narrow or close the perception gap. I shall assume that the agent is acting rationally and in self-trust, namely that there is a (prudential) reason for his action and that he acts because of that reason even when he does not take part in conscious reasoning and deliberation about that reason. Meanwhile, non-humans may be aware or unaware of some of their health interests, but the latter

[27] Dante, Paradisio, XIII, 88–104.

often are beyond the control of their wants. Small wonder, then, that veterinary medicine often must force prudential interventions on them.

First-Person Indexicalised Account: Naturalistic Inferences

The scientific naturalist believes the world to be in all respects open to science. He furnishes an agenda for the view that acting prudentially is acting according to desires depending on a person's goal in the context of the network of our beliefs, desires and competing interests.

Naive naturalism assumes that we can construe a justification of the requirement for medical care in an idiom from which any value-laden or normative term have been eliminated. Basic needs are basic wants (in the case of sentient beings) and are mainly defined by sheer biology resulting from unlearned primary drives and motives and from empirical self-interests and social conventions. The naturalist argues that the gap which divides personal impulses, wants, and wishes from objective self-preserving interests might be a mere matter of their prevalence level in the population: health needs, as it were, are mere wants shared by a large proportion of the population: occupying oneself prudentially merely describes some ubiquitous patterns that are statistically characteristic of how human beings presumably preserve themselves from objective harm. What is good is what is being desired and what is bad is to be averted: both are apt to be reduced to those utilitarian aims conducing to a decreased preponderance of suffering and harm. There are no values but mere valuing activities involving goal-directed processes.[28] Part of the attraction of this way of thinking lies in its objectivity and simplicity; it appeals to facts alone.

But if we venture this naturalistic conjecture, we must then deal with the following question: do people or sentient beings always, or often enough, act from a prudential standpoint and upon their own best interest?

First-person statements of want, like the Cartesian argument, suffer from worrying difficulties since they do not necessarily follow from what the agent says, believes or knows to be beneficial. Within the course of his deliberations, the patient may voice some appraisals as well as his feelings of attraction or aversion for some health care choices, and use various valuing words, expressions, or sentences to that effect. If a patient says that he wants something, which from his point of view is beneficial, this defines and expresses his attitudes, even though these may be things of the fleeting moment.

'Indexical' is a term introduced by C. S. Peirce to refer to words that relate statements to the spatio-temporal co-ordinates of the act of stating. This includes demonstratives such as 'I', 'you', 'here', 'there', 'now'. Colin McGinn says: "*... the use of indexicals involves treating oneself as somehow a* centre, *as a privileged*

[28] Kurz (1965, 211).

coordinate; an objective description should not be thus invidious in its depiction of reality."[29]

Perhaps it is enough for the present purposes to set aside an alternate possibility that one of the cognitive biases, described by Daniel Kahneman and Amos Tversky, that systematically cause human beings to deviate from an ideal model of rationality.[30]

But some further difficulties come here to light since the first-person account crosses the treacherous use-and-mention gap.[31]

The patient can be rather confident about the knowledge he has of his own wants or preferences. However, when the doctor refers to what he says and quotes his words, he encloses them in inverted commas.

The difficulty here is that this first premise hinges on the first-person privileged position: it is said or thought by the patient and is stated in reference to him; it cannot rid itself of its quotation marks; it will not follow that the first premise said within quotations, i.e. from the vantage point of the patient, will be true without quotation marks i.e. from the physician's point of view.[32] *"It would be a mistake,* writes David Copp, *to equate the good 'from an agent's standpoint' with what would be good for the agent.*"[33]

This use/mention distinction is of importance because, in the process of removing the quotes, we may be crossing the fact/value barrier. Actually, something beneficial within the slot of quotation, i.e., from the private standpoint of the patient, may not be so when divested of quotation marks, i.e., from a neutral point of view. A rational soliloquist may express desires, wants and alleged needs of medical relevance but he soon gets bogged down in the egocentric predicament of how to 'get out' with the caprices of first person implicatures and the thorny quandary of *disquotation*[34]or deindexicalized discourse.

What clearly emerges from this is that what is good from an agent's standpoint is not necessarily what is good for the agent. Expressions of mere wants in combination with factual means-end relations might provide reasons why, not reasons for it, but they cannot justify prudential behaviour.

Furthermore, it is granted that human beings have other interests and goals, which may eventually conflict with their health-related prudential wants leading to compromises and trade-offs. People often perform acts that they most want, and not acts that they believe are best for them: many cigarette smokers are very much aware of the risks of smoking. Reasons of self-regard depend on what expectations the agent nurtures, what believes he endorses or discards but also on interfering primary urges, present desires, and disinclinations.

---

[29] McGinn (1983, 16).

[30] Kahneman and Tversky (1979).

[31] Searle (1969, 73–76).

[32] Quine (1951, 23).

[33] Copp (1993, 112–113).

[34] David (1996).

Finally, prudential reasons can be jettisoned without contravening any biological principle, in specific situations such as in indolence and weakness of the will, risk-taking activities, consensual sadomasochism, rational sacrifice or not fearing what Aristotle called honourable death, pain or damage.

Returning to the naturalistic account, it now seems that something is being lost in translating prudential discourse in purely descriptive terms. The naturalist must prove us that his descriptive account will commit the agent to some corrective medical intervention, and that the prescriptive terms used for that purpose can be defined away in the indicative mode: assuming, with the naturalist's putative claim, that the positive valuing of prudential reasons can be totally defined in terms of behaviour, to what extent do they warrant prudential behaviour? There is a difference between a natural necessity and a practical necessity: facts are no reason. A statement that a hysterectomy is necessary may either mean that it results from some causal process (to survive, she needs a hysterectomy) or express strong reasons for action (the surgeon thinks she needs a hysterectomy). But without wishing to belabour this matter, one must drive home the point that medical reasoning and decisions transcend the first-person perspective and its limitations, and that we cannot account for medical prudential decisions on the mere ground of patient wants, preferences and observation of human behaviour and within an idiom clear of prescriptive terms.

First-Person Indexicalised Account: A Prescriptivist Account

The scientific-naturalistic[35] account leaves us with the following contention:

I want to maintain a high respiratory capacity since this will be beneficial to myself
Refraining from smoking is necessary to maintain a high respiratory capacity

---

I want to refrain from smoking

The motivation is housed in the first premise, which is here an expression of prudential wants. The second premise meets an empirical condition, and the conclusion is a desire or a quasi-resolution. If the agent accepts the premises, which he should do since we assume here that he is an ideal rational being, he cannot legitimately eschew the conclusion.

The argument is not merely descriptive and can be analysed into:

- A verb expressing favourable or adverse attitudes (I want, or I do not want, I wish, I prefer), which have an illocutionary force, that is to say, that, in uttering it, a type of speech act is being performed over and above its cognitive content.
- A proposition expressing that which satisfies the want if realized and that which leads to harmful consequences if unmet; this is the locutionary aspect of the argument, namely the mere saying of a string of words.

---

[35] MacArthur (2010, 124).

– The concluding statement is conclusive but not prescriptive.

Since some of our prudential wants may be pre-reflective, we may thus delete the explicit mention of want, and we now have:

Maintaining a high respiratory capacity is beneficial to oneself
My refraining from smoking is necessary to maintain a high respiratory capacity

---

It is beneficial for me to refrain from smoking

As the naturalistic view is relinquished and regarded as diversionary, we now must sketch the prescriptivist argument.

Introducing a presumably more important and more general higher-order prudential principle in the inference under which the second premise will be subsumed, we now validly deduce the concluding singular imperative:

One ought to bring about what is beneficial to oneself
Maintaining a high respiratory capacity will, in the long run, be beneficial to myself
Refraining from smoking is necessary to maintain a high respiratory capacity

---

I ought to refrain from smoking

Once we accept this point, we see that first-person prudential judgements now have four components:

– A high order normative component: prudential principle
– A descriptive-predictive component: fact
– A justificatory component: reason for action
– And a prescriptive conclusion: conclusive reason for intervention

The above inference moves now, under the cover of prudential considerations, from description to injunction, namely from naturalist descriptive statements about what the case is about the psychological necessity of what the agent is doing and of his motives for doing what he is doing– to action-guiding utterances about what he ought to do, i.e., to a normative and not a mere resulting conclusion.

For one thing, resulting beliefs are predictive and have a causal intention: they formulate the result of deliberation and usually end up in action. But for another, prudential reasons do not merely guide us from one truth to another; they involve natural and practical necessity; they provide grounds for moving, through a means-end relation, from a harmful or potentially harmful situation to a remedial intervention. They convey information that is both factual and commending. Thus, certainly, reason does not merely indicate us what to do to satisfy our wants but also what we must do to provide for the future. Plainly, needs are rational means of guiding behaviour.

Second-Person Indexicalised Account: The Clinical Encounter

"A person makes his appearance by entering in relation with other persons," write Martin Buber.[36] The meaning of real or potential suffering embraces a drive to ask for help from another person; to obtain remedial help, the threatening condition must be conveyed to a caring person with whom the patient finds himself to be engaged in conversation.

First-person prudential reasoning is guided by inclinations and relies on the agent expressing some want and stating of himself some requirement on the basis of his own want. Second-person inferences are of a different kind: they are used by health care personnel within the clinical dialogue to advise and persuade the patient of some medical prudential requirement and are of the form "You ought to do such and such". The patient, it is assumed, assents, tacitly or not, to the premises; they serve as a rational justification for their normative conclusion upon which it confers an illocutionary force.

Prudential second-person inferences often have the following hypothetical form:

Since we both want to cure your subacute bacterial endocarditis,
Then, it is advisable to administer IV penicillin G

It is advisable to administer IV penicillin G

The physician asserts some sort of therapeutic requirement based on the patient's own wish. But for a doctor to do his utmost, he needs more than the bare patient's persisting wants to support the cogency of his prescription.

To obtain a statement which is more binding than a mere advice, both doctor and his (allegedly autonomous rational) patient must implicitly agree on some general prudential principle issued in the imperative or commendatory mould: we need hypothetical imperatives of the form: *If you want ... you ought to...*

This kind of statement is closer to a categorical claim, and we now have:

One ought to seek care for one's illness.
Both of you and I want to cure your endocarditis.
A penicillin regimen is necessary for this cure.

If you want your endocarditis to be cured, then you should receive a penicillin regimen.

The first premise is assumed to be an intuitively self-evident principle in the absence of which there is no medical quest.

This practice of asserting medical advice in the categorical mode could be extended to all medical decisions but it might infringe on the patient's autonomy. In most cases, doctors' advice needs to be endorsed by the patient's want, since medical advice or recommendations addressed to a patient seek to direct or redirect his behaviour. And it is plain that doctor and patient may differ in their attitude to treatment, or they may disagree about their beliefs about it. A patient may take issue with

---

[36] Buber (1958, 62).

the doctor's advice and if he refuses the proposed treatment, it is the doctor's responsibility to talk him into it, or to propose some alternative medical intervention, should there be one. Medical inferences in the second person are thus hypotheticals of the form "If you want *E*, you ought to do *M*", contrary to first person inferences which are categorical.

Third-Person De-Indexicalised Inference: From a Prudential to a Moral Account

If medicine must be a scientific discipline, it needs to rely on intersubjective verifiability rather than on mere first-person perspective.

There are third-person inferences with singular (or plural) first premises or conclusions with which the clinician may refer to his patient (or the public health officer to a community):

I (or we) want him (them) to attain E
Doing M is necessary for E

He (They) ought to do M

It should by now be clear that first- and third-person accounts are not on the same footing despite a tricky tie-in between them. A patient cannot meaningfully declare: "I believe in acupuncture although it is thoroughly ineffective" but it makes sense to say: "he believes in acupuncture although it is ineffective".

Since medicine assumes that each patient and each person have a common stock of health-related prudential valuing, it implies a shift from personal to neutral claims, i.e., from a descriptive psychological (or psychobiological) account to a normative one. Medicine is written in neutral language. Clinical case histories are built on signs as well as on symptoms defined through behavioural and observable criteria relative to some scale of measurement. In other words, medical ratiocination, arguments, and interpretations of clinical case histories, are expressed from a third-person vantage point.[37]

Crudely, the process of becoming a patient starts with construing the latter as a neutral, medicalized individual, i.e., a patient, with some core of experience, which is invariant as to both first- and third-person standpoint. Writing a medical record or presenting a case at a clinical conference, yields to an impersonation of his patient as he becomes regarded as a mere locus of medical attention. To acknowledge, should the occasion arise, that his patient acts according to some prudential values is a statement of fact: it does not express but ascribes a value judgement to his patient; the third-person account mentions the patient's prudential reasons as objects of description but cannot use their prescriptive force.

[37] Hunter (1991, 132‹w@).

Furthermore, may perhaps a patient's self-referring prudential well-grounded claim for the first-person, not be so justified when translated into the third-person account? Do we have to cross the rift between first and third personal account? This may sound like a benign jump, but it is not. If one wants to avoid the accusation of naturalistic fallacy, some additional premise needs to be introduced here. The naturalistic fallacy is the claim that evaluative properties can be causally reduced without remain to empirical properties or are fully definable by empirical terms, and that agent-relative prudential goals can be worded in straight descriptive terms.

In the light of this, third-person prudential claims—granted that they are applicable to all sentient or human beings under similar descriptions (with similar medical conditions) and endorsed by those to whom they apply—can be reduced neither to some avowal beginning with the first-person singular pronoun, nor to mere empirical statements.

How do we patch up this problem?

The strategy is this: if we forfeit this reductive move, the line of reasoning gives rise to another position according to which the economy of prudential reasons should announce itself at the very beginning of the clinical transaction in a shared public language: with this move, each instance of medical decision commits the doctor as well as the patient, to a rather general principle expressed as a tacit additional premise. The upshot is that when a physician acknowledges that something must be done for prudential reasons, he thereby undertakes an obligation to do it.

Plainly, in the case of prudential resolutions about someone's medical concern, health care personnel are in a better position to make an appropriate judgement, since they are more unprejudiced and uninvolved than the individual patient. *The logical priority of the third-person account is thus genuinely befitted to the analysis of prudential reasoning*. And the second person, then, depends on the primacy of the third-person account.

Conclusion

Prudence establishes the logical link that relates the initial or first medical negativities with the need of medical intervention. At the clinical level, it legitimates what connects the patients' illness or suffering with the needs of help and of medical care.

1. Prudence or practical wisdom is one of the basic virtues in medicine and it canvasses the sequential pattern, which leads from a patient's ailment to a medical intervention. Prudence requires being able to conceive and empathize with the future stages of one's body, and it turns out to be the same as one's obligation to any other person. Medicine is concerned with any present or future state or change of states of one's own body or mind to which one is averse, which one wants to avoid. It issues its prudential advice either in the form of imperatives or hortatory statements, or in terms of therapeutic interventions.

2. Medical needs are objective and extensional and respond directly to suffering or indirectly to long-term forms of prudentially significant incapacities such as malfunctions, diseases, and handicaps. Painful, harmful, and pathological features or incapacities afflicting individuals and their inherent needs for help—ideally backed by personal rational and informed desires—represent the epistemic point of departure, which sets about the medical diagnostic and therapeutic process.
3. Medical needs, then, forestall discounting the future: each one of us may have his own curve of concern about forthcoming adverse life events, but medical prudence turns a blind eye to time-discounting. To the extent that medical judgements and decisions depend on squaring with prudential medical norms, they are being set in some neutral vantage point, a kind of timeless no-man's land, which Nagel proposed to christen, '*the view from nowhere*'. It thus seems that agent-neutral reasons can be justified without necessary reference to the person who may either hold them or express conflicting preferences.
4. Personal wants, wishes or preferences are subjective and intensional, and provide no foundation for choices of therapeutic or preventative action: yet they may initially have a hand in inducing them, and when the medical process is about to conclude, they may be partially or fully endorsed by our prudential ordering.
5. It might be appropriate to examine medical decisions from a first-person, singular self-reflexive account, from a second-person account, namely the clinical encounter, and from a third-person intersubjective public health account; these three steps are genuinely befitted to the analysis of prudential rationality.
6. Prudence requires being able to conceive and empathize with one' own future concerns. But the reasons being moral, this obligation to the future stages of one's body, turn out to be the same as one's obligation to any other person.

Learning how people become diseased and what their needs are is integral to trying to help them, and this how prudence leads to the core of medicine.

Chapter 8
Diagnosis Clinical Epistemology

A diagnosis usually does justice to only one part of the facts and is merely a convenience of nomenclature.

Alfred Meyer

Contents

"*Physicians,* wrote Immanuel Kant, *believe in helping their patient by giving a name to his disease"*.[1] Diagnostic procedures are methods and appreciative judgements made in the hope that a patient's clinical facts will converge on some single or a limited number of disease labels. Such procedures fashion a clinical epistemology since diseases are logical constructions out of *indicants,* namely either signs and symptoms or evidence elicited by tests. But they are also semantical exercises insofar as they attempt to relate medical language and disease labels to the clinical reality.

With medical diagnosis, as we move away from preliminary considerations on prudential, harmful, detrimental, abnormal, or pathological features or situations, we are entering the area of health care. We now quit the first position and we move to the second position, namely the clinical one, as we leave a conventional, conceptual account, to reach the surface of the world.

It is with the medical propædeutics, explaining the patient's complaints and their history, their signs, and symptoms that medical norms enter clinical medicine: physicians have the problem of deciding which of those data belong to normal or

[1] Kant (2000, 1905).

L. Karhausen, *Analytical Philosophy of Medicine*, Philosophy and Medicine 152,
https://doi.org/10.1007/978-3-031-81257-6_8

abnormal, and which among those abnormal ones are to be defined as being pathological. And it is with those data, once categorized, that scientific medicine ultimately develops shifting concepts of functions, malfunctions, diseases, health, treatment, or anything else of that ilk.

A disease, physical or mental, according to J. G. Scadding, is nothing more than "*a convenient device by which we can refer succinctly to the conclusion of a diagnostic process which starts from recognition of a pattern of symptoms and signs, and proceeds, by investigation of varied extent and complexity to an attempt to unravel the chain of causation.*"[2]

Signs and Symptoms

John Brown (1810–1882) held that *"Symptoms are the body's mother tongue; signs are in a foreign language.*" Diagnosis depends on the patient's own account—i.e., indexicalized manifestations or symptoms—on objective clinical manifestations (signs) and on the hidden evidence detected by special investigative instruments and procedures. Patients are not suffering from diseases, but they manifest symptoms and signs that are supposed to indicate potential sources of suffering.

Most clinical manifestations are positive, i.e., those that are being added to the normal repertoire such as a rash, pain, or arterial hypertension, or to the language of folk-psychology, such as hallucinations or compulsions. Other signs or symptoms are negative or deficits and are manifested by the absence of some clinical feature normally present, such as dyspnoea (shortness of breath), physical handicap, anhedonia, volitional disabilities, or signs of depression. They may either be unpleasant, e.g., in the case of pain, nausea, itching, vertigo, unwanted cognitive intrusions and bouts of anxiety, or pleasant e.g., in the case of hypomania sometimes referred as "pathologized happiness".

Symptoms and signs may be temporary or evolve into harmful or hurtful clinical features, with malfunction, impairment, or incapacity. Both are quantifiable although signs are more apt for measurement. Yet, they neither need to be troublesome in themselves, nor to be a manifestation that needs to be curbed. It is only when a physician construes them as evidence for some disease process that clinical manifestations become signs and symptoms: what then characterizes them is their aboutness. If the patient ignores the meaning of his complaints, they remain merely descriptive.

Symptoms express immediately given *sensations*. Wittgenstein said that words as "pain" and "feel" have different meanings in first- and third-persons statements. "I am in pain" or "I feel dizzy" are not sense-data since they are not representative of external things. Knowledge in the absence of the possibility of doubt is meaningless, so that pain, dizziness, and symptoms give no way for doubts. As a result, such

---

[2] Scadding (1996).

"inner processes" stand in needs of outward criteria: and when you realise that there are outward, public criteria for being in pain, you can see that these are *not* inner activities. Wittgenstein rejects the model of language that allows that words such as "pain" or "tickle" stand for private objects, which is a misconception about *how* a connection is established between a name and an alleged private sensation.[3]

They are either intentional, propositional states, i.e., beliefs, desires, hopes and all the rest; or else, they are qualitative and non-propositional states, i.e., kind of qualia in that there is something that it feels like, a qualitative feel such as feelings of panic, pain, or delusions. Finally, symptoms are part of a unified conscious experience, albeit this condition is not always fulfilled in the case of psychiatric symptoms. Consequently, symptoms profiles that are part of the world, contrary to diseases that are constructs, are culture-bound: they are environmentally and culturally variable.

Symptoms are named by non-count nouns, i.e., terms which do not provide a principle of counting: one cannot have two headaches, and even if one suffers from two teeth one usually does not complain of having two toothaches; moreover, if I have been suffering from a toothache for 2 days, I do not claim to have had two toothaches. We cannot count pains or nauseas although we may be able to count episodes of pain or of nausea ascribed to an individual. Symptoms require, to be explained, a local perspective, since they depend upon an individual's personal conceptual experience; they are not necessarily walled off from one another since they may have an effect one on another and even shape each other. Finally, symptoms may be culturally and environmentally variable.

The patient's narrative also includes a chronological ordering (anamnesis) describing the development of the ailment and its contextual features. Besides, the language of symptoms assumes that patient and doctor are cooperating, i.e., that the patient is supplying information that is correct. "*Listen to the patient,* writes William Osler*; he is trying to tell you the diagnosis.*"

Clinical *signs*, by contrast, represent hard data: they are *objective* manifestations construed in a visual space: they are publicly observable by the physician and obtained from physical examination and from endoscopy, imaging techniques, electrocardiographic or spirometric tracings, tissue examinations and biochemical findings, the microbiome and suchlike. They are independent of the patient's beliefs. What's more, in psychiatry, signs include concrete features that can be observed, namely behavioural manifestations, as well as what the patients say, namely, the rational content of what patients tell us, their written expressions and ideas, which includes delusional ideas, hallucinations and suchlike. To be sure, they are part of the world just as much as symptoms, they are 'out there', but contrary to symptoms, they are explained not by a local perspective but by a non-local perspective.

[3] Wittgenstein (1969a, 1981, 2009).

The Semantics of Diagnosis

Symptoms and signs are *prima facie* merely descriptive. However, when fed into the diagnostic process, they acquire a semantic dimension and point to something else. Inasmuch as this relates the name of a disease to a patient's signs and symptoms, the diagnostic process is a semantic relation with extensional and intensional contexts.

Signs and symptoms are caused by and truly refer to some relevant diseases when the diagnostic process is successful. They are extensional since the clinical manifestations are real descriptive features expressed by factual statements*:* it is a matter of fact that Molly's flu explains her fever.

But are signs and symptoms *of* or *from* a disease, that is, are they caused by or internal to the disease process? If causal they are extensional, but if constitutive of the disease concept they are intensional. Thus, clinical manifestations are extensional linguistic features, but once they become reasons for the diagnosis, they become intensional; reasons are reasons *that:* they are over and above their descriptive content; when they turn into diagnostic items or criteria, they become constitutive of the concept of the disease they refer to. Exuberance veers into, but also overlaps with mania. And cough is a non-specific symptom or sign, but in the presence of acid-fast bacilli it becomes a diagnostic criterion for lung tuberculosis and internal to its concept.

In short, the rules that relate signs and symptoms to a disease depend on our empirical experience. But it is also of semantic nature since it relates medical language to medical facts. So, when a diagnosis is established, clinical manifestations become diagnostic and sufficient for diagnostic decision.

Facts or Events

Are signs and symptoms events or facts? This again merits a digression before we return to the question of diagnosis.

Events are anything that happens, individual occurrences, temporal entities or episodes that take place in the world. Just like objects, two events are identical if they occupy the same places at the same time. Furthermore, events are extensional: if two extensional expressions each of them referring to some event, have the same meaning, they must refer to the same event. So, two terms referring to the same event can be substituted for each other with the assurance that truth will be preserved.

Even so, there is an alternate possibility: symptoms and probably signs are not events but facts or states of affairs. What do we mean by this?

Is there anything in the world apart from events? Moore, Russell, and Wittgenstein showed that *facts are that which makes statements true or false*. Facts are truthmakers. If so, then, is the world of events coextensive with that of facts?

To begin with, facts are not events or worldly occurrences: they are not in space or time. They are identified by some convention or shared assumption about how we use words and through relative clauses that connect language with reality. Facts are thus of semantic nature: they do not properly belong to the world but rather to our representations and the way we describe the world; they are *about* things, events or conjunctions of events which take place in the world. Facts cut out bits of the world through the statements that capture them. They are consistent with, and derived or derivable from statements or other states of affairs. They may be actual or not, ascertainable, or deniable, possible, or impossible although they are sometimes defined as states of affairs that obtain.[4]

The sentences or the expressions, which refer to events, are thus extensional: different descriptions may describe the same event. But facts are propositional as they are identified by clauses that begin with 'that', e.g., the fact that Julie has fever.

Events and facts may both be elements of our representations of the world. But they differ in the following way: when we talk about events, our language is being fit to the event that is being described; but when we talk of facts, the world (or the set of relevant events) is being spliced so that it fits our language and make our statements intelligible and either true or false. Thus, Strawson writes: *"Of course statements and facts fit. They were made for each other. If you prize the statements off the world, you prize the facts off too; but the world would be none the poorer."* On this score there is, as it were, a semantic move from the world towards language in the case of events, and the reverse in the case of facts.

The Meaning of Clinical Manifestations

Physical examination of a patient may ascertain some purely descriptive features cognate to events, occurrent states or standing conditions, e.g., short stature, fair complexion, arterial blood pressure, grey hair, which represent some features of the world just like the colour of the sky or the temperature of the sea. Clinical manifestations are autonomous and are initially unrelated to the presence of a disease process.

However, the inductive or abductive diagnostic process consists in increasing the cognitive role and content of such initially sheer descriptive characteristics, which thereby mature into clinical evidence. Diagnostic procedures rely on explanatory chains, which are themselves canvassed with true propositions, and true propositions correspond with facts not with events.

Before the clinical encounter, signs and symptoms are extensional and are mere events. But at the clinical bedside they turn into facts. From a pure phenomenological standpoint, signs are like sounds; they become meaningful during the diagnostic process just like sounds become tones when they become part of a melody. The process of diagnosis relies on theory-laden features on facts. Not in the mere sense that facts are opposed to opinions and beliefs or mere psychological features,

[4] Vendler (1967, 122–146).

neither that facts are opposed to events or mere worldly phenomena. Clinical manifestations are facts in that they are pieces of understanding and, in their relationship to one another, they have meaning.

Paul Grice[5] distinguished conventional (intensional) from natural (extensional) meaning. Signs, such as traffic lights, are radically arbitrary since they are devised or stipulated to mean a specific thing. (And so are those signs that are parts of human languages such as words or sentences.) They are conventional insofar as the meaning relation that links those signs to what they stand for, *to what they are about*, is not part of the causal order of nature but result from some human choice and common agreement.

It is one thing that, in textbooks of medicine, dyspnoea may suggest the presence of some circulatory or respiratory insufficiency. Shortness of breath is a natural event within an extensional context. Such a relation between a sign and its meaning has not been devised by human beings—like traffic lights— but it is a description of non-linguistic causal order of nature conveying information. Such relation once acknowledged makes it possible to predict, anticipate or infer for instance that 'these spots mean measles'.

But it is quite another that physicians provide their significance since clinical signs are clinical conventions and are not in themselves pointing to anything. They are like clouds, which mean rain, ashes that mean fire or fever that mean illness. Clouds are not *about* rain, neither is fever *about* illness. They are meaningful because we know they are causally related to the condition, which they signify. Natural meaning expresses a correlation between two states of affairs[6]: physicians provide their significance. What clinical manifestations mean is brought about by medical understanding, and in the absence of persons endowed with medical knowledge and skills, they may be devoid of meaning. Lester King writes: "*A sign ceases to be a sign when you cannot read it. The meaning has disappeared, and the alleged sign becomes any corporeal object (or in more technical terms, merely a phenomenon).*"

In like manner, as long as the patient ignores the meaning of his symptoms, they remain sheer natural events. It is only when a physician or an informed person construes them as evidence for some disease process that, instead of being mere happenings or perceived occurrences, they become sign and symptoms *of some condition,* and that they become intensional concepts.[7] Nausea is a mere feeling, a negative feeling until a physician calls it a symptom and endeavours to investigate its meaning, e.g. a manifestation of a gastro-enteritis or of a vestibular disorder. Within the cooperative context of the diagnostic process, symptoms eventually call attention to or point at some diagnostic hypotheses.

Even then, are diagnostic processes always converging to some organic or mental disease? The answer is negative. Except for some signs or symptoms (e.g., the

[5] Dretske FI, 205, 242.

[6] King (1982, 73).

[7] Cassell (2004, 93).

exophthalmia of someone suffering from hyperthyroidism, or palmar erythema suggesting liver cirrhosis), do not necessarily indicate the presence of a disease process: they merely point to the presence of some abnormality, some irregular or atypical physiological function. Symptoms are not characteristics that merely add up until some cluster emerges and yields some discrete disorder. A large bulk if not most people presenting symptoms cannot be fitted in any disease niche. One might be tempted to call those symptoms in the vernacular language of clinical medicine, *idiopathic, agnogenic, functional, psychogenic, somatoform* and suchlike. But using these terms fails to make a crucial distinction between being abnormal and being ascribed a disease label. Patients may have various complaints or clinical manifestations, some of them anodyne and other more significant, which turn out to be either medically irrelevant or of no consequence to the patient. In most cases, seeing visions spells out disease although it may occasionally be medically insignificant (in someone suffering from sleep deprivation).

Summing up, symptoms and signs in clinical medicine point to, or at least purport to point to, some kinds of disease entity, they are not, in and of themselves, defining pieces of evidence, that is necessary and sufficient conditions for a diagnosis; they are instead set of features, groups of which may tend to be sufficient, and few of which may be necessary for the applicability of the disease term to particular illness cases. They provide inductive or abductive evidence for some relevant disease entity.

Plainly, clinical manifestations are natural features and medical science supplies their aboutness.

The Nature of Diagnosis

Medical textbooks are written with the tacit assumption that all clinical complaints and manifestations can be ascribed to some disease. Hence, patients have complaints, and doctors construe them as symptoms and by inductive or abductive reasoning may reach a diagnosis.

First, just because the number of diseases is assumed to be denumerable, doesn't mean it is not infinite. Emanuel Goldberger wittingly writes, *"The number of diseases equals the number of patients times the number of physicians who examine and treat these patients."*[8] Furthermore, diseases are intended to be, but often are not, discrete entities and may overlap with one another. Finally, the definitions of some diseases may very well shift.

What clearly emerges from this is that the diagnostic process with its uncertainties endeavours to convert descriptive signs and symptoms into stipulated (though revisable) diagnostic criteria of a disease. It converts empirical features, i.e., clinical evidence, into conceptual ones, namely disease entities. Knowing how to diagnose

---

[8] Goldberger (1965, 45).

a disease along its own biological gradient is but knowing how to interpret clinical manifestations under a set of rules. It is in this sense that the rules are neither true not false; they may be correct or incorrect; they are conventions that we must accept; for without such rules, the statement 'Molly has all the manifestations of a flu' might be true although 'Molly suffers from the flu' is false. The rules are what transforms the clinical manifestations into a diagnosis: they constitute the evidence but remains outside it, since they are not part of it. Just like treatment, diagnosis is a procedure which consists in following rules one happens to adopt, to alter or to make up as we go along, or choosing between sets of rules, i.e., some correct way of going on. It attempts to combine a body of adequate evidence with semantic rules of evidence.

On this score, 'to diagnose' is an assertive verb specified in a propositional content, as well as a directive status pointing to an individual (just as 'classify' or 'identify') even though it is not used in the form of a 'that' clause: one does not say: "I diagnose *that* he has the flu" but 'I diagnose him as having the flu". This is an *ascription*. A diagnosis is a proposition uttered, written, signalled, or silently held within the clinical encounter.

A wrong diagnosis is still a diagnosis; it is mistaken when it fails in descriptive fit, whether it is merely being held or when it is affirmed.

A diagnosis is thus a matter of truth or falsity, but it can also be described in terms of felicity conditions namely specifications for appropriate usage: when completed it is an accomplishment and when successful an *achievement:* the diagnostic process must be accomplished correctly and completely.

It follows that once ascribed to a patient, a diagnosis is not merely a doctor's belief: it is a matter of acceptance rather than belief, and of assertion rather than a statement. Contrary to beliefs, a diagnosis and its acceptance is a voluntary commitment based on evidential and prudential reasons. It is posited for deciding what to do in a particular clinical situation and it assumes that anyone should accept it, even though it might, in some cases, be a makeshift manoeuvre provisionally accepted *faute de mieux.* Assenting to a diagnosis for our clinical deliberations is a way of acting as if it was true, but not necessarily a way of declaring that it is true insofar as it tacitly assumes our intention that it will stand up to examination.

The Limits of Diagnosis

In an ideal world, all patients without disease should be free from any signs and symptoms, and all individuals with the disease under study should present a decisive set of signs and symptoms as well as positive values for diagnostic tests; finally, signs, symptoms and test-results should either be those of the diseased or those of the disease-free group.

Even so, tests can be falsely positive or falsely negative.

If we aim for a diagnosis, which leaves no room for serious doubt, for instance in clinical medicine, a highly specific definition is required using stringent criteria that might miss individuals that clinicians would diagnose as having the disease. But if, contrariwise, our aim is to detect all the persons who suffer from the disease at the risk of including false positives, for instance in the case of population-screening for cancer of the ovary, weaker criteria can be used, though screening did not demonstrate a real benefit.[9]

There are two separate aspects of the accuracy of a test. A *sensitive* diagnostic test misses few true cases of the disease it tests,[10] so that one must increase the sensitivity of a test (the true positive rate) if one wants to identify all cases of the disease, at the cost of including false-positive and questionable cases. Sensitivity is the degree to which disease is detected when it is present.

And the more *specific* a test, the more certain one is that positive cases have the disease (the true negative rate).[11] This will tend to keep the disease group free of non-cases at the cost of missing a few borderline cases. The specificity of a symptom, sign or clinical test is an estimate of its uniqueness in its relation to the relevant disease. The specificity is the degree to which the test is negative when the disease is absent.

Highly specific signs and symptoms are loaded with meaning so that their presence indicates beyond reasonable doubt that some specific disease is present. Exceptionally, there may be a one-to-one concordance of disease and symptom in the case of so-called *pathognomonic* manifestations. Inguinal hernia is such a '*spot diagnosis*' identified through learned pattern recognition since it presents specific and exclusive manifestations.

The *positive predictive* value is the probability that a person who is identified by a given test as diseased result is correct: it is the probability of a "true positive". The *negative predictive value* is the probability that a person identified by a test as healthy, is healthy: it is the probability of a "true negative"?

Sensitivity and specificity are strongly influenced by the intensity of the search for disease: conducting the diagnostic workup with varying intensity leads to different test results. Positive or negative predictive values are usually more useful for clinicians, since they indicate the "predictiveness" of a given test in a given patient. Initial results on a new test are usually tried on groups of patients with a high disease prevalence, so that the initial assessment of their predictive value is often inflated.

Sensitivity and specificity reflect the proportion of positive and negative test results in homogeneous populations of patients, that is *with* the disease (for sensitivity) or *without* the disease (for specificity). It follows that these two criteria are unaffected by the prevalence in the study population. In contrary manner, predictive value is strongly affected by the prevalence in the study population.

[9] USPSTF (2018).

[10] Sensitivity is the percentage of people with the disease who are detected by the test.

[11] Specificity is the percentage of the people without the disease who are correctly labelled by the tests as not diseased.

But there is a further consideration of immediate interest concerning the incidental and unexpected findings discovered during a routine diagnostic process or during a surgical intervention. The incidence of "incidentalomas", incidentally found asymptomatic tumors has been growing during the past decades, due to improved imaging capabilities. Incidentalomas raise numerous tricky questions: what to do with silent small masses, e.g., a renal, liver, pituitary, or lung masses of 5 mm; 50% of patients have thyroid nodules and 20% of patients have pituitary incidentalomas. Most of these findings are benign, will never cause symptoms, and do not require evaluation. Could such unexpected observations trigger the ordering of unnecessary tests and cause untold anxiety? Could it raise legal issues resulting from the physician's failure to act on early evidence of a problem, initiating a cycle of disclaimers and protecting test-ordering?[12]

Finally, diagnostic uncertainty is a fact of life, and diagnostic errors are not completely unavoidable. In 2015, the US National Academy of Medicine (NAM) published a historical report *Improving Diagnosis in Health Care,* which designated diagnostic errors as common problems relating to patient safety. The NAM report showed that diagnostic errors arise in 10–15% of the clinical encounters. History taking and assessment are the most common origins of diagnostic errors. This diagnostic error rate should not be used as an argument for philosophical nihilism.

Medical errors are also common. Art Papier, a dermatologist specialising in medical informatics and a professor at the University of Rochester, stated that "*We don't know how many diagnostic errors we make*". Diagnostic errors have become and are accepted as part of the medical process. Many of them concern common diseases such as cancer. Patients usually suffer from fairly trivial conditions, but a clinician is often fooled by variations on common diagnoses.

Paul Epner founded the SIDM (Society to Improve Diagnosis in Medicine) in the US in September 2018, which brings together 40 organisations, including the major medical societies and academic medical centres. Marc Graber, president, and co-founder of SIDM, says that the first and most important step is for "everyone to understand that we have a problem". The measures to be taken concern doctors, nurses, care organisations and patients, who must also understand that the diagnostic process must "help co-produce the diagnosis".

The American College of Physicians (ACP) believes that internists make errors in 10–15% of cases. The ACP is now introducing continuing education courses to encourage thinking about diagnosis and decision-making, how problems can occur, and how errors are likely to affect patients and clinicians.

Fortunately, new tools are available to streamline the diagnostic process and help clinicians more reliably find the right answer. Some of these tools use artificial intelligence that can mimic human behaviour, or they use machine learning that uses artificial intelligence and statistical methods that allow devices to improve by solving new problems.

[12] Stone (2006).

Diagnosis as an Ampliative Procedure

Diagnostic processes are achieved on the basis of motley of tactical and diagnostic procedures and their various combinations.[13]

However, the process of diagnosis is apt to consist in a strategy, which does not follow a one-dimensional genus-species logic leading to a stopping point before treatment, but rather some complex decision in which descriptive and therapeutic considerations are intermingled. The diagnosis and demarcation of diseases may rely on manifestations of various sorts: symptoms in the case of myasthenia gravis, physical signs in the case of most dermatological disorders, morphological abnormalities for polycystic kidneys, physiological aberrations such as atrial fibrillation, biochemical defect in case of diabetes mellitus, chromosomal abnormalities in Down's syndrome, intranuclear inclusions in cytomegalovirus infection, behavioural disorder in schizophrenia, aetiological agent in tuberculosis. Furthermore, specific symptoms and signs must be considered in their context and, among other things, among the constellation of clinical manifestations of which they are a phenomenological part, and within the clinical natural course of the patient's illness.

Prima facie, the diagnostic process has two separate stages. First, collection of facts: anamnesis, physical examination and diagnostic tests should be thorough and systematic though some branching from the usual routine may be necessary. The second phase includes interpretation of the facts, as working diagnoses of progressively higher plausibility are being adopted.

The logical parsing of great many diverse diagnostic processes is an ampliative approach, namely *inductive or abductive,* whose conclusions go beyond their premises, proceeding by hypothesis-generation springing out either from intuitive guesses or else from some standard complete history and physical. Bayesian considerations, namely probabilistic diagnosis or decision trees displaying the proper temporal and logical sequence of a diagnostic decision problem may provide some theoretical or evidential support.[14]

This is tantamount to saying that, in clinical decision analysis, diagnostic statements are usually not true or false, but should be assigned truth-values that exhibit probabilities from zero to one.[15] These truth-values do not characterize any kind of uncertainty on our part but rather reveal some features pertaining to the diagnostic statements themselves. Evidence for tuberculosis is whatever makes tuberculosis more plausible, which shows the central role played by probability in the diagnostic process.

The theorem of the English clergyman Thomas Bayes (1714–1762) is an expression of the posterior probability—i.e. the degree of belief—of a diagnostic hypothesis conditional on some clinical sign or symptom (its probability after some clinical

[13] Wulff and Lennard-Jones (1981, 80–119), Albert et al. (1988, 181–211), and Macartney (1995, 59–99).

[14] Cohen (1980, 45–62) and Good (1983).

[15] Feinstein (1991), Sackett (1991), and Weinstein et al. (1980).

evidence is obtained).[16] This is a product of the diagnostic probability before the clinical evidence, or the prior probability, and the probability of the evidence given the diagnostic hypothesis, divided by the prior probability of the evidence (the probability of the evidence in the light of all possible diagnostic hypotheses). The modelling of the inductive process of clinical diagnosis by the theorem of Bayes describes the way by which the probability of the diagnosis increases by the addition new information, signs, and symptoms. Clinical investigation suggests that physicians are natural Bayesians,[17] although Bayesian inference is most often used in the field of decision theory.

The process of diagnosis, quite often if not most of the time, sounds more like 'muddling through' rather than an exclusive and undue reliance on the rigorous and sequential procedures of formal interpretation. We cannot draw up rules for such non-argumentative reasoning. Even though various rival schemes of reasoning may be part of the process of diagnosis, a diagnosis actually emerges from a process of mutual adjustment by means of confirmation and refutation; in this matching procedure, clinical manifestations suggest hypotheses which in turn prompt further tests yielding new data which themselves modify the original hypotheses. The diagnostic process may proceed with a coherence method now termed *reflective equilibrium* which has been described by Nelson Goodman and John Rawls[18]: it portends a kind of shuttling back and forth with diminishing return as we approach certainty, to-ing, and fro-ing through several cycles of mutual adjustment between narrative features as well as clinical data and hypotheses, until equilibrium and an acceptable coherence among them is achieved and a final diagnosis is confirmed. *"Clinicians often unconsciously use multiple, combined strategies to solve clinical problems, suggesting a high degree of mental flexibility and adaptability in clinical reasoning".*[19]

Diagnostic Criteria

Disease entities are defined by constellations of manifestations, i.e., signs and symptoms; some of them may be necessary, few of them are individually sufficient, and most of them are neither necessary nor sufficient.

For reasons of medical research and of public health practice, disease entities are sometimes defined, somewhat arbitrarily, by sets of signs and symptoms, which, conjointly with some rules of application, constitute necessary and sufficient diagnostic criteria. Diagnostic criteria introduce a certain rigidity and arbitrariness in our disease concepts, but they are needed in order to reduce uncertainty in medical communication. Since diseases are clustered concepts, their diagnosis may require

[16] Hacking (2001, 68–78).

[17] Gill et al. (2005) and Chitty (2005).

[18] Goodman (1955), Rawls (1971), Albert et al. (1988, 190–197), and Johnson (1991).

[19] Bowen (2006, 2220).

the presence of a sufficient number or a specific combination of criteria. A diagnosis can be ascribed by a cluster of characteristics associated with the name of a disease and no matter what manifestation is missing if all or most of the remaining ones are present. There may be no manifestation that all cases have in common. *Pace* Ludwig Wittgenstein, by analogy with the ways members of a family resemble each other, this is the kind of similarity shared by patients classified into certain diseases: this is termed the *quorum feature of language*.[20]

By way of illustration, the modified Jones criteria for Acute Rheumatic Fever (ARF) include five major clinical manifestations and six minor ones. Major clinical manifestations include migratory polyarthritis, chorea, carditis, subcutaneous nodules and erythema marginatum. Minor manifestations include fever, arthralgias, prior history of ARF, elevated ESR (erythrocytes sedimentation rate) or C-reactive protein, elevated white blood cell count and prolonged PR intervals on the electrocardiogram. The diagnosis of ARF is made in the presence of laboratory evidence of recent group A streptococcal infection (positive throat culture or rising antistreptococcal antibody titres) together with either two major manifestations, or one major and two minor manifestations. Such criteria were initially devised for classifying or for measuring patients for research purposes. For most diseases and despite great efforts by expert committees and WHO's efforts towards the use of standard definitions, medical science is still plagued by a lack of satisfactory and widely accepted working definitions.

Being able to understand and measure the severity of mental illness is important in psychiatry, and how distressing or disabling they are. The Present State Examination (PSE), and the Schedules for Clinical Assessment in Neuropsychiatry (SCAN) are tools aimed of diagnosis and dimensional approach.

Consider the word "*schizophrenia*". Several signs and symptoms are associated with this word. Symptoms may be positive (hallucinations and delusions) or negative (diminution or loss of normal function or affect), disorganised (though disorder ad bizarre behaviour) or cognitive (deficit of information processing and problem solving). Patients may show no symptoms or may manifest impaired social competence, mild cognitive disorganisation, and some general coping deficiencies. There are no known specific causes, but schizophrenia has a biological basis as shown by alterations in brain structure, and changes in neurotransmitters (dopamine and glutamate).

Quite often pathognomonic signs or symptoms avoid an inductive approach since they allow a "spot" diagnosis, such as stare and eyelid lag in hyperthyroidism, facial puffiness, and plethora of the cheeks in myxedema, facial rounding in Cushing syndrome, coarsening of the facial features in acromegaly or palmar erythema in liver cirrhosis.

[20] Karhausen (2023, 8–10).

Conclusion: Constructing an Epistemology of Medicine

Following preliminary philosophical considerations on the path leading from medical negativities and the normal/pathological split to the need for medical help, diagnosis is the port of entry in clinical medicine.

1. Signs and symptoms are atoms of clinical intelligibility and are the entry port into medical care. But they need to be explained and understood.
2. Symptoms and signs are *prima facie* merely descriptive. They may be sensitive or specific. However, when fed into the diagnostic process, they acquire a semantic dimension and point to something else. Inasmuch as this relates the name of a disease to a patient's signs and symptoms, the diagnostic process is a semantic relation with extensional and intensional contexts.
3. When the diagnostic process reaches this second position, signs and symptoms are no more mere effects contingent upon the disease process but are constitutive of, and internal to the disease altogether. Caroline Whitbeck[21] countenances that fever is constitutive of flu since fever is part of what we mean when we ascribe the diagnosis of influenza. In such a case, the ascription of signs and symptoms is no more causal or inductive but semantic; disease and its diagnostic manifestations are ligated through the very meaning of the terms. Explaining fever as such is one thing, but explaining a fever under a certain description, namely as a symptom of flu, in another matter.
4. The perspective afforded by the two positions should not be construed as the contention that these are two successive positions, a descriptive-extensional one and an instrumental-intensional one, neither that they correspond to some historical features of personal medical history. Medical thinking is canvassed from these two apparently opposing arguments describing a reasoning that proceeds through revision towards some greater adequacy. We are merely exploring the Janus faces of a single idea.
5. Such clinical consensual conventions on the use of diagnostic categories assume either that a given creature is or is not a case. Nailing the name of a disease to an individual is not a mere semantic exercise of labelling through predicate ascription, but the issuance of an entreating statement in an actual context. It pertains to the meaning of the disease category as well as to additional assumptions securing a definite response in an audience over and above what is recognized and understood. A diagnosis should be taken as an assertion paired with a context constituting a convention-governed practice and inducing an interpersonal and therapeutic transaction in an audience, principally the health care team.
6. In clinical decision analysis, diagnostic statements are usually not true or false, but should be assigned truth-values that exhibit probabilities from zero to one.

[21] Whitbeck (1977b, 634–639), Freedman and Hollingshead (1971).

Chapter 9
Diseases, Injuries, and Impairments

Instead of being a relatively inferior abnormal phenomenon, the empirical reality maybe that illness, defined as the presence of clinically serious symptoms, is the statistical norm.

I.K. Zola (1972, 37)

Contents

It took centuries even millennia to develop a first rough understanding of one of the pivotal concepts of medicine. Medical thinking emerged centuries before the conceptualisation of the notion of "disease". On another note, "disease" is a term that is mainly used by physicians and health care personnel, so that its medical use may be more or less remote from the different philosophical conceptions in which it is understood.

People have been using the word "disease" for centuries without knowing what it stands for. With the advent of scientific medicine and the emergence of physical examination and its accompanying technology, the hospital created a new viewpoint

L. Karhausen, *Analytical Philosophy of Medicine*, Philosophy and Medicine 152,
https://doi.org/10.1007/978-3-031-81257-6_9

disposed to think of diseases as objective processes: it contributed to *a shift from 'dis-ease' to diseases*.[1] The concepts of disease in traditional medicine were concentrated on individual sufferers and their symptoms.

For the Hippocratists, the disease was a property of the person. The idea that disease was just an observable set of signs and symptoms, a combination of physical and pathological examinations and diagnostic tests results, with a predicable course date back to Sydenham in the late seventeenth century.

The recognition of specific conditions—with a physical diagnosis, a thermometer, blood, and urine chemistry and with a characteristic clinical course and pathophysiological mechanism—represents a profound conceptual change, which initiated at the end of the nineteenth century and preceded the germ theory of disease.

At this point, it becomes clear that the concept of disease turns out to be the end point of a long history, and the outcome of centuries of research and reflexion.[2] Historically, if not logically, the concept of disease is not and cannot be the starting point but the end point of medical thinking. The philosophical literature about disease attempts to find sharply outlined definitions, but the situation is far from neat and tidy. None of these definitions cover what they wanted it to cover. Countless disputes arise, new definitions keep coming in and they are becoming increasingly far-fetched.[3]

Diseases are biological processes independent of the person observing them. They have "*actual concrete out-in-the-natural world existence*".[4] Even then, diseases are a product of our thought and pieces of understanding: they are not natural kinds, and they are not items in the basic inventory of the world. "*They are not things like lungs or livers… they are conceptual entities*".[5] Although a disease has a biological reality, according to Charles Rosenberg, "*it does not exist … until it has been named and explained.*"[6]

A huge variety of defining accounts have been proposed and are the object of endless controversies. Some years ago, in a book called *What is a disease?*,[7] several authors attempted to reach some sort of definition. What was surprising, was that each author had his own definition and succeeded to defend it. But none of them tried to find out why they disagreed with one another.

So that philosophers nowadays have to deal with the need to better define the concept of disease and how diseases can and should be classified in kinds. Be that as it may, philosophers has no monopoly on the *concept of disease*, and it is the role and responsibility of physicians to manipulate, split, alter, delete, control, or develop

---

[1] Porter (2006, 83).

[2] Rosenberg and Golden (1992), Cassell (2004, 77–87).

[3] Schwartz (2017).

[4] Cassell (2004, 98).

[5] Cassell (2004, 98).

[6] Rosenberg (1989, 2002).

[7] Humber and Almeder (1997).

the use of the term "disease". For clinicians or epidemiologists, "disease entity" is a weaker expression since it makes no metaphysical assumption beyond its use which amounts, according to Brian MacMahon, to *"grouping ill persons in categories that are believed to have utility in the management of their illnesses or in understanding the circumstance that led to it"*.[8] Whether an ankle sprain or an attempted suicide is a disease is not a matter of linguistic or philosophical preferences since our health care systems take charge of them: both of them are indeed included in ICD-11 and are being stored away in medical texts.

What is "disease"? What are "diseases"? What are the basic things required by medical theories and out of which "disease" is composed? How are these things related to each other?

The basic medical concepts of medicine, such as signs and symptoms, function and malfunction, disease, or syndrome mental or physical, and their classification, and treatment are neither natural kinds, or natural events, nor social constructs.[9] They are brought into being by the medical act of categorizing them as specific kinds. Disease is an *explanatory construct* that refers to some hypothetical *medical*—not biological social or cultural—tacit *convention.*

A disease is a condition described in textbooks of medicine and in the medical literature and which exhibits a pattern of interrelated symptoms. A disease is assumed to display a stable, characteristic structure allowing to make predictions and to legitimate preventive or therapeutic interventions. It is *pathological* in various degrees, in that it is actually or potentially harmful as compared to a default position, namely the counterfactual situation in which the individual would not suffer from the relevant condition: it assumes that it could and should have been otherwise. A disease is framed contrastively in a given sick population, by subtracting what belongs to its natural history from the patients' clinical background such as pre-existing disabilities, intercurrent diseases, co-morbidity and any feature resulting from aging: this is one of the main tasks of epidemiology.

To sum it up, he term 'disease' refers to a theoretical, conventional construct concerning biological processes and ranging over creatures, which is thought up for the purposes of understanding and remedial intervention, and which resists explanation by the methods characteristic of the natural sciences. Medically, it could be defined by five jointly necessary characteristics that do not depend on the existence of other things.

---

[8] MacMahon and Trichopoulos (1999, 31–32).

[9] Margolis (1976; 247).

Defining Diseases

1. In the minimal sense of the term, 'disease' intends to refer to a discrete pattern of organic or mental, statistically recurring pathological processes or states undergone by a creature, acknowledged by medical science, lying within the boundary of individual organisms, and which may have variable degrees of severity, namely intrinsic suffering, incapacity and/or excess mortality.
2. A "disease" is a conventional construct but there are no set of criteria that fully determine the meaning of the term "disease."; being abstract, it is instantiated by a list of token conventional disease entities, but no list of such entities, however long, can fully determine the meaning of the term "disease". There is no respect in which all diseases resemble each other since they are *cluster concepts* illustrating the *quorum feature of language.* Most diseases are shifting, overlapping processes, which provide a whole spectrum, a continuum coming in all degrees and ranging over all grades of severity, and not a dichotomy since abnormal features usually exist on continua with normal features.
3. There are two kinds of types of disease, monofactorial or multifactorial. The first ones are natural kinds, and the other are not natural kinds but medical constructs, and do not have independent existence. Yet, individual instances of diseases have objective, real, concrete manifestations, and existence, they are part of the furniture of the world but the boundary that sets them apart and isolates them among all biological processes is not part of the real world.
4. Diseases have a temporally spread natural history that may be acute, chronic, or static.
5. Diseases have known or unknown causes and may be either organic or psychosocial; their mechanism is internal to the body—except for mental disorders which may extend beyond the boundaries of the body—and constitutive of the disease process, albeit there may be external contingent, contributive factors.
6. Diseases embody an a priori prudential concern for prevention, remedy, redress, palliative or appropriate care. Either a cure or some effective treatment is available, but if it is not, it is assumed that it ought to and may, in principle, be found.
7. The best provisional definition of "disease" in clinical medicine and in public health identifies its extension by listing, as in the case of ICD10 or DSM5.

The Nature of Diseases

Claude Bernard[10] held that: *"Diseases are part of the milestones of life like birth, christenings, weddings, pregnancy, and burial."* The concept of 'disease' results from our need to put order and to classify related but diverse pathological processes

[10] Bernard (1872, 17).

and sick people for the purpose of explanation, communication, prediction, and control.

In a striking formula, Galen suggested to *"assign the term 'disease' only to those unnatural dispositions which impair an activity"*.[11]

We can then distinguish two sorts of basic constituents of our cognitive grasp: on the one hand, material bodies or *things*, literally material objects, living organisms and people, in Berkeley's phrase "*all the choir of heaven and furniture of the earth*"[12]; and on the other hand, *changes or alterations* and the absence thereof, i.e., events, processes and states. Changes are ascribed to things: they are *of* things since things are constituents of events and processes.[13] Diseases belong to changes and disabilities to states.

Processes are more complex than events: they are temporally extended features that stand in relation to the events which compose them like the way a symphony stands in relation to the notes, chords, or themes it is made of. Moreover, processes are not mere change of order and structure, but the changes are themselves ordered. On this score, a process is an order of change. Events and processes occur but only processes endure. Since the beginning and ending of a process is an event, disease incidence is concerned with events, but prevalence with processes. Molly's recent flu may be seen as an event. But Molly's current bout of flu is a process; it is going on; it can be interrupted or completed; it can proceed at normal or abnormal rates, be mild or severe, brief, or protracting, simple or complicated.

Processes are not mere sequences of events causing one another because the causal relation belongs to the process as it is part of it; processes have products, results or outcomes, the latter of which are not caused by, but constitutive of the process since a whole is not a cause of its parts; hence, diseases have a clinical course; their process is not completed until the outcome has been achieved.

Events as well as processes happen, but processes enfold an internal order and a direction in which changes happen. While events may have temporal parts, processes have stages. If we break down an event we obtain other events, i.e., a set of separate events. Processes are protracted events: they are going on in time and are described by verbs that possess continuous tenses, as they consist of successive phases following one another in time.[14]

Diseases are described or related in narrative manner, but it is as if we did not discover diseases to be true, rather we make them true by completing and refining their description.

But disease *qua* disease, type of diseases, are dependent on the person observing them: diseases are conventional constructs based on clinical, and sometimes social or cultural values. If so, epistemologically, clinicians perceive instances of diseases

[11] Galen (1991, 36).

[12] Berkeley (1983, I, 6).

[13] Waddington (1977, 18–25).

[14] Mourelatos (1978).

and medical conditions directly: their seeming and aspects are anchored to external objects.

ICD-10 represents a definition by listing of what medical scientist mean by diseases and the use they make of these concepts. The next edition, ICD-11 was adopted at the 72nd World Health Assembly, and it will be implemented as planned in 2022. However, the inclusion of Traditional Medicine in the ICD puts into question its scientific validity. That prescientific mythological concepts now have gained a serious position in the WHO morbidity and mortality classification, and statistics can be regarded as a direct failure of the political consensus-strategy.

Diseases, Type and Token

The Lancet's 1926 ground-breaking philosophical paper by Francis Graham Crookshank was giving an account of the theory of diagnosis. It argued that it is necessary, as Galen said, to distinguish between words and things, and between the two great doctrinal viewpoints, the *Coan* writings about organism and disease and the *Cnidian* tradition, which was more concerned with the diversity of diseases.

Coan or *Conventional* diagnosticians proceeded in terms of some rationalized type or universals, that were not the object of experience but were mental constructs. The Hippocratic Collection relied on judgment but not discrimination between diseases. They framed their judgment or *diagnosed* by paying attention to what was common to every and particular patient.

Cnidian or *Natural* diagnosticians proceeded in terms of empirical experience, which Galen defined as "*clear cognition of things present*". Briefly, Hippocrates' and Coan's school were Platonist, while Galen and the Cnidian scriptures were Aristotelian. The struggle between the Hippocratists and Galenists emerged at the Paris Faculty of Medicine in the sixteenth century.

It is one thing to say, from a *Coanian* standpoint, that diseases *qua* clinical states of affairs are pigeonholed according to clear cut criteria. Even though they do not describe some natural order, they are scientific constructs. The concept of disease entities reflects a binary Cartesian way of thinking and fall within the normative project of understanding medical suffering: you have a gastric ulcer, or you don't; you have a cerebral haemorrhage, or you don't.

It is quite another, from a *Cnidian* standpoint, that definitions of 'diseases' are supposed to be reportive, that is to say that they report how physicians and health care personnel use the word. The yardstick of medical customary use, rather than theoretical or ontological considerations, must assess any definition of the term "disease".[15]

Philosophers are attempting to identify the defining characteristics of disease, namely a true report of what we mean by, or how we use the word 'disease'. Is it a

[15] Campbell (1979) and Editorial (1979).

definition not of a word but of a thing, of an entity or of something for which the word stands for? Is it a lexical definition or a real definition? And what is its mode of existence? Nordenfelt claims that "*medicine has no monopoly on the concepts of health or related concepts. Health, impairment, disease, and disability are concepts which are well embedded in ordinary thinking, and which has a long non-scientific tradition*".[16]

When talking about diseases, a distinction must be made between types and tokens: "Disease" is a type, and "diseases" are token. This division in two philosophical approaches— evident enough even nowadays—is independent of medical doctrine and represents a tendency that has been indicated throughout the whole history of medicine. All in all, this *type-token ambiguity* is still at the very heart of medical philosophy about the concept of disease.

For one thing, the concept of "disease" is a highly abstract concept, albeit it comes from outer objective experiences. Medical research might increase the scope of ICD-11 listing of classes of diseases, but it would have no, or practically no effect on meaning of the general word itself. The conventional construct is fenced-in and impervious to additional empirical experience. This notion does not imply any form of ontological assumption or of metaphysical realist thesis: the ultimate reality is not at issue since it does not refer to any real essence. "Disease" is a "thick term", an expression coined by Bernard Williams that has a descriptive and an evaluative component: it is action-guiding and guided by the world.[17]

For another, particular disease categories depend on the assumption about the general concept of disease. Such individual "diseases" are subject to the techniques of clinical and analytic epidemiology as legitimate objects of scientific investigations. Admittedly, in clinical medicine, diseases and their taxonomy are observational and epistemically objective features of the world, namely biological processes. Categorization or subcategorization of diseases may spring from purely descriptive characteristics, from insight into the mechanism of a particular disease, from searching for selected genes or novel serum proteins, from the presence of specific immunoglobulins, from imaging techniques or from therapeutic response potential. As Dennett puts it: "*There will always be a few transitional things, missing links, …that defy definition, but the fact is that almost all real (as opposed to mere possible) things in nature tend to fall into similarity clusters separated in logical space by huge oceans of emptiness. We do not need "essences" or "criteria" to keep the meaning of our words from sliding all over the place.*"[18]

However, if each of these diseases is a biological process, the boundary that surrounds diseases and that identifies and separates them from biological processes is not of biological nature and does not belong to the world, since it is of conventional nature. Our notion of diseases are medical constructs, and the process of selecting specific intrinsically negative clusters among those biological features is not

[16] Nordenfelt (1995, 1).

[17] Williams (1985, 142–145, 152).

[18] Dennett (1991, 421).

biological, as it presupposes the concept of pathological processes. The sorting out of such pathological processes that sets a gap within biological processes and events is not a natural exercise.[19] Concepts of diseases are thus *value-laden* like those of 'weed', 'disaster' or 'floods'. Disease categories are constructs that serve both normative and scientific purposes, or prescriptive and descriptive practices, which cannot be cleanly separated.

Since natural kinds exist independently of medical mind, it appears that diseases are not *natural kinds*, inasmuch as their existence depends on medical cognizers.[20] In the medical reality, diseases are not natural kinds and do not come out already sorted out and given.

Diseases Have Causes

The *scientific* concept of disease presumes the existence of some known or yet unknown organic or mental causal factors that are being discovered in terms of what we can observe, usually by epidemiologic investigations. Statements about causality are modal and not logically necessary. Those causal factors act within or at the boundaries of the patient's body.

Even then, it may be a matter of *meaning* that some diseases have causes, so that the link between such diseases and their determinants is then *semantic*; causes are part of what is meant in medicine by "causal disease", so that causes are *essential properties* by *definitional characterisation.*[21] If one knows that trisomy-21 is a cause of Down's syndrome, trisomy then pertains to the understanding of the meaning of 'Down's syndrome', but this is essentially a lexical question.

Diseases Have a Natural History[22]

"Disease will have its course," wrote Thomas Moffett (1553–604). Diseases are unified temporal wholes: they have a course and temporal parts, a feature already mentioned by the Hippocratics. Sydenham taught us that they have a *natural history* with a beginning and an end, which may be cure, chronicity or premature death.

The natural history of a disease is a statistical concept, which refers to age of onset, course, evolution, phenomenology, complications, laboratory findings, the way it occurs in a given community in the absence of medical care, and the response to treatment. It includes effects of environmental factors, social and physical, and

[19] Culver and Gert (1982, 64–85).

[20] Dupré (2017).

[21] Fine (1994).

[22] Ryle (1936).

the response of the patient to them. Thus, the natural history of schizophrenia is usually less severe in the equatorial areas with low levels of anomie than in occidental countries.[23] Incidentally, comorbidity, i.e., the presence of other statistically correlated diseases, may raise tricky questions about a disease's definition and cut-off points, although such conditions are in clinical theory not seen as part of the natural course of a disease.

Prima facie, there are four stages in the course of a disease:

(a) *Biological onset.* At this early point of its natural history, we cannot detect the presence of the disease. For some diseases, biological onset may conceivably precede the early stages by several decades. Genetic diseases begin at conception; atherosclerosis begins early in life, and infectious diseases with the internalization of some infectious agent. Some diseases do not necessarily move up into the stage (b), such as cervical cancer *in situ*. Identifying people at this level for instance by measuring biomarkers might allow early treatment before damage is irreversible.
(b) *Presymptomatic stage.* At this subclinical stage, although the affected creature remains free of any symptoms, there are histo-anatomical or physiological abnormalities; early diagnosis is possible so that screening tests may result in early detection such as in the case of cancer *in situ*. Some diseases such as prostate or breast cancer may not evolve beyond this stage.
(c) *Clinical disease.* In the absence of intervention or spontaneous regression, the disease gains ground to the point when signs and symptoms are manifest, and the patient becomes *ill*. This is the stage of clinical diagnosis. This phase may be asymptomatic (hepatitis A or B, arterial hypertension) although it is usually detectable by appropriate clinical, imaging and laboratory diagnostic procedures.
(d) *Outcome.* The disease arrives at its resolution, namely recovery, remission, relapse, chronicity, permanent disability, or premature death.

Some diseases, especially cardiovascular diseases, Lyme disease or some cancers, may have some critical point in their natural history for which therapeutic interventions are either easier to apply or more effective and may improve or reverse the course of the disease. Early detection supposedly improves the prognosis of a disease if this critical point is located within stage (b). Spontaneous reversal of cervical dysplasia or cervical cancer in situ may occur in some individuals.

Rational therapeutic decisions depend on our knowledge of the natural history of diseases in the absence of therapeutic interventions. Unfortunately, such information is too often unavailable. Clinical experience cannot provide valuable prognostic data because cases coming to clinical attention represent a biased sample, and because of poor or incomplete follow up.

This model has its limits. Prostate cancer screening might cause more harm than good. Cochrane meta-analysis showed no reduction of mortality and recommended

[23] Jablensky et al. (1992).

against screening regardless of age.[24] However, USPSTF concluded that the benefits and harms of screening for men aged 55 to 69 are balanced and recommended an individualised decision after a careful consideration of potential benefits and harms.[25] Likewise, half of breast cancers detected by screening would never cause problems even if undetected and untreated.

Syndromes

A *syndrome* is used somewhat loosely to signify a broad spectrum of concurrent, intercorrelated, clustering signs and symptoms, a symptom complex, presumably having some undiscovered underlying connection; either a constellation, i.e., a mere statistical or presumably statistical association of symptoms and signs; or the sum of manifestations of a disease entity. Still, the two words 'syndrome' and 'disease' work equally well in common medical language.

Diseases are supposed to have an enduring individuality which syndromes lack, being mere assemblages, although the line between the two is often blurred. The aetiology of syndromes is multifactorial: a variety of causal determinants may result in the same syndrome (e.g., migraine, Down syndrome), while a single factor may give rise to several syndromes (syphilitic infection). The cases of Banti's syndrome or gastroptosis which were still fashionable in the 1950s, show that syndromes may dissolve, split, or crystallize again in distinct disease diseases.

Meanwhile, there need not always be something that holds concepts together: non-existent clinical diseases often arise from selection bias. Doctors and hospitals are more likely than GPs to encounter people presenting two abnormal manifestations than one, and even more likely to meet patients with three than two. Those clinical observations give way to syndromes or diseases that may not be supported by proper statistical evidence. For example, the Paterson-Brown-Kelly also called Plummer-Vinson syndrome or sideropenic dysphagia is characterized by dysphagia with iron-deficiency anaemia and the presence of an oesophageal web. Such cases exist but so do cases of each of the three separate manifestations of the syndrome as well as of each of the three possible pairs. The presence of the three manifestations in a single person does not occur more often than would be expected from the chance association of three unrelated conditions.[26] It follows that the syndrome does not exist.

The relation of a disease to its characterizing signs and symptoms is like that of a melody to the sounds it is made up of. A same melody can have different tones if one moves to a different key. A disease is a Gestalt perceived as an identifiable and reidentifiable unity that supervenes on and is organized into structures and

[24] Illic et al. (2013).

[25] Fenton et al. (2018).

[26] Barker and Rose (1979, 107).

processes. By contrast and pursuing the metaphor, a syndrome would be like a sequence of sounds devoid of any configuration. What supposedly distinguishes a disease from a syndrome is that the former is perceived as a whole and the other as the sum of its signs and symptoms, the difference between a sequence of noises and a melody.

Enduring Pathological States[27]

Acute or chronic diseases differ from static abnormal conditions or incapacities such as congenital heart defect, paralysis or amnesia, a distinction already made by Galen.[28] They are characterized by what Ducassé aptly termed some *unchange*.[29] Bentham wrote: "*By bodily imperfections may be understood that conditions which a person is in, who either stands distinguished by any remarkable deformity, or wants any of those parts or faculties, which the ordinary run of persons of the same sex and age are furnished with: who, for instance, has a harelip, is deaf, or has lost a hand.*"

One usually distinguishes three kinds of pathological standing conditions or static events. *Impairment* is a temporary or permanent *state* of deviation—namely imperfection, failure, loss, or absence—from some anatomical, physiological, or psychological norm resulting in reduced functional capability such as urinary incontinence or personality disorders.

Disability refers to any restriction or lack of functional or psychological ability to adjust to impairment or some health condition, and to carry out the activities of daily life in the manner or within the range considered normal: this is a rather objective notion, which measures the basic need satisfaction.

Handicap represents the social consequences of impairments and disabilities that limits or prevents the fulfilment of a normal role (depending on age, sex, and social and cultural factors) defined by the expectations of society and the needs of the concerned creature.

Impairments, disabilities, and handicaps, such as blindness or personality disorders, are standing conditions. Once being acknowledged that they have special needs, they call for remedy, medical treatment, care, social help, or self-help, or else becoming inured willy-nilly to one's disabilities.

Even then, disability depends on societal, economic, and environmental factors that are as important as the biological constraints. Being "deaf" (with a small *d*) is the medical condition of deafness as something to be treated. But being "Deaf" (with a capital *D*) refers to people who are culturally deaf, as part of the deaf

[27] Susser (1973) and Whitbeck (1978).

[28] Galen (1991, 44, 3.4).

[29] Ducassé (1968).

community and using the sign language. In this way, disabilities, more so than diseases, are in both the medical and the social and systemic frameworks that contribute to it.

The Need for Intervention Is Constitutive of Our Notion of Diseases

A *non-contingent concern for prevention remedy, redress, or care is built into the concept of disease*[30] due to an intrinsic coupling of diagnosis and intervention: preventive measures or restoring interventions, even if not presently available, may in principle, it is assumed, be found. Caroline Whitbeck countenanced that: "*a case of disease …is an instance of the sort of psycho-physiological process that people wish to be able to prevent or to terminate*".[31] The need for intervention is internally related to the disease construct: they are *logically* but not ethically tied up to one another.[32] Yet, if '*ought does not imply can'*,[33] the concept of disease in itself and of itself provides morally justificatory reasons for the need for medical help.

But surely, when a diagnosis is ascribed to a patient, it necessarily implies that the person who has that knowledge, usually the physician, should intervene, even though in many cases such as local prostatic carcinomas and in several situations, watchful surveillance, conservative management or therapeutic abstention might be the best response; management of cardiac valvular disorders commonly requires only periodic observation with no active treatment for many years; intervention in only needed when a moderate or severe valvular lesion causes symptoms or cardiac malfunction.

A physician who assents to a diagnosis and denies the therapeutic imperative or does not acknowledge the need to help his patient does not properly understand his diagnostic move. If so, then, *the need for care* is part of the characterization of what a disease is and lies at the root of clinical care and public health.

On the other hand, contrary to the case of need, *the demands for care* or the patient's expectations are not inherent to the disease concept. A malingerer asks for care but has no disease; and some patients adamantly refuse treatment, so that the further demand overtakes need, the further away we find ourselves from the core of the disease concept.

However, since the actual clinical concept of disease is, rightfully so, not the object of any medical conceptual analysis, the demand for care may in practice dissolve into the need for care in variegated conditions such as abortion, aesthetic surgery, euthanasia, or human enhancement.

---

[30] Whitbeck (1977b, 127) and Nussbaum (1994, 33).

[31] Whitbeck (1973; 211).

[32] Culver and Gert (1982, 70).

[33] Margolis (1971, 60ff).

Illness

Moreover, when talking about diseases, a distinction must be made between types and tokens: "disease" is used to mean either a type of disease, namely a disease entity, or a case of disease such as a case of flu. In the light of this, several authors, for example Boorse[34] and Margolis,[35] have introduced a core distinction between "disease" construed as a medical notion in the doctor's third-person perspective, and illness as a subjective indexical experience, namely the patient's perspective, the individual experience of a diseased person described in the first person.

Disease is supposed to be an objective biological or psychological disturbance while illness is the patient's subjective experience of sickness.[36] Furthermore, the term "illness" might be diffuse and vague, and it has psychological, moral, and social dimensions. It supposedly stands for the experience of ill health and how it affects the patient's behaviour, his relationship with other people and his interpretation of the origin and significance of his behaviour.

Patients approach their physician because they feel ill. But, although an illness may be a case of disease, diseases and illnesses are not coextensive since patients may feel ill but have no disease, and conversely.[37] And, although some patients suffer from one or several identifiable diseases, most of them are merely suffering from formless sets of symptoms. It follows that there is no one-to-one relationship between illnesses and diseases.

Symptoms and signs are facts, and individuals' illnesses are states of affairs, while diseases are mere constructs, which refer to conditions people ought not to be in. If so, then, a disease and an illness should be identifiable, but only a disease is reidentifiable. Diseases are theoretical biological events, while illnesses are human events. Diseases codify and simplify clinical reality. Illnesses are factual brute given, while diseases are man-made, they are medical models construed as a succession of gradual changes that follows one another in a relatively fixed way and lead to a particular end. Diseases, as it were, are more conceptual, while illnesses sound more phenomenological and encompass the patient's experience of the condition. Hence, plants may be diseased but never ill. Diseases are of third-person account while illnesses are of first-person account. Diseases are thus once remote and rather meta-clinical while illnesses are about sets of feelings that people experience. The recent interest in narrative medicine portraying the individual clinical manifestations of patients illustrates the importance of this distinction.

Even so, the day-to-day and the clinical use of the term 'illness' is fairly chaotic. Illness is an all-purpose term: being ill means being in a diseased state.[38] In familiar usage, 'illness' and 'disease' are interchangeable. The medical literature as well as

---

[34] Boorse (1975).

[35] Margolis (1976).

[36] Boorse (1975), Field (1976), Eisenberg (1977), and Barondness (1979).

[37] Weiner and Fawzy (1989) and Stewart et al. (1995, 32).

[38] Fabrega (1991), Boorse (1997), and Sackett (1991, 3 & 189).

physicians and medical care personnel often use the words disease, malady, sickness, disorder, condition, ailment, and illness synonymously, although they may not always be substituted for each other without change of meaning. Chronic conditions (rheumatoid arthritis) are usually not illnesses. In the medico-philosophical literature, some authors have introduced subtle distinctions between them, which usually do not correspond to actual linguistic use.[39]

Sickness Behaviour and the Sick Role

Diseases often evolve against a background of unspecific signs and symptoms such as malaise, fatigue, tiredness, anorexia, withdrawing from one's surroundings, fever, or weight loss: what undergirds them all has been termed *sickness behaviour*. This is more than a mere manifestation of some malfunctional condition since it is also a non-verbal form of social communication which elicits, in human or animal attention and caretaking behaviour or aversion and avoidance. The upshot is that, in human societies, there is a bundle of norms of behaviour, social rights and duties which surround and are part of the connotation of sickness and health, which Talcott Parsons termed the *sickness role*[40]: sick persons occupy a special role in society.

Firstly, a person in the sick role has both the title and the obligation to be exempt from some or all his familial, professional, and social responsibilities. Most diseases soon or later interfere with the patient's psycho-physiological capacity to perform those tasks that persons commonly wish and are expected to be able to achieve.

Second, the sick person is not held responsible for his condition and its consequences; he is not expected to get well by a mere act of decision or will.

Third, being sick is held to be inherently undesirable and the patient should want to recover as promptly as feasible.

Fourth, the patient should seek and accept therapeutic help from and cooperate with health professionals.

The modalities of help and support provided by the community are not part of the notion of 'disease' but a consequence of the disease process. They are socially defined and thus open to cultural variation. However, Talcott Parsons countenanced that the sick role is universal to all cultures and, within limits, to all sentient beings, as altruistic responses of heeding a diseased member are often observed in non-human organisms.

A person may be installed in the sick role without being sick in the case of factitious diseases or of some of the so-called functional disorders.

The upshot is that the disease role is external to the disease process: it is epiphenomenal as it is not part of its concept. A diabetic who refuses to acknowledge his medical condition may be deviant as he breaks with the social norm, but a patient

[39] Fulford (1989).

[40] Parsons (1951, 428–479), Parson (1964, 274–291), and Herzlich (1973, Chapter Eight).

who occupies the sick role is not deviant within his new role; actually, the principal function of the sick role is to prevent deviance; a patient in the sick role is deviant only if he keeps from conforming to that role and if he refuses a much-needed treatment.

The Yielding Concept of Disease

The health systems uses, creates, abandons, recasts, splits, and decides on diseases. It groups them into various categories according to their use, be it clinical, administrative, legal, physio-pathological, or anatomo-pathological. Doctors do not give diseases a metaphysical reality: diseases do not exist in an indefinite sense, independent of diseased organisms. They may use the name of a disease with different defining characteristics: it may be a clinical syndrome, a specific anatomical change, a specific dysfunction, or various combinations of these. The practical nature of medicine is not conducive to theorization. Some diseases may be multiply realized.

Traditionally, the concept of disease has arisen to explain the occurrence of processes of suffering, and disability, actual or potential, in the absence of any obvious injury.[41] Even then, when it comes to some of the mental disorders described in the DSM 5, doctors accept much more heterogeneous characteristics, since personality disorders or psychoneurotic disorders are such that the conventions by which diseases are defined no longer apply: the definition is then reduced to the need for treatment.[42]

Classification of diseases is essentially made of compromises between different classification principles. We are still far away from a disease classification that would be etiologically oriented, even though it is only through their causation that we can hope to control diseases. On this score, our lack of knowledge is compounded by the fact that most diseases have multiple causes.[43]

For all that, countless disputes arise because of the vagueness of the term "disease". It is vague, since it depends on what criteria we decide to use to separate pathology from physiology, and, in any case, there is usually no precise cut-off point between their applicability and their non-applicability.[44] There is a multiplicity of defining criteria as well as a multiplicity of senses for the use of disease terms, and there is not a proper single definite set of conditions governing their use. Small wonder then, that this assumption on who is deemed sick and who is not, works well in the hospital ward, but is not properly applicable on populations where diseases exist as a qualitative continuum.

---

[41] Campbell et al. (1979).

[42] Kendell (1975).

[43] Mackenbach (2004).

[44] Hofmann (2017, 134–135).

Philosophers of medicine provide essentialist definitions of the notion of disease. But how to adjudicate between what is stored away in medical texts and what is in constant use in clinical care. Granted then, the final account becomes unwieldly, and leaves us with a shuttling back and forth between the philosophical view that diseases have a set of attributes that are necessary to their identity, and the clinicians who strive toward moving targets. Diseases are primarily not what philosophers, but what doctors, nurses and health care officers describe as such. R. Saracci proposed to define disease by regarding as cases "*those subjects that have been so diagnosed by a doctor*".[45]

Yet, if the notion of 'disease' is one of the chief junctures of medicine, its usefulness is in no way impaired by its vagueness. Following Ludwig Wittgenstein, by analogy with the way members of a family resemble each other, this is the sort of similarity shared by diseases and their classifications. Medical concepts are often what Ludwig Wittgenstein called "*family-resemblance concepts*", which manifest a similarity shared by things classified into certain groups in the way members of a family resemble each other: each shares characteristics which many but not all the others, and there are no necessary or sufficient conditions for belonging in that classification.[46]

Considering a given disease each patient shares characteristics with many but not all of the others. The categories of diseases are neither mutually exclusive, not jointly exhaustive.[47] Hilary Putnam[48] christened such concepts *cluster concepts* based on what has sometimes been described as the *quorum feature of language*[49]: a cluster of characteristics associated with the name of a disease is such that they do not all have to be there; each of them can be absent as long as some of them are present; the more of them there are, the more secure we feel about applying the term.[50]

Meanwhile, Carl Hempel who had been invited to present a paper to the American Psychopathological Association wrote[51]: "*In scientific research (diseases) are often found to resist a tidy pigeonholing of any kind ... Some of the objects under study will present the investigator with borderline cases, which do not fit unequivocally into one or another of several neatly bounded compartments, but which exhibit to some degree the characteristics of different classes.*" Hilary Putnam ventured the suggestions that "*the principle of bivalence of classical logic is simply a useful idealization, which is not conformed to fully and cannot be conformed to fully by any actual language, natural or artificial, that human beings could possibly use*".[52]

---

[45] Saracci (2010, 13).

[46] Karhausen (2023, 8–9, 39–42).

[47] Wittgenstein (2009, 67–77).

[48] Putnam (1976, 52t).

[49] Hospers (1990, 122–124).

[50] Wittgenstein (2009, 67–77).

[51] Hempel (1965, 151–152).

[52] Putnam (1992, 123–124).

From the vantage point of population, that is for the epidemiologist, diseases, and more particularly mental disorders, give rise to a fuzzy boundary between sickness and wellness. Diseases are what Gilbert Ryle called "topic-neutral". The epidemiologic analysis of the population reality is not constrained by the clinical need for intervention. It follows that disease in the population is a quantitative not a qualitative issue.

The problem of deciding between a broad or narrow definition of diseases inherits the need for therapeutic intervention. This therapeutic imperative defines the condition's perimeter and cuts off categorical pathological disorders from scalar abnormal processes. The boundaries of pathological/normal suffering, of appropriate/over-diagnosis, treat or not to treat, beneficial or harmful effects of diagnosis, are for all physicians complicated and contested. Both sides have valid points to make. Each side is probably thinking of different kinds of case. Our inferences about what is pathological are thus epistemically modified by the pragmatic appeal of need for care.

Expanding disease definitions are causing more and more previously healthy people to be labelled as diseased.[53] This apparent ambiguity is often misunderstood. With a narrow definition of a disease, few healthy people will be unnecessarily diagnosed and overtreated, but a substantial number of diseased people will be missed. Conversely a broad definition will cover most if not all diseased people, and at the same time many healthy people. In the case of arterial hypertension, the risk of future complications shows an exponential relation to blood pressure. Hypertension is defined as the level above which intervention is necessary, namely above which the risk is unacceptable, namely the level that is being medically decided to be pathological. The diagnostic threshold may change under the pressure of new evidence. With a systolic BP <110 mm Hg, very few persons with putative hypertension would be missed, but many normal individuals would be wrongly classified as hypertensive; conversely if set at 140 mm Hg. systolic, a smaller proportion of putative hypertensives would be misclassified as normal, and a greater fraction of normal would be classified as hypertensive. Thus, setting up a diagnostic threshold implies trade-offs: to avoid misclassifying normal individuals, one must accept failure to include all abnormal ones; conversely to make sure of identifying all pathological individuals, we must accept misclassifying as abnormal and increased fraction of normal ones. Opting for the best boundary value for hypertension, becomes quite tricky when *the normal and the pathological population distributions overlap*.[54]

Preventing and treating have now the same end, the removal of disease. Medicine has two different aims, and the goal of prevention is identical to the goal of treatment.[55] But treatment is a continuum, a gradable stance. This stance has had the

---

[53] Moynihan et al. (2019).

[54] Blood Pressure Lowering Treatment Trialists' Collaboration (2014), Brunström and Carlberg (2018).

[55] Broadbent (2019, 53).

effect of dissolving the theoretical concept of disease entity and to blur the distinction between disease and pre-disease, diabetes and pre-diabetes, hypertension and pre-hypertension and the capacity to create protodiseases and disease states that shape medical practice and public health policy.

Patients with so-called 'sub-threshold' disorders suffer from significant distress and disability with signs and symptoms, which do not fulfil the accepted diagnostic criteria. By way of illustration, in sub-threshold depression, psychosocial disability (household and financial strain and limitations in physical and job function) increases linearly in stepwise fashion with each escalation in the level of depressive symptoms. The relative risk of negative outcome is lowest in subjects with an asymptomatic status, higher in those with subsyndromal depressive symptoms, even higher in those with minor or intermittent depressive disorders, and highest among persons with major depressive disorders.[56] Furthermore, the population distribution of these categories of depressive illness is continuous and unimodal. It follows that it does not make much sense to ask what fraction of the population suffers from depressive disorder. All depends on where we decide to draw the demarcation line.

Archie Cochrane[57] observed that as a student he was taught that there was a simple dichotomy between ill and healthy people. And it was not until he got involved in measuring quantitative characteristics of random samples that he learned that there was nearly always a continuum and no dichotomy. Leukaemia's may seem to be yes-or-no conditions but have various degrees of severity, from asymptomatic, to moderate, major and catastrophic manifestations[58] Even an autosomal dominant genetic disease such as acute intermittent porphyria showed a skewed but unimodal distribution of total faecal porphyrin among 150 members of a family in which there had been a case.

In other words, the bimodality implied by clinical descriptions is often the result of selective biases inherent to clinical or hospital samples rather than to their actual distribution in the general population. Most diseases present us with a smooth and continuous distribution of attributes: their manifestations shade gradually in a linear order into 'normal' features and there is no point where one can draw a boundary line except very arbitrarily.

Textbook clinical entities may be overlapping and not only in the case of mental disorders. J. Hoey contends that: "*As physicians… we want to base our actions on valid evidence, but we know that this evidence — when it exists — will rarely be directly relevant to the case of Mrs. Jones. We cannot expect to find a study of even a small number of Mrs. Jones. There is only one Mrs. Jones.*"[59]

Kendell and Jablensky wrote: "*Thoughtful clinicians have long been aware that diagnostic categories are simply concepts, justified only by whether they provide a useful framework for organizing and explaining the complexity of clinical*

---

[56] Judd et al. (2002).

[57] Cochrane (1972).

[58] Morley (1998).

[59] Hoey (1998).

experience in order to derive inferences about outcome and to guide decisions about treatment. Unfortunately, once a diagnostic concept such as schizophrenia or Gulf War syndrome has come into general use, it tends to be reified. That is, people too easily assume that it is an entity of some kind that can be invoked to explain the patient's symptoms and whose validity need not be questioned.[60]

Hence, diseases qua processes provide a whole spectrum, a continuum coming in all degrees and ranging over all grades of severity, and not a dichotomy since abnormal features exist on continua with normal features. Except for causal diseases, there is an ultimately imperceptible grading from the diseased to the non-diseased.

Two Types of Diseases and a Semantic Digression

Brian McMahon has identified two ways to classify ill individuals into disease categories: diseases can be construed either on purely descriptive terms, in the case of manifestational diseases, or through their aetiology, i.e., causal diseases.[61] Hilary Putnam captured a similar intuition.[62] Morton Beckner called them *polytypic* and *monotypic,*[63] but the terms *polythetic* and *monothetic* might be more appropriate.[64]

Manifestational diseases are based on manifestational criteria because there are few causal criteria to follow. Ill persons are grouped according to similarity of signs and symptoms, body chemistry, behaviour, or prognosis.

In causal diseases, ill persons are grouped according to their similarity with respect to one or more experiences believed to be the necessary cause of their illness.

Manifestational Diseases

Manifestational disease such as rheumatoid arthritis, schizophrenia or cervical cancer have multiple criteria and no single necessary one. Their aetiology is multifactorial. They display overlapping recurrences or similarities of signs, symptoms, course, behaviour, or prognosis, without any common feature they all share, and in the absence of available linear (one-to-one) single causal chain. In other words, the meaning of the term is mainly descriptive. They are not given to us by nature but have been identified either by some intuitive Gestalt or by methods of multivariate statistics and cluster analysis, which pick out a frequent correlation of symptoms.

---

[60] Kendell and Jablensky (2003, 5).

[61] MacMahon and Trichopoulos (1999, 335).

[62] Putnam (1975, 310–311).

[63] Beckner (1968, 21–22).

[64] Sokal and Sneath (1963, 13).

Our relation to this kind of disease is many-stranded: we relate to them through memory, clinical experience, epidemiological data, or consensual criteria. Such a definition is contingent: it could be overturned or abandoned under the onslaught of further medical research, as it was the case with gastric ulcer and *Helicobacter pylori*.

Single-Criterion or Causal Diseases

Single-criterion diseases are a quite different group of disease, which have a universally accompanying characteristic that is some specific and necessary identification criterion, such as diseases resulting from host-parasite interaction. Causal diseases are a subcategory of single-criterion diseases, which have been known ever since Galen.[65] Their definition is given in terms of a single, particular, *a posteriori* necessary, proximal cause (or causal process). Something is a scar if a wound has caused it. And it is through this internal causal structure that the scar is an effect. Stress and adjustment diseases, injuries, vitamin deficiency, substance-use diseases, bacterial or viral infectious diseases, prion diseases, molecular diseases, genetic defects, or anything of that ilk, are causal diseases since they are defined by their aetiology or pathophysiological mechanisms. Freud's theory, now abandoned, that hysteria is rooted in unconscious sexual conflict or memory through the mechanism of conversion had the earmarks of the causal entity model. Acute porphyria or general paralysis of the insane are single-criterion causal mental disorders.

It is hoped that manifestational diseases will gradually be converted to causal diseases, though this might be mere wishful thinking. As an example, a gene has been identified by Judith Grolleman, which is involved in colorectal cancer, and which also causes breast cancer.

Are Diseases Natural Kinds?

Natural kinds are groupings that are supposed to reflect the structure of the natural world, cutting nature at the joints. In *Phaedrus,* Plato writes that we should "*divide into forms, following the objective articulation; we are not to attempt to hack off pars like a clumsy butcher.*"[66]

Since natural kinds exist independently of medical mind, it appears that diseases are not *natural kinds*, since their existence depends on medical cognizers.[67] In the medical reality, diseases are not natural kinds and do not come out already sorted out and given. Natural kinds are supposed to exist independently of our minds, and

[65] Galen (1991, 42).

[66] Plato Phaedrus 265d–266a.

[67] Dupré (2017).

independently of or classificatory practice: they are discrete and categorically distinct, separate entities. Science is usually successful in revealing those kinds such as physical particles or chemical elements. However, in biology, many philosophers recognized the inadequacy of natural kinds to account for species.

Even so, a causal disease such as tuberculosis is out there the way a liver is out there, i.e., it is *defined by denotation*, so that it can be considered a single natural kind. The physician is correct to assign such a disease type in view of the necessary causal presence of Koch's bacillus.

Against this background, it is not difficult to appreciate that identifying a manifestational multifactorial disease involves, alternate, overlapping, conventional, shifting itineraries so that it is being *defined by description,* and it is not a natural kind.

The question is whether there is an invariant core of essential properties persisting over time, or whether such core is subject to change or to radical modification of their meaning as a reflection not of the nature of the world but of the nature of our own mind. The choice is between external realism and value-loaded conventionalism, or between ontologically or intrinsic objective entities and epistemically subjective entities. There is no privileged way to divide nature into objects or diseases into 'natural kinds', but there are many natural ways to assort those objects into kinds. The partitioning of nature and the medical taxonomies accord with our interests and, in less obvious ways, with our capacities. The world is not made of obscure metaphysical processes, but we are dividing up nature on account of our goals of inquiry. The medical world is partly the world as we make it.[68]

Grades of "Naturalness"

Causal diseases illustrated by Koch's postulates are supposed to be 'natural kinds' since they exhibit a feature of essentialist thinking.[69] They are to be explained by the internalisation of their own invading microorganism and it is assumed that a different microorganism would define another disease. Causal diseases are obvious taxonomic schemes that partition nature according to our interests.

On the other hand, chronic diseases such as atherosclerosis, hypertension or autoimmune diseases arise from a wide range of genetic, metabolic, behavioural, or environmental risk factors.

Natural kinds are contextually defined kinds with explanatory value, and they admit of degrees. Disease categories may, as it seems, be lined up in a spectrum at one end of which we find causal and single criterion disorders, at the other end of which are pure manifestational multifactorial disorders, but in the middle of which we cannot say which is which; moving along this scale from one item to the next,

[68] Kitcher (2001, 43–53).

[69] Kendler et al. (2011, 1144).

from causal to manifestational conditions, brings out a progressive dissolution of what supposedly holds patients together, and a succession of disorders which seem to have less and less of a unifying causal mechanism and increasing referential indeterminacies. In this view, naturalness might be less of a ontological dichotomous and more of a spectrum feature. I want to illustrate this issue with the following listing. The deployed progression is dialectically suggestive:

(a) Causal disease disorders defined by the internalization of a specific necessary causal agent: infectious diseases or genetic defects are natural kinds.
(b) Single-criterion hereditary disease disorders, which are genetically complex. Women who carry mutations in genes Brca1 or Brca2 have a 30 to 80% lifetime risk of developing breast cancer.
(c) Diseases having one non-necessary contingent causal factor which carries a great weight: Bronchial carcinoma and tobacco smoking, infection with human papilloma virus and cancer of the cervix, hepatitis B and C and liver cancer, asbestos and mesothelioma, or gastric ulcer and *Helicobacter pylori.*
(d) Multifactorial disorders, the cause of ischaemic strokes is heterogeneous: atherothrombosis accounts for 20% of cases of ischaemic strokes; 40% are due to atherothromboembolism, 20% to emboli from the heart and 20% are due to other causes.[70]
(e) Unexplained diseases. This includes unexplained clinical conditions with symptoms out of proportion to their physical examination findings, laboratory results and with absent pathological findings or clear physiological malfunction: irritable bowel syndrome, fibromyalgia, and tension headache, to name but a few.

We saw that causes can be evaluated in terms of tendency towards necessity. It is by now plain that causal necessity is also a scalar concept: it is as if there were grades in membership in the set of necessary causes: a scale of intelligibility, a sort of *sfumatura* or a shading-off of meaning, with a resulting loss of a clear separation between causal and manifestational disorders, natural and non-natural kinds. Naturalness might be a matter of degrees and some diseases are a fair amount more natural than others.[71]

Does this represent inadequate grouping of ill persons due to our ignorance? Or should we accept as a brute fact that causal factors can be ranked by order of decreasing necessity?

---

[70] Hankey (2001).

[71] Dupré (1993, 63).

Real Illnesses and Fictitious Illnesses

There are two possible approaches to illness: one is medical, the other sociological. The first approach is medical-biological and seeks to identify clinical entities in order to prevent or treat them. The second looks at the place of the sick person in society: the role of the sick person means that, whether in Western or non-Western society, a patient is exempt from some of his or her family, professional and social responsibilities, and obligations. What's more, being ill is an undesirable thing, and those who suffer from it generally seek to recover or leave this state as quickly as possible.

However, sometimes a person settles into the role of a sick person in the absence of illness, as is the case with malingerers, but also with fictitious illnesses, and this is a form of social deviance.

The pressure of public opinion, encouraged by the media, creates new needs, and seeks to manipulate and shape new diagnoses. The lycanthropy epidemic in the Middle Ages, *tarentism*, a dancing mania attributed to a spider bite (hence *tarantella*), Charcot's hysteria and spasmophilia in France, and so-called hypersensitivity to magnetic fields are all manifestations of epidemics that rely on abusive and inappropriate medical interventions created and maintained by the combined efforts of enthusiastic therapists, curists, and reflexologists of all stripes, follicular specialists, and malleable patients. These fallacious and inflationary diagnoses are exaggerated by speculation, the media, conjectures, and the prejudices of experts; they were all the rage at a certain time and usually go out of fashion, like neurasthenia, chronic appendicitis, Banti's syndrome, arthritic diathesis, ptosis of the stomach or of the kidneys, which required 'nephropexy', i.e., surgical fixation of the kidney.

One example: public opinion was recently gripped by panic over the alleged carcinogenic effects of mobile phones. If it is true that the IARC has classified mobile phones as category 2B, that of possible carcinogens—whereas 2A would have meant 'probable carcinogen' and 1 would mean 'proven carcinogen'—this means that according to this body, the carcinogenic risk of mobile phones is comparable to that of pickled vegetables or caffeic acid in coffee.

The World Health Organisation writes: "*When a journalist chooses to report a story, he or she does so for all sorts of reasons, which may be anything but technical: journalists compete for time and space, while newspapers and magazines seek to increase their circulation at the expense of others. To achieve this, they need sensational headlines on new kinds of subjects likely to interest as many readers as possible. Bad news is not just the big news, it's often the only news you hear about. For example, the press gives scant coverage to the many studies suggesting that electromagnetic fields are harmless. Science cannot guarantee absolute safety, but progress in research is reassuring on the whole*".

There is a lot of money to be made if, driven by alarmist pressure from the media, the European Parliament, and an armada of lawyers, we succeed in convincing a court of the danger we are trying to conceal from ourselves. The alarmists, not to mention the firms of 'engineers' and technical advisers specialising in the sale of anti-wave products, the apostles of alternative medicine and charlatans of all stripes are faced with the fact that insurance companies are no longer insuring the risks attributed to radio frequencies, for fear of being sued. One of the world's largest insurance companies has this to say: "*We regard the risk of change as extraordinarily explosive, not because weak electromagnetic fields may, contrary to expectations, turn out to be dangerous after all. We consider the risk of change so dangerous because it is clear that a large number of pressure groups have both a political and a financial interest in making people believe that electromagnetic smog is dangerous for society*".

Barnabas Kunsch of the Austrian Research Centre in Seibersdorf sums up the issue: "*There is no convincing evidence of any adverse health effect attributable to electromagnetic fields*". "*The absence of proof of a harmful effect does not seem to be enough for modern societies. What is increasingly demanded is proof of its absence*".

Caught up in a fictitious illness, and in order to escape the anxiety that plagues them, patients, despite the absence of disease, take on the "*role of the sick*". As a result of contagion, these syndromes become epidemic through mimicry and the amplifying and multiplying effect of the media. The effect of a contentious mood, of inevitable and legitimate conflicts with the health insurance systems and a medical profession that is necessarily powerless, of appeals to pressure groups and the intervention of consumer defence bodies that are all too often uncritical, but supported by the media, play a major role in the attribution, shaping and consecration of these syndromes. All this leads public, political and medical opinion to treat the difficulties and existential problems of everyday life in terms of illnesses, a trend encouraged by the pharmaceutical industry, which pushes the consumption of tranquillisers.

In *À la Recherche du Temps perdu*, Marcel Proust writes: "*For every ailment that doctors cure with drugs (at least, we're assured that this has happened a few times), they produce ten in healthy subjects, by inoculating them with this pathogenic agent, more virulent a thousand times than any microbe, the idea that we're ill*".

Doctors and patients become victims of a self-fulfilling prophecy, with a vicious circle of unnecessary medical interventions and a growing sense of frustration on both sides. Patients tend to abuse the health services with which they are dissatisfied, so that they often change their GP; they do not benefit from their treatment and are dissatisfied with it; they may suffer prolonged incapacity which can, in some cases, last for years. Karl Kraus (1874–1936) once wrote that "*diagnosis is the most common disease*". Patients and doctors find themselves caught up in a vicious circle of unnecessary diagnostic and therapeutic interventions, a source of medical inflation and a feeling of mutual frustration that reinforce each other. Let Goethe conclude: "*If you start thinking about your physical or moral condition, you usually end up believing yourself to be ill*".

The Magic of Alternative Medicine and Its Lobbies

Throughout its history, medicine has always been closely linked to magic. Didn't Pliny (23–79) write that magic has its source in medicine? Two things characterise alternative medicines: on the one hand, they are not based on any coherent doctrine or on any sensible evidence except for the enthusiasm of their advocates; on the other, they have hardly been subjected to any kind of evaluation, and if they have, the conclusions are negative, even though they claim to improve a very large number of disparate and often ill-defined ailments.

Unfortunately, not all doctors practise medicine in accordance with reason. But what makes alternative medicine so successful is the largely unrecognised or unacknowledged power of the placebo effect, and the *vis medicatrix naturae*. What is essential is the doctor's belief in the efficacy of the treatment, and the patient's parallel faith. Addressing doctors, Richard Asher (1912–1969) wrote: "*If you believe fervently in your treatment, even if controlled investigations show it to be totally ineffective, then your results will be infinitely better, and so will your income*" In these days when supposedly correct diagnoses and therapeutic suggestions can be found on the web, it is not surprising that myths are among us. This wealth of information is so widespread and so familiar that we take it all as truth without worrying about its scientific value. Using this information to make decisions about our health, how to treat a cold or how to care for our child can be harmful or even fatal.

Homeopathy, developed by Samuel Christian Hahnemann (1755–1843), is a giant among myths, as recounted in a very fine and comprehensive monograph on the subject. Its principles have hardly changed despite the considerable progress we have made in our knowledge of chemistry, physics, biology, and medicine: there is not a shred of evidence to justify any of the indications of homeopathy, which consists of selling and administering water with no other ingredients. The difficulty lies in the fact that their dilution is such that there is often not even a single molecule of the active product in the dose administered. In general, homeopathy has neither positive nor negative effects, but homeopathic products sometimes contain active agents, such as thyroid hormone, which has been the cause of deaths published in the medical literature.

The public, encouraged by too many doctors, is demanding homeopathic products manufactured by a company that rakes in more than 300 million euros a year to sell gullible patients a product that costs nothing to manufacture; for a long time, these products were reimbursed by the social security system and therefore paid for by the public, both those who believed in them and those who did not.

Death

It is not death that we fear, but the tragedy of aging, and of physical and mental decline. The word death, like truth, cause or disease will have different meanings in different contexts. *Prima facie,* death is the end of life. But as soon as one attempts to agree on some definition or some criterion of death, it becomes a moving target. Giving a satisfactory definition is difficult and a tricky business. From a philosophical or a legal standpoint, a criterion of death cannot have exceptions: it can yield neither false-positives, nor false-negatives.

Gilbert Ryle drew attention to a class of episodic words, which he labelled 'achievement *words'*. Construing the difference between task verbs and achievement words, appeals to the difference between the progression and the denouement, between looking and seeing, between treating and healing, and between the process of dying and the moment of death. And does the word 'death' stand for the process of dying or for its product: death? Dying is a process (which can be protracted), being dead is a state (which lasts forever), and, according to Wittgenstein, death is not an event of life or a fact of life[72]: we do not live to experience death.

But does death occur at some definite time? There are several possibilities. We may define death as a permanent cessation of the organism as a whole, or as the cessation to be a person, i.e., the loss of consciousness and of cognition, or else as the irreversible loss of that which is essentially significant to the nature of man, namely, a permanent vegetative state. Each new definition is adding some different or additional meaning to the previously stipulated one.[73] The dispute is more verbal than factual. No degree of clarity will settle this dispute in an acceptable way. "*Death has so many doors to let out life.*"[74]

The development of life-saving technology has important consequences for clinical medicine and philosophical theory. Our ability to maintaining a patient's respiration and circulation after cerebral functioning has ceased raises several questions, e.g., about the boundaries of life and the definition of death, of the ethics of euthanasia and of our obligation not to bring death voluntarily.

Incidentally, passive euthanasia is accepted in France, Chili, Finland, Germany, Ireland, Israel. On the contrary, voluntary active euthanasia called "physician assisted dying" is legal in Belgium, Canada, Columbia, Netherlands, and Luxemburg.

Useless to say, passive euthanasia, i.e., death by omission, is ethically totally unacceptable since we never know for sure whether the patient is still conscious: switching off a respirator, disconnecting a feeding tube, stopping the infusion, means letting the patient dying of thirst and of hunger, or suffocating to death.

One of the major difficulties in medical research is to differentially diagnose a vegetative state from minimally conscious state and from locked-in syndrome. A permanent *vegetative state*, PVS (also called *unresponsive wakefulness syndrome,*

[72] Wittgenstein L, GH. Von Wright & GEM Anscombe, (eds.). 74 (18).

[73] Beecher (1968), Luper (2017), Lizza (2011), Holland (2017), Walton (1979), and Taylor (2017).

[74] Beaumont F and Fletcher J, Act II, Sc ii.

UWS), is a state in which patients *lack any consciousness or awareness*, but, unlike a comatose state, they can open their eyes and exhibit sleep-wake cycles. Then, beyond this, a *minimally conscious state,* MCS, is a state in which fragments of meaningful interaction with the environment are preserved: such patients may establish eye contact, purposefully grasp at objects, respond to commands in a stereotypic manner, or answer with the same word. Finally, a *locked-in syndrome*, LIS, is a rare neurological disorder in which there is complete paralysis of all voluntary muscles except for the ones that control the movements of the eyes. Patients are *fully conscious* and awake, they are fully alert and aware of their environment, but they have no ability to produce movements (outside of eye movement) or to speak. Cognitive function is usually unaffected. Communication is possible through eye movements or blinking. Locked-in syndrome is caused by damage to the pons.

The accurate diagnosis of a patient's state of consciousness–in particular, the differentiation of PVS from MCS (and sometimes from LIS)–remains a major challenge in medicine. The rate of misdiagnosis has been estimated at around 20%, since many patients are wrongly diagnosed as PVS, even though careful evaluation would in fact disclose clear evidence for MCS. Frequent clinical misdiagnosis is a serious matter, because even minimally conscious patients might suffer from the lack of personal attention that results when their state is not properly recognized: furthermore, an incorrect assignment could result in the premature termination of life-sustaining care.

Until 2022, it was taken from granted that vegetative state was a condition of complete unconsciousness, when AS Nilsen from the National Institute of health[75] wondered whether we are really unconscious in "unconscious" states?

The fMRI scanner, developed in the early 1990s, allows to detect brain activity associated with thoughts, emotions, feelings, and intentions. Active areas of the brain receive more oxygenated blood, and the scanner can detect this and can identify the location where the activity is occurring. Owen put his patient in a magnetic resonance imaging scanner. When she was asked to imagine a play of tennis, a significant neural activity was observed in one of the motor areas of her brain. When asked to imagine walking through her house starting from the front door, there was significant activity in the parahippocampal gyrus (for scene recognition), in the posterior parietal cortex (for intentional movements and spatial cognition) and in the premotor cortex (for bodily motion).

Even then, it was shown beyond any doubt that she was consciously aware of herself and her surroundings. Small wonder, then, that the finding that this patient in such a terrible situation could be aware of what goes on around her, thinking and imagining, was an astonishing breakthrough.

In contrast with the PVS, MCS and LIS syndromes, the vegetative state is a condition of complete unconsciousness.

The most important tool to infer consciousness has been and still is to probe the brain with the electroencephalogram (EEG), which is considered as the safest signal

[75] Owen et al. (2006); see also Andre Sevenius Nilsen (2022).

of a conscious subject; magnetoencephalography (MEG) tracks the magnetic field around the brain, and another way is to evaluate the dynamics of blood flow inside the brain with the functional magnetic resonance imaging scanner (functional MRI and fMRI).

The question before us is how to avoid misdiagnosis, how to adjudicate amongst behaviourally unresponsive patients and how to evaluate disorders of consciousness in severely brain-injured patients. The problem consists in avoiding the uncertainty of clinical evaluation that relies on the patients' abilities to connect with their surroundings and to manifest their private experience through motor behaviour; even then, disconnected consciousness may result from motor or executive function impairment or because of sensory disconnection from the environment. Silvia Casarotto of the University of Milan, and her colleagues utilized a metric of consciousness, the *Perturbational Complexity Index* (PCI) in patients with disorders of consciousness and in healthy controls. The PCI was determined by the quantification of the EEG responses to transcranial magnetic stimulation. PCI has been shown to distinguish between conscious and unconscious subjects: the authors identified an empirical cut-off point that discriminates with 100% sensitivity and specificity between conscious and unconscious conditions, irrespectively of connectedness, responsiveness, and presence of brain lesions, namely between, on the one hand, *unresponsive, unconscious patients,* and on the other, *unresponsive conscious* and *responsive conscious.*[76]

Digitally processed EEG recordings may offer a more precise method of determining consciousness in clinically unresponsive patients in the days following a severe brain injury. A subset of patients has brain activation in response to verbal commands, that can be detected by EEG recordings analyzed by machine learning, raising the possibility of consciousness. The dissociation between behavior and brain activation is known as *cognitive-motor dissociation.* Sixteen (15%) of 104 unresponsive patients had brain activation detected 4 days after an injury. Detection of cognitive-motor dissociation by EEG could inform prognostication of acute brain injury and provide means of communication with patients who seem unresponsive based on conventional clinical examination.

These findings clarify ethical issues that arise surrounding end-of-life care including withholding and withdrawing of treatment, terminal sedation, determination of death, 'do not resuscitate' orders, artificial nutrition and hydration, euthanasia, assisted suicide, as well as the availability and opportunities associated with palliative medicine.

Against this background, it is not difficult to appreciate that, beyond the forensic dimension, unconsciousness raises a major philosophical issue. To investigate the nature of consciousness, why not compare the conscious state with the unconscious one. This has been the task of researchers from the University of Michigan's Center

[76] Casarotto et al. (2016).

for Consciousness Science.[77] What is going on in the brain when it is drifting into unconsciousness?

Anesthesia is not a process that turns off the brain, but that isolates processes in certain areas of the brain. During unconsciousness, disrupted connectivity between different areas of the brain are creating an environment that is inhospitable to this kind of information transfer and integration. The authors propose a measure of integration of information, they christen Φ, and results from studies suggest that it shrinks as the timing of communication across the brain's disparate areas, and the brain gets modular and has more local conversations. The authors think that Φ is closely correlated with the role of the brain in consciousness. They developed a method to estimate the change of Φ and showed it was able to differentiate all states of consciousness. The prediction holds true for pharmacological (anesthesia) and pathological (coma) states of consciousness.[78]

John Searle claims that consciousness is a property of the brain. But when a person's brain is well correlated with this person being conscious, this does not imply that his brain is being conscious too. The brain is not conscious, but it can be said to be the vehicle of these abilities.[79] Persons, dogs or birds are conscious, not their brain.

Conclusion

1. The concept of "Disease" is a conventional construct that is not a natural kind: it involves grouping ill persons into categories that are believed to have utility in the management of their illness and in the circumstances that led to it. The concept of "disease" embodies a necessary and prudential concern for prevention, remedy, redress, or appropriate care.
2. The "Disease" concept is descriptive and prescriptive, scientific, and evaluative, informative and action guiding. It has a spread: it is a cluster of overlapping diseases related to one another by family resemblances.
3. Diseases are instances of "disease" discrete patterns of organic or mental statistically recurring pathological processes or states, undergone by a creature, acknowledged by medical science, whose mechanism is internal to the body of individual organisms, and which may have variable degrees of severity, namely intrinsic suffering, incapacity and/or excess mortality.
4. The sick role is a term regarding sickness and the right and obligations of the affected. A person being ill is temporarily relieved from social responsibilities; he could stay off work and he try to recover as quickly as possible.

[77] Huang et al. (2018) and Kim et al. (2018), and Mashour (2018).

[78] Kim et al. (2018).

[79] Bennett et al. (2007, 136–137).

5. Diseases have a natural history as well as known or unknown causes. Diseases are either manifestational entities or causal entities according to the criteria that are used to categorize ill persons. When feasible, a disease process incorporates a single *a posteriori* necessary cause that defines it. Such diseases are quasi-natural kinds having some real existence in nature although stripped from their time-honoured essentialist flavour. But this is the exception to the rule: most diseases have a multifactorial and open aetiology; their categories are overlapping.
6. At the population level, diseases are scattered on a broad spectrum of severity and the question is not: 'who has got it?' but rather 'how much of it has he got?'.[80] But for clinical purposes, diseases are extreme events, and their ascription is a yes-or-no conventional and instrumental decision forced upon physicians.
7. Diseases are like constellations in the firmament, grouping stars—i.e., patients' illnesses—which seem to form a pattern to which we give a name. Our search for a unified concept defined in terms of some shared property among all individuals afflicted with some disease is essential but it often turns out to be a piece of confusion. Diseases are nothing but our ways of grouping sick people and there are multiple legitimate schemes of classifying afflicted individuals. They depend for their utility on the explicit enunciation of the conventions and the explanatory hypotheses governing their constructions: they are often provisional and merely conventional ways of grouping sick creatures into overlapping categories. Their demarcation lines are drawn on the surface of the world and in that sense, diseases are not part of natural sciences.
8. Death and passive euthanasia raise important issues between on the one hand, unresponsive unconscious patients, and on the other unresponsive conscious and responsive conscious patients.

[80] Rose (1989).

Chapter 10
Mental Disorder

Canst thou not minister to a mind diseas'd,
Pluck from the memory a rooted sorrow,
Raze out the written troubles of the brain,
And with some sweet oblivious antidote
Cleanse the stuff'd bosom, of that perilous stuff
Which weighs upon the heart?

Shakespeare Macbeth 5.3....

Contents

L. Karhausen, *Analytical Philosophy of Medicine*, Philosophy and Medicine 152,
https://doi.org/10.1007/978-3-031-81257-6_10

"*Psychological disorders... I shall define as conditions which–to beg no question–cause a person to seek, or need, or be directed towards the care of a psychiatrist*", writes the Nobel-winning Peter Medawar. Psychiatric diseases are not properly diseases of the mind any more than so-called physical diseases are diseases of the body. Mental disorder, for the philosopher and psychopathologist Derek Bolton, is not a naturalist notion, but one that is focused on harm and suffering, and in which the personal, the social and the biological are hardly clearly distinguished.

Defining psychiatry as the branch of medicine that is concerned with diagnosing and treating diseases of the mind is quite misleading. "*Cartesian dualism is the source of misguided oppositions and assumptions*", writes Eric Matthews. He adds that: "*...human experience and behavior, including disordered experience and behavior, and their explanation are more complex and subtle than is allowed for by the abstract metaphysics of "mind" and "brain"*.

There are a considerable number of books on mental disorders, psychiatry, and the philosophy of psychiatry. However, there is one book that stands out from the rest, because it covers everything not only about psychiatry and the philosophy of psychiatry, but also about the nature of these disciplines and what links them together. This is what makes *Alien Landscapes*[1] by Jonathan Glover a book that should be read before all the others: it covers both psychiatry and its rationality.

The definition of mental disorder used in ICD-10 and DMS-5 acknowledges the existence of patterns of *behaviour* associated with present distress and disability but avoids using the term "mind".[2] So-called mental diseases are diseases or disabilities that affect minded beings, sentient beings, but mostly human beings, even though animals may also suffer from depression, obsessive-compulsive disease, substance abuse and hyperactivity. They are diseases in the same way as non-psychiatric diseases: they are processes that are harmful, have causes, a natural history and they imply a commitment to medical intervention.[3]

Thomas Szasz held that the concept of 'disease' fastens the clinical manifestations with some biological explanation. He assumed that clinical signs in the absence of some patho-physiologic foundation do not pertain to the medical sphere but belong to social and valuation judgements. Yet, schizophrenia is now known to be accompanied by pathological brain structure and function. For example, the ventricles are somewhat larger, there is decreased brain volume in medial temporal areas, and changes are also seen in the hippocampus.

The term 'disorder' "*is used to imply the existence of a clinically recognizable set of symptoms or behaviour that in most cases are associated with distress and with interference with functions, always at the individual level and often at the group or social level (but not the latter only)'.... 'The mental disorder is not,*

[1] Glover (2014).

[2] Matthews (2013, 531–544).

[3] Culver and Gert (1982, 87–108).
Nesse (2019, 128).

therefore, to specify the presence of a disease, but to recognize the presence of the designated syndrome. It does, however, allow hypotheses concerning a pathology or other biological abnormality to be tested."[4]

Mental disorder may actually result from or result in some physical disease: depression often aggravates recurrent or chronic physical diseases and increases the risk of diabetes, osteoporosis, alcoholism, and cardiovascular disease. Conversely, most physical illnesses have a psychological component: 50% of Alzheimer's disease and of multiple sclerosis, 33% of diabetes, cancer and chronic renal diseases suffer from depression. But the demarcation line which divides them looks less like a split than like some wrinkle or crease in the seamless fabric of the world and the differences between physical and mental disease are in all respects gradual as shown by functional disorders.

Two national epidemiologic surveys revealed that 28% of the adult population in the United States meet the full criteria for a mental disorder, addictive disorder, or both in any year.[5] They create a heavy burden on individuals, their families, and the health services. They consist of a heterogeneous group of conditions, which are treated by general practitioners or by psychiatrists and which have no common physiological or psychological features.

Psychoanalysis has been a major step in twentieth century psychiatry. It has had a widespread influence on human thought and conduct. Not only have Freud's major contributions influenced modern medicine and related sciences such as psychology, sociology, and criminology but they also have been reflected in the seven arts. Even so, ever since the historic paper *The Decline of Psychoanalysis* by Jacob Conn published in 1974, psychoanalysis has been progressively discarded from academic psychiatry in most occidental countries, except for France. According to Peter Medawar, Freudian psychoanalysis is a glaring example of the corruptive influence of what he calls *the literary syndrome*. With Lacan and Kristeva, an unsubstantial literature supported a world on the brink with wild proliferation of unchecked speculations which have been competing with science and took over the territories claimed by psychiatry[6]; Wasn't Lacan "*a fraud to extort money from the bourgeoisie injured by patriarchy while building a reputation as an untouchable genius?*"[7] These procedures, according to Peter Medawar, are "*highly mischievous ... because they represent a style of thought that will impede the growth of our understanding of mental illness*".

Psychoanalysis is the lost tribe of psychiatry and will not be discussed any further in this chapter.[8]

---

[4] Wing et al. (1998, 1).

[5] Reger et al. (1993), Kessler et al. (1993).

[6] Grant (2000, 255).

[7] Camiret (2023, 193).

[8] Paris (2017).

Defining Mental Disorder

A mental disorder is a disease or a syndrome, and often a mere set of symptoms, and it shares all the common elements that define a disease, namely a clinically significant construct that is either a process or an enduring state. It supposes victimization of a certain sort by causal forces over which the patient has no control.

On this score, what distinguishes mental disorders from disorders that are not mental, and from mere problems or disturbances of living? To be qualified as a mental disease, a patient must display the following sine qua non characteristics:

(1) *Mental disorders are purported identifiable recurring patterns of presently or potentially intrinsically harmful disturbances of minding, distress, and disability, which befall the whole of a creature, and which occur in clusters and are acknowledged by medical science; they imply distress, mental suffering, disabilities and or excess mortality.*

This usually implies:

- *mental suffering* (present or potential distress such as excessive and irrational anxiety, phobias, compulsions, anxiety or depression in certain situations, obsessive thoughts, irritability, low self-esteem, dissonance, panic, paranoia, pervasive guilt, self-accusation, suicidality, or pain) or absence thereof (anhedonia, de-affectualization, or perverting of pain perception).

Derek Bolton claims that there is something distinctive about the nature of mental suffering but also a distinctive kind of response to that suffering.[9] The problem of mental suffering that people bring to the clinic is distinctive in that it is typically unmanageable or intolerable so that they seek outside expert help, the appropriate help would be healthcare. This can be contrasted with distress that people can manage that can be called 'normal'.

- *Disability or defects*, i.e., limited, or systemic incapacitation hindering the patient's well-being, achievement, growth, and fulfilment, and rendering his life or his relations with other people difficult, i.e., delusional beliefs, illusions or hallucinations and loss of the sense of reality, i.e., *usually incorrigible false beliefs held despite countering evidence or arguments to the contrary.*
- *Comorbidity* as well as *excess mortality* (suicide, homicide, accidents, alcohol, tobacco or drug abuse and the increased risk of immune system disorder and death from cardiovascular diseases). Patients' acts may contravene their own prudential interest, and one per cent of them are very dangerous and poses serious risks to their family and to others, though usually not because of their own wishes or rational choice.

Mental disorders are not harmful malfunctions because "*psychiatric diagnosis lacks a perspective of normal useful functions that physiology provides for the rest of*

[9] Bolton (2017).

medicine".[10] The signs and symptoms of the disorder are not the effects of the disease process but integral parts of it; and so are the disturbances between the individual and society that are not a mere culturally sanctioned deviant behaviour.[11]

(2) *Mental disorders amount to interpersonal disturbances, which befall an individual, a person, or a sentient being seen as a subject and as an object as well as the relation with his social surroundings, and which disrupt the sufferer's relationship with others and the world around him.*

Psychiatry extends the threshold of medical care beyond the boundaries of the body. Mental illness differs from physical illness in that *the perimeter of a mental illness is not limited to the patient's body* (as is usually the case in physical illness), but it includes the relations of the patient with his unique environment and personal life history. Psychosocial impairment, as it seems, is consequential of the disease process in the case of physical disorders but is constitutive in the case of mental disorders.

Mental disorders are thus not a historical invention resulting from power relationships within society and manifested in particular families as held by Foucault. Madness is not the effect of a mere social definition wished by society on a fraction of its population. Madness, as it were, is intrinsic to a person and his relationship with society but it is not an expression of this relationship.

(3) *The concept of mental disorders presumes the presence of both a restricted range of causal, i.e., physical, or mental factors, operating within or outside the boundaries of the individual, and which may be internal or external to the disease process.*

Causal factors result from a plurality of underlying biological, i.e., presumably neurophysiological, and / or psychological malfunctions, and not a mere disharmony between the patient and society. In some cases, a mental disorder may be a self-inflicted condition such as alcohol-induced disorder or drug addiction. Kant thought that mental illnesses take their roots inside the patient's body, even though antecedents in the social environment may play a role in initiating, maintaining, and enhancing the process. Vulnerability to mental disorders is partly genetic: genetic influences and epigenetic factors susceptible to environmental influences might have a role.[12]

But, prima facie, it is the presence or absence of causal factors or of causal chains operating within the boundaries of the individual that set out the departure from normality and that delimitate mental illness from deviant behaviour.

Except for a few specific conditions, mental disorders are not causal but manifestational disorders, and are being classified *pro tempore* according to symptom-orientated criteria and independently of their causes.

---

[10] Nesse (2019, 27).

[11] Whitbeck (1977b), von Wright (1963a), Glover (1970, 118–125).

[12] Hollingshed and Redlich (1958).

Stegenga claims that medicine generally employs pathophysiological or aetiological nosology, while psychiatry employs a symptom-based nosology. This represents a very optimistic outlook of medicine: actually, except for some common conditions, the aetiology, and the pathophysiology of most physical or psychiatric diseases is poorly known and understood, and often out of reach.

In medical care, symptoms and signs are usually evidence of an inner disease process. By contrast, in psychiatry the symptoms of distress are more constitutive of the illness, particularly in distress-related conditions such as anxiety and depression. The mental pathology has a defining role in mental illness. Beyond this, much is known about causes or risk factors in psychiatry: genetic risks, prenatal environment, birth complications, child rearing practice, socio-economic determinants, and the complex role of neurotransmitters, etc.[13]

(4) *Mental disorders imply a need for prudential concern, for help, counselling, remedy, redress, prevention, or care, which is built into the concept of mental illness, even more so, since patients are unable to adjust their behaviour or mental states by any forthright choice.*

This suggests that effective psychological and pharmacological therapy, if not available, ought to, or may in principle be found. It calls on others (health and social services, physician, family, occupational environment) to help those in need of intervention. Very rarely, may the sufferer be unaware of the harmful consequences of his disease and some coercion might be needed to prevent harm to self or to others. Living experiences may be therapeutic.

(5) *Mental disorders, just like physical illnesses, have a temporally spread i.e., a natural history; in their broadest sense, and they include enduring states, i.e., personality disorders and mental handicaps.*

So, *in the first place*, mental disorder as a process may manifest a succession of well-defined stages.[14]

The patient may go through a progressive course with clinically recognizable stages.

(a) A premorbid phase with no evident psychosocial impairment.
(b) An early prodromal syndrome, namely a phase of anomalous subjective experiences with transient feelings of depersonalization, self-perceived disturbances of thought, concentration or attention and initial psychosocial impairment.
(c) A late prodromal phase which manifests attenuated psychotic symptoms (increased suspiciousness, early feeling of change in the sense of the self) and brief intermittent psychotic episodes.
(d) An overt psychotic phase with full-blown symptoms of psychosis likely to develop into schizophrenia.

[13] Schaffner (2013, 1003–1022).

[14] Raballo and Laroi (2009, 366).

All in all, the concept of natural history substitutes a taxonomy of risk to a categorical nosology of all-or-nothing psychiatric diseases.

Secondly, and next to disease processes, we have another possibility that of enduring states or standing conditions such as personality disorders, psychological infirmities and enduring mental malaises characterized by the presence of enduring traits which portray rigid and often stereotyped dispositions to subjective experiences as well as to overtly behavioural or cognitive processes and they may be exaggerations of normal personality traits.

Personality disorders are the most frequent disorders treated by psychiatrists with a prevalence estimated to be in the range of 10% to 13% in community surveys. They seem to have a genetic basis, but they may have their origins in childhood or adolescence. These conditions are chronic, usually lifelong diseases, even though 50% of them improve with time either due to therapy or through the reparations that adult life offers them.[15]

Personality disorders centre on extreme traits or various combinations of those extreme predicative traits: paranoid, dependent, narcissistic, avoidant, obsessive-compulsive or schizo-affective or antisocial personality disorders present themselves as continuous or dimensional concepts comparable to the regularly reviewed boundaries of arterial blood pressure above which hypertension is being diagnosed.

(6) *Contrary to non-psychiatric diseases, mental disorders are characterized by the fact that the patient's attitude to his disease, his feeling or is awareness of being ill, or the complete absence of both, are an integral part of the disorder itself, and is not something additional to be easily corrected (as in pure somatic diseases.*[16] Psychiatry is the only branch of medicine that necessarily has recourse to the whole of an individual, a person, namely a 'self-interpreting animal'.[17]

Breakdown of the Self[18]

What do we mean by a self, by the sense of '*myness*'?[19] David Hume contended that the identity we attribute to the human mind is '*something that really binds our several perceptions together, or only associates their ideas in the imagination*'. We only have access to a succession of states—occurrent thoughts, sensations, emotions &c.—but the idea of an enduring, determinate self is fictitious. There are no real ties between a person's perceptions and experiences, merely a felt one. This is usually called 'the bundle theory of the mind'. The 'self' as the private aspect of

[15] Skodol (2008).

[16] Jaspers (1963, 788).

[17] Nussbaum (1994, 51).

[18] Sims (2003, 209–229), Glover (2014, 369–370).

[19] Glover (1988).

personhood is a bearer of meaning but it is not an entity: you would be wasting your time trying to discover it as an inward experience.

But Hume was grappling with two different problems: the question of unity and the question of identity.

For one thing, he refers to the self as a source of internal unity: this is the question of unity of integrated, *'I-ness'*. He writes, *'We are every moment intimately conscious of what we call ourselves'*. One uses the term 'I' (the non-bodily, mental 'I') to refer to oneself and to refer to one's inner unity: I am a single individual, a particular person. Each one of us lives through a variety of experiences; all these experiences have one thing in common in that they are united by being all in one mind.

But for another, Hume had in mind something very different, namely the question of identity. How are people identified? To identify someone, we may ask him who he is, take account of his facial and bodily appearance, or consider what he claims to remember. What is meant by 'the self', in the present view, i.e., the experience of temporal endurance, is regarded as a source of one's uniqueness, of what makes him identical with himself through inner diversity, and what makes two experiences the same person's experiences.

Introspection does not give a meaning to feeling words, i.e., a special sense which each of us can learn on his own account. *'The use of "I" in such soliloquies,* writes Peter Geach,[20] *is derivative from, parasitic upon, its use in talking to others; when there are no others, "I" is redundant and has no special reference'*. The identity issue supposes one has the capacity to differentiate oneself from others and to objectify one's own experiences. If so, the question of identity presupposes our knowing the existence of other people since this helps to account for a person not being identical with another person.

In the light of this, there are two *prima facie* possibilities of breaking the bundle, that is, of some breakdown of the self.

In dissociative disorders the patient experiences a temporary aberration or departure from identity, i.e., a severe or significant modification of one's sense of self or character, a loss of conscious self-sameness or a fragmentation of consciousness. Dissociative disorders consist of various symptoms such as amnesia (when some areas of memory are being split off or dissociated from conscious awareness) realization (when the world and the others seem unreal), depersonalization (the subjective experience of disturbance in self-image with a feeling of unreality), identity confusion or alteration, or any temporary divergent identity. There is an alteration of the patient's self-conception, or dissolution of what he thinks he is, of what one might call his 'epistemological self'.

Contrariwise, schizophrenics may suffer from ego boundary loss and have difficulties experiencing the 'I'—who they are, their ontological self—and of accepting that their conscious and mental activities are their own. Some of them feel themselves influenced like automatons, with no self-control or self-expression. Self is

[20] Geach (1957, ch 26).

disturbed due to alteration or misinterpretation of self-perceptions with delusional elaboration and loss of reality judgement. They may lack an integrated system of reference: some mental states such as judgement are being divided from another, such as emotions, with resulting incoherence of speech and thought. Yet, there is no splitting of one entire personality from another, with different identities succeeding one another. What is being split or segmentalized is his inner unity of the self, the interconnection between emotional and cognitive states, with various personality fragments coexisting simultaneously.

Beyond this, Julian Jaynes raised the provocative question whether the unitary self could be considered as an illusion created by Western cultural and social ways of viewing reality, developed after the Homeric ages, and philosophically reinforced by Plato's account of mind-body dualism.[21]

The Paradoxes of Solipsism

"*I am sitting*," writes Wittgenstein, "*with a philosopher in the garden he keeps repeating 'I know it's a tree', pointing to a tree near us. Someone else comes along and hears this, and I say, "This guy is not crazy. We're just doing philosophy.*"

Although Wittgenstein was not particularly interested in mental disorders, his philosophy played a major role in the development of the notions of schizophrenia and psychosis in both psychiatry and the philosophy of medicine.[22]

Schizophrenia has traditionally been described as impossible to understand both because of the profoundly strange content of the delusions that are among its main crucial diagnostic symptoms and because of the inconsistent epistemic and practical behaviour of patients. In 1913, the psychiatrist and philosopher Karl Jaspers had declared that many symptoms of mental illness (and delusions) were "incomprehensible" and therefore hardly worth considering except as signs of some other underlying primary disorder. Then, in 1956, Gregory Bateson and his colleagues, Donald Jackson, and Jay Haley, formulated a theory of schizophrenia as arising from double-bind situations in which a person receives different or contradictory messages. Laing put forward a similar explanation for psychosis as early as 1959: the strange behaviour and seemingly confused speech of people experiencing a psychotic episode was ultimately understandable as an attempt to communicate worries and concerns.

In saying this, a new approach surfaced in an academic debate between two philosophers on the following question: what light Wittgenstein's philosophy can shed on the most elusive, the strangest and the most philosophically charged of the 'mental illnesses': schizophrenia.

[21] Jaynes (1976).

[22] Fulford et al. (2006).

On the one hand Louis A. Sass[23] famously commented on the memoirs of Daniel Schreber, a legendary schizophrenic and author of a notorious autobiography. Sass observes that while somewhat milder deviations or insufficiently severe cases are, to a large extent, understandable, it seems strange to speak as if somehow there is some absolute line beyond which all understanding wavers, beyond which all interpretation can only be complete madness or total projection.

Wittgenstein comments: *'One of the main sources of our inability to understand is that we do not have a synoptic view (**übersichtliche Darstellung**) of the use of our words. Our grammar lacks synoptic character. A synoptic representation produces precisely that understanding which consists in "seeing the connections*". Hence the importance of finding and inventing intermediate cases.[24]

Wittgensteinian approaches develop a concept of delusion as a disruption of the enactive and intersubjective constitution of a shared reality. The alteration of experience is a subjectivity of perception, which results in a global experience of egocentrism and derealization. The delusion then converts the perceptual disturbance into a reframing of the perceived world, namely an alleged persecution by ordinary enemies. Through this, a new construction of meaning is established, but in a way that is fundamentally decoupled from the world ordinarily shared with others.[25]

Thus Louis Sass, professor of psychology at Rutgers University and a leading exegete of Wittgenstein, considers that the word '*delusion*', if applied to many cases of schizophrenia, has a quite different grammar from that of 'error': delusions are no longer mere cognitive errors, since they are not univocally believed, but form an alternative delusional system to that of everyday life.[26]

Consider, he tells us, the sequence of Wittgenstein's notes:

> *The greatest happiness for a human being is love. If you say of the schizophrenic: he does not love, he cannot love, he refuses to love, what is the difference?*
>
> *He refuses to love" means it is in his power. And **who** wants to say that?*
>
> *Well, what do we say, "it is in his power"? - We say it in cases where we want to make a distinction. I can lift that weight, but I won't lift **it**; I can't lift that weight.*[27]

The diary of a schizophrenic published in 1950 by the Swiss psychoanalyst M.A. Sechehaye, was written by a young patient diagnosed with schizophrenia. Perhaps one of the most remarkable passages in Renée's autobiography is the one in which she describes a transformation of the perceptual world in which the functional meanings of objects seem to disappear. "*Objects*," she writes, "*are stage props, placed here and there, geometric cubes without meaning*."[28] Such passages capture a perceptual transformation that many individuals with schizophrenia may experience but are unable to describe with much precision.

---

[23] Sass (1994).

[24] Wittgenstein (1968, §122).

[25] Fuchs (2020).

[26] Sass (2007).

[27] Wittgenstein (1980a).

[28] Sechehaye (1970, p. 33).

Moreover, Sass' main claim is as follows:

> *Schreber's mode of experience is strikingly reminiscent of the philosophical doctrine of solipsism, according to which the whole of reality, including the external world and other people, is but a representation appearing to a single individual self, namely the self of the philosopher who defends this doctrine. Many details, all the complexity and contradictions of Schreber's delusional world can be understood in the light of solipsism.*[29]

Solipsism is defined as the position that the self is the only thing one can know or, more extremely, that one's own mind, i.e., one's inner self, is the only thing that exists in the universe. This does not mean that one is the only human being, or mind, that exists, but that all that exists can only be either oneself (conceived as a mind or subject of experience) or a part of oneself (the content of one's mind).

Louis Sass thinks that the *'analogies with solipsism suggest that schizophrenia may be less a Dionysian disease than an Apollonian, or perhaps even Socratic, disease: a matter of the perverse triumph of the mind over the body, the emotions and the external world'*.[30]

During his argument, however, Rupert Read raises difficulties for the project of interpreting people with schizophrenia, difficulties based on the apparent impossibility of a (self-)description of these people. The utterances of schizophrenics are usually interpreted as mere word salad, and it seems that any apparent order is completely illusory.[31]

A profound psychic disturbance may deprive the sufferer of the resources necessary to enable us to make the usual distinctions in our social-psychological interactions with each other. Our conceptual faculties reach a limit, a limit of meaning, but not because of a poverty of concepts on our part, nor even on the part of the sufferer. We are not dealing with a situation like that of understanding an animal, but we are dealing with a system of non-clarity. The task of understanding cannot be "completed" because there is nothing that can be considered completed; even a limited understanding is of much the same sort as the understanding of a nonsensical poem that runs up against a hermeneutical limit that cannot be overcome.[32]

Commenting on Sass, Rupert Read noted that solipsism as characterized by Wittgenstein in the *Tractatus* and the *Philosophical Investigations* is impossible to describe, and incoherent as a concept:

"*There is nothing by which solipsism can be understood. The very idea of solipsism is ultimately a delusion of meaning. One can think of understanding it, one can think of having a clear idea of what it means to think that "only I exist". Wittgenstein's great achievement, in the wonderful therapeutic details of his later work, was to show that we don't have a clear understanding of all this; or rather, to show that*

---

[29] Sass (1994, p.8).

[30] Sass (1994, p. 117).

[31] Read (2001).

[32] Read (2012, p. 138).

there is no "it" here."[33] It is a mistake, as his opponent does, to accept a solipsistic interpretation of schizophrenia.

Schopenhauer, who influenced Wittgenstein, wrote: "*Theoretical egoism (solipsism) is only to be found in a madhouse and is not so much in need of refutation as of treatment.*"[34]

It is significant that the schizophrenic type of world corresponds so closely to the description of solipsism - the illusory doctrine which, for Wittgenstein, is the quintessential example of the philosophical disease, that tendency to overvalue and reify abstract and contemplative thought and to lose touch with the true sources of wisdom that lie in a life of engagement and activity.

Wittgenstein writes: "*'I' does not denote a person; "here" does not denote a place; "this" is not a name. But these words are related to names.*"[35] In other words, "I" can be a subject or an object. Or: "*In any case, I have only this. - What are these words for? They are useless. - Can't we just as well say: "There is no question here of a 'seeing', nor consequently of a 'having', - nor of a subject, nor consequently of an I"? Could I not ask: What you speak of and say that you alone have it, - to what extent do you have it? Do you possess it? You don't even see it. Shouldn't you say that nobody has it? And this too is clear: if you logically exclude that other people have something, it loses its meaning to say that you have it.*".[36]

"*The world is my world; the world and life are one. I am my world (The microcosm)*" writes Wittgenstein.[37] He adds: "*What solipsism* intends *is quite right, only it cannot be* said, *but shows itself*".

Solipsism is thus, according to Wittgenstein, either empty or self-contradictory. Wittgenstein knew very well that the logical impossibility and incoherence of solipsism as a philosophical doctrine can hardly exclude its existence in the real world. He insisted that, despite their absurdity, sentences like 'the only reality is my present experience' correspond to a deep intuition of the centrality of the experience of the self to the world. Although this intuition cannot really be said (because it is absurd), it can somehow be shown.

Let us add that Wittgenstein did not believe that the meanings of concepts could be reduced to a finite set of necessary and sufficient conditions. Even if there are rules of grammar (proper use of language) to be made explicit, these rules should not be rigidly applied. Thus, instead of trying to erase differences and achieve an artificial generality, the conceptual analysis should highlight differences where they have not been seen before.[38] Lilienfeld and Marino link Wittgenstein's notion of *family resemblance* to a special theory of diagnosis: rather than seeking the instantiation of concepts with rigid necessary and sufficient conditions of use, they

---

[33] Read (2003).

[34] Schopenhauer (2009; pp. 251–252).

[35] Wittgenstein (1968, §410).

[36] Wittgenstein (1968, §398).

[37] Wittgenstein (1961, 5.621f).

[38] Aucouturier V, Demazeux S.

approach mental disorders through the characteristics that lead to the choice of one category over another.[39]

Elisabetta Lalumera suggests, in a very lucid essay, that Wittgenstein clarified the phenomenon of schizophrenia in at least three different ways: for the purpose of empathy, scientific explanation or philosophical clarification.[40]

A philosopher and a psychiatrist from the University of Leuven, Van Duppen and Sips, in a study of a Wittgensteinian approach to psychosis, have shown that the delusional mood and the onset of psychosis involve the dismantling of the language game, the way of seeing things. The authors cannot distinguish a clear chronology or etiology and, therefore, find no argument to support one of the two-stage models of delirium formation that claims that perceptual disturbances generate cognitive disturbances or vice versa.[41]

Language games and background beliefs constitute social reality, and they structure our perception of and interaction with others. Yet the habits we have learned from others and incorporated into our most personal ideas, beliefs and behaviours can suddenly lose their evidence in mental disorder. While the experiences of psychosis are often thought to be strange, bizarre, or incomprehensible, the aim of this article is to offer a new step towards a better understanding of how the psychotic process affects a pre-reflective background. The authors use concepts from Wittgenstein's philosophy to clarify the first-person perspective on psychosis. They describe the early psychotic process as a disruption of the "*nest of propositions*", shaking the scaffolding of our language games: "*What I hold fast to is not a proposition but a nest of propositions*", writes Wittgenstein.[42]

Thus, the pre-reflective context that forms our existential orientation in the world is fundamentally altered. This psychopathological process transgresses the boundaries of language games, imposing a multiplicity of perspectives on reality, leading to the experience of groundlessness and blind spot biases. According to the authors, the perceptual, cognitive, pre-reflective and reflective aspects of psychosis are closely interwoven. Together these alterations lead to a radical reorientation of the experienced world.

Wittgenstein describes the flow of thought of the "philosophically perplexed man": "*The philosophically perplexed man sees a law in the way a word is used and, in trying to apply this law consistently, he comes across cases where it leads to paradoxical results. Very often, the discussion of such a conundrum proceeds as follows: First, the question is asked: "What is time?" This question gives the impression that what we want is a definition. We mistakenly think that a definition is what will remove the problem (like in some states of indigestion, we feel a kind of hunger that cannot be removed by eating). The question is then answered with a wrong definition, e.g. "Time is the movement of celestial bodies". The next step is to*

[39] Lilienfeld and Marino (1995).

[40] Lalumera (2017).

[41] Van Duppen and Sips (2018).

[42] Wittgenstein L. On certainty. Blackwell. (1969a, §225).

see that this definition is not satisfactory. But this only means that we are not using the word "time" as a synonym for "motion of celestial bodies". Now, by saying that the first definition is wrong, we are now tempted to think that it must be replaced by another, the right one."[43]

Recognizing and exploring the depth and impact of this process on a person's world can be a first step towards resolving isolation and suffering. Philosophy can facilitate such an exploration, while interpersonal therapeutic activation can provide structure and confidence in the world, helping the patient to find a solid foundation in action and interaction.

The treatment of psychosis focuses mainly on acute positive symptoms, such as hallucinations and delusions, and therefore mostly neglects possible antecedent alterations that may go beyond positive symptoms. If, according to Wittgenstein, the foundation of our language games is an unfounded way of acting, and if we consider that the psychotic process is able to break the "nest of propositions" that forms the language games, then therapy for psychosis should focus in particular on shared interpersonal activities - activities which Wittgenstein regards as unfounded but which, by their interpersonal character, can rebuild fundamental trust in others, in the world and in oneself, and thus offer at least a basis for recovery.

What precedes combines a philosophical approach with a first-person perspective that illuminates aspects of psychosis that have not been described or elaborated before: psychosis involves an experience of existential groundlessness.

Rational Breakdown

Hilary Putnam[44] argued that rationality and reasonableness couldn't be reduced to formal rules such as those of logic or mathematics. Rationality is an ability, which does not proceed as exactly determined by formal or by general rules. It is not defined by some set of rigid, ahistorical, algorithmic, standardized canons or principles by which the premises or some facts entail the conclusions. Yet, this does not mean that anything goes!

The notion of rationality sounds like those of health, function, physiology, or normality, in that it is a default state. Rationality is a privative term and irrationality, or a-rationality have logical priority. Admittedly, there is a lack of strict dividing line between perfect rationality and extreme non-rationality. It follows that rational actions must be defined in terms of a-rational actions, and rational beliefs or desires in terms of a-rational beliefs or desires.[45]

Most a-rational or irrational ideas or delusions are rather mundane. We may all be sometimes incoherent, senseless, be fatigued, believe foolish things, fail to

[43] Wittgenstein L. The Blue and Brown Book. Blackwell. 1968.

[44] Putnam (1981, 1983).

[45] Bortolotti (2013).

perform, be down in the dumps and act in unreasonable ways, but generally such ways of behaving respond to reasons and are susceptible of an explanation to make them understandable and intelligible, and make them seem rational. An angry person is temporarily fully irrational. Irrational acts or beliefs are acts that are not based on reasons making a self-consistent whole. Anyone is entitled to hold irrational beliefs or to behave in ways other people might judge to be irrational and to be irresponsible even though this might be at the cost of ending up in prison: sane human beings are responsible, and this supposes a minimum of rationality.

Irrational behaviour is scalar: one may be more, or less irrational, and severe psychotics lie at one end of the continuum. Hence, there are degrees in this a-rationality, and it is not a one-dimensional concept. Practical reason assesses the means to realize one's goals, but a-rationality seems to have some normative foundation as it also pertains to the worth of goals. The frequency of a-rational traits, their severity, their duration, and the extent to which they impair the individual's ability to maintain interpersonal, familial, and professional relationships defines the severity of the disease.[46]

Kant identified three kinds of psychiatric a-rationality (*Aberwitz*): first, a person with a disturbed brain may manifest perceptions that have no basis in objective reality (hallucinations: *Verrückung*) and in which he properly dreams while being awake; second, he may have delusions, namely disturbances of the imagination and of the ability to appreciate, evaluate and judge those false perceptions (*Wahnsinn*); finally, there are thought disorders (*Gebrechen*) regarding general judgements (*Wahnwitz*), such as hypochondria, melancholia, depression (*Trübsinniger*). What's more, Kant contends, a patient suffering from psychiatric disequilibrium (*Gestörtes Gemüth*) believes that his delusions and his false sensory perception are the truth as he resists all arguments.[47]

A deficiency of deductive reasoning, namely of the proper use of the laws of logic was suggested by Von Domarus.[48] Carl Schneider[49] described some aspects of the nature of this 'pathology of logic': *verschmelzung* (fusion), *faseln* (muddling), *entgleiten* (snapping of) and *entgleiten* (derailment.) It follows that one of the manifestations of mental disorder is not that of being a-rational, but rather of being non-rational or partially rational. They constitute an intrinsic disturbance of thinking, which encompasses concrete thinking, self-reference, and loosening of associations. Cognitive-behavioural therapy is an essential adjunct to medication as it attempts to help the patient to enhance rational capacities, reorient and enhance his self-knowledge and understanding of his situation.[50]

Even so, the delusions of people with schizophrenia cannot be properly explained in terms of impairments or truncation of formal reasoning: actually, schizophrenics

[46] Radden (1985, 43–71).

[47] Kant (1905).

[48] Von Domarus (1944).

[49] Schneider (1930).

[50] Rathod et al. (2010).

might be more logical than healthy volunteers. In a case-control study, Owen, Cutting and David[51] have shown that under conditions where common sense and logic are in conflict, people with schizophrenia reason more logically than healthy individuals. The authors suggest that it is because they are worse at common sense: they might have a bias towards theoretical reasoning over and above practical rationality. With this move, this exploratory study portends that the plight of schizophrenia pertains to the breakdown of practical reason rather than that of mere cognitive theoretical rationality.

Summing up, Culver and Gert subscribe to the view that a person's action, desire, or belief is a-rational if it is harmful to himself and if he has no adequate reason for doing it (or holding it) that explains why he is doing it (or holding it).[52]

Syntactic Breakdown

Psychotics experience a breakdown of the integrity and the self-consistency of intentional processes and failure of such intentional processes may occur at different levels which spread from desires, motives, fear, actual decisions as well as what links decisions with action; disruptions at different points of this course of events allow for a variety of mental diseases: suffering from diminished ability to engage in the social rituals of everyday life, anhedonia or loss of feelings, bulimia or anorexia, attentional malfunction, apathy and aboulia, hallucinations and thought insertion or any kind of inability to initiate and persist in intentional activities; schizophrenia is manifested by a conflict of perceptual experiences or a persistent failure in the perception of reality resulting in disarray in action resulting in unpredictability and loss of control.

King Lear chronicles the dissolution and the disorganization of 'higher' processes and a psychic reorganization at a 'lower' level. Schizophrenics often manifests fantasies, magical thinking and illogical ideas and they may be drowning in sensory stimuli, with hallucinations, auditory or visual, unconnected experiences, shifting content, attention deficit, incoherent and chaotic system of beliefs, which represent the breakdown of perception and thought processes, which unravel on an underpinning emotional background; these experiences may be threatening, helpful, religious, ceremonious, musical; they may make the patient appear confused as he develops a gross distortion of reality.

Rochester and Martin[53] examined schizophrenic speech, looking at what he called *cohesive ties*, namely the links between sentences in the discourse. Such patients use much less of these links and use weaker bonds that make it more difficult to follow the meaning of the sentences: they use less *reference* ties, namely

[51] Owen et al. (2007).

[52] Culver et al. (1982, 20–41), Martin (1986).

[53] Rochester and Martin (1979).

linking through meaning, and more *lexical* ties, connecting words. Compared with manic or normal controls, schizophrenic patients showed more syntactical and semantic mistakes and less fluency.

Semantic Breakdown

We know that there are natural barriers to empathic leaps across the boundaries of the self into the world of the other, namely between two horizons of relevance or from a third person into a first-person account. If relations between the subject and the external world regulate mental states, a mental disorder may be recognized from any comprehensive breakdown or temporary impairment or truncation of the connection between mental states and the non-mental world, and of the ability to understand one another through sympathetic identification.

Autistic people may have difficulties to understand the world in ways other people do. They may find it strange and difficult to understand. The utterances of psychotics may be so weird that they flout even empathic understanding.

Actually, if one fails to understand or misunderstands someone else's speech or behaviour because it is unintelligible, this does not, in itself, imply that the person is mad. Misunderstanding does not imply madness. A psychiatrist's first-person ascription of madness to someone is legitimate only if this lack of understanding is generally shared. Incomprehensibility is conceptually related to third-person ascription of madness only if the unintelligibility is public.

"*If lions could speak, we would not understand them*", wrote Wittgenstein.[54] It may very well be that we could not understand the lion because even if he could utter grammatically correct sentences, his way of understanding the word would presumably be too radically different from ours.[55] Wittgenstein's remark points out the incommensurableness of our lives with the phenomenological world of other species and our consequent inability to imagine what it is like for a lion to be a lion or for a bat to be a bat.

For Thomas Nagel, having conscious experiences means that there is something that it is like to be a creature of a certain type with that experience: the subjectivity of experiences is perspectival, as it can only be understood from the standpoint of relevant creatures, human beings, bats or lions. We have not and cannot have any access to another sentient being's subjective experience.[56] Yet, to point this out is to wonder whether our stance towards the lion is comparable to our standpoint towards the incomprehensibility of an individual with schizophrenia or delirium. A severe psychotic and we are strangers. But there is a further consideration: contrary to Nagel's lion, and, since a patient shares with us the same language, his mental

[54] Wittgenstein (1953, 223).

[55] Pitcher Gn (1964, 20).

[56] Nagel (1979a, 165).

disorder may take the form of being unintelligible or incomprehensible to himself; this makes him incapable of rational self-examination with consequent loosened emotional or mood connections.

Two consequences follow from this analysis.

First of all, Wittgenstein writes that "*human beings ... agree in the language they use*"[57] and that sharing a language, "*is not agreement in opinion but in forms of life*". Understanding a language quite strange to us presupposes a convergence not so much of beliefs, but rather of patterns of pre-intentional features which we take for granted and which include shared emotions, perceptual and understanding abilities, tacit assumptions, dispositions, an implicit minimal rationality, what Wilfrid Sellars called a *space of reasons*. It is only within the framework of some common implicit agreement, of some way of seeing things that we understand each other.

But in the second place, the point of Wittgenstein's remark is that even if the lion were using a language we know, he would still be an enigma to us. The intentionality evinced by animals and mentally ill humans, relates these temporary bedfellows to the reality of the world.

In what way does our difficulty of contacting a madman differs from that of a lion? When a person's behaviour seems to make no rational sense is unintelligible, we are inclined to call such person and such behaviour incomprehensible. We cannot reduce the first-person account into a third-person account. But it is at this juncture that the lion and the patient diverge from one another.

In one way, the patient is maladapted because of the chaotic texture of his interrelated intentional states, a quasi-syntactic feature, and a persistent misrepresentation of his own self and the world, a semantic feature. This springs from the intrinsic disruption and conflicts of cognitive and intentional features, as well as propositional awareness, which are constitutive of mental illness.

The behaviour and verbalization of patients suffering from severe mental disease may be incomprehensible, incoherent, or unpredictable and seem to have no rational meaning to observers. Disordered language and speech include derailment, tangentiality, neologisms, breaks or interpolation in the train of thoughts, resulting in incoherent or irrelevant speech. Laing writes: "*One had for long periods that uncanny feeling, described by the German clinicians, i.e. of being in the presence of another human being and yet feeling that there was no one there*".[58] Karl Jaspers thought that the quintessence of schizophrenia is its un-understandability.[59] Kathleen Wilkes writes that "*it is difficult if not impossible to see the world through the mind of a schizophrenic, in terms of which the way he behaves seems, and might be shown to be quite rational; and he for his part has temporarily, or perhaps permanently, lost the ability to see the world as we do*".[60]

57 Wittgenstein (1953, 241).

58 Laing (1965, 195).

59 Jaspers (1962, 577).

60 Wilkes (1988, 90).

But in another way, it is the great merit of Ronald Laing[61] to have shown that within the cognitive disarray that characterizes patients with schizophrenia, their symptoms and signs and the content of their delusions might be more socially intelligible and understandable than usually supposed and could be decoded once their sources in the patient's conflicting experiences of others are recognized.

What clearly emerges from this is that the incomprehensibility of psychotic patients arises from a twofold failure, internal and external. But not the lion, and not the bat, I should argue. Thomas Nagel points out that lions and bats are probably aware of facts "*that do not consist in the truth of propositions expressible in a human language.*" But schizophrenics are human beings suffering from a disrupted semantic access to the facts. "*The mystic sees the ineffable,* writes Somerset Maugham, *and the psychopathologist the unspeakable.*"[62]

Epistemic Breakdown

In everyday life, most mental occurrences or dispositions have an intrinsic tendency to represent actual or hypothetical states of affairs related to the world or themselves, which may be present or future, probable or improbable, desirable, or undesirable. Intentionality is this cognitive ability to represent to ourselves our present as well as possible worlds. Content (or epistemic appearances) is what is specified by the content sentence, a 'that' clause or a sentence complement in the ascription of some mental state, what someone feels sad, happy, amused, or angry about.

Psychotic disease such as schizophrenia can thus be seen as the realization of pathological epistemic processes. What characterizes misperceptions, delusions and hallucinations is that they are about objects that do not exist or else that they seem not to arise from missing objects but by the way real objects are perceived, pictured, described, or thought about. Hallucinations, just like illusions, belong in the same category as mental pictures or afterimages: they have descriptive content and are not distinguishable by internal criteria from perceptions, which have an outside cause. Yet they deceive the patient since their content lacks semantic value: they deceive their bearer because of the absence of a proper relation of signs to the objects they describe whether or not the experience is being taken as the perception of an external object.

The upshot is that mental illness is a 'pathology of reality', which pertains to an externalist standpoint: it consists in this unstructuring of the mental states which precipitates the patient 'really' and totally into the unreality of the imaginary, a fall which will tend to be lived and thought of as real, namely not abnormal as far as the patient is concerned. "*Therefore, the subtle and pervasive adjustments of thought that allow us to keep track of events and to use interactions with others to update*

---

[61] Laing (1965).

[62] Maugham (1919, Ch 1).

and correct our lived narratives of experience are disconnected from the guides and signposts that usually orient the mind."[63]

In genuine mental disorders such as schizophrenia, mental experiences may tend to become kind of a screen between the subject and the world in such a way that their truth or falsity becomes an external feature as if the context of intension progressively was taking over the context of extension. The subject is like being enclosed in Plato's cave progressively unable to know whether having this or that experiences tallies to anything happening in reality. For Jaspers, delusionality is the core feature of psychosis. The patient then tends to see himself as a condition of the world which becomes his world, and he becomes unable to think of the world in abstraction from his own relation to it; other people's experiences are not completely real like his own; he finds himself not quite able to bring his subjective first-person conception of the world in harmony with the reality of the third-person account.

Solipsism is the unwieldy philosophical theory that there is nothing in the world other than myself and my perceptions and my ideas. Bertrand Russell writes that the world of a solipsist will be one of disjointed fragments, which change completely from moment to moment.[64] Philosophically, solipsism is inconsistent as it relies on some assumptions, which entail its falsity.

Mental illness, e.g., depression, attention-deficit disease, or schizophrenia, often pushes one towards solipsism. Only exceptionally, i.e., in severe psychotic conditions, does it become what Russell called 'drastic'. This kind of solipsism in not a question of knowledge or the outcome of a Cartesian doubt, wondering about how much of what we believe might really be illusion. This drift towards a kind of solipsism is rather a way of looking at the world, an epistemic stance, inasmuch as it is not a *seeing that* but rather a *seeing as*.

Breaking the Rules of the Game Whilst the Game Is in Play

> *I am sitting with a philosopher in the garden; he says again and again 'I know that that's a tree, pointing to a tree that is near us. Someone else arrives and hears this, and I tell him: 'This fellow isn't insane. We are only doing philosophy.*[65]

There is a tradition, which started with Descartes, which holds that the nature and extent of human knowledge is the fundamental topic of philosophy. So how can we be sure that our claim to knowledge or our beliefs are true? According to the standard analysis of knowledge, one knows that *p*, if one believes that *p* is true, and if there is good evidence for believing *p*.It follows that one cannot claim to know that a proposition is true, if there are any grounds for doubting it. But are there such

[63] Ey (1969, 136), Gillett (1999, 33).

[64] Russell (1973).

[65] *OC* refers to Ludwig Wittgenstein numbered last notes collected under the title *On Certainty*.

things about which we are infallible? Aren't we living in a world of uncertainty? In order to escape sceptical objections about the bounds of knowledge, our statements and beliefs need to be justified, namely they must be derived from a fulcrum of propositions that are immune to rational doubt.Is there something wrong in such characterization of knowledge and certainty? And if so, is there any alternate possibility? Enters here Wittgenstein who undermines a long tradition and contends that we have been misconstruing the concept of knowledge: the difficulty before us is that we have been working with a wrong position about knowledge which leads us to seek situations about which we are infallible; but plainly, the realm of knowledge and justification is the fallible area where a mistake is possible. It follows that there is a fundamental difference between knowing and being certain.

What is the point at issue?

Wittgenstein contrasts knowledge with certainty. Not subjective certainty, which is personal conviction but objective certainty, i.e. certainty not based on grounds, even were they compelling grounds. The only objective certainty distinct from knowledge is a certainty, which would not depend on justification, for which we do not need to provide grounds. The truth of certain empirical propositions, he avers, belongs to our frame of reference. At the foundation of any well-formulated belief lies a belief that is not founded. With this move, certainty is groundless, logical, non-epistemic.

But what are these propositions, which are held to be certain?

The first thing to notice is that if one did believe that the earth had been in existence some short time ago, then he would not only be bringing into question all the facts of history, geology, and much else besides, but he would be doubting the whole system of evidence and verification in these areas. The contention that the earth existed long before my birth isn't supported by evidence, yet it is unshakably certain: it is not a matter of knowledge. No experience, no observation could ever justify them or count as grounds for them or evidence against them. In like manner, our certainty, our assurance of existing, of having a body, of having parents, of not being alone in the world, or that the world exists, is not due to reasoning: it cannot be justified since it was never verified. Such claims are not falsifiable propositions or empirical observations but logical certainties that unquestionably and ineffably stand fast for normal human beings.

But there is a second consideration: although these propositions look like being empirical, they constitute the ungrounded necessary pragmatic basis of our knowledge since they belong to the framework of our thinking and our behaving. In learning to speak about tables and chairs, a child doesn't first have to learn that they exist or assume that they do. "*Remember, they are learning to do something...Any explanation has its foundation in training*".[66] So the natural starting point of thought is not in language but in our instinctive actions and reactions. Goethe wrote: '*in the beginning was the deed*', i.e., the beginning is what we find natural to do. So, at the end, there are no unfounded propositions but rather ungrounded ways of acting. The

[66] Wittgenstein (1981, 416, 419).

system does not rest on what we can know to be true, but on what we do. In this sense, knowledge requires that the instinctive or animal part of man, as opposed to the ratiocinative part, be given a proper place. The propositions lying at the foundation of any discourse are not indubitable or self-justified statements but are either animal certainties, non-reflective, i.e., instinctual certainties (I have a body), or else, they have become acquired automatic certainties.

Our primitive ordering of the world, which John Searle calls *the Background,*[67] is non-conceptual, but belongs to the realm of instinct and action. The world is not primitively embraceable in words or thought we embrace it non-intellectually and then move to a more sophisticated grasp. "*I want to say: it' s not that on some points men know the truth with perfect certainty. No; perfect certainty is only a matter of their attitude*". So, the relation towards these certainties is not of an epistemic nature: it springs not from something we know but from something we do. Certainty, then, has the nature of a propositional inarticulate trust.

Normal Malcolm[68] holds that "*The framework' within which one's thinking occurs is constituted in part by propositions that are contingent and, in a sense, empirical; Propositions of the form of empirical propositions and not only propositions of logic belong to the foundation of all operating with thoughts (with language).*"[69]

But if someone were claiming that the earth is 100 years old, the situation would be categorically different, since the world would then be unintelligible to us. Not surprisingly, if somebody believes he is dead (*Cotard's syndrome*), or doubts he has a body,[70] doubt is misplaced inasmuch as a mistake is not possible. In like manner if he thinks that his body is made of stone or of glass, this string of words is not an error but is unintelligible.

So, accordingly, there is a difference between a mistake and an anomaly such as a mental disturbance: there are no grounds that can justify such beliefs as mistakes, although there is a cause, which explains them, i.e., a mental disturbance. It follows that in some cases, a doubt, a mistake, any hesitation is a sign of madness. "*But the absence of doubt and the unintelligibility of doubt are very different things*" writes Malcolm[71,,72]

It is logically impossible to doubt or be wrong about some beliefs while remaining within the ken of normal human understanding. The point at issue is thus that an individual whose thoughts and action earnestly cross the threshold of these bounds of sense, that is, shrugs off or ignores those 'framework' propositions against which all other propositions are being judged, is not manifesting some sort of scepticism or of uncertainty but a pathological delusional belief i.e. a mental disorder. If

[67] Searle (1992), Wittgenstein (1953, 206).

[68] Malcolm (1977, 194).

[69] Stroll (2007, 39).

[70] Malcolm (1977, 82).

[71] Malcolm (1977, 186).

[72] Moyal-Sharrock (2007, 197).

someone claims to be Jesus Christ, this is not something, which may be questioned, something that may be doubted, but it is evidential of mental disorder. Such doubt then arrests the game: it is a sign of the speaker's being demented.

Mood Disorders

Mood is somehow quite distinct from intellectual understanding. Mood disorders refer to a sustained emotion that colours the way we view life. Acknowledging when mood is disordered is extremely important because as many as 20% of adult women and 10% of adult men may have the experience at some time during their lives. The prevalence of mood disorders seems to be increasing in both sexes, accounting for half or more of mental health practice. Mood disorders are more common among those who are single and who have no "significant" other. It is also more frequent in someone who has relatives with similar problems. Mood disorders include depressive disorders, bipolar disorders, personality disorders, schizoaffective disorder, adjustment disorder with depressed mood and uncomplicated bereavement.

Moral Identity, Responsibility, and the Sick-Role

The essence of disease, be it physical or mental, is helplessness and victimization.[73] We say that someone presenting with distressing and uncontrollable obsessions and compulsions that he suffers from obsessive-compulsive disease. We see the patient as laid low by forces, which he does not control. Such a disease is thus, in a critical respect, caused and undergone rather than decided upon.

But surely, the distinctions between the objective and the subjective stance, between the third person and the first-person epistemic access, are drawn by pointing out that our ways of describing and discussing normal, everyday personal interactions are quite different from the ways in which we talk about abnormal reactions and their appeal for a causal explanation. When faced with some deviation from normal functioning, we either speak the language of *choice and responsibility* in intentional or motivational terms, or else we abandon it completely and speak exclusively in *causal* terms: these are two logically distinct reckonings of human behaviour. Both kinds of accounts can answer for many of our psychological traits although not everything that is explainable by the first stance may be adduced by the second one and conversely. We thus need both logical pleas for a full account of interactive transactions.

[73] Margolis (1966, 11).

Mental illness is the paradigm of those appropriate exempting conditions by which someone is temporarily or permanently discharged from the basic demands made in the first place.

In a certain sense, the sick role penetrates mental disorders. Psychiatric diseases are diseases of mentation and the interconnectedness with the social world is part of mentation, hence of its disorders. It follows that the sick role is an intrinsic feature insofar as it is constitutive of a mental disorder. Siegler and Osmond differentiate between the sick role and what they call the psych role. "*The sick role*, they write, *deals with what one "has"; the psych role with what one "is"*." They argue that this dichotomy is much more useful than the corresponding body / mind dichotomy as it substitutes a choice of role to an ontological split.[74]

Responsibility is a scalar concept in psychiatry. Persons under the influence of alcohol or of certain drugs, persons with severe confusional or obsessional conditions, or some psychotic patients may be more or less unable – not merely unwilling – to discriminate between right and wrong and to evaluate appropriately the nature and the consequences of the acts they commit. On this score, Nagel subscribes to the view that there is continuity between the familiar interactive attitude and the objective critical viewpoint just as there is a continuum between normal and severely mentally ill persons.[75]

Mental illness is thus intimately connected in clinical medicine with prudential interests touching on the use of one's body and person and capacity to choose, i.e., whether one could have done otherwise if other choices had not been interfered with: we see the patient as "*incapacitated in some or all respects for ordinary personal relationships*"[76] because he may be deficient in appreciative judgement, rationality, or self-control.

In contrary manner, the forensic concept of insanity is not a medical condition but an altered legal status, which may allegedly affect moral reasoning, under which some mentally ill persons are not legally responsible for otherwise criminal acts: it is an all-or-nothing feature (although impaired responsibility may be seen as disculpating).

The term "forensic" is derived from the Latin term *forum* and means 'public'. The famous nineteenth century *M'Naghten Rule* provides a rule bearing on the legal protection of personal and institutional interests against the transgress of others. For exempting someone from legal responsibilities, it requires that it be "*clearly proved that at the time of the act, the accused was labouring under such a defect of reasoning as not to know the nature and quality of his act, or, if he did, that he did not know that what he was doing was wrong*".

---

[74] Siegler and Osmond (1974, 48).

[75] Nagel (1979a).

[76] Strawson (1974, 12).

Breakdown of Autonomy and Enforced Treatment

Autonomous agents (animals, human beings or children responding to the direction of adults) are beings who act on reasons they have chosen, and which are their own: their cognitive and volitional states are free from extraneous influences and are not restricted by constraint or external factors. Autonomous behaviour is thus a biological feature of the animal kingdom (and more characteristically of sentient creatures that is incipient persons), which supposes a minimal rationality and purports to the capacity of self-control and procedural independence leading to choosing for oneself what would be the best course of action.

To be sure, there is some confusion in the use made of the term 'autonomy' in moral and political philosophy[77] and it is often used in normative philosophical examination: autonomy and liberty should be kept separate. Autonomy refers to self-determination, the ability to select one's path through life in a range of circumstances, while liberty is concerned with freedom from any coercion, restrictions, and external constraints.

But surely, even though autonomous behaviour evolves in an interpersonal context (as the cultural penetrates the mental), it allows some freedom from social constraints and the ability to decide for oneself. Autonomy holds that human beings are the best judges of their own best interests and should accordingly mix with other people and engage in activities willingly, without coercion. "*Autonomy* writes Dworkin, *is conceived of as a second-order capacity of persons to reflect critically upon their first-order preferences, desires, wishes, and so forth and the capacity to accept or attempt to change these in light of higher-order preferences or values*".[78]

It follows that mental or emotional disability, depressive states, delusions or gross inconsistencies in thought patterns may affect a patient's capacity for judgement regarding his need for treatment; if he is too incapacitated, such features tend to intermittently impair his autonomy in various degrees, resulting in a "*disposition to behave contrary to one's own presumed prudential interests and / or contrary to those of society at large*".[79] On this score, a person in the grip of delirium or of some mental disorder is not responsive to reason, and is then not responsible for what he does.

For one thing, the therapist's role is that of enhancing the patient's autonomy, his reason responsiveness, by fostering his ability to decide independently as a self-directing person by means of informed consent. But for another, there are rare cases—such as some instances of anorexia nervosa, cases of dangerousness to self or if the patient is hopeless at understanding his condition—where autonomy of thought, or will, may be sufficiently and comprehensively impaired, that involuntary psychiatric treatment through persuasion or even coercion may be justified by considering either the patient's best interests or his self-harming resolutions.

[77] Dworkin (1988, 5–6).

[78] Dworkin (1988, 20).

[79] Margolis (1975, 114).

The Interactive Stance: Breakdown of Interpersonal Relations

What is an appropriate relation between a physician and his patient? Surely, objectivity is unavoidable since it stems from shared, intersubjective experiences and knowledge. A physician often merely attends his patient's physical disease and pays little attention to the person. This is bad practice, but it is not uncommon. In most cases this has little consequences on the outcome of the illness in question. However, a psychiatrist who sees his patient as the mere holder of a role, and who would not complement his objective attitude by some reciprocal personal relationship will neither cure nor heal; he is likely to fail if he does not see his patient both as object and subject, and if he does not connect with his patient empathically, because the disease process is not limited to the patient's body but spills over into interpersonal transactions.

We have seen that one of the things, which distinguishes mental from physical processes, is that the social environment is constitutive of an individual's mental life. For Thomas Szasz, mental illness is essentially a problem of abnormal communication. Strawson held that a world without any interpersonal responses would be a world without persons: if we were to constantly surrender our inter-personal attitudes, we would give up holding others as well as ourselves as subjects. We would eventually find ourselves in a world of connected objects instead of relating subjects. Even the medical notions of abnormality, disease, treatment, or control would suffer loss of sense since there would be no more subjects and objects, doctor and patient, psychiatrist, and madman. In a similar spirit, Kant writes, "*a person is the subject whose actions are capable of being imputed*".[80]

Human beings come to understand each other both as subject and object. One way of being a human being, namely the personal attitude, implies being party to a convention: it involves patterns of intelligible attitudes between individuals in which each considers the other as a subject and in which each has a disposition for the cognizance of new facets of the other. We expect and demand a reaction of some predictable kind will occur in our daily encounters with other people unless something prevents it from happening. P.F. Strawson calls *reactive attitudes,*[81] those range of feelings and reactions that we are subject to in our dealings with each other such as approbation and indignation, pride, guilt, and shame, hurt feelings and resentment, gratitude and indifference, love and repulsion, sympathy, and antipathy, praising and blaming which are essential parts of the weave of our life.

Meanwhile, in some circumstances, we may be related to a person in an impersonal relationship for instance by seeing him or her as an object of social policy, of public health programme or of some forms of medical treatment. What distinguishes the mentally ill individual is that he defuses our reactive attitudes and forces us to adopt to such fellow beings the objective attitude: this is a criterion for mental illness. In such cases, as well as if someone is either morally incompetent,

[80] Kant (1964, 23).

[81] Strawson (1974).

psychologically abnormal or under a heavy bout of drinking, we suspend this inter-subjective way of dealing as well as our attitude of involvement and participation in interpersonal relationships. If so then, our resentment and blaming attitude are relinquished and we envision our interlocutor extensionally as some object to be worked upon and opt into a detached attitude Strawson calls *objective*. This objective attitude, a kind of moral distancing,[82] a primitive 'reaction of recoil',[83] cannot include reactive attitudes: our indignation, our puzzlement or our resentment may come to be mitigated by excuses e.g., if we become aware that someone 'didn't know' or 'didn't realize', 'couldn't do otherwise' and suchlike. "*I sometimes permit myself to feel anger and outrage at normal criminals, whereas I cannot help feeling some pity (mixed, perhaps, with repugnance) toward those whose conduct appears bizarre and unnatural*" writes Feinberg.[84]

Two consequences follow from this analysis.

It is one thing that, at that point, a psychiatrist adopts the objective attitude and sees his patients as objects of knowledge. He looks at signs and symptoms and at coexisting biological, biomedical, psychological, and social characteristics. The suspension of his reactive attitudes calls forth some causal explanation of why the person displays some abnormal, non-conventional and clearly pointless behaviour and reactions. But it is quite another that the psychotherapist cannot escape some degree of involvement and participation in the interpersonal relationship. Surely, the success of his enterprise supposes that in handling his patient, he establishes with him a kind of empathic understanding. It implies a genuine acceptance of how the patient feels. Dieter Wyss[85] writes: "*The difficulty facing psychotherapy here, the need to strike a balance between perceiving and being devoted, the need to see through the patient whilst yet warmly accepting him, the need to regard him not only as an object of transference but also as a human being in his own right, should not be underestimated. The point of balance, it would seem, is no sooner established than it is lost again.*"

Seeing Faces and Making Faces

In 1563, Giambattista Porta (1540–1615) published a work on *physiognomy.* The caricatures of the human head presented faces explained through analogies with those of animals, suggesting a similarity of natural predispositions in both.

Johann Kaspar Lavater (1741–1801) distinguished between physiognomy and pathognomy.

[82] Glover (1977, 286–297).

[83] Wittgenstein (1980a, 190).

[84] Feinberg (1970, 283).

[85] Wyss D, in: Havens LL, 359.

Physiognomy is the study of stable facial and bodily features, such as the cranium, the forehead, the nose, and the eyes and mouth in a state of repose, as signs of lasting traits of character. *Pathognomy* is the study of passing facial expressions and bodily motions as signs of passing emotions and states of mind, such as anger or fear or shame, as well as the ability to mimic emotions. "*Physiognomy therefore teaches the knowledge of the character at rest,*" as Gilbert Austin put it in 1806, "*and pathognomy, of the character in motion*".

The American psychologist Paul Ekman (1934–) focused on visual illustration techniques and the measurement of non-verbal communications such as hand movements or often faster and more complex facial movements. According to him, body language and the visual representation of facial expressions activate subtle signs that express emotions. Emotions are a kind of '*thought without thought*', their expression universal: people from different cultures agree on which feelings, emotions or moods are shown in each expression. Human beings can assume more than 10,000 facial expressions, 3000 of which are related to emotions.

"*The idea ...,*" writes Stanley Cavell, "*is that expressions, the human figure, in order to be grasped must be read ... the human body is the best image of the soul, and not primarily, I might add, because it represents the soul, but because it expresses it. The body is the field of expression of the soul'*.

For Wittgenstein, the perception of appearance depends on sight, not interpretation. "*We see emotions,* Wittgenstein writes: "*Contrary to what? We do not see facial contortions and make inferences based on them (like a doctor creating a diagnosis) to arrive at joy, mourning, boredom. We immediately describe a face as sad, bored, even when we cannot provide any further description of its features. One might say that mourning is an integral part of the face. This belongs to the idea of emotion.*"

The perception of appearance, '*seeing as*', denotes what Wittgenstein calls the '*dawning of appearance*' or "*change of appearance*", i.e., the ineffable features of ordinary perception: a vocational and spontaneous reactions of intimacy to what we see. One important thing we do when we notice an appearance is to give it what we call context: when we see the same thing in a different context, we make new connections or comparisons. What changes in the perception of the appearance is not what we perceive, but our attitude towards it, how we react to it, and what we do with it.

Alan Tormey analyzed the logical relations that link behavioral expressions and thoses stares of a person that are said to be expressed.[86]

When we describe a face as sad, we are not saying that his face looks sad, but that it is. We immediately describe a face as sad. The description of a person by the expressions on his face belongs to a given attitude towards the person.

There is no privileged introspective access that provides more basic knowledge than the recognition of emotion in another person's face: Wittgenstein emphasizes the '*myth of the inner*', according to which the meaning of a given segment of

[86] Tormey (1971).

human behaviour is a manifestation of the inner (*Innen, Inneres*), of hypothetically private events that make up the special way that only you can discover the content of your mind.

Jonathan Glover contends that *"It will now be clear that the familiar physiognomy of a human face manifests our sensitivity to the unique identity of the person we observe to the sense that it creates a world with which we are already comfortable; in short, it presupposes a recognition of that world as the world of a person.*"

Béla Balázs's Film Theory and Pathognomy of Human Emotions

A film should not be analysed as a language because there is no language of the image: cinematic images are signs or representations. The problem of *'visual expression'* marks a relationship between human behaviour and certain characteristic states of a person. The expression of a feeling or belief, or whatever, is the expression of an intentional state.

One could say that the actor expresses his own emotions and represents those of the character; all a person's movements and words are symptoms from which to infer something about emotions and characters. If this is so, then we can speak of the expressive function of movements and words.

On the other hand, acting should not be regarded as a kind of expressive behaviour, but as an activity that appropriates the surface of expressive behaviour for representational purposes.

Bela Balázs (1884–1949), Hungarian writer, director, and film theorist, spent part of his university studies in Berlin and Paris, where he was a student of Bergson; he also took part in Georg Simmel's courses in Vienna.

The progressive increase in the dominance of the word in culture led the content of texts to be increasingly expressed through intellectual notions detached from inarticulate spontaneity and lively nature. The body had lost its content, and 'devoid of soul' became empty. Before, 'man was visible' with his whole body, but in the culture of speech, the soul (since listening to it became so simple) was now almost invisible. Today, thanks to the cinema, man regains his visibility, and the soul is consequently able to express itself as before it was atrophied by the process of cultural development.

Balázs tolerated a film depicting the individual experiences of a character and an objective component, i.e., what is represented in a film as the historical and social reality that shapes these experiences. He believed that it was the close-up of the human face that distinguished film from other acting arts, especially theatre. In contrast to stage performance, the camera brings us closer to a face and allows us to peer into the depths of a gaze, examining every twitch of muscle in the most intimate detail. Balázs called this a '*polyphonic*' play of features. "*The appearance, on the same face, of contradictory expressions". "... facial expressions, isolated from their context, seem to penetrate a new dimension of the soul*".

These components, objective and subjective, interact and synthesize to produce what he called *microfisionomy* in the case of close-ups. The qualities of the soul belong to the various thoughts and emotions, and the innate conditions of human experiences of space and time produce the physiognomy in film. If this is so, then cinema transforms the particular into the universal. Put more simply, Balázs focused on the power of expression in film through the face and the body.

Pathology of Face Perception

Face recognition processing is complex and not currently understood. It depends on a combination of anatomical structures, cognitive functions, and psychopathological processing.

Visual working memory allows us to hold a visual picture in our mind for a few seconds after it disappears from our sight. During this time span of a few seconds, a small subset is transferred into visual working memory.

Our brains are exceptionally tuned to perceive, recognize, and remember faces. The neurons responsible for our "*face sense*" lie the fusiform gyrus. Findings implicate dysfunction of posterior brain areas that mediate visual perceptual processing and the prefrontal areas involved in the active maintenance of information during delay intervals.

Why do we see faces everywhere we look: in the moon, in Rorschach inkblots, in the interference patterns of oil spills? Doris Tan, a neuroscientist at Bremen University, had an inkling that people might process faces differently from other objects. "*Her suspicion was that a particular area of the brain gives faces priority*" she said. "*So, we started questioning whether there really might be an area in the brain that is dedicated to face recognition*". She discovered that three areas of monkeys' temporal lobe seemed to be strongly attuned to faces. Even so, areas of the brain she had identified as face-specific occasionally lighted up in response to objects that bore only a passing resemblance to faces.

The systems that govern object and spatial visual perception and working memory appear to be affected differentially by schizophrenia. Schizophrenia has been linked to impaired face and emotion perception. Individuals with schizophrenia typically demonstrate marked deficit on such complex tasks, as well as on more elementary working memory task.[87]

Thus, it has been proposed that an impairment of working memory may be responsible for much of the observed cognitive disturbance of the illness. Persons with schizophrenia get the emotions right only 60% of the time, a highly significant

[87] Gambarota and Sessa (2019).

difference statistically.[88] Facial emotion processing has been extensively studied in schizophrenia patients while general face processing has received less attention. Patients had a hard time distinguishing between familiar faces with new expressions and unfamiliar faces. The findings could be interpreted to indicate a broader, more global perception of faces occurs in schizophrenia, and that individuals with schizophrenia have an impaired ability to retrieve the global perception when perceptual nuances, such as expressions, occur. People with schizophrenia can experience facial recognition deficits, which can have a dramatic impact on social interactions, even when they try to identify faces familiar to them. Whether the deficits in psychosis concern recognition of all emotions, or whether there is differential recognition ability across emotions, still requires further investigation, but increasing evidence, from meta-analysis and from the largest studies to date, suggests the most prominent deficits are in fear and/or anger. Furthermore, there is a positive correlation between self-face recognition and other-face recognition difficulties in schizophrenia patients who often report feelings of strangeness when looking at a mirror.[89]

Individuals with borderline personality disorder features may be related to decreased accuracy toward detecting neutral faces, but increased accuracy in detecting negative emotion faces negative emotions face that are not or barely present.[90] Anxiety has been widely studied and generally correlates with a low visual working capacity. Autism spectrum disorder is a comprehensive neural developmental disorder that produces difficulties with facial identity recognition and recognizing emotional expressions, which may be derived from abnormalities in the fusiform face area and amygdala.

Prosopagnosia (from Greek *prosopon*, meaning "face", and *agnosia*, meaning "non-knowledge"), also known as *face blindness*,[91] is a cognitive disorder of face perception in which the ability to recognize familiar faces, including one's own face (self-recognition), is impaired, while other aspects of visual processing (e.g., object discrimination) and intellectual functioning (e.g., decision-making) remain intact. Prosopagnosia is highly heritable and surprisingly common, affecting 1 in 50 people, more than 5 million in the U.S. alone.[92] In congenital prosopagnosia, the individual never adequately develops the ability to recognize faces.[93] The brain area usually associated with prosopagnosia is the fusiform gyrus,[94] which activates specifically in response to faces.

Acquired prosopagnosia results from occipito-temporal lobe damage and is most often found in adults. Within those sufferers, the condition varies widely. Some have

---

[88] Tek et al. (2002).

[89] Gaebel and Wölver (1992).
Guillaume et al. (2012).
Tripoli et al. (2022).

[90] Meehan et al. (2017).

[91] Davis (2006).

[92] Grüter et al. (2008).

[93] Behrmann and Avidan (2005).

[94] "Face blindness not just skin deep". CNN. Retrieved 19 November 2015.

trouble recognizing their own face. For most, the problem is not so much as detecting a face, as it is about recognizing the same set of features when meeting them again. Acquired prosopagnosia is subdivided into apperceptive and associative prosopagnosia. Mild prosopagnosics can train themselves to memorize a limited number of faces by learning to distinguish one from another, based on hairstyle, voice, gait, or body shape. Most of them learn to cope early on.

Are Mental Disorders Natural Kinds?

Could psychiatrists brace themselves to confer natural kind status to some mental disorders? Even in a minimal sense of the term, 'natural kind' is extremely controversial and there are different understandings of what it is to be a natural kind.[95] At that point, it is up to us to select the taxonomy, which is most useful for the diagnosis and treatment of mental disorders. What's more, mental disorders often merge into one another through lack of any natural boundary between them.[96] Small wonder then that despite their success, signs of dissatisfaction have progressively arisen within the legacy of mental nosographies. Categorical classifications attempt according to Plato's analogy "*to cut nature at the joints*",[97] but the same is true for the dimensional approach: the difference rests in how wide the slice is.

Then, beyond this, medical thinking remains tainted with the view that medical research ought to identify for every disease its single necessary cause. The search for *a posteriori* necessary cause is the grail of medical science. The ability to define a natural kind ought to rely on some necessary criterion, or a single cause, the presence of which exhausts what groups together all the members of that category.[98]

A less telling argument banks on the assumption that specific conditions require specific drugs: depression if it is a natural kind must be targeted with one drug or a family of drugs acting on specific targets.

This purview helps to classify diseases by their aetiology rather than by their symptoms. As a matter of fact, a variety of abnormal behaviours form some sort of unity and represent natural kinds in the minimal sense of the term since they have a single underlying necessary cause: nutritional (pellagra, cretinism due to lack of iodine), metabolic (acute intermittent porphyria), toxic (delirium tremens or heroin abuse), infectious (general paresis of the insane), experiential (post-traumatic stress syndrome) or genetic (Huntington's disease).

However, the majority of objects classified in psychiatry are polygenic, aetiologically multifactorial, and clinically heterogeneous. They are usually not grouped into mutually exclusive and jointly exhaustive categories and do not represent

---

[95] Wilkerson (1998).

[96] Dupré (1981, 2000).

[97] Plato Phaedrus 265d-266a.

[98] Samuels (2009).

biological entities. It seems that the same environmental factors and the same genetic predispositions can each contribute to the genesis of seemingly unrelated disorders. There is usually no known specific linear chain of development between molecular genetic determinants and clinical features. For example, the category of schizophrenia is probably too heterogeneous, and needs to be split into a number of subtypes. It is as if there was no unique ideal classification but countless legitimate ways of categorizing and classifying mental diseases.

As soon as one admits borderline instances, one has to accept scales, or some graded series of naturalness and the concept loses its bite if not the core of its meaning. The major categories of mental disorders, for now, are syndromes, i.e., unnatural constructs since they fail to emerge from single causal processes. Kraepelin's ambition of classifying mental disorders as homogeneous conditions attributable to some common cause failed to materialize. Contrary to many somatic disorders, mental disorders lack defined biological mechanisms, although one should not exaggerate this discrepancy. Isn't the exact cause of a very common ailment, low back pain, never found in 85% of patients? The theoretical perspective afforded by biomedical science has got itself in a quandary and we are often in a growing muddle trying to clarify the mechanism leading to the common chronic diseases most of which are multifactorial.[99]

Do mental disorders really exist in the world or are they socially or medically constructed categories? These questions are not unique to psychiatry: mental disorders are medical diseases. Just like diseases, they are medically constructed real processes; they are real, and stable across time and space in the way that atomic elements are real.

In short, if most mental disorders are not natural kinds, they are real kinds, as real as they can be.[100] It is possible that disorders might merge into one another with no clear boundary. There is evidence that there is a similar genetic predisposition to schizophrenia and bipolar disorders, as well as seemingly unrelated disorders such as autistic spectrum, intellectual disability and possibly epilepsy.[101,102]

According to A. Jablensky: "*Variations in psychiatric symptomatology might indeed be better represented by an orderly matrix of symptom-cluster dimensions than by a set of discrete categories.*"[103] These clusters are similar not only in their clinical manifestations, but also because the co-occurrence of these properties from individual to individual is explained by causal mechanisms, which ensure that these properties are instantiated together.[104] These mechanisms span several levels.

For one thing, the individual symptoms interact to sustain the other symptoms characteristic of the illness. In a case of depression, insomnia predisposes to

[99] Cooper (2007).

[100] Kendler et al. (2011).

[101] Tabb and Schaffner (2017, 343).

[102] Borsboom (2017, 88–90).

[103] Jablensky (2012).

[104] Kendler et al. (2011).

tiredness, and guilt predisposes to suicidal ideation. Phobias led to avoidance which prevents habituation to the feared stimulus. The same cluster of symptoms might arise from different etiological, underlying or sustaining mechanisms in different cases, in philosophical terms these kinds are multiply realizable.

For another, at a more basic level, a state of depression through its impact in cognitive biases, can create self-fulfilling expectations, which further exacerbate the depressive state. In this case, we have a series of causes that interact with each other to produce an underlying state that reinforces old symptoms into existence over time.

The stability of these kinds is maintained by mechanisms at multiple levels, including symptoms themselves, in addition to mechanisms sustaining or underlying the imperfect cluster of symptoms that characterize our best-codified diagnostic categories. The symptoms of a psychiatric disorder are not effects of a common cause at all; rather they stand in direct causal relation to each other.

Ian Hacking: Representing and Intervening[105]

The influential argument proffered by Ian Hacking in a series of publications contrasted *interactive kinds* to *indifferent kinds* and defended the notion of *looping effect*. Contrary to indifferent kinds, interactive kinds—schizophrenia, childhood autism, or mental retardation—are conditions whose instances are people who change their properties, namely their behaviour, emotions, and self-concepts, which consequently may affect their status as instances of the kind in question or affect the kind of classification itself. Patients become self-aware of being classified in a certain way. It follows, Hacking writes, that "*What was known about people of a kind may become false because people have changed in virtue of how they have been classified, what they believe about themselves, or because of how they have been treated as so classified. This is the looping effect.*"[106] Classifications may influence what is classified.

There is another looping effect of kinds. Social media and special websites illustrate how psychiatric labels are publicly disseminated with a risk of self-harm and self-medicalisation. Neither psychiatry, nor the pharmaceutical industry are responsible for this situation.[107]

All in all, the scientific value of a theory does not depend on the mind of the scientist who develops it or understands it, but it relies on objective ascertainable facts: science seeks from a position outside the world; it is actually about the world and as such it is not within the world. It usually brackets its own existence or treats itself as scientifically out of sight. So, in a way, natural scientists are *externalists*

[105] Two texts of Arthur Danto have been drawn upon widely as the source of the views presented here.

[106] Hacking (1999, 104–105).

[107] Charland (2013, 167–170).

since, for them, meaning and descriptive meaning, sense, and reference are one and the same: a statement should be either true or false, otherwise it is nonsense.

The scientist, as it were, casts no shadow over the world, give or take a few minor exceptions, such as quantum theory or the method of observant participation in sociological research.[108] Small wonder then, that for one thing, as scientists, psychiatrists form beliefs, makes observations and inferences.

To that extent, *psychiatrists are externalists*: like biologists or physicists, they seem not to be themselves part of the object of the exercise. Psychiatry—as well as medicine in general is a kind of natural science covered by causal laws and as such it is not supposed to affect the disorders observed. Wittgenstein here says: "*'Observing' does not produce what is being observed.*"

But there is another consideration of immediate interest. Surely, a therapist needs to cross the barrier between his patient's outward verbal self and his inner, subjective, and elusive inner world. If, as scientists, psychiatrists are externalists, but *as therapists they are internalist*, allowing them to have meaningful exchanges and taking the patient's perspective and his narrative self-awareness. As internalists their language is a performative, causal shaping force: their way of understanding the world is to change it.[109] To understand normal or pathological behaviour, a psychiatrist has to ponder the way people represent the world around them, his patient and himself included. It is as if the medical language of a psychiatrist describing mental disorders, which was about the world and thus outside it, was now, in the therapeutic situation, housed inside the world.

It should now be plain that the concepts of mental illness as well as the concerns of the therapist and the anticipations of the patient are of self-intimating nature. There is something special about mental symptoms. Mental symptoms elicited in the doctor-patient relationship have a self-reflexive and self-referential character. Every clinical judgement is framed by expectations. The relationship between the psychiatrist and his patient is instrumental in the effectiveness of the psychotherapy, and the need to break the self-defeating and self-sustained and self-confirming vicious circle, which characterizes mental disorders, and which perpetuates, reinforces, self-modifies, and entrenches his patient's ongoing mental illness. A cardiologist may have a bad rapport with his patient, and this will have no consequences (within limits) on the efficacy of his treatment as long as he is competent. But for a psychiatrist to help his patient, he needs to understand him, to empathize with him, and to understand him he needs to understand himself and to bring his own experiences, his feelings, and his emotions into the process of psychotherapy; this is why fifty years ago, the training of a psychiatrist in the United States impelled him going through a didactic psychoanalysis.

Beyond this, the Rosenhan study showed that whether a psychiatrist is expecting to see a person in distress or a normal individual, strongly affects diagnostic judgement. The accuracy and usefulness of a psychiatric diagnosis may be compromised

[108] Danto (1989).

[109] Danto (1968, 232–240).

or modified by both the background in which it occurs and by expectations. The process of diagnosing may affect the diagnosis as it opens the loop. And anticipation may significantly modify the act of labelling of a patient, thereby closing the loop.

To introduce a concrete example, Charcot's assistants at l'Hôpital de la Salpêtrière artificially produced the symptoms of hysteria described by him: they were coaching the patients on how to act and prepared them every morning before the medical round on the wards. *'Physicians need sick people'*, writes Thomas Szasz. And people who need help, who desire assistance are being accepted in their sick role and become a genuine occupant of the sick role, again, closing the loop.[110]

In our globalizing world, several mental illnesses have been spreading like epidemic contagious diseases: depression, hyperkinesis, chronic fatigue syndrome and various so-called functional disorders. This process of 'disease mongering' has been promoted by multinational pharmaceutical companies, by the mass media as well as by the health-care systems, putting normal people under the canopy of the 'sickness role'.

Meanwhile, some mental disorders, such as catatonic schizophrenia have progressively disappeared over the last hundred years at least in Western but not in non-Western countries. In like manner, there has been a decline in the incidence rate of hysteria in the Western World from the nineteenth century and during the two World Wars, as psychoneuroses now take the form of anxiety and depressive disorder. Yet there is no evidence of a similar decline in India, Lebanon, Egypt, and Sudan.[111] Why so? Mentally, writes Leff,[112] the emergence of the individual weakens the cohesive bonds that are so evident in traditional societies. Individuals become more important than the roles they perform, freedom of action in relationships increases, and a variety of emotions begins to be explored in the context of relationships with others. The unique qualities of the individual become prized, and introspection consequently flourishes. The effect of these profound changes on emotional life is to shift the expression of emotion from the somatic to the psychological mode, and to increase differentiation between emotions. Psychiatrists both overtly and covertly encourage patients to present their distress in psychological rather than somatic terms and thus constitute one of the forces behind the continuing process of emotional differentiation.

[110] Szasz T,

[111] Leff (1981).

[112] Leff (1981).

Psychiatric Nosology

Kant ventured the following suggestion: "*I see nothing better than to imitate physicians who think they render a great service to their patient by giving a name to his illness, and thereby I propose a small nomenclature of mental disturbances.*"[113] When Emil Kraepelin (1856–1926) proposed the first widely accepted classificatory system for severe mental disorders, he emphasized the discreteness of mental disorders.[114] This influential view based on the hypothesis of the nosological entity, allowed the subsequent development of a psychiatric nosology based on observable criteria and consensus among experts.

All in all, the development of classifications of mental disorder over the past century, the ICD, and the DSM, has been a fruitful exercise. Assen Jablensky indicated in 2009 that the two approaches of DSM and ICD are complementary and compatible. These manuals, though overinclusive, are indispensable instruments for clinicians, mental health services, researchers, insurance companies, and for all those who need facts about occurrent risks, frequency, replicable diagnoses and aids for treatment, teaching, and training, as well as in data reporting and public health practice. Inter-rated agreement, that is, diagnostic reproducibility, for most categories indicates excellent consensus. This classification has its uses in clinical medical or psychiatric care: ICD, the International Classification of Diseases, now in its eleventh revision (ICD-11) presents 21 major sections, each endeavouring to provide a degree of aetiological homogeneity, without quite succeeding. The broad groupings are composed of aetiological disorders (organic disorders or those due to psychoactive substance use), patterns of clinical presentations (schizophrenia and allied diseases, affective, stress-related and somatoform disorders), typical of age at first manifestations (disorders of development of with onset at infancy, childhood or adolescence, mental retardation, disorders of adult personality and behaviour).

There is an alternate possibility, namely DSM, The US Diagnostic and Statistical Manual of Mental diseases (DSM-II, DSM-III, DSM-IV, and DSM-5) approved by the American Psychiatric Association. DSM is essentially descriptive, supposedly atheoretical and generally free of aetiological assumptions; it results from the gradual, ever-increasing accumulation of provisional clusters achieved by splitting rather than by lumping and that have no natural boundaries, such as single symptoms, habitual behaviours, or personality traits, rather than clinical syndromes. Such lists of names, conditions or behaviours come with elaborate rules as to how those names should be used.

There are two approaches for psychiatric taxonomy.

DSM-III developed explicit *operationalized* criteria, which sort out patients into homogeneous groups identified by those symptoms that have great significance on their own right, and for which this or that treatment will work. The operational approach defines schizophrenic syndromes by exclusion. So, schizophrenia is

[113] Kant (1905).

[114] Kraepelin (1913).

defined by what it is not rather than by what it is. Mario Maj argues that the DSM-IV criteria do not capture a proper understanding of the diagnosis of schizophrenia and fail to acknowledge the intuitive ranking of symptoms, which have equal status in the DSM account.

Another taxonomic strategy represents a category by its *prototype.* Psychiatric signs and symptoms are free standing items, but are meaning-laden and mutually interdependent, they constitute a *Gestalt,* a salient unity of symptoms and signs: European psychiatrists agree about something phenomenologically distinctive and characteristic about schizophrenia, the *whatness,* namely a prototypical essence whose features are not casual surface symptoms, but a deep phenomenological reality that brings in the clinical flavour of this complex condition.[115]

Clinicians usually prefer a phenomenological description, while researcher favour a symptom rating system.

For one thing, mental pathology in contemporary psychiatry is *wishfully assumed* to be subdividable into well-demarcated, mutually independent, discrete diseases or discrete categories characterized and delimitated from other mental disorders by a tendency to cluster together, by their clinical description, their biology, or by the identification of some homogeneous group of patients defined by specific treatment.

For another, the classes of diseases are neither mutually exclusive, nor jointly exhaustive, and there is no empirical evidence for natural boundaries; they are convenient instruments and they do not map into some conceptual framework; they do not classify people but disorders people have.[116] Most mental disorders are actually mere provisional syndromes, namely groupings of statistically co-occurring signs and symptoms with different levels of severity from mild to severe. The scope of such syndromes is restrained to some prototypal symptom clusters, which are presumed to represent real maladies, such as obsessive-compulsive, paranoid, or depressive syndrome, as well as schizophrenia, which occur globally with certain uniformity although tainted by individual or cultural differences. By and large, DSM conceives diseases as processes occurring in clusters of recurring patterns, based on the assumption that any individual can be classified as a case or not. These conditions should tend towards some form of aetiological sorting for want of which they are considered provisional. Even though provisional, they possess practical utility for the clinician.

The upshot is that our disease classifications, medical or psychiatric, are slipping back and forth from diagnoses made on the basis of descriptive features into those made on the basis of any particular aetiological theory; they represent a patchwork, a motley of conditions, which often overlap one another, some of them being causal entities (general palsy, post-traumatic stress disease), some being probable single criterion ones (Alzheimer's disease) and some being mere syndromes (panic disease) or aggregates of clinical entities (schizophrenia).

---

[115] Parnas and Urfer-Parnas (2017).

[116] Jablensky (2002, 11).

The process of categorization leads to unavoidable arbitrary thresholds, *fuzzy boundaries,* and consequent difficulties of validation due to the lack of diagnostic standards[117]: the actual function of DSM is to provide a uniform system for keeping records, namely an inventory of all conditions that are considered diseases as well as to provide an exhaustive system of many anomalies even if they are not impairments.

Beyond this, some authors claim that DSM categories, or at least some of them, tend to medicalize features that are part of the human condition. But there are different sorts of differences, and medical science arises in connection with keeping apart normal from pathological, for which there is no objective scientific criterion, but rather some sets of levels or degrees that separate what is medical from what is not. It follows that a given category may be a manifestation of some mental disorder for one patient, but not for another.

Since the boundaries between disease categories are more fluid over the life course than DSM-IV recognized, and since many symptoms assigned to a single disorder may occur at varying degrees of severity, suggest that, like many other medical diseases, DSM-5 introduced a dimensional approach to mental disorders, including dimensions that cut across current categories. This approach allows a more a more detailed description and an increased validity of a diagnosis, namely the degree to which diagnostic criteria reflect the comprehensive manifestations of an underlying disorder.

I want here to drive some key points home in an attempt to pinpoint the pitfalls of this purview of enhancement. Those who propound second-order enhancement might be settling in a conspicuous cul-de-sac. There are two groups of propounders in front of the consumers: for one thing, the pharmaceutical industry, and the medical community, and for another, the population in quest of some earthly paradise with its romance with pharmaceutical chemistry.

The fact is that the evidence base is highly imperfect and that the research was done in the context of larger context of social, economic, and industrial forces. Is psychiatry turning normal human traits into diseases merely to promote the interests of the pharmaceutical companies? American psychiatry is allegedly largely determined by profit. Even the leading medical magazines are cautious about avoiding antagonizing Big Pharma or upsetting pharmaceutical advertisers.[118]

Drug discovery is a lengthy, difficult, and inefficient process with a high attrition rate: new therapeutic promising candidates turn out to be ineffective or toxic. At present, research and development cost of each new molecular entity is about 1.8 billion dollars.

Carl Elliott writes: "*My worry is that we will ignore important human needs at the expense of frivolous human desires; that dominant social norms will crowd out those of the majority; that the self-improvement agenda will be set not by*

[117] One might be tempted to say that the individual patient suffers from illnesses or mental illnesses. Their illnesses are then tentatively assigned to some mental disorder.

[118] Kassirer (2005).

individuals, but by powerful corporate interests; and that in the pursuit of betterment, we will actually make ourselves worse off." On this score, and in the context of a preoccupying dwindling number of new medications, drug makers need to widen their net, lower the threshold of their interventions, and extend the scope of people's needs beyond the limits of medicine and, instead of targeting only patients, go for consumers who are already victims of an epidemic of pill-popping. In a market-driven health care system, the strategy is then, for one thing, to create demands cobbled together out of people's disposition to strive for self-improvement and cancel out their social disadvantage. But for another, it implies instilling in the masses both the belief that their stressful life situations are requiring medication as well as the hope that medications will soon be found that will easily increase efficiency, improve some bodily and mental trait, strengthen their cognitive abilities, equip them with a better personality, and, at the end of the day, meet the whole society's existential *angst*.

Since medical antidepressants must be ordered by a physician, manufacturers medicalize their technological fixes and present them not as enhancement, but rather as treatment for newly found or not yet acknowledged disorders: SSRIs, for instance, were marketed with a copious supply of publicity for clinical depression as well as for concocted diseases such as 'premenstrual dysphoric disorder' or shyness now upgraded and christened 'social anxiety disorder', with the risk of inappropriate short and long-term use of antidepressants.

What's more, DSM-III and DSM-IV have been accused of putting the bar of authentic pathology too low and consequently of laxity in separating vague discontents from real disorders. Of course, it would be a misjudgement to aver that the collaborative development of DSM-5 and WHO's ICD-11 has wilfully shaped the psychiatric nosology to tip the balance if favour of drug companies. Yet, most of the psychiatrists who prepared the manual have financial ties with the pharmaceutical companies. By the way, one of the contributors acknowledged that the manual lacks scientific evidence. It is, according to Marcia Angell,[119] "*the product of academic politics, personal ambition, ideology, and, perhaps most important, the influence of the pharmaceutical industry.*"

Louis Charland concludes that diagnosis has been dismantled and appropriated by interest groups, since psychiatry, historically to the present, has been promoting commercialisation of psychopharmacology.[120]

Be that as it may, chapter V of ICD-10, which covers mental disorders, has been accepted in countries with different degrees of development of medical education and mental health services. It has been prepared under the chairmanship of A. Jablensky and N. Sartorius, two leading world psychiatric epidemiologists, without being financially involved with the pharmaceutical industry. The process started in Geneva in 1982, involved hundreds of experts, numerous meetings, comments and criticisms, and the intervention the World Psychiatric Association. The resulting

[119] Angell (1985a, 12).

[120] Charland (2013).

draft underwent extensive field trials in 1987 and 1988. A series of four last meetings between 1988 and 1992 were attended by WHO advisers and some of the DSM-IV Task Forces. The result was that there are surprisingly few differences between the final versions of the two classifications.[121]

In his wonderful book on mental disorders, Jonathan Glover's detailed analysis of Van Gogh mental disorder epitomizes the difficulties of psychiatric diagnosis better than lengthy discussions concerning DSM. Few patients, if any, present manifestations that are unique to, peculiar to, and restricted to some single category.[122] Van Gogh's 'madness' exposes the ambiguities and the complexity of diagnosing people long dead, even when we have ample and detailed information on the course of their condition. Glover indicates that the various diagnostic hypotheses that have been offered, work on the premise that Van Gogh suffered from a single clinical entity that should be covering his tortuous clinical course. His illness has been the subject of much speculation, with some people suggesting he suffered from bipolar disorder, or schizophrenia, highlighting incidents such as when he cut off his own ear.

Yet, there is little doubt that he has suffered from partial temporal epilepsy, from bipolar disorder with some psychotic manic episodes with some paranoid traits, and from alcohol abuse. Even so, just as in the case of Mozart, this led to a free-for-all of diagnoses including schizophrenia, acute intermittent porphyria, saturnism, syphilis, Meniere disease, malnutrition, digoxin poisoning. One may now propose new diagnoses at will since there is no way to decide between them. Van Gogh's death has become a grabbing of hypotheses.

To sum it up then, the neurocognitive evidence, the clinical course, the biomarkers, the criteria of genetic risks, the environmental factors and the treatment responses do not add up yet for unequivocal distribution of patients within the available nosological categories.

The Spectrum Perspective

In 1913, E. Kraepelin[123] wrote: "*Wherever we try to mark out the frontier between mental health and disease, we find a neutral territory, in which the imperceptible change from the realm of normal mental life to that of obvious derangement takes place.*"

But the question now before us is: could mental disorders amalgamate with one another? Prototypes described in textbooks are actually pretty rare: atypicality is the statistical mode.

---

[121] Cooper and Sartorius (2013).

[122] Glover (2014).

[123] Kraepelin (1913, 295).

To begin with, most patients who meet the DSM-5 or ICD-10 criteria for some mental disorder present themselves with the criteria of at least two clinical disorders. This psychiatric comorbidity tends *to blur the distinction between clinical categories*. For DSM-IV, a patient, in principle, who belongs to one category is unrepresentative of other categories. But extensive comorbidity makes this principle rather shaky.[124]

For instance, 18% of the total population, or 60% of those with at least one disorder, had at least two mental disorders in their lifetime. Or 50% of cases of schizophrenia have at least one comorbid psychiatric disorder (such as substance abuse, obsessive-compulsive disorder, depressive or obsessive-compulsive disorders) or medical condition (e.g. heart disease, cardiovascular disease, osteoporosis, obesity, diabetes, autoimmune process).[125] Co-morbidity indicates shared symptoms, shared risks, or that one condition causes another.[126] Beyond this, comorbidity is not limited to mental disorders: 20% of Medicare beneficiaries have five or more chronic conditions.[127]

But the second question, then, is that psychiatric epidemiology has shown that *mental diseases are dimensional* and not categorical, whether or not a zone of rarity suggesting some sort of valid boundary is separating them.[128] Clinical observations indicate that, even more so than in non-psychiatric medicine, mental disorders are, with rare exceptions, not categories but continua.[129] Most symptoms mentioned in DSM-5 are expressed as dimensions or continua: obsessions and compulsions, panic attacks, anxiety, depression, impulsivity, or anything else of that ilk. Adaptive styles and coping strategies are not all-or-nothing features but lack strict dividing lines.[130] In other words, dimensional relationships—within and between clinical syndromes—underlie more and more apparently discrete clinical syndromes, in psychiatry but more generally in disease concepts such as those of diabetes, heart failure or hypertension.

DSM-IV has a *'Not Otherwise Specified'* category, which reflects the vague non-diagnosis conditions or those, which do not seem to warrant a full-fledged diagnosis. DSM-5 provided another way to list a diagnosis that seems uncertain: '*Other Specified Disorder*' or '*Unspecified Disorder*' in the case of depressive disorder (F.32.9). There is, to be sure, a temptation to attribute the vagueness inherent to the notion of mental illness either to our ignorance, or to some chaotic nature of the universe. But it might actually very well lie in the nature of things.

It follows that there is a growing interest in the so-called sub-threshold disorders since unmanageable disability and harmful malfunction, associated with mild

124 Buckholz and Meyer-Lindenberg (2012).

125 Sartorius et al. (2015).

126 Robins et al. (1984).

127 Tinetti (2004, 2870).

128 Sneath (1975).

129 Jablensky (2001, 11).

130 Maj et al. (2002).

depression (which differs in degree not in kind with DSM-IV defined disorder) may exceed the disability associated with common medical conditions (threshold conditions) in primary care and in the general population (F31.89; F31.9; F32.8). For example, nearly 20% of the adult population reported one or more depressive symptoms in the previous month in the ECA study. And the 1-year prevalence of two or more symptoms (11.8%) exceeded the 1-year prevalence of all mood disorders combined (9.5%). Sir Aubrey Lewis held that "*a gross blatant psychosis may do less damage in the long run than some meagre neurotic incubus: a dramatic attack of mania or melancholia, with delusions, wasting, hallucinations, wild excitement and other alarms, may have far less effect on the course of a man's life than some deceptively mild affective illness which goes on so long that it becomes inveterate.*"[131]

Beyond this, if *each disorder consists of a continuum between 'normality' and overt psychosis*, it also constitutes a spectrum or continuum with other diagnostic categories, that is to say those mental disorders might merge into one another with no valid boundary in between, and without being separated by a zone of rarity. For instance, dysthymia and schizotypal personality may be forerunners or extensions of major depressive disorder and schizophrenia, respectively. And there is some evidence that schizophrenia is a spectrum that comes in degrees. About 1% of the cases stem from the deletion of a small stretch of DNA on chromosome 22. One percent of the cases might be due to an encephalitis caused by autoimmune disorders. In addition to influenza, a wide variety of maternal viral infections (measles, rubella, varicella/zoster) during pregnancy are associated with increased risk of schizophrenia. Exposure to *Toxoplasma gondii* when young, a parasite transmitted by cats may double the risk of developing schizophrenia. There are other factors such as childhood viral infections of the central nervous system, and some abnormality of the normal process of pruning connections between brain cells that happens during the adolescence. Some neurotransmitters such as dopamine and glutamate may contribute to the disorder. Different pathogenic pathways may need different treatments. Schizophrenia is turning out being very different things.

To sum it up then, our psychiatric nosography faces great challenges to take account of the ubiquity of comorbidity, the failure of the categorical system to capture the full spectrum of subclinical, subthreshold and atypical manifestations of a given disorder, the fact that separate disorders might exist in a continuum with one another without clear boundaries, and the potential usefulness of a thorough dimensional assessment of the full spectrum of psychiatric symptoms that might be experienced in a lifetime. Classifications cut across one another, and none of them is properly primary. It is up to us to choose the provisionally most useful one with a creditable predictive success.

[131] Lewis (1936).

The Fragility of Diagnosis

Mental disorders just like somatic diseases are initially identified phenomenologically, i.e., by their signs and symptoms, natural history, prognosis, outcome and exceptionally by their aetiology. Of course, there is more to rheumatoid arthritis or to schizophrenia than "what it is like to suffer from rheumatoid arthritis" or "what it is like to suffer from schizophrenia", but this phenomenological dimension, subjective and objective, provides the initial material with which we construct our concepts of diseases, or at least our central prototypes of a category. The starting point of mental disorder identification should include not only third person behavioural symptoms but also whatever we can initially learn from first-person awareness of the patient's conscious mental life.

Diagnostic reliability in physical medicine is far from being perfect but it is much more valid than psychiatric diagnosis. In psychiatry, there are usually no specific physical data and no biomarkers to support the clinical judgement and often no biomarkers.

A challenging experiment has been carried out under the supervision of David L. Rosenhan, the late professor of law and psychology of Stanford University who was the co-author of one of the best books on abnormal psychology.[132] A group of perfectly normal people simulated a single presenting symptom, hallucinations, and successfully gained admission to psychiatric hospitals; after admission, they behaved in a cooperative way and in normal manner. They remained undetected until being discharged within a few weeks with a diagnosis of 'schizophrenia in remission'. Interestingly, about a third of the real patients, inside the hospital, detected that they were frauds.

Another experiment was then conducted in which the staff of a hospital was told that at least one pseudo-patient would be presenting himself in the next three months. The psychiatrists became much more conservative in their diagnosis. Among the 193 patients admitted, one doctor was firmly convinced that 41 were fakes, while another doctor suspected 23. Actually, no fake patients have arrived at all.

This experiment could, of course, not be repeated for several reasons: fewer patients are now being admitted in a hospital and the concept of schizophrenia was still excessively broad in the United States in 1973.[133]

If so, then, what does Rosenhan's study mean?

Perception is surely an active process. The meaning and value we attribute to a figure are in some part contributed by the ground. "*A hand in the air,* writes Rosenhan, *has different meanings whether you are sitting in a classroom, making a right turn in your car, or marching in a German parade during the 1940s.*"

For one thing, in Rosenhan's experiment, the hospital settings as well as the diagnoses themselves constitute contexts that let a diagnostician to accept some types of information and reading, and to reject some others. The situation controls

---

[132] Seligman et al. (2005).

[133] Wing et al. (1974).

the diagnosis. For example, the fake patients spent much of their time taking copious notes on their observations. As many as a third of the real patients detected that they were frauds and concluded that they were journalists or college professors. But the staff made no such conclusion. "*Patients engage in writing behaviour*" was the nursing comment on one patient. Nobody troubled to ask him what he was writing. But this did not apparently raise any suspicions. The labelling process itself – 'patient', 'schizophrenic in remission' – produce a background for perceiving the person and for explaining his behaviour.

But for another, whether a psychiatrist is expecting to see a normal person or a person in distress is important in moulding diagnostic judgement. Once a person is single-out as abnormal, his entire demeanour is coloured by that label: Rosenhan mentions the 'stickiness' of psycho diagnostic labels. In Rosenhan's second study, each staff member was asked to rate each patient who sought admission or who was already on the ward, using a scale that indicated how likely it was, in fact, a pseudo-patient. Among 193 patients admitted to the psychiatric wards, more than 20% of the patients admitted were judged with high confidence to be pseudo-patients by at least one staff member and nearly 10% were firmly thought to be pseudo-patients by two staff members.

Physicians gingerly tend to call a healthy person sick more often than calling a sick person healthy. They play safe, which is legitimate since it is more hazardous to misdiagnose illness than health.

Symptoms of mental illness are dispositional traits, i.e. dispositions to behave in certain ways under certain circumstances. But unlike other epistemic tasks like those of science, the epistemic task of diagnosis involves a framing of the relevant context. It is exercised within the borders of the clinical context. Friedrich Waisman observed[134]: "*Try as we may, no concept is limited in such a way that there is no room for any doubt. We introduce a concept and limit it in some direction; ... We tend to overlook the fact that there are always other directions in which the concept has not been defined. And if we did, we could easily imagine conditions which would necessitate new limitations.*"

The hospital staff in Rosenhan's study overlooked the fact that people presenting to mental hospitals might exceptionally be journalists pretending to be insane. Psychiatrists are well aware that criminals may pretend to be insane to avoid being imprisoned and they have developed techniques for identifying them. But they are not prepared to rule out pseudo-patients.

A similar situation might arise if a fake patient presented himself complaining of severe, intense precordial crushing sensation occurring spontaneously at rest and progressing in nature, which radiates from the chest to the shoulders, neck, or left arm. Even in the presence of a normal EKG, such a patient would be admitted with a provisional diagnosis of myocardial infarction.

[134] Waisman (1953, 120).

There is however another reason for the surprising results of this investigation: the diagnosis of schizophrenia in the United States in the 1970s and 1980s was somewhat erratic in sharp contrast to the British standardized interviews.[135]

Rosenhan's study illustrates, in the provisional absence of biomarkers, some the theoretical limits of psychiatric diagnosis. 'And it suggests that if you do not want to be locked up, you should not claim to hear voices', concludes Rachel Cooper.[136]

Diagnostic Reliability and Validity

Diagnosis in psychiatry is a conceptual quicksand. In order to arrive at a diagnosis, psychiatrists use clinical interviews, observation, and psychological tests. But to arrive at an acceptable diagnosis, a good assessment instrument or method of diagnosis must possess two quantifiable characteristics: validity and reliability. *Validity* refers to the extent to which a diagnosis successfully differentiates patients in one category from those in another, and eventually helps to make predictions (e.g., will the patient respond to a particular treatment?). *Reliability* refers to the measure of agreement of a diagnosis between skilled users.[137]

To be considered for inclusion in a diagnostic manual, a disease entity has to undergo some validating procedures involving epidemiological field trials, clinical description, and treatment response and, as much as possible, it should be culturally transferable. For example, in the 1950s there still was a wide discrepancy in the meaning of 'schizophrenia' between New York and London so that the differing prevalence between the two cities reflected differences in the definition of the condition being used unwittingly in the two countries.[138]

Be that as it may, some categories of mental diseases (schizophrenia, depression, mania), may reach a level of diagnostic reliability, which is, comparable to that obtained in many physical diseases, albeit other categories such as personality diseases do not reach such a high reliability rate.

The advent of the operationalist approach inaugurated by DSM-III had implications for a balanced appraisal of reliability and validity, and particularly on interrelated reliability (diagnostic agreement), and construct validity. It may however have contributed to proliferation of mental disorders and a distorted conception of some forms of psychopathology.[139]

[135] Cooper et al. (1972).

[136] Cooper (2007, 14).

[137] Jablensky and Kendell (2002, 9–14), Wing et al. (2017).

[138] Pull et al. (2002).

[139] Haslam (2013).

Strawson and the Span of the Concept of Mental Illness

How do we know what is meant by a 'depression'? What is, in general, our procedure for answering questions about whether someone like *Richard II* or *Ophelia* suffer from depression? Shall we answer that such a person manifests depressed behaviour? Not at all: actors can simulate depressed behaviour. Or should such a person rather be in the grip of feelings of depression? Suppose we say of someone he feels depressed. Are we describing his feelings or are we describing *him*? Of course, someone who is depressed has feelings of depression and conversely, if someone has feelings of depression, we can declare that he is depressed.

One of the leading analytic philosophers, Peter Frederik Strawson proposed a rough division of the kinds of predicates ascribed to persons. M-predicates are those observed by a clinician, those applied to material bodies. P-predicates include things like 'is in pain', 'is anxious', or 'has an auditory hallucination' that are perceived by the patient. Strawson claims that there is not one primary process of learning to ascribe a depression to someone else on the strength of behaviour criteria, and then another process of ascribing a depression to oneself.

It is one thing to speak of feeling depressed (of a feeling of depression) and it is another to speak of behaving in a depressed way (of depressed behaviour).[140] One is inclined to argue that feelings can be felt, but not observed, and behaviour can be observed, but not felt, and that "*therefore there must be room here to drive in a logical wedge*".

Strawson adds: "*But the concept of depression spans the place where one wants to drive it in. We might say, in order for there to be such a concept as that of X's depression, the depression which X has, the concept must cover both what is felt, but not observed, by X and what may be observed, but not felt, by others than X (for all values of X).*" It follows that it is essential to the character of psychiatric diagnoses that they at once have both first-person ascription on the basis of observation of the behaviour of the subject, and third-person perspective on the basis of behaviour criteria.

Depression is not an object, not a thing, but a relatively permanent disposition to feel a certain way, which spans first-and third-person reports.

[140] Strawson (1959, 110).

The Two Explanatory Roots of Mental Disease: Hippocrates and Samuel Tuke

Eisenberg maintained that psychiatry is neither mindless nor brainless.[141] Yet, psychiatry has until recently been deeply divided into two grand traditions, each of which went its own separate way. Both have a philosophical interest and importance beyond the sphere of psychiatry itself.

Kant combined both strategies and contended that mental illness is a disease of the soul *(Gestörtes Gemüth)* manifested by realization when a patient takes his phantasms for real findings or by distorted judgement such as the hypochondriac "*who feels in himself the illusion of all the diseases he just heard talking about*". For one thing, it may spring from a *'disturbed brain or some defect of the body'*. But for another, social environment and social condition may induce madness.[142] This articulation between internal and external features characterizes the chief junctions of mental disorders.

The first member of the philosophical tradition is biological and pertains to neuroscience including pharmacological treatment. "*Men ought to know,* wrote Hippocrates in the celebrated passage from the Sacred disease, *that from nothing else but the brain comes joys, delights, laughter and sports, and sorrows, grief, despondency, and lamentations . . . And by the same organ we become mad and delirious and fears and terrors assail us, some by night, and some by day, and dreams and untimely wanderings, and cares that are not suitable, and ignorance of present circumstances, desuetude, and unskillfulness. All these things we endure from the brain when it is not healthy.*"[143]

The goal of identifying specific topographic mappings to localized brain regions or to district neural networks has not materialized and may be untenable for instance in schizophrenia.[144] Or else, a systems neuroscience approach may be more fruitful, i.e., pertaining to brain structural, functional and connectivity alterations, although a fully mechanistic description remains elusive.[145]

This means that causal laws, should they be found, are bidirectional, i.e., interactional, like those of the stock market. Input and output may be mental or physical: if the disruption of brain circuits may cause mental disorders, mental states may in turn give rise to mental disorders with their aberrant relevant circuits.[146]

Even so, Karl Jaspers[147] wrote that "*Diagnosis is expected to characterize in a comprehensive manner the whole morbid occurrence which has assailed the person...*" and: "*The statement that all mental illnesses are cerebral illnesses... is*

141 Eisenberg (1986).

142 Kant (1905).

143 Hippocrates (1939, 357–358).

144 Lende (2014).

145 Frangou (2014. Fig. 1), Mirsky and Duncan (1999).

146 Gur and Gur (2010).

147 Jaspers (1963, 496).

nothing but dogma ... To do so is to imply that we are making an absolute of cerebral events, taking them as the very substance of man and considering all human events as brain events."

Jaspers added: "*From the psychological point of view cerebral diseases are just one of the causes of psychic disturbance among many. The idea that everything psychic is at least partially conditioned by the brain is correct but is too general to mean anything.*" Mental and behavioural events and corresponding physical events are thus linked in whatever way cause and effect are liable to be connected: causality, should it be ascertained, may proceed from physical-to-physical events, from physical to mental events and conversely from mental to physical events, as well as from mental to mental events.

However, Plato[148] held that although "*nerves and bones and sinuses*", i.e., biology, are a necessary medium to explain human actions, they are insufficient to account fully for human actions. In sharp contrast with the biological standpoint, clinicians have nurtured a psychological viewpoint. It is an outgrowth of the '*moral treatment*' approach originally backed by William Tuke (1732–1822) and his grandson Samuel Tuke (1784–1857) in the eighteenth century at the York Retreat and promoted by Philippe Pinel (1745–1826) and Jean Etienne Dominique Esquirol (1772–1840) in France. Moral treatment was based on a general humane, intimate relationship between psychiatric care personnel and patients in which useful activities were brought into play as a vehicle for re-education.

"*If we adopt the opinion,* Samuel Tuke writes, *that the disease originates in the mind, applications made immediately to it, are obviously most natural; and the most likely to be attended with success. If, on the contrary, we conceive that mind is incapable of injury or destruction, and that, in all cases of apparent mental derangement, some bodily disease, though unseen and unknown, really exists, we shall still readily admit, from the reciprocal action of the two parts of our system upon each other, that the greatest attention is necessary, to whatever is calculated to affect the mind.*"[149]

Treatment now meshes these complementary approaches: chronic depression can be treated either by cognitive behavioural therapy or by antidepressants. The rate of remission was almost twice as high among patients who received both nefazodone and psychotherapy as among those who received either drug alone or psychotherapy alone. Their effects seem to be additive, not synergistic. The superior effectiveness of the combination approach suggests that psychotherapy and pharmacotherapy might act at different levels of the causal process that leads to the disease.[150] It seems that Greek philosophers such as Plato and Democritus were aware of the distinction between these two therapeutic dimensions, between the practical and the rational, or between working by being causes or by being reasons.[151]

---

[148] Plato (1975, 98 c-d).

[149] Tuke in Goshen (1967, 490).

[150] Keller (2000).

[151] Nussbaum (1994, 51–52).

Crucial to explanation is the contrast between cause, which is something one discovers and reason, which is something one understands. So, billiard ball Humean causation belongs to the biological standpoint while reasons and intentional causation pertain to the psychological standpoint.

Quine concludes: "*Each occurrence of a mental state is still, we insist, an occurrence of a physical state of a body, but the groupings of these occurrences under mentalist predicates are largely untranslatable into physiologic terms.*"[152] If so then, we supposedly have on the one hand, a neurophysiological explanation locked into self-determinism and natural wiring, and on the other hand, kind of understanding nested in psychodynamic motivation.

Body and Mind. The Mereological Fallacy

One of the Hippocratic texts told us: "*Men ought to know that from the brain, and from the brain only, arise our pleasures, joys, laughter, jests, sorrows, pains, grief and tears.*' It added that: '*It is the same thing that makes us mad and delirious, inspires us with dread and fear…brings insomnia…aimless anxieties …*" If so, then, for the *Corpus Hippocraticum* and for Galen, madness was a physiological disturbance of the brain, due to an imbalance among humours with physical reactions. For Galen, the psychic *pneuma* that he equated with the soul, was formed in the brain, and distributed in the nerves.[153]

With the Romans, the picture changed. Lucretius contended that the study of nature, of everything that is existing, can reveal to men the way they can count and trust: for him it illuminates life, as already marked by the name of Epicurus, whose philosophy excludes all kinds of survival after death.[154] Furthermore, Cicero wrote that "*the word* insania *indicates a discomfort and a disease of the mind, namely a morbid condition and a suffering of the soul that I shall call* insania".[155] Seneca called it *affectus animi,*[156] a *morbus* that rattles the soul and attacks its stability.[157]

According to the Greeks such Galen---, brain states that lie behind sensations and emotions are what is primary and important. For the Romans, mental events and behaviour can be studied and theorized in their own right. On this score, mental illness was a brain disease for the Greeks, and a disease of the Mind for the Romans[158]: this was an important consideration since they traced the future lines of force of neurosciences. The challenge for philosophy and the neurosciences is to find the

152 Quine (1995, 87).

153 Galen (2011, 7.3 (K5: 606)).

154 Lucretius (2014, III, 1070).

155 Cicero (2014, III, 4, 8; See also: 2014, III, 10, 22).

156 Seneca, 2, 1.

157 Seneca, 2, 10, 22.

158 Puliga (2017; Wikipedia).

right balance between the Greeks and the Romans without falling into Descartes mind-body dichotomy: most psychiatrists and neuroscientists deny a Cartesian substance dualism. To be sure, psychoanalysts are the last dualists since they are completely uninterested in the brain.

Besides there are various kinds on monism: broadly, they are either reductive or eliminative. Francis Crick,[159] the Nobel prize, is a reductive materialist, since he believes that consciousness and our mental states will, someday, be *explained*, without residue, in terms of neural events. Eric Kandel,[160] another Nobel prize, is a nonreductive materialist, and he believes that, at some level, mental phenomena and brain phenomena *translate* into each other. The American philosopher DC. Dennett believes that mind is *dependent* on brain but not reducible to brain. Mental phenomena have their own symbolic language analogous to the language of a computer. Gerard Edelman,[161] the Nobel prize, dismissed the hypothesis that the brain functions like a computer and he is antireductionist inasmuch as physics could not account for consciousness. Yet, he argues that the mind and consciousness are purely biological phenomena, arising from complex cellular processes within the brain. Patricia and Paul Churchland are proponents of *eliminative materialism*: the world of the mental is not entirely physiologically explicable; our desires, fears, states of experience such as pain cannot be properly described in physical terms and are illegitimate and should therefore be eliminated. "*Nervous systems'*, she contends, *are information processing systems*." A future scientific psychology will eliminate the language of *folk psychology* since it can play no role in any future explanation of behaviour. Actually, "*nervous systems are information processing*".[162]

Even though psychiatrists and neuroscientists tend to believe that the nervous system and especially the brain is the seat of thought and memory, reductive and eliminative theories of the mental/cerebral problem are confusing since mental events are not physical events occurring in the brain.

Neuroscientists, just like any physiologist, work in laboratory *closed systems*. Systems insulated from their surrounding causal background are indispensable in physiological research. It follows that the mind and the brain are being equated.[163] However, such artificial isolated systems, that are necessary for methodological reasons, are being cut off from the real world, from the *open systems* with its *ceteris paribus* features, in which animals or human beings are living, and in which clinicians and epidemiologists operate.

Against this background, brain imaging has been used in numerous studies endeavouring to visualize functions of discrete brain regions while subjects are engaged in various activities. Efforts to link brain regions to feelings and emotions have confirmed previous studies about cognitive and mood alterations associated

[159] Crick (1994).

[160] Kandel et al. (1995).

[161] Edelman (1994, 109f., 130), Ghaemi (2007, 296).

[162] Churchland (1989, 36).

[163] Malafouris (2013).

with brain lesions, epilepsy, and the use of various drugs, although researchers are still far away from being able to read out anything specific.

Neurophysiology relies on high-quality images of the brain. Any increase in focal neural activity coexists with an increase in oxygen/glucose consumption and consequently in blood flow in those areas of the brain that are activated. The increase in blood flow alters the local magnetic environment, and this is why neural activity can be delimited with functional magnetic resonance (fMRI) scanners. Neuroscientists then compare such located MRI imaging with conscious experiences, perhaps or sometimes finding their neurological correlate, and thereby allegedly identify more or less precisely what area of the brain is associated with some conscious experiences. Research on the working of the brain as well as the limits imposed by brain functional magnetic resonance imaging, have by necessity imposed isolating the experimental subjects from their usual environment and their culture, for the sake of studying them. For neuroscience, there is only behaviour and physiology.

Even then, must we accept the irreducible reality of the mental or the psychological? Mental events such as *qualia* are ineffable, and they are unreachable by human understanding. Moreover, much of our thinking, our intelligence, our knowledge, our memory, our minds spread into our social and physical surroundings, into our culture, and are engaged in complex experiences outside our heads and our body. What makes our conscious experiences the kind of experience they are is not the neural activity by itself but the dynamic interaction to our external world, that itself depends on our neural openness to changes in our relation to things. Mind crosses out of the skull in a shared socio-linguistic environment.

An epiphenomenon is an occasional product of some process that has no effect of its own. In attempting to solve the difficulty of reconciling the existence of mind with the notion that only physical events can cause physical events, epiphenomenalists such as C.D. Broad, contend that mental events or states of consciousness and experiences, have no causal powers and produce no effect on the physical world. They accept one-way causation whereby either physical events produce mental events (functionalism), or else they are two systems running in parallel (for Malebranche), on order to avoid causation between mind and body or between body and mind.[164] But this is missing the point, since it ignores the intentionality of mental events, our awareness is *of* or *about* something other than themselves. If we are looking at a garden of roses, 'intentionality goes *upstream,* it points *from* effects (nerve impulses in the higher levels of visual pathways) *backwards* to their causes (the interference between the object and the light).'[165] And then it points further backwards to the roses bathed in the light and producing the effect.

Consciousness does not have its locus in the brain, just as digestion happens in the digestive tract. Mind is more than a brain.[166] Talking of the brain perceiving,

[164] Jackson (1982).

[165] Tallis (2011, 105).

[166] Noë (2009, 87–89).

knowing, believing is common language among the neuroscientists, namely inserting consciousness into the matter of the brain, but this is incoherent. Pretending to ascribe psychological attributes to the brain is not a scientific question. Psychological powers, perceiving, thinking, feeling emotions, planning, and intending, are attributes of human beings, not of their brain or of their heart. It makes no sense to replace those ordinary psychological explanations by neurological explanations: brains are not conscious, persons are.

Gilbert Ryle argued that some *category mistakes,* for instance the separation of body from mind and its behaviour are made by misunderstanding what categories their concepts belong in, just like talking about the nutritional value of prime numbers. Neuroscientists often represent the facts of mental life as if they belonged to one logical type or category (or ranges of types and categories) when they actually belong to another. The confusion springs from what one might call the different *fields* or *categories* to which belong the two arguments that are either those of closed or of open systems. Neuroscientists, confined in their closed system that suggests to them that you are your brain and that you stop at the limit, have uncritically adopted a language of cognitivism where 'mind' is largely replaced by 'brain'.

P.M.S. Hacker who made a major contribution to this question, contends that it is nonsense to ascribe to the brain psychological concepts that make sense when ascribed to persons (and other animals). Wittgenstein averred that "*Only of a human being and what resembles (behaves like) a living human being can one say: it sees, is blind; hears, is deaf; is conscious or unconscious.*"[167]

Philosophers have argued for ages over the nature of persons. How do we decide what determines the numerical identity of someone? How do we commonly use this concept? To be a person is to possess the abilities needed to qualify for the status of a moral agent.

The best answer seems to be what philosophers call constitutive reductionism according to which persons are constituted by their physical and psychological continuity, and nothing over or above these continuities. This rule out the claim that persons are separately existing entities over and above their physical and psychological continuity. Whether physical continuity takes precedence over physical continuity or conversely is a matter of debate in academic philosophy.[168]

All in all, the brain makes no decision, but it is not indecisive. The brain is not conscious: only living creatures can be conscious or unconscious. The brain, brain states and brain processes are not intentional either.

This way of explaining of neuroscientists and cognitive psychologists is termed the *mereological fallacy.* Mereology is the logic of relationship of part to whole, and the mereological fallacy consists in ascribing to a part of a creature attributes that logically can be ascribed only to the creature as a whole.[169] Explaining the intentional stance is to engage in explanatory practices at the *agential level,* rather than

[167] Wittgenstein (1953, §281).

[168] Parfit (1999, 2007).

[169] Bennett and Hacker (2003).

describing some feature at the cerebral level, a category mistake. We do not need some sort of glue between mental science and brain science.

Some neurophysiologists opposed to this neuroscientific received view, characterize the closed approach of neuroscience as *neuromania* or *neuromythology* or as *the grand illusion,* and are arguing for the extension of our mind schema and our capacity of thought beyond the internal private resources of an individual: they may be partly but not all inside us.[170]

Even so, this is not to deny or to question the neural basis of cognition. Surely, the brain is a *necessary* condition and vehicle for experience or for the mind, but it is not a *sufficient* condition.[171] Neurosciences do not explain consciousness, but consciousness is unthinkable without neural processes.

More on Neuroscience and Neurophilosophy

So, the final account leaves us with Wittgenstein claim in *Zettel*:

> *No supposition seems to me more natural than that there is no process in the brain correlated with associating or with thinking; so that it would be impossible to read off thought processes from brain processes. I mean this: If I talk or write there is, I assume, a system of impulses going out of my brain and correlated with my spoken or written thoughts. But why should the* system *continue further in the direction of the centre? Why should this order not proceed, so to speak, out of chaos?*[172]

If brain activities can be mapped in some detail using the various methods of brain imaging, does the ability to locate some neural patterns imply that correlated belief, desires, hopes, obsession, anxiety, or fears are being therein structured, somewhere within the nervous system? Actually, the majority of brain imaging studies is correlational so that they tell us nothing about causation and even less about identity. Neuroimaging does not answer this crucial question since where something happens is no answer to how it happens. Even if each mental event is paralleled with a physical event, this does not imply that mental talk can be reduced to neurophysiological talk. What the neurophysiologists observe is that a certain brain change coincides or is immediately followed by something the patient feels or does. But there is nothing in this observation to correspond to the brain alteration causing some feelings or behaviour. For exempla, a seven-year-old Australian boy missing the visual processing centre of his brain, has been described, and despite the fact that the primary visual cortex is crucial for sight, the child had a normally preserved vision, except for a strong short-sightedness. His brain still had a great neuroplasticity because he lost his sight the first two weeks of his life, when visual information could still be rerouted via other centres.

---

[170] Noë (2009, 87–89), Tallis (2011, 2004a), Legrenzi and Umiltà (2011).

[171] Tallis (2011, 90).

[172] Wittgenstein (1981, §608).

For all that, the fact that different areas of the brain are in control of different mental functions is something physiologists and neurologists have known for a long time, using evidence from lesions and other forms of brain damage. If so, then, fMRI images and brain scans are telling us something we already knew. Neuroscientists are hoping to be able to read off answers about the mind. But knowing more about the brain will not help, since, in order to know how to interpret images, we must already apply our knowledge or real psychological structures.

Perhaps it is enough for the present purposes to quote Ludwig Wittgenstein when he is asking[173]: do, "*certain physiological processes correspond to our thoughts in such a way that if we know the correspondence we can, by observing these processes, find the thoughts ...?*"

> *I suppose we imagine the correspondence to have been verified experimentally. Let us imagine such an experiment crudely. It consists in looking at the brain while the subject thinks .... assuming that the subject is at the same time the experimenter, who is looking at his own brain, say by means of a mirror' ... then, I ask you, is the subject experimenter observing one thing or two things? ... The subject experimenter is observing a correlation of two phenomena. One of them he, perhaps, calls the thought .... The other experience is one of seeing his brain work. Both these phenomena could correctly be called 'expressions of thought'; and the question, "where is the thought itself?" had better, in order to prevent confusion, be rejected as nonsensical.*

Wittgenstein here tells us that it is nonsensical to talk of a locality where thought takes place. "*An object may exist and yet to be nowhere*", wrote Hume. Danto put it in another way: '*We may, for example, distinguish what goes on in my thoughts from what goes on in my brain if we but acknowledge the difference between an extensional and an intensional "in". The thought itself may be a half-inch back from my forehead, though what is in the thought may stand in no spatial relation to my forehead (even when it is my forehead I am thinking of).*"[174]But there is a further consideration of immediate philosophical interest. If we are identified with our brain, namely with a piece of matter, we probably cannot escape the laws of material nature. If our actions were mere physical events, it would follow that we are deprived of volition and free will and of personal responsibility.[175]

There is an alternate perspective afforded by ontological *naturalism,* according to which mental properties, consciousness and pain are neither neuronal processes, nor non-neuronal processes, and they are not ghostly processes either. This standpoint avoids dualism similar to materialist philosophy. Ontological naturalism contends that everything that exists in the world, every object, every property, or event, is a scientifically describable and measurable entity, including mathematics or logic that can be explained by the natural sciences. This allows to get rid of any appeal to mysterious, supernatural entities that would obstruct or obscure our ontology, so that it avoids dualism. It is perfectly compatible with our cognitive abilities or with our normative concepts if we endorse a non-descriptivist stance towards our mental

[173]Wittgenstein (1969a, 7–8).

[174]Danto (1999, 27).

[175]Tallis (2011, 51–57).

abilities: they do not stand for facts that can be characterized in some scientific discourse; yet, they are ineliminable, not because they refer to corresponding properties in the world, but because they express commitments and evaluations that people acquire over the lifetime, and they occupy no natural or non-natural space at all. Gilbert Ryle defended this non-factualist dispositionalism, which rejects the idea that if something is considered a genuine cognitive process, it is the product of *internal* processing or computation of representations. Dispositions, according to Ryle, are neither witnessable nor non-witnessable since they are not actually existing properties.

To sum it up then, mental, and behavioural events do not stand in spatial relations to anything: they lack mass, momentum, and velocity. And assuming that the content and meanings of our mental states are encoded in the brain does not imply that they are actually in the brain. If so, they cannot be fully understood by focusing exclusively on what goes inside our heads.

Splitting Mental Disorders from Neurological Disorders

Clinical medicine usually distinguishes between two broad groups of abnormal behaviour: neurological syndromes which are the behavioural expression of some specific structural or functional pathology in the brain, while psychiatric syndromes are harmful mental conditions supposedly due to abnormal structural or functional pathology of the brain and to experiential factors.[176] One needs both Ryle's ghost and his machine, empathetic understanding and neurophysiological explanation, cognitive therapy and antidepressant, reasons ad causes, movement qua movement and movements qua signs, first and third-person account, the body that is the body and me that is mine, or folk psychology and neurophysiology. Albeit W.V. Quine believed that "… *We must perhaps acquiesce in the psychophysical dualism of predicates, though clinging to our effortless monism of substance*"[177] and we have to conclude that "*considerations of scientific parsimony and theoretical strength suggest that physiological accounts will be the ultimate victors in this rivalry*".[178]

What distinguishes a mental from a neurological disorder? The dividing line is not as clear-cut or ascertainable as it might seem. The borderline is unavoidably blurred since it is not conceptually well-grounded: ICD-11 classifies encephalitis as neurological disorders but postencephalitic syndrome and post concussional syndromes as mental. Amnesic syndrome and dyslexia are borderline conditions related to damage to a set of interconnected brain structures.

Besides, 'proper' mental disorders are characterized by a lack of specific diagnostic neuropathology, with the exception of conditions resulting from brain

[176] Kendell RE., 110–116, in: AL. Caplan et al. (2004).

[177] Quine (1995, 87).

[178] Clark (1980, 17).

damage such as Alzheimer's disease, Lewy Body dementia; brain tumours, arterio-sclerotic vascular diseases, multiple sclerosis or general paresis of the insane (a dementia syndrome caused by syphilis) and anything of that ilk, which, with their specific neuropathology may cross the boundary of neurology and psychiatry.

Next, the question of immediate interest, perhaps, is whether psychiatry will, someday, be absorbed in neurology.[179] Even if the distinction between organic and functional disorders remains useful, at least at the opposite ends of the spectrum—e.g. agoraphobia at one end and Alzheimer at the other—psychiatrists pin their hope on producing syndromes, in some utopian future, that will mostly or exclusively reflect some single underlying pathophysiology for each category. Should schizophrenia be broken up into clinically homogeneous mutually exclusive and jointly exhaustive subgroups with relevant specific neurobiological data, would these resulting mental disorders be absorbed into neurology? Would we reclassify such diseases as neurological? Are mental disorders mere complex genetic brain disorders and is psychiatry meant to slowly fade into clinical neuroscience?[180]

On the other hand, McHugh & Slavney[181] offers grounds supporting the distinction between psychiatric and neurological disorders: "*In the everyday world of the clinic, psychiatrists are distinguished from other medical specialists not because they are concerned with "minds" rather than "bodies", but because they focus on complaints appearing in people's thoughts, perceptions, moods, and behaviours rather than their skin, bones, muscles, and viscera.*" For a psychiatrist, mental suffering and harmful behaviour features are dominant and operative in the construal of mental disorders; they have to be acknowledged at the onset of the analysis since they are epistemically prior and take precedence over associated patho-physiological brain malfunctions, although recent developments suggest that mental disorders might have by and large a significant organic component.[182]

To sum it up then, Jablensky and Kendell claims that an eliminativist "mindless" psychiatry that discards psychopathology and that relies on biological models is a possible situation "*It is much more likely in our view that clinical psychiatry will retain psychopathology (i.e., the systematic analysis and description of subjective experience and behaviour) at its core.*"[183]

---

[179] Fizgerald (2015).

[180] Insel and Quirion (2005).

[181] McHugh and Slavney (1983).

[182] David and Nicholson (2015).

[183] Jablensky and Kendell (2002, 22).

Mental Disorder and the Brain

Neuroscientific and genetic research will be increasingly tied to explanation, understanding, and treating mental disorders. Disease-specific mappings of local brain regions or of some distinct neural networks have failed to materialize and could well be unrealistic. The goal of identifying disease-specific mappings to distinct brain regions or to specific neural networks is probably untenable even if neuroimaging has provided evidence correlating symptom dimensions to various types of malfunctions in the brain. A gap necessarily remains between what the patient experiences, and the correlated brain abnormalities, if any.[184]

Even so, 'pain' or 'anxiety' or 'obsession' are correlated and may assumedly be partly caused by neurophysiological processes, even if they cannot be defined by, or identified with them. We must already know what 'pain' or 'anxiety' or 'obsession' mean, to know what kind of experience the neurological process leads to.

Extensive progress has been made in neuroscience devoted at the neurobiological correlates of some mental disorders, mainly in cases of schizophrenia or of bipolar disorder.[185] Understanding how the physiology of neural networks becomes malfunctional holds great promises for the development of new approaches to diagnosis and treatment. It may involve several large-scale transdiagnostic neural networks as well as small local microscale circuitry. For example, in schizophrenia or in bipolar disorder, cognitive disturbances (disorganization, distractibility, disinhibition, or psychosocial disability) are correlated with spatially distributed structural and functional anomalies in the prefrontal, the anterior cingulate, and the parietal cortex; and perceptive disturbances (hallucinations, delusions) are correlated with the prefrontal, anterior cingulate, parietal, and visual cortex. The goal of identifying specific topographic mappings to localized brain regions or to district neural networks has not materialized and may be untenable for instance in schizophrenia. However, a systems neuroscience approach may be more fruitful, i.e., pertaining to brain structural, functional and connectivity alterations, although a fully mechanistic description remains elusive.[186] The measurement of such impairments, initially proposed as a test of brain damage, emphasize the role of brain dysfunction in schizophrenia.[187]

In recent years, large-scale research projects at Washington University in St Louis, such as the *Life Span Human Connectome Project* (HCP), led by David van Essen and Kamil Ugurbel, and the Life Span Human Connectome projects, have focused on defining and mapping the functional connections of the healthy brain across the full human life span.[188] These include networks involved in cognitive

---

[184] Gur and Gur (2010).

[185] Mundt and Spitzer (2001, 13–19), Strik (2001), Frangou (2014), Zorumski and Rubin (2014), Editorial (2016).

[186] Frangou (2014. Fig. 1).

[187] Mirsky and Duncan (1999).

[188] Van Essen and Glasser (2016).

(thinking) function, attention to internal emotional states, and the 'salience' of events in the environment. The Life Span Connectome initiative is setting up research projects to address various diseases including schizophrenia and Alzheimer disease. The Brain Activity Map Project in the United Kingdom proved catalytic for President Obama's Brain Research through the *Brain initiative* (Brain Research through Advancing Innovative Technologies). Launched in April 2013 it aims to achieve over the next 10 years a comprehensive mapping of the activity of single neurons and their connectivity by applying nanotechnologies and large-scale computation techniques.

The National Institute of Mental Health conducted the *Research Domain Criteria* (RDoC) project as a strategic counterpart to the DSM/ICD classification, based on primacy of translational research, integration of neuroscience and behavioural science, a quantitative dimensional approach to psychopathology.

Shall we be able to identify types of mental disorders, which would coincide with some distinct neurologically definable abnormalities? It is to be hoped that the answer is yes. What we lack are biomarkers. Even then, that some disruption of neural circuits or some disease-specific marker might map neatly onto some newly defined disorder is one thing. Quite another would be to bring all categories of mental disorders under the canopy of some sort of unified causal mechanism. Contrariwise, forfeiting the concept of natural kind, or accepting that it is not an all-or-nothing feature, would bring mental disorders nearer to somatic diseases.

There is a big gap between statistical associations of common genomic variants with specific disorders, symptoms or manifestations, and the evidence of causality. Since most of such associations have minute effect sizes, it follows that several hundreds of genes make statistically significant but negligible contributions to the disease risk. In a contrary manner, improved neurophysiological investigations and neuroimaging are more likely to yield reliable measurable biological, behavioural, or cognitive markers (*endophenotypes*), thus being of pragmatic utility in the evaluation of patterns of individual pathogenesis. Genome-wide association investigations on data from more than 100,000 individuals have found out common polymorphisms shared by apparently unrelated disorders, such as schizophrenia, bipolar disorder, autism, attention deficit/hyperactivity disorder and possibly certain forms of intellectual disability and epilepsy. All of this provides strong justification for pleiotropy as a rule, instead an exception in the role genetic plays in psychiatric disorders.

In the light of this, it seems feasible that neuroimaging might help differentiating between patients and healthy groups of individuals and among various mental disorders. This might start with the identification of undifferentiated homogenous groups of clinically identified patients, in terms of complex neural features, who would hopefully partake similar anomalies at the level of microscale circuits.

On the other hand, another research approach could consist in selecting groups of patients according to their neural, genomic, cellular or molecular, instead of their clinical profile, and in deliberate ignorance of the latter; this method could lead to new medically biological nosologies for mental disorders.[189]

Katharine Thakkar, a neuroscientist from Michigan State University believes that psychotic symptoms might be related to a defective signalling between regions of the brain that are involved in producing movement (agency) and regions of the brain involved in sensations (sentience). These signals allow to predict sensations resulting from actions, and the match between them contributes to our feeling of agency. These processes probably go awry in psychoses, be it of schizophrenic or of bipolar nature, with an impaired contact with reality and a disruption of the concept of 'I'. This is an interesting attempt to link psychotic experiences with physiological processes.[190]

If so, then, could some disruption of brain processes give rise to mental disorders? If physical variables predict the occurrence of mental variables, their correlation may suggest a causal link between them. Surely, brain activity is a necessary condition for consciousness. And if brain malfunction or brain abnormalities do not foreordain mental disorder, they are a necessary though not sufficient condition for some mental disorder.

But do mental events or mental disorders merely emerge from brain activities the way stock markets emerge from the collective behaviour of individual investors, or do they emerge like elasticity emerges from the mechanisms of atoms? Actually, the stock market and the collective behaviour of individual investors are interacting kinds, since they are bi-directionally causally connected with one another by a (binary) symmetric relationship, the emergence, and its converse; meanwhile, elasticity is related to atoms in an asymmetric one-way relation. So, mental disorders are echoing the emergence of the stock market, not that of material elasticity.

This means that causal laws, should they be found, are bidirectional, i.e., interactional, like those of the stock market. Input and output may be mental or physical: if the disruption of brain circuits may cause mental disorders, mental states may in turn give rise to mental disorders with their aberrant relevant circuits.

Not surprisingly, clinical advances in psychiatry based on neuroscience have not been rapid.

Even so, Karl Jaspers[191] wrote that "*Diagnosis is expected to characterize in a comprehensive manner the whole morbid occurrence which has assailed the person...*" and he added: "*The statement that all mental illnesses are cerebral illnesses... is nothing but dogma ... To do so is to imply that we are making an absolute of cerebral events, taking them as the very substance of man and considering all human events as brain events.*"

One does not call the TV repairman because the programme is bad.

Jaspers adds: "*From the psychological point of view cerebral diseases are just one of the causes of psychic disturbance among many. The idea that everything*

[189] Lende (2014).

[190] Thakkar et al. (2017).

[191] Jaspers (1963, 496).

psychic is at least partially conditioned by the brain is correct but is too general to mean anything."

Depression is one of the least heritable psychiatric conditions, while bipolar disorder has a heritability of about 85%. Genetic studies showed a significant relation between risk of depression and polymorphisms of a serotonin transporter gene that is responsible for serotonin uptake. Yet, the risk of depression is also associated in dose-dependent fashion with a number of stressful life events: one, two or three events, are respectively associated with 30%, twofold and threefold higher risks for depression. We know that stress increases dramatically the release of the neurotransmitter glutamate, the excess of which can damage or kill nerve cells, a process to which the hippocampus is particularly vulnerable.

A study of the Centre for Addiction and Mental Health (CAMH) of Toronto shows that untreated depression lasting more than a decade leads to increased inflammation in the brain. Positron emission tomography (PET) brain scans were carried out on 50 depressed patients, half of whom had suffered the condition for more than 10 years. Images were also taken from a comparison group of 30 non-depressed individuals. Participants with long-lasting untreated depression were found to have 30% more translocator protein (TSPO), an inflammation marker made by microglia than those who had experienced depression for shorter periods. Is depression an inflammatory process or does the inflammatory result from the disease itself?[192]

Mental and behavioural events and corresponding physical events are thus linked in whatever way cause and effect are liable to be connected: causality, should it be ascertained, may proceed from physical-to-physical events, from physical to mental events and conversely from mental to physical events, as well as from mental to mental events.

Is neuroscience still like a Ptolemaic geocentric cosmology, assuming that the earth is stationary at the centre of the Universe. It is a closed system that does not go beyond nature/nurture to look at what interferes with the function of the brain: it assumes that mental disorders are biological disorders involving brain circuits. Mental health research centres are brain research institutes not mental health institutes.

However, then mental activity may be influenced by forces outside the brain. Brain states can be what they are, details and objections aside, independently of what happens outside the skull, but mental states are essentially relational and world involving. It is nonsensical to locate them anywhere but even more so within the skull.

There is a bidirectional communication between the gut and the brain between peptide hormones released from the gut and intestinal bacteria. This communication is regulated by different systems including the endocrine, immune, autonomic, and enteric nervous systems. The gastrointestinal tract is the largest endocrine gland. Gut peptides can bind cognate receptors on immune cells and vagus nerve, enabling

[192] Centre for Addiction and Mental Health (2018).

indirect gut-brain communication. The body's microbiome or bacterial system are increasingly recognized environmental factors that influence brain development, behaviour and affect how people think and feel. Microbiota influence stress-related behaviour.

It follows that the gut-brain axis has an important role in anxiety and depression, since the gut microbiota are regulators of anxiety and depression, and gut-derived peptides are signalling molecules in microbiome-gut-brain communication. Bacteria in the gastrointestinal tract–be they commensal, probiotics or pathogenic– can activate neural pathways and brain signalling systems and have an effect on the development of anxiety and depression. The macrobiotic compositions of healthy controls are indeed significantly different with major depressive disorder.[193]

But the question now before us is whether all mental disorders must be brain disorders. The mereological fallacy suggests that since all mental processes take place in the brain, all mental disorders are brain disorders. But surely, this argument starts from a false premise. One can at least conclude that whether a given condition listed in DSM-5 is judged as to be a disorder, requires a criterion independent of brain functioning or malfunctioning. What clearly emerges from this is that mental disorders are harmful psychological dysfunctions.

Can Mental States Be Multiply Realized?

It is often assumed that mental properties are multiply realizable.[194] The semantic content of mental states, occurrent thoughts, desires, anxiety, or disposition to addiction, cannot be read from brain states even if they are foreordained by them. But surely, it is to be hoped that every type of mental disorder will someday be shown to be strictly correlated with some relevant specific type of brain state, although it might seem more plausible that any particular (token) mental state might be correlated with some particular brain state.

Even then, the assumption of one-to-one structure-function mappings is a disputed issue. 'Degeneracy' is defined as the capacity of elements that are structurally different, to realize the same function or yield the same output.[195] "*Thus, broadly defined, degeneracy is found at all levels of biological organization, from the molecular, cellular, and genetic levels up to the level of organism: different nucleotide sequences encoding the same polypeptide, different antibodies binding the same antigen, different patterns of muscular contraction yielding the same movement, and different encodings of the same message.*"[196]

---

[193] Foster and McVey (2013), Smith (2015), Zheng et al. (2016), Lach et al. (2017).

[194] Putnam H, 429–440, in: H. Putnam 1975ctive.

[195] Edelman and Gally (2001).

[196] Figdor (2010).

So different brain abnormalities can correspond with the same mental disorder. Furthermore, a single type of brain activity may realize more than one mental state. This pluripotentiality seems to license the view that mental disorders cannot be characterized in terms of shared physical properties; mental disorders might not be numerically individuated since two of a same brain state might perfectly well fail to have the same manifestations or the same semantic content. "*There is token identity, to give it the jargon, but type diversity*".[197]

A look not at the genes, but at the relative activity of genes expressed in samples of the cortex of patients suffering from one of the five mental disorders (autism spectrum disorder, alcoholism, schizophrenia, bipolar disorder, major depressive disorder), was compared with samples of patients who had no diagnosis. A CLA study failed to find any similarity between alcoholism and the four other conditions. Bipolar disorder and depression shared important similarities, and so did bipolar disorder and autism spectrum disorder, as well bipolar disorder, and schizophrenia. Such studies show how twisted are the roots of phenomenally seemingly diverse mental illnesses.[198]

Clinical appearance does not fully map on the neurobiological associated mechanisms[199] although sometimes it does when such as deficit in cortical plasticity leads to impaired cognitive performance observed in patients with schizophrenia.[200] If mental disorders of the same type can be physically distinct, can brain abnormalities be used as a criterion to determine whether a proposed disorder is a valid condition in its own right?

Could the assumptions of adopting single or multiple validators such as molecular genetics and molecular biology, neurochemistry, neuroanatomy, and neurophysiology, as well as cognitive neuroscience and response to treatment, by the dominant research programme remain a moot point? On this score, the model consists in a rejection of the requirement that malfunction is a definitional criterium of the notion of disease. A disease may be mono-malfunctional or multi malfunctional, so that malfunction cannot be a necessary feature of its definition, since multiple realizability would mean multiple functions, which would mean multiple diseases.

May Mental Disorders Be Adaptive?

Although there is nothing globally good to be said about mental disorders, when a patient's signs and symptoms are not too out of control a mental disorder may play a positive role in a patient's life. Kay Redfield Jamison psychiatrist who suffers from bipolar disorder has provided evidence that suggests an aetiological

[197] Quine (1995, 87).

[198] Gandal et al. (2018).

[199] Danielyan and Nasrallah (2009).

[200] Kantrowitz et al. (2016).

relationship between mania and artistic creativity as illustrated by Coleridge, Byron, Schumann, Van Gogh, and many others.

"*I believe*, writes Sandison, *that extreme and persistent depression is one of the conditions to which man is heir, and that its absence in a person is rare and more abnormal than is a liability to it*".[201] The suffering of mental illness can be unspeakable but it may also "*bring greater joy over long periods than any state of normality*".[202] Did not melancholia come to have a positive connotation in the 1600s as an affliction of those of superior intelligence? Mentally ill individuals may also adapt better than healthy individuals in certain difficult environments such as concentration camps or combat.[203]

Furthermore, most of us tend to believe that fewer bad things will come about to ourselves than to others. Most human beings are overoptimistic and tend to overestimate their potentialities and effectiveness and to believe they have more control over the circumstances and the way things are than they actually do. Comparing the beliefs and inferences of depressive patients with those of normal people, suggests that the attitudes of depressed individuals are often more realistic and more rational than those of non-depressives, who exhibit more misjudgements and errors.[204]

Depressive conditions might thus have a selective advantage.

All in all, one may wonder whether severe depression over a substantial period might not be justified, given the world we live in, for instance the tragic twentieth century, or given the circumstances in which some persons find themselves, the global warming, and the destruction of our planet? If so, then, if despondency may be an appropriate response to one's circumstances, is it malfunctional? If we take it to be a disease, does it mean that it results from some brain pathology? Laing, himself suffering from schizophrenia, writes, "*Schizophrenia is a special strategy that a person invents in order to live in an unliveable situation*".[205]

In like manner, people with autism spectrum disorder may have better musical pitch recognition, superior visuospatial skills, greater attention to details and better rational decision-making due to a lower susceptibility to emotional factors.

Taking stock, mental disorders are not modules, which can be taken out of the realm of biological processes; they are not clear-cut, distinct features which stand out against the background of people's mental lives either, but they are being pieced together from abnormal features, the psychological makeup of particular clinical cases includes normal and abnormal, positive, and negative and maladaptive traits.[206]

Human beings, as well as all sentient beings, foster an inner thrust guided toward fulfilment. This existential thwart, namely whatever gives shape to our lives and makes them creative, budding, dynamic, and sprightly, may be partly or completely

---

[201] Sandison (1972, 1228).

[202] Jameson (1985).

[203] Tas (1951), Ripley (1951).

[204] Alloy and Abramson (1988).

[205] Laing (1965, 195).

[206] Nesse (2019, 89–94).

blocked or diverted by the disease process and its cognitive distortions. But what is primarily pathological, that is what serves to advance the disease, cannot be disentangled from what remains of the patient striving to achieve maturation and recovery, and the progression of his *Eigenwelt* under adverse circumstances.

Is Schizophrenia Adaptive?

Schizophrenia reduces rates of producing offspring so that it is not adaptive in the sense of natural selection. If so, then, why are its incidence constant and uniform all over the world?

Even though schizophrenia substantially decreases fertility rates,[207] its incidence seems to be more or less constant all around the world (around 12 per 100,000 population per year) with a prevalence oscillating between 0.5 and 1.5% of all populations: this suggests both that the disease dates back from prehistory and that it might also give some selective advantage.[208] This situation might be comparable to balanced polymorphism already mentioned in the case of sickle cell anaemia in areas of high malarial prevalence. It is often assumed that relatives, parents, and siblings of individuals with schizophrenia, the heterozygotes, if not psychotics themselves, have some selective advantages in the cognitive and emotional system as well as an increased innate resistance to some infectious diseases.

Moreover, loci associated with evolutionary markers suggest that several gene variants that went through positive selection are linked with cognitive processes. There are basic identities between acute schizophrenics and marginally adjusted shamans of certain primitive cultures.[209] There are no significant differences in the sequence of underlying psychological events that define their abnormal experiences. The major difference lies in the cultural acceptance of a life crisis.

For one thing, shamans are men who communicate directly with his 'spirits' in the other world, and who exhibit the most blatant form of psychotic behaviour, i.e., non-reality-oriented ideation, abnormal perceptual experiences, trance-like ecstasy states, profound emotional upheavals, and bizarre mannerisms. They are usually accorded great prestige and the faith in their power is total.

For another, in the psychosocial environments of our occidental culture, which do not provide reference guides for understanding such types of crises of experience, an acute episode of schizophrenia is being boosted since it has absolutely no cultural significance beyond the proof of his insanity.

Sullivan concludes: "*The essential difference between the psychosocial environments of the schizophrenic and the shaman lies in the pervasiveness of the anxiety that complicates each of their lives. The emotional supports and the modes of*

[207] MacSorley (1964).
[208] Crow (1995).
[209] Silverman (1967).

collective solutions of the basic problems of existence available to the shaman greatly alleviate the strain of an otherwise excruciating painful existence. Such supports are all too often unavailable to the schizophrenic in our culture."

Then, beyond this, whole communities, for example among the Tanala of Madagascar and the Mohave, a Native American people indigenous to the Colorado river of Southwest United States, have been known to live their lives under what could be qualified, from a psychiatric point of view, as altered states of consciousness. They dream of supernatural experiences often confused with everyday activities so that these communities could tentatively be labelled as psychotics.[210] A Mohave coming of age must consume parts of Datura plants, a potent hallucinogen, in a rite of passage in order to enter a new state of consciousness.

There is a further consideration, which Daniel Dennett believed to be important. Julian Jaynes, a Princeton University psychologist is famous for one single book.[211] Julian Jaynes and more recently David Horrobin[212] have argued that the genetic change that brought about schizophrenia and related serious mental illness like bipolar disease was a necessary condition for the development of civilization and the reason why we became human. It is this genetic change that was associated with the development of human higher order consciousness and self-knowledge and drove the explosive flowering of symbolic thoughts and activity, arts and religion around 3000 years ago. And it is the same genetic change that brought in the dark side of humanity, the agony of mental illness but also the nemesis of paranoid dictators. Much of the discussion is fascinating but highly speculative.

Beyond this, Jaynes' main idea was that our modern form of consciousness and the suppression of hallucinatory experiences began no more than 3000 years ago but can be dated after 2000–1000 B.C. when the mental separation between gods and people gradually dissolved.

In earlier times—and this might apply to the Neanderthal who became extinct about 40,000 years ago—human mental function was characterized by auditory as well as visual hallucinations, with which people heard the voices of the deities talking to them and telling them what is to be done. The hallucinated voices reflected right hemisphere activity channelled via cerebral commissures to the left or language dominant hemisphere and construed as linguistic utterances. Schizophrenics present indeed a decrease language lateralization. Modern conscience began only when this psychological process became internalized and acknowledged as coming from within the observer's own minds.

But before that, human minds were split in two, *the bicameral mind,* resulting from a dissociation between the two hemispheres of the brain. Jaynes identified strong evidence for this in Homer's *Iliad*, where the poet's heroes are continually being provided by instructions and advice from the various gods. This, according to Jaynes, is not merely figurative language, but are actually an authentic description

[210] Benedict (2002).

[211] Jaynes (1976).

[212] Horrobin (2001).

of how people really perceived the world around them. Beyond this, the characters of the *Iliad* lack of insight, of introspection and self-reflection, and do not have our soliloquy and our interior stream of consciousness. In a contrary manner, their resolutions, plans and their inventiveness evolved at an unconscious level and were then divulged to them by a voice or by the ghostly figure of a friend or a god. These voices are comparable to those of a schizophrenic today. The difference is that they were not taken to be something abnormal, but a command for the gods. Jaynes described the same process at work in the art and literature of the civilizations of Mesopotamia and of the Hebrews. Tethys came to Achilles and Yahweh up to Moses. The first serious indications of collapse of the bicameral mind appeared at the time of Egypt's Middle Kingdom, around 1700 B.C.

Jaynes did not equate schizophrenia with bicamerality but he did explore the parallel between the bicameral mind and schizophrenia. Hypnosis and schizophrenia, according to Jaynes, are reversion to bicamerality. He writes, "*volition came as a voice that was in the nature of a neurological command, in which the command and the action were not separated, in which to hear was to obey*". This ancient control structure manifests itself in the schizophrenic hallucinations when the patient is invaded by admonitory voices that pass condemnatory judgements and command behaviour, just like three thousand years ago, when men had no consciousness and were automatically obeying the voices of the deities.

Disease Mongering: The Case of Mental Illness

Disease only exists after "experts" have agreed that it exists as a clinicopathological phenomenon, a medical construct. Redefining diseases and deciding to expand them by lowering thresholds value for diagnosis seems to be an encouraging trend. It captures more people at lower risk of future illness and extends treatment benefits to patients with early disease. Overdiagnosis does not imply misdiagnosis. Even so, such technologies can be harmful and helpful. Welch summarized the growing concern in the medical community: "*The biggest problem is that over-diagnosis triggers over-treatment, and all our treatments carry some harm.*"[213]

Disease mongering assumes that the selling of sickness that broadens the boundaries of illness in order to grow markets for those who sell and deliver treatments. To be sure, pharmaceutical companies have a vested interest in enlarging categories of mental diseases. It is also an informal alliance with doctor's groups, patient's advocates, using the media to push certain views about a particular health problem.[214] The media themselves, independently of the pharmaceutical industry, play an important role in the creation of new syndromes such as electromagnetic field sensitivity.

---

[213] Welch (2015).

[214] Moynihan et al. (2019).

Most panels which proposed changes to disease definition had a major part of members disclosing financial ties to pharmaceutical companies. As a result, a 2011 report from the US Institute of Medicine (IOM) recommended that, whenever possible, guideline developers should not have conflicts of interest, that a minority of the panel members involved in guideline development should have conflicts of interest, and that the chairs of these panels should be free of conflicts.[215]

For one thing, disease-mongering refers to an intentional commercial strategy of drug companies consisting in tinkering with the definition of a given disease or creating a new one, to promote the sales of their drugs; they have a vested interest in enlarging categories of mental diseases.

For another, this does not mean that the pharmaceutical industry is simply concocting diseases or that psychiatrists are being misled into diagnosing well people as sick. People really suffer from depression, ADHD, panic disorders and are treated with the adequate medicinal products. Surrounding the kernel of many of these disorders is a wide zone of ambiguity that can be gouged out or expanded. A given condition once recognized as a disease needs to be treated with the medication that the companies produce. The larger the diagnostic category, the more patients who will suit within its, and the more psychoactive drugs they will be prescribed.

Evidence suggests that specialist guideline groups are lowering thresholds to capture more people at risk of future illness. Truth lies between these two extremes and shows two quite different faces. From the health care perspective, there is another way of looking at this issue. There is no doubt about the unhealthy conflict of interest held by guidelines panel members, but is it also a matter of *ad hominem* circumstantial fallacy, an argument which makes appeal to the special position of the person? That a medical researcher has financial ties with a pharmaceutical company does not show that his medical decisions are incorrect or unfounded. A speaker's race, sex or sexual orientation almost never give us any good reason to challenge the truth of what he says or the soundness of his argument. To evaluate an idea or an argument, one should focus on that idea, not on its origin.

Oddly enough, if the literature devoted to overdiagnosis is considerable and it has been the subject of multiple surveys indicating that it deals with a real problem, the question of under-diagnosis received surprising little attention, although its prevalence is considerable[216]; community surveys have shown that epilepsy, chronic bronchitis, psychoneurotic illness, and rheumatoid arthritis are more prevalent than suggested by morbidity statistics.[217]

I shall now turn to some concrete examples.

Ritalin (Methylphenidate) was first made nearly seventy-five years ago, and was identified as a stimulant in 1954. The FDA approved it in 1955. Beginning in the 1960s, it was used to treat children with ADHD based on earlier work starting with the studies by the American psychiatrist Charles Bradley who gave Benzedrine and

[215] Moynihan et al. (2019), Gonzalez-Moreno et al. (2015).

[216] Falagas et al. (2007).

[217] Last (1963).

psychostimulant drugs to then called "maladjusted children". The medical literature became filled with scientific papers on ADHD although it was called by different names over the years. It was widely regarded as safe, and it appeared to be effective for children with ADHD. Then, something happened. In the 1990's, the rates at which Ritalin was being used hit the ceiling. Production and prescription of methylphenidate rose significantly. The annual US production of Ritalin increased by 300% from 1990 to 1995 and it tripled during the early 1990's. There have been numerous papers addressing epidemics of ADHD and "*a definitive explanation for increased diagnostic rates over time is impossible leading to rancorous debates regarding the significance of such increases*".[218] Does it indicate an increased recognition of overlooked illness, a broadening of diagnostic criteria resulting in an increase in the overdiagnosis of normal individuals, or a diagnostic mislabeling of ill individuals? To be sure, patients diagnosed as suffering from ADHD often fulfill the criteria for schizophrenia.[219]

The alarming rise in ADHD prevalence from 2003 to 2011, occurred over a period of time in which the DSM definition of ADHD remained unchanged, so that the definition played a minor role in determining prevalence. Several current studies estimate the number of schoolchildren suffering from ADHD at anywhere between 3 and 5%. Joseph Biederman from the Harvard Medical School Psychopharmacology Clinic believes that it is around 10% and that Ritalin is not overprescribed. In the late 1990's, and in response to public alarm and a debate that made headlines about Ritalin, the American Medical Association set up a taskforce to look at the question whether ADHD was overdiagnosed. Despite the alarming statistics, the task force concluded that ADHD was so wildly diagnosed because physicians and the public know much more about ADHD than they did several years ago and are more prone to recognize it. Doctors also know more about Ritalin itself and are not preoccupied about the risk of abuse and more likely to prescribe it for long periods. Furthermore, adults were made aware that stimulants by increasing attention can help them concentrate at work.

ADHD is a syndrome manifesting a well-established heterogeneity. Evidence suggests that three neurobiological distinct circuits serving distinct neurocognitive functions become involved, and the malfunction of these circuits give rise to separable neuro-cognitive deficits, the malfunction of which underlie three distinct types of ADHD that can be distinguished in terms of distinct neurocognitive deficits.[220] There have been major advances in the treatment and the understanding of ADHD during the last twenty years. Although stimulants remain the US-FDA-approved medical treatment of choice with an exceptional response rate as indicated by findings of the Multimodal Treatment of Children With ADHD, a combination of behavior and medical therapy may benefit most patients. The diagnosis,

---

[218] First, MB, 131.

[219] Parnas (2017).

[220] Wigal (2009).

characterization, and standard setting for quantification of ADHD are crucial to assessing treatment effectiveness.

Panic disorder, another issue, was not listed in DSM-III. Until 1960, panic was understood as part of anxiety neurosis. In 1964, Donald Klein, a New York psychiatrist from Columbia University, suggested in a paper (partially funded by Geigy and Smith Kline) that panic is different from anxiety. This is what Klein called 'pharmacologic dissection'. What differentiated panic disorder from those with other form of anxiety, was a rapid response to relatively low doses of imipramine. We now know that panic disorder and general anxiety disorder (GAD) differ meaningfully in etiology and in pharmacologic and psychotherapeutic efficacy. Klein was part of the preparation of DSM-III, and he convinced his colleagues that 'panic disorder' should be a distinct diagnosis separate from 'generalized anxiety disorder.' The Upjohn Company then shifted its anxiolytic medication Xanax (alprazolam) as a treatment of panic disorder and funded an extensive clinical trial to show that panic is distinct from anxiety. In 1998, 2.4 million adults between 17 and 54 were treated for panic disorder.

Even so, diagnostic splitting has proven fertile in medicine, for instance the splitting of diabetes mellitus in type 1 and type 2, or the splitting of Down's syndrome into Trisomy 21 or into a translocation. Similarly, Kraepelin's 'manic-depressive insanity' includes what we now call major expression and bipolar disorder, or else separating anxiety disorder into panic disorder and generalized anxiety disorder (GAD).

It goes without saying that the ways in which and the extent to which DSM has been affected by direct or indirect pressures of the pharmaceutical industry is a central question. The pharmaceutical industry played a key role in establishing disorders and the global marketing contributed to strengthen the tendency to ever more disorders to be included in DSM. Rachel Cooper who wrote the best, the deepest, and the most authoritative report on this issue concluded that "*DSM has been shaped by the use of psychoactive drugs albeit to a lesser extent than has often been claimed.*"[221]

Stegenga *Medical Nihilism* is an extraordinary book and probably the best one and the deepest one on these issues. One doesn't need to agree with Stegenga to acknowledge that his account is always extremely challenging. Its shortcoming is to accept without any reservation his fundamental tenet. When he discusses problematic practices regarding disease definitions, he is encumbered by his formal premises: a disease, as he puts it, "*is a category, or what philosophers call a kind*". He probably means that they have essential properties, but most diseases have no essential properties. This question raises again the issue of the *quorum feature* analysed by Wittgenstein.[222]

---

[221] Cooper (2007, 126).

[222] Incidentally, Stegenga claims that the "bereavement exclusion criterion" mentioned in DSM-IV has been excluded from DSM-5, a surprising statement since it truly included in this last version. (DSM-5, 125–126, 134, 161, 194).

Some kinds have a natural existence independently of us, such as the individual elements of the periodic table, gold for instance, other such as the game of chess or psychiatric disorders share some properties that depend entirely on us: they are not natural, they are not true categorical entities in nature. In other words, Stegenga's strategy favours a reification of diagnostic categories[223] even though the authors of DSM-IV pointed out that: *"there is no assumption that each category of mental disorder is a complete discrete entity"*. However, the fact that a diagnosis is listed in an official classification and provided with precise operational definitions tends to promote tacit reification. Unfortunately, this reification becomes inevitable and may lead to a form of narrow-mindedness by the researchers and funding sources: new drug development tends to focus their interventions on treatment of DSM-5 defined categories despite ubiquitous evidence that pharmacological treatments are liable to be effective in treating a relatively wide range of DSM disorders. From this, how can one help researchers to break the diagnostic shackles resulting, rightly or wrongly, from DSM-5? A major conundrum posed by ICD and DSM systems is that the same document is being used by clinicians, educators, researchers, and administrators to serve the same purpose.

By contrast, clinical psychiatry's 'up' schedule needs an agreed way of describing the meaning of diagnostic terms, to know exactly what is being discussed, since some symptom rating schedule is certainly better than merely stating a diagnosis. All those symptoms currently assumed to indicate mental disorder and several other non-diagnostic symptoms that may be salient features for the patient such as worrying were combined in an interview to form the *Present State Examination* (PSE), that is probably the most comprehensive rating schedule. The PSE does not produce a diagnosis but a list of symptoms that are founded on the experience of the patient interviewed in his native language. The researcher determines how to use these symptoms, and whether they will be used to make a diagnosis. A comparison of complaints and symptoms in field studies may be full of insight. Complaints are what the subject says in replying to questions about how he feels or what is bothering him most. They reflect the unpleasantness and suffering of the illness experience. Symptoms in psychiatry are subjective experiences reported by the patient such as pain, anxiety, or physical signs such as body deformities. The clinician treats some but not all these complaints as symptoms that indicate the presence of a disease, so these will be given precedence over mere complaints in conceptualizing what is wrong. It is not possible to decide whether the complaints or the symptoms should be preferred as a clinical basis. Psychiatrists are likely to give more weight to the symptoms, but social workers probably give more importance to the complaints.

Anthony et al.[224] studied DSM-III diagnoses in a two-staged morbidity survey made by the lay Diagnostic Interview Schedule (DIS) method, in relation to a standardized DSM-III diagnosis by psychiatrists, in a sample of 810 Baltimore subjects:

[223] Hyman (2010).

[224] Anthony et al. (1985).

after the DIS interview by a lay interviewer, the subject was interviewed again by a psychiatrist using a PSE-type interview.

The psychiatrist did not know the results of the first interview.[225] The table is the one-month prevalence of a selected DSM-III diagnosis.

DSM-Diagnoses	Lay DIS	DSM-III diagnoses
Phobic Disorder	11.2	21.3
Alcohol Misuse	3.6	6.9
Major Depression	2.3	1.1
Obsessive Compulsive	1.3	0.3
Panic Disorder	0.8	0.1
Schizophrenia	0.7	0.5

The psychiatrist found twice as many persons with the Phobic Disorder as the lay interviewer, and similarly for alcohol abuse. On the contrary, the lay interviewer found twice as many subjects with Major Depressive Disorder than the psychiatrist.

Most psychiatric disorders are medical constructs since they would not exist without the activities of social but mostly of medical conventions. In the history of psychiatry, many diagnostic categories that were once accepted have been abandoned such as drapetomania, hypermania, demonomania, monomania, fanatic psychopathy, phonemic paraphrenia, parakinetic catatonia, hysteria, and homosexuality. Isn't likely that someday a substantial proportion of current categories will noti be represented in ICD or DSM manuals?[226] Mental disorders have no essence, since to have an essence, they would need to have a set of properties necessary and sufficient for them to count as mental disorders. It is much more plausible that our psychiatric syndromes arise from inter-connected neural networks that are manifesting themselves at the level of mind (severe anxiety) and at the level of brain. Psychiatric disorders can then be understood as emergent clinical clusters of symptoms arising from brain-based networks rather than from 'essences' located somewhere in the brain.

In closing, we may conclude with J. E. Cooper and N. Sartorius[227]:

1. That "... *we shall have to live with a variety of classifications, each serving the needs of a particular profession or a particular purpose.*"
2. *"In contrast to deciding whether a disorder is present or not, an alternative way is first to describe a number of dimensions which contribute to the disorder, and then state the position of the subject in each dimension"*.

[225] Anthony et al. (1985).

[226] Kendler et al. (2011), Kendler (2016).

[227] Cooper and Sartorius (2013), 54, 68.

Conclusion

1. Mental diseases or disorders are diseases in the same way as non-psychiatric diseases, except that the missing piece is the interplay between the individual and the environment and that there are no biomarkers. They are the result of bidirectional brain-mind determinism. Mental disorders may be the object of widening their diagnostic boundaries to expand market of those who deliver and sell treatments.
2. The clinical manifestations include fragmentation of the self, syntactic, semantic, and epistemic breakdown, a-rationality, and breakdown of autonomy, breakdown of interpersonal relations. They may sometimes be adaptive. There are two types of predicates: Diagnosis may be either valid or reliable. There may be degrees of severity: *the spectrum perspective.* It includes the pathology of face perception and the physiognomy of human emotions.
3. The classification of mental disorders, DSM 5 and ICD-11 are atheoretical and generally free of etiological assumptions. Clinically, psychiatric comorbidity tends to blur the distinction between clinical categories so there is a serious question whether they represent distinct comorbid disorders or spectrum disorders? Assessment tools and procedures must be reliable and valid. The accuracy and utility of any diagnosis must be demonstrated.
4. Psychiatric diagnoses, for instance of depression, have both first-person ascription (based on observation of the behaviour of the subject) and third-person ascription (based on behaviour criteria).
5. There are two serious levels of analysis of mental disorders: the psychological and the biological approaches. Psychiatry assumes that both conscious states and brain events are part of the causal structure of the world.
6. Mental disorders are interactive kinds since once aware of their diagnosis, a so-called *looping effect* arises, that, being recursive and self-applied, affects and biases the diagnosis.
7. The brain-behavior relationship is increasingly tied to explanation of mental disorders, although it seems unlikely that the neurosciences might someday eliminate the role of clinical psychiatry. Mental disorder may be multiply realized.
8. Depression may have a positive role: it seems that the prevalence of depressive disorders is somewhat higher among writers, painters, or musicians, unlike that of scientists. Similarly, schizophrenia reduces the rate of offspring, even though its incidence remains constant.
9. Schizophrenia might have had a seminal role in the birth of our civilization.

Chapter 11
Socially Deviant Behaviour

Physicians are not competent to take care of social distress.

Ivan Illitch

Contents

Robert Merton contends that: "... *social structures exert a definite pressure upon certain persons in the society to engage in nonconformist rather than conformist conduct...we should expect to find fairly high rates of deviant behaviour in these groups, not because the human beings comprising them are compounded of distinctive biological tendencies but because they are responding normally to the social situation in which they find themselves.*"[1]

Prima facie, deviancy is a divergent or abnormal behaviour, a variation from the average. Being deviant means being different from the neighbouring cultural and social norms. It may or it may not be associated with or caused by some mental disorder. And conversely a mental disorder may or may not be associated or be the cause of a deviant behaviour.

In other words, deviancy is the failure to keep to *social* norms, while mental disorder is the failure to keep with *medical* norms.

To be sure, every one of us may, every now and then, deviate from society's expectations. One must thus distinguish between occasional and permanent deviant behaviour; occasional or situational social nonconformity (e.g., drinking in excess, stealing or promiscuity), which may be considered as normal, and may or may not

[1] Merton (1949, 26).

L. Karhausen, *Analytical Philosophy of Medicine*, Philosophy and Medicine 152,
https://doi.org/10.1007/978-3-031-81257-6_11

be the beginning of a full-fledged deviancy in which society casts the actor in his deviant role. We need a definition that does not include accidental or brief sidestepping that are part of the complex games of social interactions.

What is of interest here are forms of deviance, which are frowned upon and called 'social problems'. *Deviance is rule-breaking*: It violates some important psychosocial, ethical, or legal norm, commendable or not, though not medical. It does not necessarily result from any behavioural, psychological, or biological malfunction.

'Deviant' is a *societal reaction to behaviour*. It is an attribute that other people confer to an act or a pattern of behaviour rather than a characteristic of the act itself: it is an interpretation of some action in terms of agreed-upon rules of the group and not in terms of its relationship to the personality structure or intrapsychic features of the person who commits the act.

It follows that labelling creates deviance. It is a label that society stamps on a certain conduct on behalf of cultural grooming and that is singled out for attention and organized into a social role. Deviance unfolds within a social career with labelling, isolation, condemnation, alienation, and in more severe cases, punishment followed or not by repetitive behaviour especially if it rekindles satisfaction or rewards. But the group externally defines the career process of the deviant: deviancy lies ultimately in the hands of the beholder, shaped by political, cultural, and religious values. Schizophrenia is perceived as deviant on occidental countries, but much less so in Nigeria or in India.[2]

For all that, once this process is set going, it becomes self-reinforcing. A society must identify some groups of persons as deviant in order to clarify, which is normal for the community. There, then, follows a reciprocal relationship, an interactive kind, between social norms and deviancy, defining one another in a given social context. The social labelling of the deviant paves the way for his self-identification. Deviance ends up being a sequence of reciprocal interactions described by Georges Herbert Mead,[3] in which deviant and labeller act and respond by turns. People who manifest some deviant behaviour may create psychological conflicts and discomfort for others.

Poor and old people in affluent countries may be relegated in urban ghettos. A Caucasian is deviant in central Africa and so is a Jew among Christians or a classical pianist among jazzmen. Being sick is thus a deviant behaviour since the sick role legitimates the deviant condition as long as the patient performs the functions assigned to that role. A stutterer is a deviant: talking with him may be embarrassing and he may very well be sanctioned and laughed at by his friends although he does not break any formal societal rule. Epilepsy was called *le mal sacré* and hysterics led to the suspicion of diabolical behaviour. Deviant attributes often feed on prejudice and tend to be seen as either positive or negative, most of them being considered negative. Ordinary people who have sought psychiatric care often become

[2] Jablensky and Sartorius (2008).

[3] Mead (1934).

pariah even though they may be perfectly sane or cured. By his ability to label people as mentally ill, the psychiatrist tags them as deviant if they were not already bandied about by society.

Deviance may remain unnoticed but when visible it includes bad manners due to poor upbringing, chronic hostility, maladaptedness, unpredictability, violation of moral standards, vengefulness, wickedness, depravity, divorce, suicide, homosexual behaviour, alcohol abuse, addiction, mental illness, delinquency, or crime. It may be labelled 'original', 'innovative', 'eccentric', 'nonconformist', 'dissident', 'odd', 'weird', 'fanatic', 'antisemitic' 'freak', 'marginal', 'misfit', 'mad', 'unconventional', 'wayward', 'maverick', 'improper', 'bad manners', 'naughty', 'rude', 'vicious', 'heretical', 'witch', 'rebellious', 'unruly' 'unnatural'; the deviant person may be christened 'perverse', 'perverted', 'depraved, criminal, terrorist, 'vicious', 'degenerate', 'wrong', 'offensive', 'threatening', 'frightening' or even 'sick'. People's actions lie, in effect, along a spectrum of modality, which ranges from rigid conformity to an overly nonconforming behaviour in which both extremes are deviant. We can thus construe deviancy as a seamless robe starting from the original creative individual, through volitional disorders such as teenage pregnancy, drug abuse, pathological gambling, pyromania, kleptomania, and exhibitionism. Further along the line, a transgressing conduct involves delinquent or criminal behaviour that violates legal codes and decisions regarding the maintenance of public order. On this score, when deviancy passes a certain point—and ruling out mental illness—moral and forensic considerations must be paid to this behaviour. There are thus degrees of transgression of conventional norms. Deviance is not a moral concept although it may end up this way when individual maladjustment or psychosocial malfunctioning fades into wrongdoing or criminal behaviour. Presumably the more deviant such behaviour, the more wicked the deviation. However, moral rules and laws change and the law-abiding individual of the days of yore may very well be the delinquent of today and conversely. Societies are culturally heterogeneous and are made up of a number of groups each of them with their own norms and cultural practices. The shift into or out of criminal deviancy of genocide, terrorism, slave ownership, marital rape, child abuse, driving under the influence of alcohol, homosexual behaviour or sadomasochism suggests that the concept of transgression and its limits are themselves at issue.

It may go through behavioural contagion[4] when the same conduct disorder triggers an imitative increase in racial prejudice, drug abuse, teenage pregnancy, or suicide, known as the Werther effect. Moreover, people engaging in deviant behaviour may enter into a deviant subculture, which differentiates them from the mentally ill: this is a system of beliefs and values spawned in a course of interactional communication in an attempt to determine how members of the group conceive of each other and of themselves, and to solve the problems of adjustment. It may then crystallize into a set of rationalizations for deviance seen as valid by the deviant, and

[4] Gould (2001, 200–221).

which serve as a mechanism for neutralizing the internal and external demands for conformity.

Deviance and its acceptable limits beyond which interventions are necessary are a question of law, a social institution enforced by social representatives, interpreted by judges, and adjudicated in courtrooms.

Social Deviance Versus Mental Disorder

The decreasing number of psychiatric beds coincided with the increased number of people in prison with mental disorders. Are prisons taking the place of mental hospitals or are we faced either by a medicalization of criminality or by a better diagnostic ascertainment in the prison system?

Thomas Szasz deliberately identified mental disorders with deviancy and claimed that psychiatric doctrine wrongly labels social misfits. Meanwhile, in occidental countries, dyslexia is a disorder but illiteracy is a deviance. *Prima facie*, any deviant behaviour socially defined as deliberate, is a boundary of mental illness. Although a mental disorder may manifest itself in deviant behaviour, psychiatric care involves at the outset the exclusion of socially deviant persons. What demarcates psychopathological features is largely found through the process of diagnosis, whereas deviance is a label socially affixed to a person, inasmuch as moral issues and their limits are being chosen.

It follows that mental disorders are essentially phenomenological; they are, in a way, 'out there'; they are primarily organized into physical behaviour, attributes or idiosyncratic characteristics that differentiate a patient from others. Diseases and mental disorders have a natural history resulting from neuropsychological factors.

By contrast, forms of deviant behaviour are being assigned and are not 'out there'; they are construed outside a time perspective. Contrary to the identification of mental disorders, which is more or less broadly shared and depends on the current medical doctrine, labelling of deviance is thoroughly anchored in its own time and place.[5] Explaining the social role and career of deviancy consists in explaining the attribution of deviance: its aetiology must be found not in the deviant person as in the case of mental disorder but rather in the social process of its ascription. Deviancy makes no assumption as to whether the person is either psychologically abnormal or intrinsically deviant or not. *A priori,* there is nothing inherently deviate or abnormal in the psychological makes up of a deviant but a mere conflict between himself and society; and this demarcates deviancy from mental illness. What's more, deviance is not a state, like handicap or disability, or a process, like mental illness, but a *value judgement about the meaning of some form or state of behaviour.*

Amusia, a deficit in musical processing, may be a handicap, but it is not a deviancy. It follows that explaining a deviant behaviour does not consist in explaining a

[5] Siegler and Osmond (1974, 116) and Freidson (1970, 217).

state so much as the meaning ascribed to a state. The question is: how does such kind of attitude or behaviour come to be categorized as aberrant or deviant? In a way, and contrary to mental illness, deviance does not exist in isolation but only in relation to those who define it and control it.

Medical norms, those that divide normal from pathological are not social norms since they have no external sanctions. In contrast, deviant behaviour violates social norms and may invite requital.

As long as the causal process, which leads to some capacity reduction, remains external to the individual the deviant condition is merely existential. But the causes of diseases are embedded in their processes. A deviancy becomes a medical one whenever the incapacities are due to the penetration of the extraneous causal chain inside the individual's mental or bodily sphere where it initiates some pathological processes. Diseases have a causal history, and their *medical* aetiology refers to the proximal causal factors, which determine them: when a patient suffers from starvation, a clinician will not attribute his condition to poverty but to a diet deficient in calories and proteins. Poverty may be an ordeal and cause a reduction of some capacities, but it is not until it causes some medically relevant condition such as a depression or poor nutrition that it becomes a causal candidate to the class of diseases. Alcohol and psychoactive substances have been used for centuries to relieve pain and alleviate anxiety; yet their abuse is deviant and often seen as immoral or evil. But the dependence and adverse effects that result from drug and alcohol abuse make it into a medical issue. What's more, the need for treatment is constitutive of the concept of mental illness just as the pressure to conform to prescribed aspirations and behaviour is inherent to what we mean by deviancy.

Sociological or Psychiatric Pathology

According to the Lawrie Reznek,[6] there is an alarming inclination for people to succumb to delusional thinking, a thinking process that is not limited to the mentally ill. Instead, there are large segments of the public that harbour a wide variety of delusions, none of which are innocent, and many of which are pushing our societies to the brink of war. This book aims to understand the nature of delusions and how they are generated.

We all believe that we can tell the difference between someone who is mad, or whom psychiatrists call psychotic, and someone who is sane. Do we really know what madness is and how it should be recognized? Have psychiatrists made a sensible distinction between the patient who believes that aliens are beaming messages to him from a foreign planet, and the religious fanatic who believes God communicates to him via automatic writing? Is there a difference between the paranoid patient who believes that the FBI is after him, and the sizeable proportion of our

[6] Reznek (2010).

normal population that believe that the US government orchestrated the 9–11 bombings?

Mental disorders are diseases because they have pathological consequences like delusions, i.e., beliefs that are firmly held on inadequate grounds and out of keeping with the person's communities. Disease status and mental disorders depend on having both harmful consequences, which is a value judgment, as well as some organic underlying abnormality, i.e., brain abnormality or histopathological lesion.

Since delusions are being a sign of mental disorder, we are committed to a value judgment. Even then, a Catholic is mistakenly convinced that when the priest performs Mass, an ordinary wafer, and a glass of wine turn into the flesh and blood of Jesus. Nevertheless, this is not considered to be a delusion because the belief is not harmful and is shared by members of the community.

Reznek sheds light on the delusions of the masses, those delusions that are common to everyday people living so-called ordinary lives. He provides an understanding of madness and the psychological processes that drive us to adopt delusions, arguing that it is a mistake to view only schizophrenic patients as delusional, while excluding large groups of society from such an analysis. On this score, delusions are either a social (antisemitism) or a psychiatric abnormality, and in both cases, they may become pathological, i.e., that causes consequences we judge are harmful.

Michael Stone, a psychiatrist, and scholar, sought to define "evil", and sharpened the distinction between mental disorder and evil.[7] Only human beings are capable of evil, although it is not a feature that people are born with. The word "evil", he acknowledged is vague, and derived from "over" or "beyond" and could apply to "*certain acts done by people who clearly intended to hurt or kill others in an excruciating painful way*". All too often, mass murderers and crazy politicians are regarded as mentally ill by public opinion and the media.

Two Complementary Views

The discontinuity that separates deviance from mental illness does not refer to two mutually exclusive realms but rather to two different ways of thinking. Mental illness is discussed in terms of causes and effects, deviancy in terms of goodness and badness. If a diagnosis of schizophrenia is a finding supported by public criteria, deviant behaviour is an appreciative judgement often disguised as a finding.

Given the medical stance of disease as pathological processes, it is quite understandable that sociological approaches to deviance are unambiguously excluded from the psychiatric purview.

So, can these two approaches be kept fully apart? Medical and legal experts are constantly confronted with the question: is he bad or mad? Should then a deviant behaviour perforce be seen either as sick or as volitional? Sure enough, symptoms

[7] Stone (2009, 2019).

of mental illness such as delusions, disorientation, withdrawal, or suspiciousness may be shared with mentally healthy individuals. The same symptoms can thus lend themselves to different interpretations, sick or deviant, mad, or eccentric, culpable or insane, psychotic, or criminal and so forth since there is nothing inherently pathological in eccentric behaviour, wickedness, or criminality.

Not surprisingly, disagreement may arise in court about whether a person is a criminal or whether in addition he presents evidence of some mental disorder. It is often the case that it is impossible to draw a demarcation line, not because we are faced with a continuum, as in the case of hypertension, but because the question raised is not properly an *either-or* question; what is being asked is: under what description are we going to judge a person's behaviour? It refers to two differing standpoints: whether we decide to classify a disorder under a health or under a moral perspective, or else under a thesis of scientific determinism vs. a doctrine of free will and responsibility. Mental disorders belong to the first category, and immoral as well as criminal behaviour, to the latter. But antisocial conduct, and deviancy may belong to both universes and correspond to an area where medical and moral concepts are not merely bordering but overlapping. Deviance, in its extreme forms, may be associated with pronounced anxiety, compulsive self-defeat and uncontrolled aggressivity.

It follows that there are two perspectives from which we may understand any abnormal behaviour, a medico-psychiatric and a sociological one, or a health welfare and a social-defence one. The psychiatric view is that no one can legitimately be considered as fully responsible for deviant or criminal behaviour, his behaviour being the consequences of internal and external causal factors over which he has no control. But the concept of deviance supposes that one is always able to act in conformity with or contrary to prevailing socially assigned rules. Quite often the answer is obvious especially in clearly identifiable mental disorders.

But there may also be room for disagreement in borderline cases, whether the person needs therapeutic help or not and whether he may invoke a sanity defence or not. Some of those deviants are mentally ill: mental disorders and social deviants have some common members. Odd mannerism, eccentric behaviour, inattention to usual social conventions, difficulties in relating to other people and observer discomfort may be signs of schizotypal personality disorder as well as of mild deviancy. This is why psychiatric treatment inevitably is bordering on social control.

But there are also forms of behaviour, which were previously identified as a crime or falsely classified as a sickness. For example, homosexuality went through the sin, crime, and disease unfolding before being excluded from the medical realm. It was deemed to be a disease until the 1960s, when it was removed from psychiatrists' lists of diseases and turned into a deviancy; by now, it comes, under the medical gaze, to constitute a mere personality trait. Since its prevalence is 3–5% in the general population, it is 'statistically abnormal', but it is less and less viewed as blameworthy in the large occidental metropoles. Medical thinking, as it were, is taking over social prejudice.

Foucault's image of psychiatry emerges from his unwillingness to clearly distinguish mental illness from deviancy.[8] Szasz also attempts to collapse the difference between deviance and mental illness, which he countenances not to be factual but psychological.[9] Both authors virtually ignore the incalculable human suffering that constitutes the process of mental disorders; they also disregard the philosophical issue of human agency: mental disorder does not merely belong in the relationship between the individual and society.

It should now be plain that these two points of view are complementary although deviancy and mental illness are not necessarily referring to a same behavioural process. They designate overlapping but not necessarily coexistential features: 40–70% of young people who come into contact with the justice system during a year have a mental disorder with high rates of substance abuse, major depressive condition and mental retardation.[10] If the psychiatrist is not prepared to deal with the primary basic 'abnormality' of the deviant, he can see to the consequent or secondary problems, although not much better than symptomatically.

We may conclude with DSM-IV: "*When the disturbance is limited to a conflict between an individual and society, this may represent social deviance which may or may not be commendable but is not by itself a mental disorder*".[11]

Is Alcohol Abuse a Mere Bad Habit?

Early in 1813, Thomas Trotter and Benjamin Rush from Edinburgh Medical College described alcohol abuse as a disease that could lead to liver disorders, jaundice, mental illness and wasting. In 1849, Magnus Huss introduced the term 'alcoholism' in Sweden to summarize the harmful consequences of alcohol abuse.[12] Contrary to tobacco, moderate alcohol consumption might decrease the risk of ischaemic heart disease and be associated with significantly lower mortality from all causes than the consumption of no alcohol or the consumption of substantial amounts.[13] Herbert Fingarette[14] has been the proponent of the theory that alcohol abuse is not the result of an underlying psychiatric disorder and that it cannot be properly regarded as a disease. This controversial view is based on a disease concept, which is defined so narrowly that it seems to be tailor-made for the specific purpose of ruling out the inclusion of alcoholism. Fingarette mistakenly assumes that alcoholics can control their intake, that medical treatment of alcoholism is ineffective, that to properly call

8 Foucault (1954).

9 Szasz (1974, 105–106, 205–206).

10 Kessler (2002).

11 Spitzer (1987).

12 Miller (1986, 111).

13 Doll et al. (1994).

14 Fingarette (1988).

it a disease, alcohol abuse needs to show an inevitable course that tallies to almost all affected individuals and that it is not a genetically influenced condition. Furthermore, Fingarette argues that alcoholism results from a culpable wilful misconduct that is not beyond the agent's control. The alcoholic is thus not the helpless victim of some alien forces. Heavy drinking is not a disease but an 'a way of life'.[15]

To begin with, nearly half of the people diagnosed with alcohol abuse or alcohol dependence in the US population had a lifetime psychiatric disorder, including drug abuse or dependence, anxiety disorders, antisocial personality disorder and affective disorder. There is also strong evidence for genetic influence on alcoholism with non-Mendelian inheritance patterns. Dependence is characterized by a cluster of symptoms such as loss of control over intake, physical withdrawal syndrome in the event of cessation, increased tolerance and continued drinking in spite of acknowledged psychological, medical, and social problems. Furthermore, drug treatment of alcohol abuse and various psychotherapeutic approaches are effective.[16]

Then, beyond this, even moderate amount of alcohol can damage the brain and impair cognitive function over time.[17] Over time, alcohol is comparable to tuberculous bacilli or to the agent of some infectious disease since it can be the source of several variegated manifestations. Alcohol is a necessary cause of alcoholic liver disease, acute alcohol intoxication, delirium tremens, Wernicke's encephalopathy and Korsakov's psychosis. It is also a risk factor for cirrhosis, various cancers, cerebrovascular disease as well as suicide and homicide, like tobacco smoking for lung cancer.

Finally, alcoholism and other drug abuse and dependence are chronic, recurrent diseases involving multiple cycles of treatment.[18]

On this score, it is perfunctory to wonder whether abusing the grape is or is not a disease, whether it should be seen as a mental disorder or as a deviant behaviour: the answer might be yes or no seconding the framework within which it is being raised. Be that as it may, the fact remains that being an alcoholic is something one does but also something one undergoes: alcohol abuse is a bad habit that people consider as unacceptable, but an alcoholic who seeks help from a physician is sick. The term 'disease' is fuzzy at the edges but if we want to know where lies the borderline, we must turn to how the word is being used by the professional who are handling it. In 1972, the American Medical Association, the American College of Physicians and the American Psychiatric Association have accepted the concept of alcohol abuse as a disease rather than a character weakness. For the American Medical Association, 'disease' does not refer to a natural kind, but it is rather a conceptual tool which describes what doctors are doing, what are their moral responsibilities and what are some of the major preoccupations and duties of public health policymakers.

---

[15] O'Connor and Schottenfeld (1998).

[16] O'Connor and Schottenfeld (1998).

[17] Topiwala et al. (2017).

[18] McKay and Hiller-Sturmhöfel (2011).

The classification of alcohol abuse as a disease does not apply to excessive drinking, but to the loss of control, which occurs in one group of alcoholics. It is a manifestation of one of the most common evil associated with mental disorders, namely volitional disability. Addictions involve intentional involuntary actions.[19] The upshot is that alcohol abuse is either a sign and an intrinsic part of some life-threatening diseases, or a major risk factor for numerous other conditions, both resulting in a high mortality; it can also be associated with a high medical and psychiatric comorbidity. Alcohol abuse may a moral flaw, but it is a medical and public health problem almost without parallel in history: it calls for medical treatment and public health intervention.

Self-Harm

Self-harm, namely substance abuse, self-poisoning, self-mutilation, suicide, or self-inflicted harm with or without suicidal intent do not describe some medical disorders but a wide range of behaviour in response to intolerable stresses or tensions. Schneidman proposed the following definition: "*Currently in the western world, suicide is the conscious act of self-induced annihilation, best understood as a multidimensional malaise in a needful individual who defines an issue for which the suicide is perceived as the best solution.*" Then, beyond this, forms of suicidal behaviour range from ideas to gestures, to risky lifestyles, suicide plans, nonfatal attempts, and realization. This definition does not medicalize self-harm but merely excludes suicide from the present discussion as a form of social protest, such as hunger strikes or its use in military tactics such as suicide bombing. Suicide may also be the endpoint of a disease and of medical treatment in case of assisted suicide in terminal illness.

Acts of suicide are often condemned either on moral terms or on religious grounds as sinful.[20] Ruling out suicide due to escape, altruism or when prudential self-interests are being drained out (e.g., in cases of terminal illness), it seems undeniable that suicide, if it is not intended to serve some other purpose than merely ending one's own life, is minimally a kind of deviant behaviour, at least in occidental countries. This results from the assumption that life is intrinsically valuable, a view, which one does not necessarily need to espouse, but that is a basic axiom of medical care.

Even so, here is what Thomas Szasz writes: "*Suicide is a fundamental human right. This does not mean that it is morally desirable. It means only that society does not have the moral right to interfere, by force, with a person's decision to commit this act.*"[21] But granted that more than 90% of suicide victims have a diagnosable

---

[19] Culver and Gert (1982, 109–125).

[20] Kant (1978, 148–154).

[21] Szasz (1990, 250).

mental illness, most commonly depression[22] isn't the decision process of committing suicide itself defective? Two thirds of individuals who commit suicide had seen a physician within the month preceding their suicide. And if suicidal behaviour is part of some disease process, it mandates some sort of medical help. Although a successful suicide does not necessarily result from some medical or public health failure, the assessment and treatment of the suicidal patient is a major medical challenge.

Suicide and suicidal behaviour are highly familial. Adoption and twin studies have shown that genes contribute to both completed suicide and suicide attempt, although there are environmental causes for family transmission such as imitation, parental bereavement or parental separation or childhood experience and abuse.[23]

Small wonder that, when Szasz was asked whether or not, if he was suffering from a temporary suicidal depression, would he wish to be rescued from it against his will and would he afterwards feel grateful to psychiatrists, he was unable to resolve the dilemma.[24] Suicidal behaviour is not primarily a moral issue but it represents a repudiation of prudential self-preservation. Only rarely is it a mere form of deviant behaviour. It may not be a disease in itself but most of the time, it is a manifestation of mental disorder.

Medicalization of Criminal Deviance

There is a chair of psychopathology of mass psychology at Rome University la Sapienza (*Psychopatologia della psychologia delle masse*). This use of the term pathological may seem illicit: 'pathological' is of individuals and though individuals may be aggregated, mental pathology of aggregated individuals is not a mental pathology of the aggregate. This looks for causes and solutions for social problems in the individual rather than in the social system.

Should deviant behaviour be redescribed as a symptom or a consequence of mental illness? In his novel *Erewhon,* Samuel Butler put scorn on this quandary inasmuch as he was imagining a country where illness and deviance stand in reverse order. This make-belief country stands the welfare system on its head. Persons beset by tuberculosis are found guilty in court and condemned to life imprisonment. But persons who counterfeit currency, set houses on fire, steal, and commit acts of violence are diagnosed as suffering from 'severe fit of immorality' and are taken care of in hospitals at public expense.

Benjamin Karpman suggested in 1949 that we ought to treat rather than punish all criminal defendants because they are led into their behaviour by strong irrational or aggressive impulses over which they have no control: we have no reason to

---

22 Maris (2002).

23 Brent and Elhem (2008).

24 Siegler and Osmond (1974, 71).

punish them any more than "*to punish an individual for breathing through his mouth because of enlarged adenoids.*"[25] Those who construe the criminal, the delinquent, the perverse, the morally unacceptable and the revolutionary as clinical cases must not lure us. In some totalitarian regimes, the line of demarcation between custodial care and imprisonment and between sociocultural valuation and psychiatric diagnosis may be vanishing. These moves debase medicine by pervading its neutral scientific dictates with moral and political considerations.

Insanity is a legal concept, which entails incapacity to distinguish between right and wrong or incapacity to control one's own behaviour. Although they may be overlapping, mental disease should be distinguished from insanity: the first one springs from prudential concern bearing on the use of one's person while the second one depends on a legal verdict and the protection of personal and public interests. The first one is medical, the second one forensic. Furthermore, the extensions of the two concepts are not coexistential: one may be declared insane in the absence of mental illness, such as Ezra Pound, and one may suffer from a mental disorder and escape the verdict of insanity.

About one in seven prisoners in Western countries suffers from psychotic illnesses (3.7–4.0%) or major depression (10–12%) and about one in four has mental retardation; one in two male prisoners and about one in five female prisoners are psychopaths.[26] Among men, 65% of them had a personality disorder including 47% with an antisocial personality disorder; among women, 42% had a personality disorder, including 21% with antisocial personality.

What's more, dangerous deviants call for restriction and repression but a dangerous psychotic needs treatment in addition to social protection. Patients with major mental disorder (schizophrenia, severe depression, and bipolar disorder) were two or three times as probable as people without such an illness to be violent: the lifetime prevalence of violence among people with serious mental illness was 16%, compared with 7% among people with no mental illness. However, since mental disorder is rare, it contributes very little to the overall violence in the general population, with an attributable risk of 3–5%. On the other hand, the lifetime prevalence of assaultive behaviour among some deviant behaviour such as substance abuse was 35% that is five times higher than those without substance abuse.[27]

Thus, criminal deviance, insanity and mental illness constitute divergent though overlapping deviations from the prudential type of rule-following behaviour endorsed by the group. A plea of insanity might mitigate the patient's accountability, i.e., his responsibility for the psychiatrist or his liability for a forensic physician or a court of law. It may call forth some form of constraints or even custodial care, while a judgement of (deviant) criminality entails culpability and some sort of punishment.

---

[25] Karpman (1949).

[26] Fazel and Danesh (2002).

[27] Friedman (2006).

But surely, this theoretical demarcation line is often quite muddy inasmuch as public health tends to extend its dominion to a therapeutic culture, yielding expanded nomenclature of DSM and ICD classifications. The boundaries between mental disorder and social deviance are constantly shifting, alcoholism, sexual malfunction, strained family relationship, vandalism, theft, or crime are, rightly or wrongly, increasingly envisioned as 'diseases' justifying treatment.

If so then, why not substitute positive preventive measures, such as education, public measures to reduce alcohol abuse, and appropriate preventative social interventions, for the negative aims of moral and legal convictions?[28] There is a trend in the Western World to conceive criminality as if it were a state capable of modification and conversion over time, by analogy with sickness. We are witnessing a growing tendency to group a range of human behaviour under the title of diseases[29] and to treat violence as a public health problem.[30]

On this score, the World Health Organisation passed a resolution in 1996 on violence as a public health problem, on the premise that it is intentional, predictable, and preventable. It calls for a plan to support violence prevention programmes around the world as well procedures on how to treat victims and to lower a society's acceptance of violent behaviour. Even if criminality is not in itself and of itself a mental illness, why couldn't epidemiological aetiological research be applied to outline some risk factors that might be susceptible to manipulation?

Research conducted during the last 40 years has shown that exposure to family and community violence during childhood, or to violence from television, cinema, the Internet, and games screens make children violent. Even so, exposure to media violence comes rather low down the list as a risk factor. Much stronger predictors include involvement in crime, poverty, family breakdown or abuse, drug use, and psychiatric illness, although further observational studies are needed to identify high-risk groups.[31]

Plainly, deviance and mental disorders are not mutually exclusive. It may thus seem pointless to wonder whether violence is a medical or a social issue: it can be apprehended both ways since these are two legitimate styles of reasoning. Deviancy, as it were, is to mental disorder what functional disorders are to diseases. It ventures a sociological look without a medical selective sieve.[32]

More recently, Gary Slutkin, professor of Hygiene and Epidemiology at Washington University observed that urban violence behaves exactly like all other epidemic issues. Like other contagions, violence tended to cluster, with one event leading to another. What was the greatest predictor of violence? The answer: a preceding act of violence. In February 2000, Slutkin started an initiative *Cure Violence,*

---

[28] Gilligan (2000).

[29] Nolan (1998, 9).

[30] Heath (2002).

[31] Somers (1977), Ingelfinger (1977), Feingold and Johnson (1977), Browne and Hamilton-Giachritsis (2005), McLellan (2001), Devine et al. (2004), and The Lancet (2008).

[32] Annals of the New York Academy of Sciences (2004).

which, where implemented reduced violence by 41–73%. Punishment does not work because behaviour is formed by modelling and copying. People care more about what their friends do than what some authority is telling them. Beyond the United States, this program has been applied successfully in several countries and among them in UK and in Scotland.

Conclusion

Social deviance is not a mental disorder, but a socially eccentric, unconventional, nonconformist, strange, aberrant, or undesirable behaviour. Only a very small fraction of deviant behaviour is associated with mental illness. The terms 'deviance' and 'mental disorders' come from different modes of discourse.

Deviant persons are labelled abnormal by other people because they are seen as a departing from the norms of the culture in which they originate. However, forms of deviancy are social constructs whereas mental disorders are medically construed such as alcohol abuse. Being deviant when applied to non-human beings, such as an aggressive dog, is being defined and understood in relation to human society. It follows that the ambiguous border between deviancy and mental disorder may lead to medicalization of deviant behaviour and tends thereby to condemn physicians to become agents of social control, which raises deep ethical questions.

Chapter 12
Unexplained Physical Symptoms and Functional Disorders

Doctors take care of people; some of whom have diseases and all of whom have some problem.

Jerome Groopman (2007, 98)

Contents

Vikram Seth wonders: "*Why do people (in a concert) have to cough immediately after a musical movement ends? If they have held back for ten minutes, can't they hold back for 2 seconds more*?" And why is coughing the privilege of the public, never of the musicians?

It is often argued that "*The pivotal concept in clinical medicine is disease*,"[1] but this is a misconception. What initially concerns a physician is the disadvantage, pain and suffering involved, namely the pathological features from which, through some inductive process, he is either diagnosed or coming up with a disease that is one step distant from the patient's lived experience or ending up with some unexplained clusters of symptoms.

Diseases are a mere part of the total medical activity. If a disease is a statistically based causal relation discovered using epidemiological and experimental methods,[2] this chapter covers another major domain of medicine and health care, that does not belong to diseases.

To begin with, are physical symptoms perforce medically meaningful? What kind of cognitive weight do they carry outside a medical context? May symptoms

---

[1] Thomson and Upshur (2017, 17).

[2] Thagard (1999, 114–115).

L. Karhausen, *Analytical Philosophy of Medicine*, Philosophy and Medicine 152,
https://doi.org/10.1007/978-3-031-81257-6_12

be bodily feelings to which a sentient being bestows meaningful status? Actually, a sign or symptoms may have no medical meaning in isolation. A precordial pain may indicate a myocardial infarction but most often it is quite meaningless. Actually, most symptoms observed in clinical practice are medically meaningless. Some symptoms or signs affecting individuals may cluster together and be turned into diseases, while many others remain randomly distributed and unexplained. The epidemiology of symptoms carrying individuals in a general population is comparable to the firmament in which some but not all of the stars are set out in constellations, but most of them remain single and uncompounded. In other words, symptoms, be they emergent or submerged, are part and parcel of reported or unreported factual clinical manifestations that time and again may refuse to cluster around our disease categories, although they may occasionally be explainable and sortable into disease categories.

Not all sick persons can be ascribed to the class of known diseases, even though ICD-10 and DSM-5 are listing a lot of symptoms as if they were diseases. This may lead, for one thing, to a surge of medical or psychiatric conditions and for another, to the proliferation of spurious studies describing comorbidity thereby hiding real, bona fide syndromes. Weren't fevers regarded as diseases until the nineteenth century?

If we could subtract from the world population all those people to whom some disease is being or could be ascribed, assuming that all medically relevant information concerning them was to be available, we would be left with a large remainder of 'sick' persons. In jargon of logic, we might say that the predicate *diseased* does not universally quantify over the predicate *sick*. Diseased people represent a mere fraction, a subclass of those suffering from medically relevant conditions. For instance, some psychiatric patients and, even more so, persons suffering from so-called "somatic symptoms and related disorders", un umbrella term, may lack one of the essential characteristics of diseases, namely a natural history: is every person who applies for medical or psychiatric help a suitable candidate? Incidentally, the term *psychosomatic disorder* does not appear in DSM and is a term no longer recommended for use.[3]

History of Unexplained Symptoms and Functional Disorders

> "*My body* wrote Lichtenberg, *is that part of the world which can be altered by my thoughts. Even imaginary illnesses can become real*".

Table 12.1 is a very instructive one, which relates the recent history of a pool of unexplained physical symptoms and functional disorders: The US National Center for Health Statistics reported that the incidence of illnesses and acute conditions has increased substantially from 1920 to 1980. The term 'functional' here does not

[3] Cooper and Sartorius (2013, 117).

Table 12.1 Time trend of briefly disabling illness episodes in the US between the 1920s and the 1980s: how the annual incidence of illness per hundred population has increased (Shorter 1985, 213)

	COMC survey 1928–1931	NSHC survey 1981	Increase 1920–1980
All ages	82	212	+158%
Males	72	202	+180%
Females	92	222	+141%
Juveniles only	83	276	+233%
Males	64	277	+230%
Females	82	274	+234%

mean 'malfunctional': functional disorder in medical language means either unexplained disorder or physically and provisionally unexplained disorder.

Shorter points out that the explanation is not that in the 1920s people were less willing to take time off work since illness rates for children under sixteen have also increased by 233% during the same period. Shorter shows that reported common diseases which are the daily bread of GP's have undergone a staggering increase: colds have increased fourfold in frequency; sinus problems more than thirteen folds; urinary tract disease has doubled, and digestive tract tripled; accidents and injuries quadrupled.

The author then makes a challenging suggestion. During the twentieth century and due to changing lifestyle, patients became increasingly prone to fresh anxieties about their health, and readier to consult their doctors. Shorter makes a powerful case for the symptom iceberg (which he calls "the symptom pyramid"), which represents a pyramid of demand not necessarily of needs. Demand substitutes for needs. The symptom iceberg becomes a new and major socio-medical phenomenon instead of the disease iceberg, which is, and should remain the central object of interest of health-care providers. Those symptom clusters are limited to the subjective feeling of being ill *in the absence of any discernible, identifiable changes in function or structure of organ systems.*[4] Small wonder, then, that, in contrast to diseases, those conditions are susceptible to placebo effect. Shorter points out that patients, men, and women, show a general increased sensitivity to body symptoms, whatever their origin, resulting in a shift from submerged to manifest status. He tentatively explains this through the intervention of various factors such as changing lifestyles, stress, the breakdown of the nuclear family, all factors which lead people to interpret their symptoms as illnesses.

Goethe countenanced that "*If you start to think about your physical or moral condition, you usually find that you are sick.*"[5] Hence, we now have a major shift from diseases to symptoms and from demand instead of needs. Health care, as a result of this process, is faced and more and more overwhelmed by a population demand that largely exceeds medical needs. This situation created what Shorter

[4] Weiner and Fawzy (1989).

[5] Goethe, Pt I, Bk I.

calls a tug-of-war between what physicians have been taught and their patients' expectations.

All in all, one of the characteristic trends of health care demand after the Second World War is an increased tendency to translate vague sensory impressions that are not medically significant, into illnesses, thereby transforming persons with existential difficulties into patients. For instance, a study at the Mayo Clinic concluded that the prevalence of sinusitis might be highly exaggerated: 14–16% of Americans are said to suffer from the disorder. But when researchers looked at confirmed diagnoses, they found that just fewer than 2% of the population has sinusitis.[6] This situation now creates a dramatic needless increase in health care expenditures and a crisis of communication in the relationship between the population on one side, physicians, and health care providers on the other side. Every medical visit must end with a prescription because each demand becomes an assumed need for care that is itself supported on shaky grounds by so-called 'patients rights'.

If functional disorders have been on the increase during the past hundred years despite striking improvement in our health, conversion disorders went the other way around. There has been a marked and rapid decline of the prevalence of conversion in the Western World where it was substituted by anxiety disorder, while no such trend was observed in the rest of the world. It follows that during the twentieth century, emotions expressed as bodily feelings have been progressively replaced by experienced emotions.[7] "*In premodern moral orders,* writes James Nolan, *pain, suffering, and injury were viewed as part of life.*" But by now, "*the tendency for individuals and groups to understand themselves as victims of their abusive pasts or of the oppressive social environment that surrounds them appears to be on the rise.*"

Our sensitivity to aches and discomfort seems to have heightened over the decades in the Occidental World. Small wonder, then, that our willingness to put up with them, our tendency to attribute them to a physical disorder and our promptness to seek medical care have increased at the same time. "*Diagnosis is the most common disease*", quibbled Karl Kraus.

To sum it up: patients increasingly present the sick role in the absence of sickness, and illness-behaviour in the absence of illness. The prevalence of functional and conversion disorders is largely influenced by cultural factors. Societal and biological factors determine, as it were, *the sick role* within our therapeutic culture, and so do, to a certain extent, the classification schemes and naming of those unexplained and unstructured conditions. By contrast, *diseases*, their naming, classification, causes, and diagnostic criteria depend much less on cultural factors than on cumulative clinical and epidemiological evidence.[8]

---

[6] Leff (1981, 54–66).

[7] Nolan (1998, 15).

[8] Simon et al. (1996), Scott and Walter (2010), Creed et al. (2012), and van der Feltz-Cornelis et al. (2018).

The Symptom Iceberg

Somatic symptoms, not diseases, are the main reasons for more than half of the clinic visits. People do not complain about diseases or discrete disorders but about symptoms, such as fatigue, backache, rash, palpitations, dizziness, weakness, difficulty breathing, pain or insomnia or anything of that ilk. Bodily symptoms, as well as perceptions of visceral stimuli are sensations, which grasp our attention. But symptoms may or not be tell-tale marks of some malfunctioning organ or system. The pattern of symptoms encountered by physicians is quite different from their profile in the community. The analogy of the iceberg is used to describe what the physician sees (the floating tip) and the hidden pass of symptoms (the submerged part). The iceberg metaphor[9] is often used to describe phenomena, which have a proportionately large, invisible, or potential component and a smaller phenomenal and visible fraction: most of an iceberg remains concealed and only its tip is manifest. Those submerged symptoms may generally be less severe or disabling than those for which people seek treatment.

The profile of symptoms observed by doctors is very different from that found in the epidemiology of the population. They are sometimes called *medically unexplained symptoms* (MUS), medically unexplained physical symptoms (MUPS) or *medically explained symptoms* (MES) when they last for more than a few weeks, but doctors cannot find a problem in the body that might be causing them.[10,11] But the philosophy of medicine ignores them completely.

They may be ascertained through health surveys or screening campaigns on representative samples of the general population, while emergent symptoms are more readily available to clinicians and obtainable from medical records. For all that, clinicians are perforce limited in their knowledge of the amount and the severity of symptoms manifested and recognized in the general population and conveyed to health care professionals. Actually, a large amount of suffering and symptoms are treated with home remedies and are never referred to a doctor. What's more, people seen by physicians, i.e., the tip of the iceberg, may differ in important ways from the large population they are supposed to represent. Patients who seek medical attention are likely to be those who are the sickest: this situation tends to exaggerate the apparent severity of diseases in the general population.

Table 12.2 shows that most patients with common symptoms do not request medical care. It shows the result of a survey made in the seventies and based on health diaries kept by 79 London women aged 16–44 years. Most symptoms shown by them were not the object of medical consultations and treatment decisions were largely a matter for the laity. The table shows that a headache was the subject of consultation in only one case for every sixty cases recorded. Actually, except for

[9] Morris (1970) and Last (1963, 2013).

[10] Marks and Hunter (2015).

[11] Klaus et al. (2013).

Table 12.2 Symptoms recorded in the health diaries of 79 London women aged 16–44 years

Ten most frequent symptoms	Ratio of consultations to symptom episodes
Tiredness, lack of energy	No consultations
Nerves, depression, irritability	1:74
Headaches	1:60
Backache	1:38
Sleeplessness	1:38
Muscle and joint aches and pain	1:18
Cold and flu	1:12
Stomach pain	1:11
Women's complaints (e.g., period pain)	1:10
Sore throat	1:9

severe illnesses, contact with the health care system is determined by demand not by need, even though they might in some cases be largely overlapping.

In Dunnell and Cartwright's study, 91 per cent of a sample of adults has experienced one or more abnormal symptoms in the 2 weeks preceding the study.[12] There is thus a large reservoir of unreported symptoms in the community that are managed with self-treatment, since only 5–34% of symptoms result in a consultation with a primary care health professional.[13] The most common symptoms are respiratory (acute respiratory infection) and musculoskeletal (arthritis, injury). Wadsworth et al. showed that 95% of some sample of 1000 adults had experienced symptoms in the 14 days preceding the interview, but only one in five had consulted a doctor.[14] Somatic symptoms, not disease, are the main reason for more than half of all clinic visits. People complain of symptoms that can be disabling, such as exhaustion, fatigue, back pain, headaches, insomnia, rashes, pruritus, paresthesia, tingling, palpitations, restless legs syndrome (RLS), dizziness, weakness, breathing difficulties, runny nose, pain, insomnia, night-time apnea, or anything of the sort. These symptoms represent a major fraction of the population's medical problems and are an important if not major part of primary care. They may or may not be indicative of specific organ or system pathology.

[12] Dunnell and Cartwright (1972, 13).

[13] Verbrugge and Ascione (1987) and Elliott et al. (2011).

[14] Wadsworth et al. (1971) and Hannay (1979).

Unexplained Physical Symptoms and Health Anxiety Observed in the Doctor's Office

More than half of the patients observed in primary care or in population-based studies are medically unexplained, so persons manifesting some kind of symptoms, whether they seek treatment or not, may or may not suffer from some identifiable disease.

Plainly, writes Irving Kenneth Zola "*It seems that the more intensive the investigation, the higher the prevalence of clinically serious but previously undiagnosed and untreated disorders.*"[15]

Physicians and nurses are very well aware of the emergent pool of pointless symptoms. When 248 doctors were surveyed in California they thought, on the average, that one patient in five, had only trivial problems with the consequent climate of resentment and distrust at being deluged with trivial symptoms. "*They spend four years in medical school learning about diseases like renal amyloidosis and phenylketonuria. They spent two years in a family practice residency learning to manage high blood pressure. Then they "treat" an endless process of colds, "treating" in quotation marks, of course, because there is nothing a doctor can do about a cold save let it run its course.*" One Welsh physician said, "*We doctors can't really treat the vast proportion of conditions that are brought to us–they're not illnesses at all.*"[16] After 20 years in his own general practice, Dr. John Fry wrote: "*The first shock and rude awakening that comes to the physician entering the field of primary care and practice is to be faced with a mass of apparently unrecognized, indefinable and unfamiliar ... disorders*" ... that "*cannot be categorized neatly or labelled or diagnosed with any accuracy or on objective bases*". This corresponds to the woolliest and major sector of the realm of primary care.[17] Several studies have shown that in 50–79% of all patients presenting to a family doctor, no evidence for a specific organic diagnosis could be found.[18]

Prima facie, we should be able to differentiate symptoms or clusters of symptoms attributable to organic diseases from those unexplained at the time of diagnosis. If one subtracted the morbidity attributed to somatic diseases, from the general morbidity in general practice setting, we would be left with unexplained somatic symptoms which show no expected or typical pattern: such episodes may be genuine sicknesses prompting medical care from the patient's viewpoint but are not diseases from the medical point of view. Such leftovers of aggregated sundry physical *symptoms* without identifiable pathology are either descriptively called 'atypical',

[15] Zola (1966, 617).

[16] Comaroff J, in Shorter (1985, 216).

[17] Cartwright (1967).

[18] Stewart et al. (1995, 33).

'symptoms-based', 'somatoform', Briquet's syndrome, or, in a form of incipient causal explanation, 'idiopathic', 'functional', or else 'unexplained' or 'agnogenic'.[19]

They may include a gamut of disparate symptoms such as fatigue, bouts of anxiety, weakness, headaches, dizziness, disturbed sleep, muscle and joint pains, backache, disabling cough, palpitations, dry mouth, running nose, skin abnormalities or lumps, nausea, heartburn, bloating, loose bowels, constipation, sexual malfunction, dysmenorrhoea, loss of memory, attention deficit, cognitive impairment, age-related memory loss, mood swings, depression, anxiety, irritability or something else of that ilk.[20] Somatic symptoms did sometimes cluster into meaningful groups such as neurological/conversion, musculoskeletal or gastrointestinal conditions. Unexplained signs and symptoms might be detected by physicians or nurses but most of those episodes will pass unnoticed since affected *persons* will not take the matter up with them.

Population surveys show that minor bodily symptoms are very common and that a mere small fraction of them are reported to physicians: for instance, a health care survey of the general population showed that 73–95% of community respondents have at least one symptom every 2–4 days.[21] And 81% of healthy university students and hospital staff reported having at least one somatic symptom in a 3-day period.

Such symptoms cause trouble and are the main reason for clinic visits. They may also be subject to a process of self-amplification and self-appropriation: the more confident patients are that their symptoms are abnormal or alarming, the more severe, protracted and incapacitating the symptoms become. Moreover, the perception of bodily symptoms may be enhanced by comorbid depression or anxiety, by suggestion, by stressful life events and by the fact of becoming a patient: the sick role may very well become a way of life.

Nathalie Steinbrecher and her collaborators studied the prevalence of medically unexplained symptoms between February and September 2008 at two primary care practices in Mainz, Germany. Both practices are part of the regular German health system with two GPs working at each practice. Medically unexplained symptoms made up two thirds of all reported symptoms with women, younger persons, and non-native speakers having the highest rates.[22] There was a strong association between somatic symptoms and psychological distress. Anxiety and depressive symptoms showed more or less the same association with somatic symptoms, and specific somatic symptoms or symptom clusters did not show any differential association with anxiety or depression.[23]

---

[19] Mitchell (1984), Angell (1985a, b), Simon et al. (1999), Kroenke (2001), White et al. (1961), O'Malley et al. (2000), McAteer et al. (2001), Creed et al. (2012), Scott and Walter (2010), Hubley et al. (2014), and Rosendal et al. (2017).

[20] Deary (1999), Sloane et al. (2001), Wessely (2001), Barsky (2001a), Council of Scientific Affairs, American Medical Association (1992), and Bytzer and Talley (2001).

[21] Reidenberg and Lowenthal (1968).

[22] Steinbrecher et al. (2011).

[23] Simon et al. (1996).

Symptoms are either *somatogenic* when the patient is the victim of his sickness, but they may also be *psychogenic* in which case the patient is the actor of his condition. If so, when we move the cursor further along this broad spectrum towards somatic distress in excess of what might be expected from physical findings, we meet more and more patients manifesting genuine symptoms that do not point to anything and are not explained by underlying organic disorders or known pathological mechanisms; all in all, these findings are less and less indicative of any medical condition; yet, they constitute the bulk of people treated in primary health care. The further we move along this scale, and as we progressively leave the realm of medical diseases, the more these bodily sensations or symptoms tend to represent a transduction of emotions. These patients are not undergoing bodily suffering, but they become authors (though not simulators) of their ailment.

When they cluster together, those independent and autonomous symptoms may turn into some interdependent systems, namely various somatic single or multiple symptom complexes. Some of those conditions pertain to some organic diseases we failed to diagnose, but most of them correspond to a range of coalescing physical and psychological symptoms with no rational evidence to support such symptom aggregates. They may become epidemic and reflect some pattern of health beliefs prevailing at the time, but they do not fit in any category of organic or psychiatric disorders, and they represent conditions we might be reluctant to recognize. What characterizes them is a common denominator characterized by the absence of physical disease and consisting in the presence of vague and unstable bodily complaints and symptoms inexplicable by the results of physical and laboratory examinations. At this end of the spectrum, the diagnostic process has lost its point, and the symptoms have gained emotive meaning but dissipated much of their cognitive weight.

Somatoform Disorders: Symptoms-Only Conditions in the Medical Setting

In 1694, Everard Maynwaring (1626–1699) was cautioning about the dangerous consequences of *"distempers and alienation of the soul"* causing *"wounded, disturbed, or restless mind" and "a diseased body"*.

First, we must rule out an earlier approach strongly influenced in the 1930's and 1940's by a psychodynamic understanding of medical patients propounded by the Chicago Institute of Psychoanalysis, namely *psychosomatic specificity*.[24] It attempted to establish some psychological causation for physical disorders such as peptic ulcers, hypertension, asthma, or eczema. According to this model, particular types of unresolved conflicts produced permanent tensions and prolonged autonomic arousal, which in turn bred-specific persistent or recurrent diseases. Psychosomatic specificity was compared to microbiological specificity since one

---

[24] Alexander (1950).

was supposed to be able to describe a personality profile or some inadequate expression or solution of emotional conflict characteristic of specific psychosomatic disorders such as duodenal ulcer, asthma, ulcerative colitis, or arterial hypertension. Medicine is still under the spell of Descartes' dualism in its way of distinguishing sharply body from mind, organic from mental illness, and psychological from physiological causes. On this score, it avails itself of a linear model of causality according to which psychological conflicts are turned into somatic symptoms and functional disorders. But, if we choose a non-Cartesian orientation position and wish to reunite bodies with minds, we may salvage the situation by abandoning a linear unicausal model and opt for a multicausal origin, i.e., a neuropsychological conceptualization. We are now seeking after factors that prompt the diagnosis of functional disorders as well as those that breed them: they cover personality, childhood experience, illness experience, knowledge of self and others, psychiatric disorder (depression, anxiety and so forth), disease of and trauma to the brain, stressful situations, social background (modish illnesses), compounding of the symptoms through routine medical examination, the patient's reaction to symptoms and symptom amplification through illness fear, yellow media coverage and the sick role with the ensuing climate of litigation. Marcia Angell answers, "*It is time, to acknowledge that our belief in disease as a direct reflection of mental state is largely folklore*."[25]

DSM-5 uses the term *somatic symptom and related disorder* that corresponds to ICD-10 *somatoform disorder.* Within that category, there are several subclassifications. *Somatization disorder* is characterized by chronic, multiple recurrent and frequently changing physical symptoms and a long and complicated history of contact with medical care services, during which negative investigations or fruitless exploratory operations may have been carried out. Included are *Illness anxiety disorder, conversion disorder (functional neurological disorder),* and *Other Specified or Unspecified Symptom and Related Disorder.* An organic physical disease may be present so that the classification includes *psychological factors affecting other medical conditions.* These diagnostic terms are used in an indiscriminate way in medical practice. In DSM-5, that the physical symptoms are medically unexplained or not, is no longer a criterion of somatoform disorder. Instead, DSM-5 focuses on the way a patient copes with the physical symptoms, emotionally, cognitively, and behaviourally. On the contrary, ICD-10 mainly focuses on medically unexplained symptoms for its classification, rather than their psychological difficulties. Somatoform disorders are responsive to placebo therapy.

Kurt Kroenke countenances that: "... *the difference between somatoform disorders and functional somatic syndromes depends mainly on the investigator: functional somatic syndromes are used widely in primary care, whereas somatoform disorders are used in psychiatric settings.*"[26] If so, then, the idea of somatisation might very well spring from the fact that somatic symptoms often serve as a ticket of admission to medical care, because the depressed patient assumes that such

[25] Angell (1985a, 1572).

[26] Kroenke and Rosmalen (2006, 14).

Table 12.3 Criteria for the differential diagnosis of unexplained symptoms suggesting physical illness

Classification	Are the symptoms medically explained?	Are the symptoms linked to psychological causes?	Are the symptoms intentionally produced or feigned?	Is there an obvious goal?
Physical disease	Always	Sometimes	Never	Never
Somatoform disorder	Never	Always	Never	Sometimes
Illness anxiety disorder	Sometimes	Always	Never	Medical help
Factitious disorder	Sometimes	Always	Always	Medical attention
Malingering	Sometimes	Sometimes	Always	Manipulative
Undiagnosed physical illness	Sometimes	Sometimes	Never	Never

symptoms are a more sensible motive than feeling sad, low, discouraged and unhappy, for seeking health care.

In October 2017, the American Board of Medical Specialties (ABMS), Acting on recommendations from the American Board of Psychiatry and Neurology (ABPN), the American Psychiatric Association (APA) and the Academy of Psychosomatic Medicine (APM) voted to change the name of the subspecialty Psychosomatic Medicine to Consultation-Liaison Psychiatry. The term Psychosomatic Medicine was problematic and had pejorative connotation. From the beginning there was controversy and general dissatisfaction with the chosen name.[27] Anne Harrington suggested that the appeal to psychosomatic disease models might follow from the "*existential deficiency*" of biomedicine's strictly physicalist approach.[28]

In a primary care practice in Mainz, the 12–months prevalence of somatoform disorders was 22.9%, for affective disorders it was 12.4%, and for anxiety disorders it was 11.4%. Somatoform disorder was comorbid with at least one other mental disorder in 43.2% of the cases, and most frequently with anxiety or depression.

Table 12.3 presents a highly schematic summary of the differential diagnosis of unexplained somatic symptoms suggesting physical illness.

Up to 31% of medical patients suffer from *illness anxiety disorder* or *hypochondriasis* though the prevalence is 4% in the general population: such patients are highly sensitive to normal bodily sensations and tend to attribute them to a serious disease.[29] Among patients attending GP's, 18.5% were classified as somatisers and 57.9% of people with anxiety or depression were also somatisers. They tend to amplify somatosensory information resulting from a lowering of the threshold level of perception of common bodily functions and minor physical symptoms such as

[27] Lipsitt (2001).

[28] Harrington (2008).

[29] Creed and Barsky (2004), Clarke et al. (2008), and Starcevic (2013).

sweating, dizziness, rapid heartbeat, precordial pain, abdominal cramps and suchlike. The hypochondriac is a person who suffers from an anxiety disorder and who manifests excessive bodily vigilance, i.e., an excessive attention to one's body for signs or symptoms, which he incorrectly perceives as threatening, but he is not a malingerer. And he is not the victim, either, of cognitive failure insofar as his symptoms are not false beliefs, in the sense of being discordant with the facts: they are no beliefs at all but value judgements. But such bodily symptoms are not medically meaningful, and they are not embodied emotions either, namely emotions experienced in the body.

Medically Unexplained Disorders

In 1893, Gowers (1845–1915) thought that a large fraction of patients suffers from "*diseases that consist only in a disturbance of function: ... they are transient and not permanent, and they are not known to depend on organic changes.*"[30] The various conditions that have been the subject of this chapter make up a large portion of primary and chronic disease management. Actually, 26–33 per cent of patients seen in a practice of internal medicine betray unexplained medical disorders. If one excludes respiratory infections and skin disorders, the incidence of unexplained conditions rises to more than 50 to 70 per cent.

These disorders often overlap with each other and are ranging along a continuum of distress and chronic pain, and from normality to extremely disabling gravity. They are poorly understood syndromes, often nested with one another, and for which no explanation is available. They fail to fulfil the normative criteria for a bona fide disease. They can be understood in several different ways, but the drift involved in them shares the following characteristics.

What characterizes them is that since, contrary to physical and mental diseases, they are largely socially constructed, they are not inevitable, and they keep changing. It follows that, for those syndromes to be eliminated, the social context would have to be altered, even though they are the most difficult of all conditions. Contrary to the paradigmatic diseases, such disorders and symptom complexes are not disjoint sets: they meet the diagnostic criteria of several conditions that often overlap one another, and they are shifting with technological, environmental, and cultural concerns. Ailments that were traditionally thought to be of a moral nature are now being medicalized.[31]

Although there might be several distinct ways of grouping clinical symptoms that hang together, the diagnostician's *medical specialty shapes the diagnostic label* given to the patient. "*Could it be,* writes Allen Barbour, *that we diagnose what we*

[30] Trimble (1982, 1768).

[31] Kirmayer and Robbins (1991), Simon et al. (1996, 1999), Barsky (2001b), and Aaron and Buchwald (2001).

"want" to diagnose?"[32] A similar polysymptomatic clinical picture may be diagnosed as chronic fatigue syndrome by an internist, alexithymia or hypochondriasis by a psychiatrist, or fibromyalgia by a rheumatologist. More and more, expert consensus is used to legitimize symptom clusters, which may then become entrenched.[33]

The physician-patient relationship is thus particularly important in these conditions. Unmet symptom-specific expectations including desires for information for a diagnosis and prognosis when satisfied may eventually alleviate symptoms and reduce their worrying over serious illness. And there is an association between multiple physical symptoms and difficulty in the physician-patient relationship,[34] which increases as the number of unexplained physical symptoms increases.

Diagnosis becomes a moving target inducing looping effect, and functional disorders mutate within a complex cultural and biomedical context. And as functional symptoms and disorders are now the main clinical and public health problems, it should not be surprising that the level of professional satisfaction has dwindled substantially during the past few decades.[35] Physicians may foster the patients' somatising fixation and reinforce illness beliefs and behaviours: actually, both patients and doctors have a preoccupation with finding biomedical reasons and both are preoccupied when no biomedical cause is being found by diagnostic tests. The initiation of medical care may start a self-sustained catch-22 process of medicalization, a classificatory looping,[36] inasmuch as it may very well enhance the severity of the patient's complaints.[37] So doctors and patients together get involved through a kind of self-fulfilling prophecy through a feedback effect, in a vicious circle of unnecessary medical interventions and growing frustration on both sides, reinforcing one another. Medical doctrine and lay opinion reflect and change one another so that what may appear as a new disease does not reflect new knowledge, but a change in the character of an unstable condition because of changes of its representation.

Such disorders are often the result of basic conditioning processes, namely reinforcing and modelling. Expression of symptoms may be reinforced by amplification of normal bodily symptoms. This either lets the patient to escape and avoid anxiety-provoking situations or allows him to enter the sickness-role when it results in social solace and medical care. Symptoms clusters develop when expected patterns and cultural stereotypes model them. By way of illustration, 'normal' pre-menstrual physical and psychological changes set in motion a conditioning process of misattribution due to women learned expectation. Pre-menstrual changes become fashioned after a medical or mental disorder, now termed 'dysphoric disorder' or 'pre-menstrual syndrome'. The symptoms are modelled by others as expected

---

[32] Barbour (1995, 64).

[33] Aronowitz (2001).

[34] Hacking (1983), Bensing and Verhaak (2006), and Hahn (2001).

[35] Zuger (2004).

[36] Hacking (1999, 10).

[37] Rippere (1992).

patterns of female behaviour and are then reinforced through Hacking's *looping effect* by medical care.

But there is another distinctive feature about those syndromes: if the need for intervention is constitutive of what is meant by 'disease', the intervention in the present case is not a medical need, but it is socially demand. Patients with such conditions have explicit and fixed illness belief and phobias about their condition. They tend to stand by their 'sick role' convictions in the face of any effort to dislodge them. Their symptoms are usually difficult to alleviate and poorly affected by reassurance, explanation, and symptomatic treatment. These quasi-delusional self-attributed diagnoses probably result from the growing influence of the media on public opinion. Such patients are high utilizers of health care and generally dissatisfied with the care which they receive.[38]

What's more, some of those syndromes considered to be functional may soon be understood in molecular terms and turn out to be legitimate diseases. Chronic fatigue syndrome (CFS) or Myalgic Encephalomyelitis (ME) is a disabling illness which often lingers for months or years with extreme weakness, inability to think straight, disrupted sleep, and headaches with no apparent connection with demographic or psychological factors. Rather than being a specific condition, it could be the end-stage of multiple conditions: infectious mononucleosis, flu-like illness and a human retrovirus have been shown to be triggering this syndrome.[39]

Taking Stock

Those patients presenting disorderly conditions are forestalling the search of patterns and the attempt to make medical sense of our clinical experience and recalcitrant unexplained symptoms or their clusters. Functional disorders and unexplained symptoms are major features that remain irreducible or insulated from mainstream medical itineraries. If the concept of disease dominates our medical thinking, functional disorders, by contrast, make the lion's share of medical care as they make up the bulk of reported and unreported medical complaints, and are responsible for a large amount of loss of absenteeism, productivity, and medical expenses. Some might like to call these socially constructed conditions "illnesses without disease", since many of those patients may assume the sickness-role without being properly sick. Unexplained symptoms or their clusters, whether reported or not, are not a mere diagnostic dustbin; they may or may not need or justify medical care, yet they reveal specific demand and expectations for care.

"*Theoretical health is the absence of disease*", writes Boorse.[40] But it should now be plain that this view is flawed since many or most of the pathological processes

[38] Karhausen (2014).

[39] Prins et al. (2006) and Lombardi et al. (2009).

[40] Boorse (1997) in: Humber and Almeder (1997, 4).

encountered in the general population and in clinical medicine are not diseases. Once patients have been classified and fitted into the appropriate niches, there is a large remainder of individuals which complain of generic symptoms suggesting some physical illness, but which cannot be neatly categorized.[41] What undergirds this recent fourfold increase in frequency of those patients is a learned anxiety and a lowering of the threshold of sensitivity to bodily signals. This exotic process is made possible by the loss of medical authority together with the corresponding increase in the power of the media. Could it be that "*In place of evil, therapeutic society has substituted illness*", as suggested by Charles Sykes?

Summary

This chapter covers patients who do not suffer from any disease entity but who attend physician or a primary care provider and who represent a sizable fraction of health care need or demand.

1. There is a large reservoir of unreported symptoms in the community that are managed by self-treatment. More than 90% of the symptoms were not reported to a physician's attention; 73–95% people have at least one symptom every 2–4 days.
2. The symptom iceberg represents the visible fraction observed by physicians. Symptoms, be they emergent or submerged, are part and parcel of reported on unreported clinical manifestations that refuse to cluster around our disease categories. More than half the patients in primary care or in population-based studies are medically unexplained.
3. A survey of acute illnesses, namely, in the USA showed more than a 158% increase of unexplained symptoms and functional disorders from 1920 to 1980. During the twentieth century, the symptom iceberg represents a pyramid of a demand not a need for health care, in the absence of any identifiable change in function or structure of organ systems. Patients show a general increase sensitivity to body symptoms resulting in a shift from submerged to manifest status. Patient increasingly present the sick role in the absence of sickness.
4. Factitious disorders are unexplained symptom clusters suggesting physical illness and include somatoform disorders as well as functional disorders and they are not known to depend on organ changes. These symptom disorders have a high degree of psychiatric comorbidity, principally anxiety, depression, or obsessive-compulsive symptoms.

[41] Lloyd (1986).

Chapter 13
Health

It is being said that a doctor is a professional man who suffers from good health.

Contents

The concept of health is of a different nature from that of illness. Sick people can be observed and described in a doctor's surgery, or in a hospital or health center. They are countable and can be interviewed, examined, described, diagnostic tests and measurements can be carried out, and an attempt made to classify them into categories of disease, and they are therefore natural and objective phenomena. On the other hand, it is difficult to identify, count and describe healthy people, because the concept of health is a subjective, global, contextual, evolving, and multidimensional qualification. What mobilizes people is suffering, unmanageable disability and illness, i.e., ill health. When expectations are not met, there is a cry for help. Contrarywise, when expectations are met, nothing is usually said. Health is one of life's great blessings, yet we are not aware of it while we have it but become aware of it after we lose it.

The Latin words "*salus*" and "malus" are transmitted into French as "salut" and "maladie". "*Salus*" means health, rescue, redemption and wealth and it derives from "solvus" which means safe, alive, healthy, unharmed. In English, "health" derives from the Old English "halts", which is related to "whole", "a thing that is complete in itself" (Oxford Dictionary), which in turn is derived from "hal" of Germanic origin, which relates health, fullness, happiness, and salvation.

Jonathan Sholl edited a book[4] in which 60 authors attempt unsuccessfully to define and/or explain the concept of health on biological grounds. None of them define what they mean by health, so the 60 essays that make up the book share the term "health" but there is no indication that they refer to the same concept. The authors also fall victim to the common mistake of basing health on the concept of homeostasis. But homeostasis is the essence of life and is found in most issues

L. Karhausen, *Analytical Philosophy of Medicine*, Philosophy and Medicine 152,
https://doi.org/10.1007/978-3-031-81257-6_13

relating to biology or disease (cancers or hyperthyroidism are homeostatic systems) and it is not specific to the idea of health!

The editor, Jonathan Sholl, tends to believe that health is a natural state, but he adds that this claim does not lend itself to observation, which seems to indicate that it is not a scientific claim, but a philosophical one. If this is so, then the various uses of the term 'health' are not descriptive but speculative. The project of this book represents a plight which consists in trying to unify on biological grounds a heterogeneous set of socio-cultural and biomedical situations, which have nothing in common with each other, except for a vague psychosocial or existential concept. It is what Gilbert Ryle has called a *category fallacy*, in which things belonging to one category are presented as if they belonged to another, so that a property is attributed to something that could not possibly have that property.

In 1946, WHO defined **health** as wellness that is to say, *"a state of complete physical, mental, and social well-being and not merely the absence of disease or infirmity"*.[1] R.S. Downie believes that this definition is a step in the right direction.[1]

Yet, the requirement of "complete" well-being means most of us are unhealthy most of the time. With this move, a person is said to be in good health when he is free from illness or injury.[2] Health is defined by default. It is thus construed as what it is not, in other words, in the shadow of the diffuse background of abnormal features, diseases, handicaps and other negativities: it is conversely anchored into suffering, disease and harmful malfunction. Being abnormal, in good health or not, is not the result of the breakdown of some natural order lying underneath a world of complexity, multiplicity and suffering, since there is no such order that can eventually be discovered, no formula, symmetry or rightness that virtually rule the biological universe.

But health allegedly is something over and above physical fitness, or the absence of disease: is a high level of health actually compatible with some degree of disease, injury, or impairment? Health cannot be defined as the mere absence of disease or infirmity, especially as a person may have an illness without feeling ill, and may experience unwanted symptoms such as headaches, fainting or vomiting, when no disease is present. It may be the normal infirmity of old age or a kind of deformity, but the person with it is not necessarily in poor health. In their seminal paper, R.N. Jadad and L. O'Grady[2] point out that health can coexist with illness: it is possible to be sick and healthy at the same time. A Canadian survey of more than 3000 people aged 65 or older clearly illustrates this point, showing that 86% of people with one, 77% with two, and 51% with three or more diseases rated their health as good, very good or excellent. These data are replicated countless times in quality-of-life studies that include self-rated health. A meta-analysis of 22 cohort studies found that people who rated their health as poor had twice the risk of mortality as those who rated their health as excellent.

[1] World Health Organisation (1981).

[2] Whitbeck et al. (1981b, 15) and Doyal and Gough (1991, 656–59).

From the vantage point of *philosophy,* the strategy is this. Plato in the *Timaeus*[3] describes health as harmony and disease or illness as disharmony, that is to say discord of the elements, which enter in the composition of the body. This eudemonistic view envisions being healthy as a condition connubial to being good, happy, thriving and blessed. Thus, health when realized is a natural good, a state of flourishing without disease or impediment.[4] We find this positive notion of health in fitness and health cults that envision health as a positive good and acknowledge the presence of a logical gap between health and the absence of disease. Plato conceived the health of an organism and the *eunomia* of a society as being, each of them, a condition of the harmonious working together of its parts. Actually, social scientists tend to favour some broad view of health while clinical and practising public health workers concentrate on disease, disability and causes of death.[5] Galen wrote that *Health is a condition in which we neither suffer pain nor are hindered in the functions of daily life.*[6] Epicures wrote: *"What produces unsurpassed jubilation is the contrast of the great evil escaped. And this is the nature of the good, if one applies one's mind correctly and then stands firm and does not go walking about chattering about the good in an empty fashion".*[7] And Jeremy Bentham added: *"Health is the absence of disease, and consequently of all those kinds of pain which are among the symptoms of disease".*[8]

In a more *pragmatic* approach, the US *Office of Disease Prevention and Health Promotion* is setting objectives concerning health promotion and the prevention of diseases. The goals of *Healthy People 2020* consist in:

1. Attaining high-quality, longer lives, free of preventable diseases, disability, injury, and premature death.
2. Achieve health equity, eliminate disparities, and improve the health of all groups.
3. Create social and physical environments that promote good health for all.
4. Promote quality of life, healthy development, and health behaviours across all lie stages.

Besides, *lay accounts of health* rest on the current, though popular assumption that the majority of people are normal. It implies that an individual is healthy as long as he or she is free of abnormal symptoms and signs. People are naturally sound in wind and limb, and they become sick through either some external interfering agent or some lack of physical fitness. Talcott Parsons[9] narrowed down slightly the preceding definition: health, for him, was a condition of physical and mental fitness, namely the optimum capacity of an individual for the effective performance of the

---

[3] Plato Timeus, 4 a.

[4] Nussbaum (1994, 30).

[5] Wilson (1970), Seedhouse and Cribb (1989), Aggleton (1990), Nordenfelt (1995), and Bowling (1997a, b).

[6] Galen *De sanitate tuenda,* I, 5.

[7] Nussbaum (1994, 102).

[8] Bentham (1996, VI, 7, 53–54).

[9] Parsons (1981).

roles and tasks for which he has been socialized. We must thus look at the social surroundings to determine how a given population or some individuals fit their optimal social capacity. Well-being, Downie suggests, requires both an essential reference to a good life and *some conception of control over one's life*.

This is tantamount to saying that "*Health is a dynamic state of complete physical, mental, social and spiritual well-being*". Health thus defined is being part of other than mere medical aspects of social well-being such as happiness, wealth, social status, and legal capacity.[10] Caroline Whitbeck defines 'health' as the psycho-physiological capacity to act autonomously and effectively or respond appropriately in a wide variety of situations. By 'appropriately', she means in a way that is supportive of, or at least minimally destructive to the agent's goals, projects, aspirations and so forth: for Whitbeck, health is the context for an adequate understanding of medicine.[11]

However, some such definitions are very theoretical and express philosophical preoccupation. Their general approach is disconnected with and doesn't address the scientific or medical or clinical issues and their need of intervention and policymaking.[12] How do the different aspects of social, economic, or psychological wellbeing relate? Are they different concepts about the same thing?

Health Status: Negative or Positive

Positive is a state of well-being and fitness, the ability to perform the tasks needed. Negative is the absence of illness.

What distinguishes the positive from the negative view, is precisely what divides Olimpicus from Galen, namely contraries from contradictories. In medical context, one cannot be at once non-diseased and non-healthy, in contrast with the lay beliefs or the sociologist's views about health.[13] Galen (131–201), friend and physician of Marcus Aurelius, criticized Olimpicus, a medical Methodist, who claimed that since between health and disease there is some intermediate state, they couldn't be contradictories. They are actually contraries.

Subtracting the negative from the positive notions of health leaves an irreducible residue made out of a cluster of characteristics: quality of life, well-being, social adjustment, physical fitness, functional ability, ability to cope with the demands of everyday life and of stressful situations, high morale and life satisfaction, or anything of that ilk. There is no limit. This remainder brings forward the looseness of the positive views. The privative perspective avoids having to fill this recondite logical gap created by WHO's definition between health and the absence of disease.

[10] Nordenfelt (1995, 76, 97–104, 145–150), Thomson (1987, 18–22), and Baybrooke (1987, 29).

[11] Whitbeck (1981b, 611, 616).

[12] Alexandrovna (2014).

[13] Whitbeck (1981b, 624).

Everything being considered, what divides the positive from the negative view of health? The answer is that the absence of disease or of any harmful conditions is a minimally necessary condition for health but not a sufficient condition for eudemonia, namely human flourishing.[14] This draws attention to the frequent confusion between these two concepts.[15] Health is not a passive process, but an active state that should be studied and maintained. Positive health, just like goodness or happiness has an intrinsic positive emotional although non-capricious meaning; and in that sense, it does not correspond to anything that is true or false: the norms of health are not discovered.[16] The 'positiveness' of health lies in our efforts to reach this unattainable goal. It is a political idée-force; health concerns states of affairs that are being and that ought to be sought after. Although it may sound descriptive, WHO's definition of health is actually prescriptive.

But there is a further consideration of immediate interest. The word "health" is often used in predominantly directive sentences whose purpose is to cause action. Health is not something which happens to people like diseases do.[17] It is more performative than conceptual. It involves the physical and mental capacity and capability to act appropriately in a variety of situations, to participate autonomously and effectively in a wide range of situations, to achieve one's own purposes and to live as a member of the human community.

Contrary to disease and to suffering, health has no ontological status and is not an identifiable state but rather, like the God of the Old Testament, a challenge, and an incentive. Health is not a goal, but it serves as a norm, that is not an average but a standard,[18] or a norm of how one's body ought to be.[19]

All in all, when it is only the perspective afforded by medical insight that is up for grab, the confines of medicine stipulate the limits of its language, so that the medical use and meaning of the term 'health' cannot extend beyond those limits: this semantic austerity thus restricts the meaning of the term, which has to be congruent with the logic of the medical edifice. Moreover, medical understanding needs concepts that are or that can be trans-culturally shared and compared.

Doctors are interested in diseases, not health, so that medical textbooks are a massive catalogue of diseases. The preferred view of health is the ability to adapt, self-manage and participate in social activities, including work. Sigmund Freud defined 'health' as the ability to love and work.

Health may be seen as a *disposition*, i.e. a source of resistance to infections and to bodily and mental deterioration, a reserve or stock of strength and a power of overcoming disease.[20] This concept of possessing a reservoir of health, a personal

---

[14] Nussbaum (1994, 15).

[15] Mill (1971, ch 1, §9).

[16] Margolis (1975, 105).

[17] Whitbeck (1981b, 616–617).

[18] Jaspers (1982, 7823) and Whitbeck (1981b, 614).

[19] Vickers (1958).

[20] Herzlich (1973, 57) and Sacks (1982).

strength and ability is cognate to a substantial view of health as an essence belonging to the world of Plato's cave, an essence which is seen as a positive good that can be captured by appropriate medical and public health procedures, supplied by doctors, purchased in pharmacy or lost when one gets sick.[21] For Galen, health was a disposition (he called a diathesis) of both behaviour and bodily functions as well as of the body itself in accordance with the absence of disease.[22]

Spector, a geneticist at King's College, London, has identified a new generation of tests, based on emerging insights into human biology. Having analysed more than 100,000 microbiomes, he identified 50 "good" bacteria and 50 "bad" bacteria, the balance of which appears to be a good indicator of health.

Spector's second recommendations is to test the "metabolome"; such tests are markers of a human's current condition and future disease risk. A recent study combined metabolite data with blood lipids from more than 118,000 participants of UK Biobank. There is a striking similarity of some patterns of biomarkers across various diseases, that might help detecting metabolites that are ideal for informing someone about has overall health.

The University of Texas Health Science Center at San Antonio in attempting to better understand health, indicated that it paly comes from immune resilience. By studying the blood cells of more than 40,000 people, Ahuja and his colleagues found that the balance between two types of immune T cells — one of them associated with resilience, and a low level the other is associated with inflammation — correlates with how someone responds to physiological challenges such as cancer or viruses. If the T-cell ratio comes back worse than expected, a 24-week exercise regime was followed by an improved immune grade.

Even so, there is no all-encompassing test of health. But everything being considered, a handful of those approaches can offer a much clearer picture of what it means to be healthy.

In the WHO definition, health belongs to categorical or non-scalar predicates. It pigeonholes individuals. WHO surveys that assess an individual's health status ask questions about mobility, self-care, pain, cognition, interpersonal activities, vision, sleep, energy, and affect? Health falls into a single metric ranging from dead (0) to perfect health (1). applying negative utilitarianism to medicine, Popper has suggested that we should concern ourselves with the minimization of suffering not with the maximization of happiness.[23]

In the biomedical sciences, however, one can understand a disease without prior knowledge of the idea of health. Being healthy means having no unfulfilled health care needs that is to say no normative needs as defined by biomedical doctrine. So, health is essentially a negative circumstance: objects of health value are all negative except for life expectancy. By becoming privative, the notion of health becomes empirical, highly specific and tends to be universalizable through the understanding

[21] Aggleton (1991, 12–14).

[22] Galen (1991, 28–30, 147–148).

[23] Popper (1966, vol 1, Ch 5, note 6).

provided by medical science.[24] The avoidance of bad things is a far more urgent task than is the promotion of the good.[25] As a consequence, being healthy, viewed as a privative notion, is a scalar predicate: it comes by degrees and can be measured. Contrary to normal vs. pathological, the words healthy vs. unhealthy shade gradually into each other. So, features that causally affect favourably or promote health refer to rank and grades of health. Nobody is really healthy, but some people are healthier than others and one may sustain and improve one's state of health. And a so-called healthy person may have diverse patterns of health risk, i.e., of potential illness. Being healthy can only be properly understood as a limiting predicate. Any one of us strives for health and when we fall sick, we know we are less healthy.

Hygeia was the daughter of Asklepios.

Beneficence is not a positive but a negative concept: it is anything that prevents or decreases harm or its effects, that corrects a pathological process. This purview is part of a family of counterfactual value-imbued concepts such as those of normality, physiology, sanity, safety, salubrity or functions which divide the biological continuum into features of positive and negative concern by leaning on their negative position.

Health is defined contrafactually. Since Claude Bernard, experimental medicine has created pathological anomalies to understand normal functions. Diabetes was discovered by removing the pancreas. Medical reasoning first defines 'health' negatively, but then swivels round and takes it as a theoretical standard or yardstick from which it assumes that the sick has departed. Health becomes henceforth an ill-defined notion, which emphasizes the process of referring to normalcy, rather than normalcy itself; *"the concept of the normal is an illusion of our own making,"* writes the physiologist Homer Smith.[26] Following this view, it is practically impossible to pronounce definitely on the question whether some individual should be considered healthy or not. The disease concept is not conceptually secondary to the concept of health and normality: it has epistemic priority.

There are many *indicators of health*: oral health, reproductive and sexual health, mental health, injuries, and violence such as deaths from injuries or homicides, environmental quality, substance abuse, clinical prevention services, etc.

If one wonders whether some categorical predicate is ascribable to some state of affairs, the answer is either 'yes' or 'no'. Non-scalar adjectives, such as good, useful, true, daily, square, perfect, or elected are binary, since they bring up some sort of clean and absolute division, without grey areas or penumbral cases. On the opposite side, being beneficial, healthy, therapeutic, or protective are scalar features. If the minima of health are categorical, there are, by contrast, gradations in harm and detriment, in disease severity, or in levels of handicap.

Health is a *graded, scalar, rather than dichotomous variable*, and it can be and has been defined in countless ways.

---

[24] von Wright (1963a, 55).

[25] Ohlsson (1979, 17).

[26] Smith (1947, 372).

Good and bad health are part of a continuum.

The assessment of health status by the Karnofsky index[27], which goes in ten steps from 100 (perfect health) to zero (death), quantifies the level of health.

Many, if not most, people are in a continuous state of moderate sub-clinical illness. They are in these states of illness to the extent that they are not in an optimal state of health. There are very few individuals that are healthy strictly speaking and there are large communities highly adapted to living with diseases. We generally only apply the label 'disease' when we are so far from optimal health that specific symptoms manifest themselves. The manifestation of symptoms of disease is simply the lowest point on the scale at which one is most ill, i.e., furthest from health.

In fact, many of the measures used as indices of health status are not measures of health at all but of health service use (visits to health care providers (hospital admissions)).

Nonetheless, Andrew Schroeder proposed a comparative approach to health with a thorough discussion of a graded concept of health.[27] Health cannot be judged by one measure, or one type of measure and it is illusory to look for a single right balance. There are several composite indicators available for health status measurement such as indices of well-being, or of disability, as well as the quality-adjusted life year (QUALY) based on morbidity and mortality.[28] Assessment of quality of life as a therapeutic outcome is an important and difficult issue. The literature suggests that doctors do not adequately evaluate the patients' quality of life. For one thing, measures of outcome should take account of the patients' self-evaluation and for another, their perception of improvement or satisfaction should be validated against levels of performance.[29]

Conclusion

If we were to apply the WHO's definition of Health *stricto sensu*, each one of us would be unhealthy. This definition raises too many difficult if not confusing issues about what is meant by the phrase "complete physical, mental and social well-being". The question is not: how to reach health? But how much health can we get?

It has vague boundaries unless one attempts to define it minimally and to identify it with the absence of negativities or abnormalities and in that case, it depends on the definition of what is abnormal. "Health" like "natural", is a slippery notion.[30] Yet, the source of this vagueness is not nature's fault but our own. No further research is going to help us finding the real area of application of the term 'health' because the indeterminacy of the boundaries of this remainder are subject to differently

[27] Schroeder (2013) and Timmermans (2012).

[28] Bowling (1997a, b).

[29] Lepiège and Hunt (1997), Brock (1993), and Griffin (1993).

[30] Nussbaum (1994, 29).

grounded, shifting, contingent ways of making judgement which include cultural, historical, social, professional, medical, linguistic, or psychological background.

Furthermore, although the collective health, in terms of increased life expectancy, declining infant mortality as well declining age-adjusted death rates for 10 of the 15 most frequent causes of death, has been improving in the past 30 years, surveys reveal declining satisfaction with personal health: respondents report greater number of somatic symptoms, more disability, and more feelings of general illness.[31] This discrepancy brings to light the fallacy of amalgamating two inconsistent definitions of health, namely an objective empirical one, with another one that depends on an ethnic, social and cultural climate of apprehension and insecurity about disease as well as decreased thresholds of awareness of bodily symptoms combined with unrealistic expectations of cure.

René Dubos described the occurrence of an ideal state of good health and eudemonic full flourishing well beyond the mere absence of disease or disability as a mirage without empirical application.[32] "*If that definition were correct,* writes Samuel Gorowitz, *and if physicians can give orders with respect to health, then physicians could give orders with respect to every dimension of life*."[33] And Germund Hesslow adds that: "*the health / disease distinction is irrelevant for most decisions and represents a conceptual straightjacket.*"[34] Health is like one's spirit: nobody can give it to you, but anybody can take it away from you. The literature is in disarray when it comes to define health inasmuch as it is surrounded by conceptual confusion. There is no general consensus and little chance of reaching agreement on its meaning and its boundaries, still less about the means by which it could be achieved. Furthermore, such terminology disputes are often mistaken for substantial ones.

Medicine writes Margolis,[35] is committed to a bifocal view of health. First, the privative notion of health based on prudential requirements preserves the neutrality of medical care. And second, the positive notion posits broader norms of human happiness, well-being, and self-realization. The first view is "*no more tendentious than our prudential concerns*" and it defines some of the least disputable, the least tendentious medical norms and objectives. On the opposite side, the second view is decidedly tendentious as it is committed to some favoured view which is more of cultural than of medical concern: it refers to whether treatment outcome is likely to lead to a life worth living in physical, mental, and social terms. Kant contended that "*A person can feel healthy and never know if he is healthy*".

Defining health recalls Saint Agustín of Hippone who said about time: "*What is time? If no one asks me, I know. If I want to explain it to anyone who asks me, I don't know anymore*. »

---

[31] Barsky (1988).

[32] Dubos (1959).

[33] Gorowitz (1982, 14).

[34] Hesslow (1984).

[35] Margolis (1975, 104–105).

Chapter 14
Preventive, Therapeutic, and Palliative Care

Doctors are men who prescribe medicines of which they know little to cure diseases of which they know less, in human beings of whom they know nothing.

Voltaire

Contents

The goal of medicine is cure, but cure is not a binary concept. It would be the same thing as to say that the scope of painting is to attain beauty or the scope of ethics to attain moral goodness. Actually, "cure" is a very generic term that could mean restoration, repairment, recovery, improvement, redress, alleviation, or healing.

In addition, cure is a scalar process or a scalar predicate: it has an outset, it may gradual, progressive, intermitting, or unremitting, interrupted, or unbroken.

L. Karhausen, *Analytical Philosophy of Medicine*, Philosophy and Medicine 152,
https://doi.org/10.1007/978-3-031-81257-6_14

Our prudential inclination to avoid errors, to avoid being wrong, expresses a human wish to minimize risk and maximize utility. Since diseases are construed as unfortunate errors blighting the smooth course of our lives, those disturbances require to be repaired to serve our best interests. "*As to diseases, wrote Hippocrates* (460?–377? B.C.), *make a habit of two things—to help, or at least to do no harm*". The aim of treatment is to shift patients from the sick towards the normal category, that is, to minimize suffering and to postpone death, and the aim of prevention is to forestall healthy individuals from all forms of sickness. To prevent a disease, we have to eliminate one of its necessary conditions, should they be found, either the *Anopheles* mosquito or the parasite itself in the case of malaria. And treating a disease implies bringing into play some intervention that is sufficient to alleviate, control or take care of it.

Diseases are processes but treatments or preventative measures are prudential procedures. 'Treating' is a task word or a try word, and treatment is a performance or activity, not an achievement because nothing may be achieved if the treatment is inefficacious; but 'curing' or 'preventing' are terms of achievement, words of success.

Treatment and prevention have two implicit dimensions: one is purely causal and relates the intervention with its intended effects; the other one is axiological and relates these effects with some prudential objectives based on a judgement of utilitarian goodness.

Rationality consists in optimizing our decisions. "*L'office du bon médecin,* writes Ambroise Paré in 1652, *est de guérir la maladie que s'il ne vient à cette fin, au moins faut-il qu'il la pallie.*" (The duty of a good doctor is to cure disease, and if he cannot do this, he must at least make up for it.) A *cure* is the effect of a treatment that erases the disease and one of its necessary causes if any, e.g., healing a fracture. When healing is not feasible, *palliative treatment* attempts to alleviate the symptoms of the disease or to positively change its course, since the responsibility of physicians is to prevent medical complications.

Treatment responds to a balanced evaluation of needs upheld from an impartial perspective, and in some cases to demand e.g., in aesthetical surgery. But ends need to be justified: to be beneficial means that the immediate or long-term advantages have to be weighed against the side effects and harmful consequences of it being brought about. When a physician decides not to continue life-sustaining procedures, this decision should result from a balanced judgement, which takes account of the therapeutic imperative and of the best interest and the quality of life of his patient.

A treatment may be needed even when no treatment is available, that is, counterfactually, were an appropriate treatment be available, it would be mandatory. Yet, if diseases call for medical intervention, heedful non-intervention may in some cases be the proper course. Therapeutic abstention, if intended and justified, that is, if it does not result from omission, may be a modality or part of a therapeutic intervention. It does not refute but illustrates the claim that the concept of disease implies medical intervention. Intervention includes acts of omission, i.e., refraining from intervening, as well as positive interventions, acts of commission. For example, segmentectomy, lobectomy or pneumonectomy is the standard treatment for a patient suffering from lung carcinoma stage I, with a cure rate of 55 to 75%. However, if the patient suffers from severe airflow limitations due to chronic obstructive pulmonary disease, none of these potentially curative treatments might be available since the patient would not have adequate pulmonary reserve after the resection.

Actually, one could paraphrase Ernest Nagel and state that we have an agent-neutral duty to prevent and treat a disease, namely regardless of whose disease it is. What's more, there is no theoretical inference to be made between the notion of disease and that of treatment. The need for intervention is what Wittgenstein called a criterion rather than a consequence, that is to say not just some empirical connection between them. *Within the realm of medicine*, the need for prevention and therapy is built into the very concept of disease. The source of the duty is squarely *outside* the patient's own will and *prior to* the patient-physician contract. This does not rule out that medical decisions might or ought to be submitted to the patient's consent.

But if treatment is part of the concept of disease, it may, by extension, be targeted at conditions, which are not *prima facie* pathological, so that, under the pressure of public opinion, medical doctrines seek to construe those conditions as pathological. It follows that treatment may house borderline procedures of non-therapeutic medical care—namely procedures that may be the object of demand but not of need—concerning some biological processes such as pregnancy or supportive, preventive or caring interventions namely contraception, circumcision, cosmetic surgery, abortion, cloning human beings, tying off Fallopian tubes against fertility, inducement of labour, euthanasia, or making babies without sex.[1] Sure enough, pregnancy is no disease but it is of medical concern since it is a risk factor and may cause medical problems. Primary aging is a quasi-universal feature of life: but secondary aging resulting from external influences (overeating, lack of physical activity) is envisioned as a disability and as something that could be delayed or avoided, and that physicians construe as abnormal, whatever the broad objections philosophers might be rising.

Finally, a physician reasons *well* if he reasons correctly about the relations between his course of action and his prudential goals, and if he looks for those relations relevant to finding means to the desired end. But he reasons *successfully* if he is looking for efficacious choices inasmuch as correct prudential reasoning may not be successful reasoning. Treatment is often a choice between a limited number of alternatives, many of them highly unwelcome. Besides, doctors may be wrong.

According to Gilbert Ryle, we must distinguish *task verbs* or *try verbs*—which are liable to be qualified by adverbs like 'erroneously' or 'incorrectly', but which can neither be success nor failure—from *success* or *achievement verb*s. 'Curing' or 'preventing' fit into sentences descriptive of achievements (or protracted achievements), but tasks such as treatment or preventive measures, by contrast, are liable to go wrong.

[1] Leaders (2017).

Besides its beneficent utility, therapeutics is philosophically important. We have seen that the splitting of biological facts into normal and pathological ones was a basic but conventional commitment, and not something empirically justifiable. But it is not arbitrary because it is ligated to the need for action. If there is a need for restorative intervention, this is not a mere contingent consideration about what we mean by being pathological but is constitutive of its concept: it is not some mere accompanying characteristic trait but a defining one: this is an *a priori necessity.*

In this sense, intervention is an essential part of medicine. John Locke writes: "*no man requires greater certainty to govern his actions than by what is as certain as his actions themselves*".

Therapeutic Doctrines

Medicine is a prudential discipline: prudence is that faculty which enables an individual to select what the correct action is in a given situation. This is how *phronesis* bridges what Aristotle called *theoria* with *praxis,* the cognitive with the intentional, or, in medical care, assessing health care needs with satisfying them, and describing a treatment with getting on with it.

For all that, the proper treatment of a given condition is relative to time and eventually to place. It may cease to be acceptable with the passage of time and the progress of medical research. Until 20 years ago, aspirin was the treatment of choice of acute pericarditis resulting from viral infection, but NSAID or colchicine are now preferable. Medicine fosters therapeutic doctrines that develop and change gradually or abruptly and that are neither true nor false but rather acceptable. There often is one dominant or fashionable doctrine for a given disease entity but there may also be several conflicting ones. Psychodynamic or cognitive behavioural psychotherapy have been opposed to pharmacological therapy, but they are now coming together. Evidence-based medicine is a way to formalize therapeutic guidelines and strengthen their evidential support in place of informal pre-existing doctrines.

Effectiveness and Efficiency

Even though the beneficence of most therapeutic interventions seems reasonable, they often lack supporting evidence. Meanwhile, doctors are deciding by themselves, often by seeing what they want to see in what most agree are poorly supported data. In assessing the outcome of a treatment, one among several questions that come up is: to what extent does a particular medical action alter the natural history for the better? Or else, how much efficacious is a treatment regimen in relation to some specific condition?

Most medical treatment are, or should be, beneficent and not detrimental. To be beneficent means to promote the good by making the bad better, which is what is meant be curing or healing. Medicine is beneficent since it is *good for* sick individuals and does good to them: it affects them favourably by protecting them from suffering disease and disability.[2] Yet, many, if not most standard therapies have not been validated and some of them such as psychoanalysis are probably inefficacious when not harmful. Furthermore, knowledge is fallible and medical guidance can be faulted. Hippocrates reminds us that "*interventions are perilous*".[3] Finally, a validated treatment will not work for everyone: when administered to an individual, such a treatment may have positive effects, no effect, or harmful consequences.

There are two ways to express the beneficent effect of an intervention: effectiveness and efficacy depending on whether it is defined and assessed in merely descriptive or in experimental set-up. Both of them are expressed in the realm of probability, that is to say that they are depersonalized.[4]

Effectiveness is a measure of the extent to which a specific therapeutic regimen does what it is intended to do in the experimental set up of a randomized clinical trial (RCT). Effectiveness refers to the causal effect of treatment.

Efficiency, as defined by Archie L. Cochrane (1909–1988), is a measure of the extent to which a specific therapeutic regimen does what it is intended to do during an observational study, which is closer, as it were, to what happens in some routine clinical or public health field.[5] It translates scientific effectiveness in clinical setting, taking account of economic cost of various treatment options, inasmuch as clinical science differs from clinical practice.

Cochrane also made a distinction between care and cure, so that cure represents the merger of effectiveness and efficiency.

We may, about any treatment, wonder whether it may work or whether it does work in clinical practice. An efficacious programme for the control of malaria in a community might or might not be effective, and if effective it will never be absolutely effective. Efficacy pertains to causal capacity whereas effectiveness pertains to causal role.

The *efficacy* of a treatment is a scalar measure of benefit, namely of the capability to improve the outcome, that is to say the extent to which the intervention has beneficial effects upon the natural history of the disease and upon the quality of life. Improved outcomes may be either subjective when reported by the patient, in which case the symptoms may or may not be observable by others (anxiety), or objective such as increased life expectancy or those obtained by physical examination (dyspnoea), laboratory or imaging results, and in the latter case they may or not depend on the patient's collaboration (examination of visual defects vs. measurement of arterial blood pressure).

Since treatment outcome is a scalar quantity, measuring the effects, i.e., the therapeutic gain, may be made in absolute or relative terms. To be meaningful, efficacy

---

[2] von Wright (1963a, 41).

[3] Hippocrates, Aphorism 1.

[4] Cochrane (1989, 2).

[5] von Wright (1963a, 47).

must be expressed in comparative terms, that is, in contrast with some reference point, often a targeted placebo. It is measured by subtracting the scores achieved by the subjects in the treatment group from the scores achieved by the subjects in the placebo group and the difference is usually called *attributable efficacy* (or attributable benefit). Is the treatment better than the placebo control and if the answer is yes, how much better? It tells us what fraction of a group of patients given a treatment it is likely to help. It reflects how many patients with a given disease might be improved in the population as a direct result of a given treatment. This is highly relevant information for health policy-makers responsible for resources allocation.

But if the disease is rare, attributable efficacy will be low. As a case in point, very few people will be saved by dialysis for end-stage renal failure because the condition is rare, whereas for the patient concerned, the treatment is lifesaving. This is why clinical care which is concerned with the individual patient's interest and queries, is tacitly guided by the logic of *relative efficacy:* it is a ratio of effects of intervention, divided by placebo: it is expressed as the ratio of the proportion improved who experience the treatment to the proportion improved under the placebo. From the patient's point of view, such as someone suffering from end-stage renal disease, a rare condition, the attributable efficacy is necessarily low irrespective of the magnitude of the relative efficacy, which is high since these are lifesaving procedures. If so, then, the relative benefit reflects the overriding concern of the patient.

Pickering held that, all in all, "*in more-or-less equal quantity, there are treatable diseases, there are diseases and human problems for which medical interventions may or may not help, and there are untreatable diseases*".[6]

The Myth of the Magic Bullet and Its Return

The German medical scientist Paul Ehrlich (1854–1915) known for his pioneering work in immunology, bacteriology and chemotherapy thought that diseases should be cured by the use of chemically specific therapy, which he compared with 'magic bullets' which zero in on their target and injure nothing else. "*A reckoning of the cause often solves a malady*," wrote Celsius (25 B.C.–A.D.50). Just as there is a tendency in each one of us to believe in the reality of disease entities or in the existence of 'true' causes, and we are inclined to think that to each condition must correspond some kind of specific therapeutic cure. One of the leading philosophers of medicine, Jacob Stegenga believes that "*the bullet metaphor … continues to serve as an exemplar of present-day medical interventions*".[7] The magic bullet is the counterpart of necessary causes. Treatment is something out there, which waits to be uncovered, even though it may not be available yet but if we keep tracking it

[6] Pickering (1996).

[7] Stegenga (2018a, 55).

down, we shall someday discover it. Put your diagnosis in the slot and wait for the ticket indicating your treatment to come out of the machine.

To be sure, infectious diseases are specific conditions that have specific causes. Their discovery led to a spectacular increase in the healing power of medicine. Vitamin B12 for a patient suffering from pernicious anaemia is a magic bullet. Likewise, the development of genomics might deliver the promised next medical revolution.

But even then, this is a mistaken view,[8] which stems from the clockwork analogy by which we tend to compare disease to a broken clock and pathological processes as some sort of breakdown of biological congruity. Even at the genetic level, diseases are usually polygenic and developmentally complex. Patients with the same disease may have vastly different responses to treatment. To pin our hopes on some large-scale extension of magic bullets supposes that the class of causal diseases will be expanding, but this is wishful thinking. Treatments are rarely precisely targeted, such as ascorbic acid in scurvy. Most diseases are determined by a combination of many factors. We can conclude with Rudolf Virchow (1821–1902): "*From the basic error that specific remedies were created for particular diseases came the notion that the whole course of a disease, or even its separate stages, could be annihilated by a single remedy*".

Furthermore, most drugs have deleterious side effects and long-term undesirable effects, both known and unknown and they do not target just the intended condition.

Postmarketing surveillance (PMS) and pharmacovigilance are practices of monitoring the safety of a pharmaceutical drug or medical device after it has been released on the market, and the reporting of adverse effects. Pharmacoepidemiology refers to the study of the use and effects of health products in populations. Furthermore, the toxicity of new agents has often been discovered only after they were in regular use, since detecting unwanted effects in advance may be an insoluble problem. Beyond this, the pharmaceutical inserts, which list all the potential side-effects, are a source of nocebo effect. Physicians and pharmacologists are perfectly aware that magic bullets are fascinating chimeras that unfortunately are shear phantasms. All medications that are efficacious are accompanied by unwanted side-effects.

There is, however, some truth in a weaker version of the magic bullet. Since medical science analyses diseases in terms of their components, physicians tend to fractionate their patients in structural and functional units, molecules, cells, and organ systems: this approach leads and will lead to effective therapeutic approaches. Meanwhile, these perfectly legitimate explanatory models should not keep medical personnel from addressing in a comprehensive mode their patients' emotional response to health care and personal capacities to deal with psychosocial problems.

For instance, immunotherapy, i.e., T lymphocyte therapy, in the treatment of cancer (e.g., non-small cell lung cancer or melanoma), consists in removing immune cells from a patient, arming them with new proteins that allow them to recognize

---

[8] Dixon (1978).

cancer, and giving them back to the patient in large numbers. These genetically targeted cells then persist in the body and become sort of living drugs.

A further example is the possibility to cure haemophilia A and B with genic therapy by a single intravenous infusion of a novel bioengineering gene therapy. The researchers used a viral factor to deliver their bioengineered payload of a gene that codes a so-called Factor IX-Padua, a naturally occurring clotting factor that is hyperfunctional (eight to ten times stronger than the normal factor) and that was discovered in an Italian family in 2009. None of the ten patients had serious adverse effects from the gene therapy.[9]

President Barack Obama announced in January 2015 the National Institute of Health Precision Medicine Initiative (NIH PMI) in an attempt to bring us closer "*the promise of precision medicine–delivering the right treatments, at the right time, every time, to the right person*". It aims to provide precise, tailored health care and treatment strategies for conditions such as cancer, diabetes, and schizophrenia. Precision medicine means more than Ehrlich's magic bullets, since it is devoted, personalised, targeted health care—preventive, diagnostic and therapeutic strategies—to take individual variability into account. Medicine should embrace more personal, versatile goals instead of catchall targets. The prospect of precision medicine has been bolstered up by the establishment of large-scale biological databases such as the human genome, various methods of identifying patients, metabolomics, proteomics as well as genomics, and by computational technology. To begin with, the initiative will focus mainly on cancer, but it will later be extended to all diseases. Precision medicine offers a bright future but is still in its infancy and remains at the edge clinical practice.

From Wants to Needs

More than half (50–80%) of medical consultations are due not to illnesses, but to symptoms, such as palpitations, dizziness, back pain, irritability, fatigue, breathing difficulties, insomnia or drowsiness, skin rashes, nausea, etc. Symptoms are comparable to the firmament, where some, but not all, of the stars are grouped into constellations, while the majority are isolated from any context. The symptoms are comparable to the firmament, where some, but not all, of the stars are grouped together in constellations while the majority are isolated from any context. When these symptoms do not correspond to any observable or identifiable anomaly in bodily functions, it can be assumed that it is a question of excessive sensitivity to certain bodily sensations or a lowering of the threshold of awareness of these manifestations, but that they have no medical significance.

A doctor undergoes many years of study, learning, for example, to understand the nature of diabetes and how to adjust insulin doses to keep the patient's blood

[9] Van den Berg (2017).

sugar within reasonable limits, and so on about complex medical conditions whose interpretation and treatment are evolving ever more rapidly. A sudden awakening surprise the GP who finds himself confronted in his surgery with an incessant stream of people complaining of pain, headaches, nausea, impatience, tiredness, and colds—unrecognised and uncategorisable problems which he is obliged to 'treat', even though he finds himself unable to help them; these diverse manifestations should be given time to follow their natural course, for they have nothing in common with illnesses. An English GP wrote after 20 years of practice that "The first shock and unpleasant awakening which strikes a practitioner entering the practice of medicine is that he is confronted with a mass of disorders which are apparently unrecognised, undefinable and unfamiliar, and which it is impossible to diagnose, label or categorise precisely on any objective basis". And one of his Welsh colleagues added: "We doctors are utterly incapable of treating a vast proportion of the ailments that come our way – they are not diseases at all." "*My body*", wrote Lichtenberg, "*is a part of the world that can be altered by my thoughts. Even imaginary illnesses can become real.*"

In short, there has been a shift from illnesses to symptoms. Since the 1920s, there has been an explosion, a veritable epidemic of unexplained symptoms, a phenomenon that is linked to many factors, such as changes in lifestyle, stress, the breakdown of the nuclear family unit, all of which led people to interpret their symptoms as illnesses, in the absence of any identifiable organic, functional, or structural change. The result is that doctors, healthcare systems and health insurance are faced with a demand for care that far exceeds medical needs, although it does make pharmacists happy, but the social security system runs a deficit.

These disorders were traditionally seen as inherent to life, whereas today people feel victimised by what they see as their abusive past or as the result of an oppressive social environment. These problems are therefore exclusively cultural in nature and have nothing biological or medical about them, unlike illnesses, which are identifiable or classifiable and result from aetiological factors either external or internal to the patient's body.

The demand for healthcare is a societal fact, reinforced by pharmacies, the media, consumer associations and the merchants of alternative medicines such as homeopathy and osteopathy, whereas needs are estimated or defined by medical science and should be by its practitioners. As a result, one of the major problems facing public health is that the demand for healthcare increasingly exceeds the need for it. Its rate of growth is outstripping that of the gross national product.

European countries where social security covers medical costs are faced with an inflation of medical care that is difficult to control and medically unjustified. The health insurance model is based on the principle that the more healthcare you offer, the better. But we now know that whatever the volume of care offered to a population, it will be absorbed! There are reasons to believe that a better approach would be one based on the quality of care: shouldn't doctors be reimbursed for quality rather than volume of care? To improve quality, we first need to measure it, and this should be the subject of research in the decades to come.

Sir William Osler wrote, "*Man has an inborn craving for medicine*". And "*the desire to take medicine is one feature which distinguishes man, the animal, from his fellow creatures. It is really one of the most serious difficulties with which we have to contend. Even in minor ailments, which would yield to dieting or to simple home remedies, the doctor's visit is not thought to be complete without the prescription.*"[10]

It lies in the very nature of clinical medicine that it is a patient-centred approach. A physician has a moral responsibility to defend and promote his patients' health interests, viz. to do one's best to fill their medical needs. But a doctor should also respect his patients' autonomy:[11] doctor and patient are equal partners in the therapeutic relationship. Hence, individuals' and patients' wants, as it were, ought to be sovereign. A conflict may thus arise here since patients expect their doctor to meet their demands and their demands may differ from their needs.

A similar difficulty may arise with the population-centred activities of community medicine: one must here distinguish the acceptance of a programme from its acceptability, that is whether the public is ready to accept the programme, e.g., vaccination, from whether it is scientifically legitimate. Physicians and health policy makers are thus saddled with moral responsibilities for their patients or for the population they administer, because they are in a better position to know what their best medical interests are.

Internationally tight radiation safety limits are being recommended because of popular concern and fear of radiation. But these international limits spring from fear of radiation. Actually, radiation exposure is much more foreboding than harmful. Small wonder then, that public acceptance (or refusal of acceptance) supplants scientific reasonable acceptability.[12]

However, in case where the individual's demand is in conflict with his interest and his needs, that is in case of conflict between the authority of medicine and the respect for patients' autonomy, should the doctor renege on his responsibility and accede to his patient's demand? Or should he momentarily bracket off his patient's autonomy in order not to frustrate his interests? Therapeutic medicine and public health give way to a constant effort to bridge these gaps.

Defining Treatment

A medical intervention ideally aims to reach some ultimate objectives such as to eradicate or cure a disease; slow its progression; relieve its suffering and other symptoms; prevent illness or injury, recurrence, complications, comorbidity, or premature death; limit disability and reduce communicability; restore functional capacity and care for those who cannot be cured. It embraces actions such as administration

[10] Osler (1951, 201).

[11] Glover (1984, 157–160).

[12] Allison (2009).

of drugs or some other specific measure, such as surgery, physiotherapy, radiotherapy, psychotherapy and advice on lifestyle, maintenance care in chronic disease, rehabilitation or reconstruction, social work, and advice about modification of personal behaviour, lifestyle, or dietary habits. In the case of chronic diseases, treatment ought also to help patients to live better. This also means that it must—at least as a general principle—focus as much on the patient's environment as on those internal problems specifically inherent to his disease process.

A *treatment*, namely a therapeutic intervention, is a need-satisfying activity conducted by a member of the health team, by the patient or by his family or friends: it is intrinsically good, not necessarily hedonically as a desire fulfilment, but rather in a desire-independent way; and it no more valuable if we sever it from its medical context since it is part of what we mean by being ill. Meanwhile, it is usually not part of the defining features of a treatment that it should be acceptable by the patient, although it is generally one of its desirable conditions.

A treatment fulfils the following seven conditions:

(0) *Primum non nocere.*

"*It is a good remedy sometimes to use nothing,*" wrote Hippocrates (460?–377? B.C.). "*One of the first duties of the physician*", wrote Osler, "*is to teach the masses not to take medicine*". And Jean Jacques Rousseau contended, "*Live according to nature, be patient, and to hell with the doctors.*"

Many diseases display a strong tendency to heal with, without or despite medical intervention. Just like restoring paintings, the first principle of medical care, is therapeutic abstention: watchful surveillance and no treatment are often the best treatment.[13] Any therapeutic decision must be a warranted exception to this rule. There is a famous La Fontaine's fable "*La mouche du coche*" about a fly that by biting the horse's rump convinces itself that it is the one pulling the wagon. Practitioners unavoidably often tend to play the fly in the wheel.

(1) A treatment is an *intended medical intervention* on the part of the doctor (although the patient may be unaware of it), namely a procedure, regimen, or service.
(2) It is *intended to be beneficial* even though it may actually not succeed to be so, even if for Paracelsus (1493?–1541): "*Only he whose remedies are efficacious is a physician*". It is *benevolent* as it is in the best interest of the patient.
(3) It is not arbitrary but based on a belief fortified by good evidence-based credentials that it is *effective*, i.e., that its administration will increase the probability of improvement of the relevant disease compared to the probability this will occur without the intervention. In other words, there is good inductive evidence that the treatment is linked with the expected improvement by a causal chain.
(4) So, therapeutic interventions being acts of prudence, ought to *do more good than harm*. They balance benefit and risk, or gain and cost, in an impersonal respect and through some acceptable trade-offs. Most treatments have unin-

[13] Shapiro and Shapiro (1997, 59).

tended effects and many of those are undesirable and constitute a spectrum of *side* and *adverse effects* besides their positive remedial effects. Adverse effects raise the issue of comparative risks: in the presence of two equally efficacious treatments we shall choose one of them on account of its less severe and less frequent overt side effects.

Patients may improve or survive in spite of medical interventions rather than because of them.

Some situations we may want to avoid, like surgery, may be beneficial. Electroconvulsive therapy consists in inducing an epileptic seizure (a disease) to treat a depression (another disease): the harmful but effective effects of one disease process are used to treat another disease. Cowpox vaccination in order to prevent variola is another example of therapeutic balancing (great) benefits and a (very small) risk.

(5) It should in its mode of application be *minimally comprehensive:* the disease should be captured as affecting the patient as a whole. In other words, treatment applies to the patient, not to the disease. Still, this first order standpoint should be kept to a minimum. Even then, acute well-defined and curable diseases (such as acute appendicitis, or a fractured bone) in previously healthy individuals do not need the type of meaningful empathic relationships that are required for long-term care in chronic diseases, or in the final phase of a predictably progressive terminal illness.

(6) *Compliance* (also adherence, capacitance) describes the degree to which a patient correctly follows medical advice. Most commonly, it refers to medication or drug compliance, but it can also apply to other situations such as medical device use, self-care, self-directed exercises, or therapy sessions. Noncompliance is the bane of clinical medicine and medical investigations. When philosophers argue, based on theoretical if not speculative arguments,[14] that statins and antidepressants "*are not nearly as effective as many people think*", they often ignore the clinical reality of noncompliance, that has been established, inter alia, among statins and antidepressants.[15] Furthermore, many adults eligible for statin therapy for cardiovascular prevention are untreated, resulting from non-initiation, refusal or discontinuation. Illegitimate concern about side effects was the leading reason for statin refusal or discontinuation, a misconception spread by the media,[16] and taken up by a current skeptical philosophical vision.

(7) It allows dying with comfort or dignity.

[14] Stegenga (2018a, 53).

[15] Chapman et al. (2005), Tierney et al. (2000), and Bradley et al. (2019).

[16] Kriegbaum et al. (2017).

Corrupted Pharmaceutical Research

There is a new pitfall, that of what might be called the pharmacological-industrial complex.

"*Is academic medicine for sale*" wrote Marcia Angell (1939-), Professor of Social Medicine at Harvard University and former Editor-in-Chief of the prestigious *New England Journal of Medicine*. She has denounced the manipulation of clinical research and the control of medical information by the pharmaceutical industry over the evaluation of its own products. It is the industry that designs the studies, analyses the data, decides whether and in what form to publish them, and keeps adverse effects under wraps. Medical and academic institutions are in a sense forced to endorse these abuses, with the resulting personal and institutional conflicts of interest. Clinical trials are thus truncated or ignored if they turn out to be negative. In short, little information is reliable. These observations are not confined to the United States but affect the entire pharmaceutical industry.

Placebos

Michel de Montaigne observed, "*There are men on whom the mere sight of medicine is operative*". Sydenham wrote: "*The arrival of a good clown has a far more beneficial effect on the health of a town than twenty donkeys laden with medicine*". And Claude Bernard added: "*A doctor who tries a remedy and cures his patient is inclined to think that the cure is due to his treatment*."

The term placebo means "I will please" and was first used in 1811 in Robert Hooper's dictionary, *Qincy's Lexicum Medicum*. It was in 1834 and in Germany that a placebo was used for the first time in a clinical trial, which demonstrated the therapeutic ineffectiveness of homeopathy.

In clinical care, a placebo is a therapeutic procedure (or some component thereof) administered by a caregiver or used by a patient for a medical condition (disease, disability, or symptom) for which it is biomedically inert and inefficacious. More precisely, a placebo "is any therapy prescribed knowingly or unknowingly by a healer, or used by laymen, for its therapeutic effect on a symptom or disease, but which actually is ineffective [non-efficacious] or not specifically effective [efficacious] for the symptom or disorder being treated".[17] But these definitions are ambiguous: placebos may first be medications or procedures such as physical manipulations or psychological interventions, which look like an active therapy and which are administered by a caregiver precisely because, unknown to the patient, they have no specific activity of their own; but there is another not uncommon possibility in which the doctor and the patient may both be unaware that the treatment is inactive.

[17] Louhiala (2020) and Shapiro (1997, 12).

But plainly, placebos should have no therapeutic capacity or at least no specific therapeutic capacity on the condition for which they are targeted: their impact should be confined to a no treatment effect. But that is a theoretical view, which may hold, although it rarely does, in close experimental settings such as clinical trials (RCT), where all or most interfering variables are being held constant. In the reality of clinical care, the matter is more complex: although they have no specific activity of their own, placebos may have some therapeutic effect, namely the so-called *placebo effect*. If so, then, placebos are causal factors that are necessary in the circumstances for this therapeutic effect: they have a causal role but no causal capacity. It follows that placebos administration is different from no therapy. In pure pharmacological terms, there should be no difference between the scores of a no-medication group and those of a placebo group since neither of them receives an efficacious agent. Even so, if we perform a clinical trial with two base rates, a placebo-controlled group, and an untreated control group the difference allows to quantify the placebo response that is to say the healing power of the placebo within the given experimental conditions.

Prima facie, placebos by definition should have no consequence. Placebo response then pertains to the residue left after subtracting the efficacy of an intervention (and its eventual side effects) from its clinical effectiveness in some target disorder observed in a given clinical setting: when it occurs, it is thus linked though not constitutive of the therapy, be it efficacious or not, be it a placebo or an active therapy.

There are perforce here two lines of reasoning.

First of all, *a placebo* is of a substantive nature: it is a biologically inert substance or a sham intervention wilfully, though often unknowingly, administered with therapeutic intent. Placebos fulfil, within limits, the treatment-defining criteria (0), (1), and sometimes (2), (3), (4), and (5). Being a treatment is, as it were, independent of the existence of anything else: medications or therapeutic procedures are described in the medical literature and stored away in medical textbooks, so that being a treatment is an intrinsic property, a universally accompanying characteristic of some interventions. Placebos, by contrast, are a pill or some sort of procedure to which the role of being a placebo is being assigned so that they are occupying the niche of a treatment.

Secondly, the *placebo effect* is a process that is of relational nature: it is the biological impact of believing in a medical treatment; it is extrinsic, contingent, and is not conferred on some intervention, but depends on the way other things are, since changes in other things could deprive it of its effect. A physician may perfectly well, and often does administer some treatment in the belief that it is efficacious while it is not; but its placebo effect, when it occurs, may be known or unknown to the prescriber. And our language becomes somewhat unwieldy when we call such treatments, 'placebos' meaning thereby that their efficacy merely results from its placebo effect, wherein 'placebo effect' melts into 'placebo': it follows that the term 'placebo' is often used when we mean 'placebo effect'.

Several though not all psychotherapies, and especially psychoanalysis, may be termed 'placebos'. Acupuncture alleviates backache no matter where one sticks the needles in the body. What's more, close to one quarter of patients taking placebos report adverse effects (the *nocebo effect*). In some placebo-controlled clinical trials, the incidence of nocebo side effects may equal or surpass the incidence of side effects observed in patients taking the active medication.[18] The nocebo effect is associated with the person's expectation of adverse effect. The placebo effect rests on the "*think well, hence be well*" tacit assumption, whereas the nocebo springs from the "*think sick, hence be sick*".

The frequency of muscle symptoms with the use of statins is hotly debated. Observational studies indicate high rates of station muscle complaints. Even so, randomized control trials in which patients do not know whether they are taking statin or a placebo, show identical rates of muscle symptoms in the treatment and the control group.[19] The same thing goes for the side effects of beta blockers such as fatigue, weight gain, headache, or impotence. A systematic review of 13 randomized control trials in more than 15,000 patients with heart failure showed that side effects were not more common with beta blockers than with placebos.[20]

The nocebo effect is an illustration of the interactive stance. Patients who did not know they were taking statin, began reporting muscle pain, real muscle pain, only when they were made aware they were on the drug.[21] Information disclosure about potential or hypothetical side effects can contribute to producing adverse effects. Among 47 suspected anecdotal reports of adverse drug reactions reported in 1963 in the *British Medical Journal, the Lancet,* the *JAMA* and the *New England Journal of Medicine*, 14 were false-positive reports.[22] Furthermore, once publicly reported, whether validated or not, whether causally related to the medication or not, the side effects are usually being reported on the package leaflets of the medicinal products with exhaustive warnings. Patients informed of real or false side effects reported these symptoms three or four times more than patients in a control group who were not informed of these symptoms. Packaging may set the expectation. The listing of side effects on the package leaflets, most of them poorly validated but mentioned for legal protection, are incentives for nocebos.

Placebos may very well have no placebo effect, and treatments are apt, despite the best intentions of the prescriber, and unknown to him, to display mere placebo effect. Homeopathy is a treatment, but its efficacy is zero, which defines its outcome if positive as a placebo effect. From a purely logical point of view, treatments and placebos can be construed as asymmetric binary relations, that is to say that they proceed in a single direction from provider to receiver. Yet, from this viewpoint, placebos differ from treatments. A treatment is knowingly either self-administered or delivered by a third party, while placebos cannot be intentionally

[18] Hahn (1997, 56–76) and Barsky et al. (2002).

[19] Pedro-Botet and Rubiés-Prat (2017).

[20] Barron et al. (2013).

[21] Gupta et al. (2017).

[22] Venning (1982).

self-administered: contrary to the first which are a reflexive binary relation, the intended administration of placebos is not reflexive.

Even in 1950, a bottle of medicine was a placebo for at least 40% of patients attending a general practitioner's surgery. Similarly, acupuncture, homeopathy and psychoanalysis are still placebos today, since they have no proven therapeutic efficacy.[23]

The Nature of the Placebo Effect

The property of being a placebo cannot be read off from the nature of the medication or the procedure. Nor is there any transubstantiation needed for the same glass of mineral water to be a plain beverage in a bar, or a placebo in a spa. Nor is it appropriate to assume an additive model by which a placebo would be an inert substance + a factor X, where factor X would be present in the therapeutic alliance and absent without it. On this score, being a placebo is not a property since it is not inherent in a procedure or a substance: the same substance or procedure may or may not be effective depending on the clinical background. For an intervention to be a placebo in clinical medicine, it must be intended to be so by the doctor, but it is the background of the therapeutic alliance or the therapeutic context, which surrounds the vitamin preparation or the sugar pill, which produced the placebo effect.

The meaning of the difference between treatments and placebos is analogous, I should conjecture, to the difference between Marcel Duchamp's urinal and a real urinal he termed a ready-made, i.e., between a work of art and an identical ordinary object. Duchamp's urinal does not differ from its ready-made in the way in which its ready-made differs from a urinal of a different make that has a different shape or colour. So, what makes them to be different? Treatment and placebo in a clinical trial look alike to the eyes of the doctor as well as to those of the patient. They look identical. What makes them different is that they medically play a different role: although they seem to have everything in common, that is to say that they are of the same natural kind, they belong to two different medical kinds.

But the most conspicuous difficulty with placebos and nocebo effects comes from their awkward consequences in clinical medicine: they are blurring the edges of the diagnostic and therapeutic tools.

Actually, administering a treatment or the absence of treatment are logical *contradictories*. But if we compare a treatment effect with a placebo effect, they are *contraries* since they cannot both be true, but they can both be false: the patient may improve by spontaneous remission that is independently from either the placebo or the treatment.

To sum it up, then, placebo seems to be one of those crossbred concepts which seems meaningful for the description of the surface of the clinical world, but which

[23] Benedetti (2009).

then turns into a mixed bag of disparate factual and methodological features. Placebo effect applies to the outcome of inactive controls in clinical trials, as well as to the outcome of medical interventions in clinical practice whether they are taken for a treatment—in which case placebo effect is part of the outcome—or intended as a placebo. Placebo effects may be positive or beneficial, absent, or null, or else negative or detrimental, be its efficacy pretended or real, in which case its benefit is additional to the efficacy of the therapeutic intervention: it is thus a component of any effective therapy.

Placebos in Clinical Trials

For Archie Cochrane, the efficacy of some therapeutic endeavour must be quantified by randomized controlled trials (RCT).

In some circumstances, the issue is quickly settled such as with penicillin in the treatment of pneumococcal pneumonia, tracheostomy to open a blocked air passage, or the Heimlich manoeuvre to dislodge an obstruction in the breathing passages, since the effect is dramatic relative to the natural course of the disease. But in the majority of cases, treatment effects at best a limitedor even marginal amelioration, the ultimate test of its therapeutic effects against side effects, can only be made by RCT. Such clinical trials are double blind: neither the patients being tested, nor the researchers are being told who is receiving the drug and who is receiving a placebo. Since therapeutic regimens have often been introduced and adopted by renown and esteem, rather than as a result of scientific assessment, the measurement of efficacy is often an "*iconoclastic exercise because, of necessity, therapies have to exist before they can be evaluated*".[24] Until some years after the end of the Second World War, pharmaceutical companies got drugs on the market without testing them in a controlled trial; but from then on, under the influence of operational research methods, new techniques turned out and, first among them, in a now classical publication on the utility of streptomycin in the treatment of pulmonary tuberculosis.

In evaluating a medical intervention, components that are not psychophysiological are regarded as 'noise' and must be filtered out. A placebo is used as a control against which some therapeutic procedure is being tested. It is accepted on good scientific grounds as biologically inert and non-efficacious in the disease or the condition under consideration and is designed to appear the same as the treatment that is being evaluated.[25] A placebo may be pharmacological (a tablet) in which case it has kind of a ghost pharmacology that includes dose-effect relationship, time-related effects, and adverse effects and even, drug dependence. Yet, it may also be manipulative (osteopathic manual therapy) or psychological (i.e., a conversation or a psychotherapy).

---

[24] Roberts (1977, 51).

[25] The attributable efficacy of a placebo is almost nil and its relative efficacy equals 1.

But in clinical practice both patients and clinicians may overestimate efficacy since borderline signs or symptoms may be classified one way or another by the patient or by the health team and produce an apparent difference where there is actually none.[26] In the light of this, a placebo is used in clinical trials for effective blinding in order to improve objectivity in the evaluation of therapy and to prevent bias in reporting, i.e. observer's bias.

Explaining the Placebo Effect

Are placebos mere artefacts or an uninvited 'noise' in a clinical trial?

Placebo in Latin means: '*I shall please*'. It was first used in medicine in 1785. A placebo is a mongrel medicine sometimes defined as a 'make-believe medicine'. In a way, placebo is defined by what it is not, namely by what differentiates it from a treatment.

Actually, the medical literature suggests that placebos produce large subjective and objective improvement in patients with a broad range of clinical conditions but mainly with those affected by pain, although they are probably useful in asthma, osteoarthritis, hypertension, myocardial infarction, and mental disorders.

It is the received view that placebo-effect works not for specific but rather for psychological reasons: merely believing that one is receiving an active drug is enough to produce some effect, be it real or not. It is not attributable to any specific therapeutic efficacy or to any pharmacological property but rather to this additional positive effect, which may occasionally be harmful in case of nocebo effect.

Ruling out cases where placebo effects are illusory, recent research demonstrates that placebo effects are no sham procedures but are real and have positive therapeutic potential.[27] Placebo administration shows a clear dose-response relationship. The magnitude of placebo response is correlated with the colour of the placebo pill, with its presentation and with the type of information given to patients about the pills; high-priced placebos and branded placebos are more effective than low-priced or generic placebos.

In other words, placebo effects are not merely in the mind of patients: they have objective powerful effects on physiological function and pathophysiology, even though all medical conditions are not responsive to placebos.

However, there is not a single but many placebo effects with differing mechanisms. It may depend on several factors[28] such as:

---

[26] Cleophas (1995). The literature usually fails to assess the success of blinding in randomized controlled trials so that one may have legitimate doubts about the information gained by incorporating a placebo comparison (Fergusson et al. 2004).

[27] Finnis et al. (2010).

[28] Harrington (1997).

1. First, cognitive factors such as meanings and symbolic meanings, attributions, beliefs and desires, anxiety, and rewards, which operate in their socio-cultural context to even induce placebo responses. More specifically, *expectancy*, positive expectation, namely anticipation of nonvolitional therapeutic responses to pain and depression play a determining role in treatment outcomes in terms of subjective as well as objective measures of effects. Beyond the expectation of improvement, that is to say the effect of faith or belief in a therapeutic method,[29] patients who think they have undergone an active treatment may have a persistent significant improvement in perceived physical function compared with those who thought they had been assigned to the placebo group. By way of illustration, the direction of the placebo effect chimes with that of the active treatment, the strength of the first is proportional to that of the second, the reported side effects of the first are often similar and so are the latent periods needed for both to become active.[30]
2. Yet, placebo effects are not the result of shear mental activities: the administration of placebos, when effective, gives rise to specific *neurobiological pathways* of action. It seems that placebos cause the body to undergo biological changes when the patient feels better in response to their administration. By way of example, placebos that relieve pain act by releasing the body's natural opioid substances, the *endorphins*; this placebo effect can indeed be reversed or partially reversed with a drug, naloxone that counteracts the effect of opiates, or inhibited by a peptide the cholecystokinin (CCK). Brain imaging techniques (PET, positron emission tomography, and fMRI, functional magnetic resonance) showed that placebo induced brain changes that were similar to those seen with opioid drug administration, and they involve a correlation with various neural regions (prefrontal, orbito-frontal and insular cortices). All in all, this suggests that placebo operates by altering the activity of both CCK and endogenous opioids.[31]

   Furthermore, the administration of placebos in cases of Parkinson's disease induced as much *dopamine* release (which is suppressed in Parkinson's disease) in the striatum of their brain, changes in the basal ganglia and thalamic neuron firing as when they received active drugs.[32] What clearly emerges from this is that placebo effects do actually exist and are not mere experimental artefacts.
3. What's more, the potential therapeutic effect of clinical attention and of the context of the clinical encounter is crucial, such as a meaningful doctor-patient interaction and the therapeutic ritual itself. It is thus a psychosocial effect since words and *the ritual of the therapeutic intervention* may change the chemistry and the circuitry of the patient's brain. Different vehicles of placebo or nocebo

[29] Kirsch (1997, 167), Frank (1974, 136), and Stoessl and de la Fuente-Fernandez (2004).

[30] Ross and Olson (1981).

[31] Benedetti et al. (2011).

[32] Rowbotham (2001) and Gibbs (2001).

rituals had different effects. The higher the price of the placebo the more significant its positive or adverse effect.

Thomas Sydenham (1624–1689) observed, "*The arrival of a good clown exercises a more beneficial influence upon the health of a town than of twenty Asses laden with drugs*". The presence of the doctor is the beginning of the recovery says an old proverb. It has long been acknowledged that diagnostic procedures or the mere initiation of treatment, even if non-specific, makes a sizeable fraction of patients feel better.

4. Furthermore, *compliance* with demand is important: compliant patients in the placebo groups of randomized controlled trials manifest far more favourable outcomes, including survival, than their non-compliant matched companions;[33] in a similar vein, the wish not to disappoint may enhance the placebo effect.[34] High compliance is thus a clue for a better outcome: controlled clinical trials often suggest that useless therapy is efficacious when administered to compliant patients. High compliance on the part of the participants of controlled trials reveals a positive attitude marked by the belief that the treatment will have beneficial effects; combined with the wish not to disappoint, this explains the enhanced placebo effect.
5. *Expectations can be modulated* to improve therapy. In a trial studying postoperative pain over several days, the question was the potential for using placebo effects in conjunction with an active therapy in order to reduce overall drug intake. The investigators used an intravenous saline added with a routine analgesic.

   One group was told that the administration was a mere rehydrating solution, and another group was told it was a powerful painkiller. The patients' analgesic intake was monitored. Surprisingly, the group who believed the solution was assisting in analgesia took 33% less active analgesic for the same pain control.

   There was a third group representing the classic *double-blind* instructions, which followed the same protocol but were given the instructions that "*the solution may or may not be a powerful painkiller*". In this group, patients to 20% less analgesic medication. In short, changes in the instructions and expectations may modulate the placebo responses.[35]
6. *Physicians expectations* also count. Research has shown that clinicians' belief can alter the therapeutic context and affect placebo effects. In one analgesic trial, the placebo effect was dramatically less in the group where clinicians believed that no analgesic therapy was given, while it was actually administered, even though patients were unaware of the different and wrong information given to the physicians.

---

[33] Horwitz et al. (1990), Coronary Drug Project Research Group (1980), Pizzo et al. (1983), and Hogarthy and Goldberg (1973).

[34] Cleophas (1995).

[35] Benedetti (2009).

7. Placebo effect may be *learned classical conditioning effe*ct acting not only on the patient's symptoms and subjective states of pain, nausea or anxiety but also producing powerful biological effects such as processes affecting immune responses. Past experiences, attention and emotions may influence the perception of symptoms by producing memory of past effects. Prior exposure to an effective treatment boosts up placebo effects.[36]
8. Beyond this, it is a persistent misconception about placebos that they are ineffective if patients know they are receiving them. The results of several studies showed that open-label placebo treatment compared with no treatment occasioned, at least in some patients, surprisingly higher mean global improvement scores, reduced symptoms severity scores, and adequate relief scores.[37]

Surgery and Placebos

Fake operations may have a much stronger placebo effect than drugs.

Two hundred patients with a blocked artery were randomly assigned to get either a real stent operation or a fake one. There was no difference in how the patients felt 6 weeks after surgery. Both reported less pain, and both performed better on treadmill tests.[38]

Sham surgery refers to a faked surgical intervention that omits the step taken to be therapeutically necessary. The problem is that sham surgery is not really a placebo. It is an invasive procedure with potential scarring or undesirable effects.

Sham surgery has shown that many other useless surgical procedures offer no advantages. For consider arthroscopic knee surgery for a torn meniscus or for degenerative wear and tear, or vertebroplasty for fractured vertebra, or surgery for shoulder impingement, which would do as well as physical therapy, weight loss and exercise. Sham surgical interventions have been published for Parkinson's disease, endolymphatic sac shunt surgery for Meniere's disease and in orthopaedics where they have shown to be as effective as actual surgery in reducing pain and disability.

The Placebo Effect Is Sometimes Fictitious

Meanwhile, besides its genuine healing power, the role of the placebo effect may in some circumstances be purely fictitious: it may then consist in or be made of various components:

[36] Ader, in: Harrington (1997, 138–165).

[37] Joplin (2013, 1218–1219).

[38] Al-Lamee et al. (2018).

1. First and foremost, the *natural history of the disease*. Ruling out confounding factors such as environmental changes including the treatment environment (supportive clinical patient relationship) or other treatment, when the administration of an alleged treatment to a patient is followed by an improvement, the patient may get better as a result of the body natural restorative processes; in the absence of a control group or a comparative element, namely in the ignorance of the natural history of the relevant disease, this placebo effect requires the a priori assumption that the outcome of no treatment would have been better, whereas most medical conditions have a natural tendency to improve. The success of medical care depends partly or largely on the natural tendency of most ailments to recover spontaneously.

   Even so, the use of placebos in clinical trials helps to control for natural fluctuations in disease course such as spontaneous recovery. Clinical conditions such as multiple sclerosis or osteoarthritis have a natural history of better and worse stages that make it difficult to ascertain whether a health upswing should be imputed to a therapeutic intervention. So, any therapy may be followed, in the absence of any causal impact, by a spontaneous remission; attributing this improvement to the therapy is common post hoc propter *hoc* fallacy.
2. Second, *regression to the mean*, i.e., the statistical tendency for subjects selected with extreme values—especially if the characteristics that are measured are unstable—to be more normal on retesting. This problem explains why patients may appear to improve through a mere statistical artefact and in the absence of efficacious treatment. On this score, patients seen in research hospitals where evaluation of medical interventions through clinical trials is usually performed are likely to be more ill than non-referred patients with the same condition. In many such conditions, a bad worsening period may be followed by a spontaneous improvement and conversely, a remission may precede an exacerbation.
3. Finally, common *confounding factors* are observer and subject biases, namely the optimistic assessment tainting the caregiver's interpretation of the patient status or given by the patient himself. Claude Bernard held that: "*A physician who tries a remedy and cures his patient is inclined to believe that the cure is due to his treatment.*"

Randomized Control Trials

Randomized controlled trials (RCT) are comparing at least one experimental therapy with at least one control therapy, vs a placebo (or another treatment). Allocations are double-masked. It is assumed that RCTs distinguish placebo from specific medication effect, and that it avoids biases of salience and confounding, namely performance bias and detection bias). In observational trials, the investigators observe what happens in patients who choose (or are chosen by their physician) whether to take the intervention themselves or not: it follows that, in theory, such studies are susceptible to selection biases, because those receiving the treatment might have systematic differences with those not receiving the intervention.

Nevertheless, James Worrall[39] and Nancy Cartwright[40] argue that RCT do not eliminate all possibilities of bias, and that even if they have a high degree of internal validity, i.e., of causal capacity, they suffer from problems of external validity, i.e., of causal role. Randomization can reduce but not get rid of selection bias, since it can eliminate most, but not all possible confounding factors. When there are too many factors, known and unknown, that give way to bias, this increases the liability that any randomization will not randomize regarding all these factors. On this score, any particular RCT will probably have at best one kind of bias. Unless it is repeated time and again, re-randomizing each time, and combining the results, RCT does not produce reliable outcomes.

Cartwright contends that external validity depends on the similarity of the two experimental populations, in terms of age, sex, severity of disease, risk factors, comorbidities, ethnicity, socio-economic level, as well as dose, timing if administration other medications, and suchlike. Then, it is taken for granted by Worrall and Cartwright that RCT have a lower degree of external validity than observational studies. Beyond this, the assumed superiority of RCT has not resulted to statistically significant differences between outcomes of RCT and those of observational studies. When results of well-conducted RCT are in conflict with those of observational studies, small wonder, then, that medical context and clinical judgement intervene to adjudicate amongst them, but RCT have a strong tendency to prevail.

Masking, of participants, study staff, and outcome assessors in RCT is considered crucial to avoid performance and detection biases, and it is a key element of the Cochrane risk of bias tool for randomized trials. However, the linked comprehensive meta-epidemiological study by Moustgaard and colleagues[41] challenges the dogma that blinding is always necessary to protect against bias in trials.

Surprisingly, this study found no evidence of a difference on average, in estimated treatment effect between randomized clinical trials with blinded and non-blinded healthcare providers, and between trials with blinded and non-blinded outcome assessors.

Even though randomized control trials are one of the best sources for evaluating the effects of treatment, they have their own limitations, and other methods, sometimes superior, of obtaining evidence are receiving increased interest.[42]

---

[39] Worrall (2002).

[40] Cartwright (2007a).

[41] Moustgaard et al. (2019).

[42] Frieden (2017).

Evidence-Based Medicine (EBM)

George Bernard Shaw (1856–1950) explained why therapeutic trials are necessary: "*In Shakespeare's time, and long after him, a mummy was a favourite medicine. You took a pinch of powder from an Egyptian corpse in a glass of the hottest water you could find; and it did you the world of good... But if you had tried a controlled experiment using hot water without mummy powder, you would have observed exactly the same effect, and any hot drink would have done just as well.*"

Pierre Charles Alexandre Louis (1787–1871) with his *Recherches sur les effets de la saignée* was first published in 1828, and later Joseph Dietl (1804–878) and Hughes Bennett (1812–1875) were the first to produce statistical evidence attacking bloodletting, the paradigm of Hippocratic and Galenic medicine. This was the dawn of scientific medicine.

Evidence is a cardinal concept of empirical science, and in medicine it means that the choice of treatment should be informed by scientific evidence. Furthermore, high-quality mechanistic reasoning should be combined with high-quality comparative clinical studies.[43]

During 2350 years after Hippocrates, physicians continued to use treatments that were either useless or harmful relying on anecdotal testimonies and individual clinical cases of effective treatments due to placebo effect. It is surprising that modern astronomy rejected astrology as early as the eighteenth century, medicine, the Hippocratic tradition remained embedded in medicine until the end of the nineteenth century with the development of germ theory.

David Wooton in his landmark book *Bad Medicine* declared that in recent years the medical profession has discovered what it calls 'evidence-based medicine'—that is, medicine that can be shown to work.[44]

Effective medicine began only when doctors began to count and to compare. Most medicine until the end of the second World War was not evidence based, or else, it relied on evidence proved to be useless or harmful when subjected randomized trials. The emblematic publication of Archie Cochrane's *Effectiveness and Efficiency* was a philosophical turning point in medical history: it was a manifesto advocating the use of randomized controlled assays as well as setting up a database capable of providing information allowing medical investigators to reanalyse and verify relevant results.

Sackett[45] has been the prominent proponent in evidence-based medicine that he defines as "*the conscientious, explicit, and judicious use of current best evidence in making decisions about the care of individual patients. The practice of evidence-based medicine means integrating individual clinical experience with the best available clinical evidence from systematic research.*"

---

[43] Howick (2011, 154).

[44] Wooton (2006, 3).

[45] Sackett et al. (1996).

It is a method dealing with the identification and the evaluation of the results of medical research, and the critical appraisal of the evidence, in order to guide clinical decision-making and medical practice. Philosophers proposed several accounts of illustrating and illuminating accounting for the confirmation of theoretical hypotheses on the basis of observational evidence.[46] EBM was a move to split from ways of deciding that were based on anecdotic clinical observations, on clinical reasoning based on biomedical knowledge and on the authority of experts, so that clinical decisions could be made on trustworthy evidence.[47]

EBM is best described in the EBM 'hierarchies'.

The rationale behind the many hierarchies is based on three central claims:

1. Randomized control trials RCTs or systematic reviews of many randomized control trials usually offer stronger evidence than observational studies.
2. Comparative clinical studies, including RCTs and observational studies offer stronger evidential support than biomedical pathophysiologic studies.
3. Comparative clinical studies, including RCTs and observational studies offer stronger evidential support than expert clinical judgment or consensus conferences.

It gives priority to evidence against pathophysiological explanation, in order to avoid bias. EBM is an approach to medical practice intended to optimize decision-making by emphasizing the use of evidence from well-designed and well-conducted research.

At what extent can statistical hypotheses be usefully connected with empirical descriptions by bridge laws? What criterion of adequacy must the evidence *e* satisfy in order that the efficacy of a treatment should ever be true? The evidence *e* cannot be adequate, as it seems, if *t* can be false while *e* is true. It follows that if *e* is evidence for *t,* then *e* is *adequate evidence for t* only if the falsity of *t* is incompatible with the truth of *e.*

When we say that *e* is a sufficient condition for *t,* the requirement of such deductive reasoning is too strong. Actually, evidence will differ in strength, and this variation in evidential strength licenses the view that a physician's intervention is justified when he possesses evidence *e* where *e* satisfies a criterion of adequacy.

A justified deduction based on adequate evidence might be false as a modal fact. Even then, abduction, a term introduced by Peirce often called "inference to the best explanation" consists in the creative formulation of statistical hypotheses to explain a given set of facts and proceeds to the most plausible possible explanation of the available evidence.

To be sure, RCT's suffer from fewer confounders than observational studies, yet the categorical ranking of RCT's over observational studies leads to several

[46] Glymour (1980).

[47] Bluhm and Borgerson (2011) and Howick (2017).

paradoxes, but most critiques against RCT's are straw men against the EBM position and EBM methodology has, with few exceptions escaped scrutiny.[48]

EBM attempts to identify the shortcomings and the weak spots of healthcare and of the medical literature and to produce and evaluate high quality evidence. It devalues the importance of clinical judgement and so-called clinical experience, as well as the reliance on pathophysiological explanation as a source of medical knowledge. The major hindrance medicine has had to deal with has been the overthrowing of its own authority, and more recently the gap between research evidence and medical practice as well as the explosion in the number of published papers. It is a misunderstanding to believe that understanding mechanisms in their entirety is necessary for medical interventions. Lung cancer could be and has been reduced by eliminating the exposure to cigarette smoke, even when our knowledge of its causal web of which cigarette smoke is one component, was still very limited.

The key concept of EBM is that of using the best medical evidence. The best evidence is derived from epidemiological methods and ways of thinking. The MRC trial of streptomycin in tuberculosis, published in 1948, is regarded as the first randomized control trial that established a gold standard for evidence of therapeutic effect.[49]

Incidentally, evidence is not proof. We would need an infinite enumerable set of experiments to prove some medical hypothesis, but a single negative one would refute it.

The Double Meaning of "Treatment"

Jonathan Bennett uses the label *compliance mechanism*[50] for a general form of the injunctional meaning exemplified in the therapeutic gesture. One of the bases for compliance involves, at least in the typical case, crediting the utterer (the doctor) with information addressed to the hearer (the patient) and which the latter lacks; furthermore, the utterer has an interest in the welfare of the hearer so that the latter will avoid some harm or get some health benefit, that is something which he counts as good independently of the doctor's wishes.

Actually, the term 'treatment' is used in two different ways. For one thing, in our texts of medicine, 'treatment' is used in a cognitive mode, in which case scientific language is external to the world: it refers to some up-to-date medical doctrine about the various aspects of how to handle some relevant condition. A specialist, it is sometimes said, is a doctor whose patients are supposed to be ill only during office hours.

---

[48] Howick (2011, 189–192).

[49] Medical Research Council (1948).

[50] Bennett (1979, 145–147).

For another, language corresponds to a different performative, emotive, hortatory, or persuasive, and sees it as internal to the world, as a set of processes interacting with the world. With this move, sentences are not merely used to say things, to report states of affairs, but rather actively to do things, that is to say to perform an act, such as the words "I prescribe" or a stat order at the bedside. Hence, contrary to scientific language, restricted to referential or indicative meaning, the language of therapeutic alliance is performing deeds and speech acts. Meaning is intentional since it is an endeavour to communicate that is directed at the patient. So, the doctor recommends some treatment to his patient with the intention that he abides by it through the recognition of this intention. In talking to his patient, the doctor is doing something, he is acting.[51]

Hence these two ways of speaking, semantically and pragmatically, correspond to two ways of relating with the world and with people. These two vantage points can be translated in terms of persons and the various contrasts we draw between medical science and medical practice, between what we do and what happens to us and between subject and object, between saying and doing.[52]

To sum it up, parts of the excess of meaning conveyed by the implicatures of the physician's utterances, spring from tacit assumptions, namely from the doctor's persuasive authority or charisma. The gap which divides these two functions of language separates the cognitive use of medical language from its performative role, the pragmatics of medical science from personalised medical care and the logic of therapeutics from that of treating an individual patient, and from medical care to public health.

Direct Realism and Instrumentalism in Health Care

In one-way, scientific realism is the philosophical view that there are entities that exist in this world independently of our minds. Scientific theories are true or approximately true description of the world.

In another way, according to instrumentalism, a scientific theory is to be regarded as an instrument for controlling events, but not necessarily capable of generating reliable true predictions or to describe the world accurately. There is a distinction to be drawn between observable and unobservable entities. For realism there is a continuity between the two ends but no real difference since according to realism a theory is true if its terms refer.

Scientific *realism*, which claims that scientific or medical theories are about things which are real as the familiar objects around us. It says that some theories really are literally and objectively true and others literally and objectively false. Furthermore, a theory could turn out to be false even though it was a very good

[51] Searle (1969).

[52] Brandom (2008).

predicting instrument. Until 1982, i.e., before identification of the link with *H. pylori*, gastric ulcers were treated with antacids and medications that block acid production, although there was a high recurrence rate.

It is opposed by *instrumentalism* which claims that scientific theories are just devices for making predictions and the objects mentioned in the theory may just be useful myths. We want to know what therapy, what medicine is effective against what diseases. The truth of a medical doctrine is not in itself practically relevant: all that matters in medical care is the truth of its observable consequences. Knowing only the molecular facts of a disease will not allow to predict how the disease looks like, neither will it predict whether this scientific knowledge will lead to an effective treatment.

Medical knowledge should then be understood as tools, and its sole value lies in its predictive power. We do not need to suppose that the terms of the theory—e.g., the concepts of function or of disease category—refer to anything: these words do not have to have any deep resemblance to the theory's description of it, as long as it works.

Clinicians believe in *scientific realism* and the background of scientific knowledge. But they also regard a scientific theory as an instrument for producing new techniques for controlling events: this second stance demonstrates *pragmatic allegiances*. Realism tends to prioritize a single form of knowledge as 'true', while for pragmatists, knowledge is not a representation of reality or a 'mirror of nature', but a tool for action. Instead of asking: "*Does this knowledge accurately reflect the underlying reality*?", the question becomes: "*Does this knowledge serve our purposes*?" This approach rejects or dismisses large amounts of empirical research since only humanly manipulable determinants or causes are the proper focus of interest. Knowledge of the genome is of great theoretical interest, but, for the moment, it corresponds to few humanly manipulable interventions. The pragmatist's principle of method over doctrine, instrumentalism over realism, practical over principled approach, sounds like a "bottom-up", inductive or abductive theory of truth.

Scientific realism, a "top-down" procedure, deducing from general propositions to concrete conclusions about particular cases, is too far removed from the intricate workings of clinical or public health context and experience, although it remains essential for patho-physiologic research. We explain things in order to be more effective in our action, to be more adept at controlling circumstances. Medical nosology is pragmatic. Our goal is to predict and if possible, successfully manipulate the world without getting hung up about the metaphysics, whether the constructs, which we construct to help us do so are "real" or not.

For instance, Kendell and Jablensky sharply distinguish diagnostic *validity* from diagnostic *utiliy*. They write: "*We propose that a diagnostic rubric may be said to possess utility if it provides nontrivial information about prognosis and likely treatment outcomes, and/or testable propositions about biological and social correlates.*"[53]

Finally, *knowledge translation* pertains to all activities involved in dissemination and moving research from laboratory, research journals, and the academic conferences into the hands of people and organizations who can put it to practical use. Knowledge transfer is highly context sensitive. The translation of knowledge to action illustrates the relative roles of realism and instrumentalism in health care. There is a distinction to be made between doing things right (as adhering to EMB-guidelines), writes Peter Drucker, and doing the thing (as adopting interventions and treatments considering this person's needs and context).[54]

Beyond this, it is one thing that suffering evil is externalist. By assigning the term "pathological" to suffering, medical language stands here in an *external* descriptive relationship to biological and clinical reality. This is what I called the *first position* (Chap. 1). Physicians and medical thinking then are not part of the reality that is being addressed.

But it is another, that clinicians then proceed to an epistemic *volte-face*, i.e., when physiology or whatever is counterfactually called 'normal' becomes the basis of comparison. With this second position, it is clear that our experience, the biologist's experience of the world, remains untouched; there is no difference between biological facts before and after the inversion, and since no experience can distinguish between these two vantage points, the difference between them thus lies outside experience altogether.

What differentiates them is that in the second position, the medical gaze is constrained by the central case of normality in the contrastive shadow of which the whole realm of medical research, clinical medicine and public health will unfold and bear fruit. This is tantamount to saying that clinical medicine, as practical knowledge, deals with contrasts and comparisons between normal and abnormal features as well as with the pull exercised by normal structures and physiological processes.

Medical language becomes a kind of complex box of tools or acts of language to be used in clinical or public health settings, which are performatives. Physicians and their language now belong to the world they study and which they attempt to modulate. Physicians are no more natural scientists, namely external observers as in the case of the first position but participant observers. Medical language is now part of or *internal* to the world it describes.[55] Medical knowledge does not merely tell us what the case in the biological world is but also how it ought to be, and what the goals are that we have to strive towards. Words used descriptively are retreating as language becomes performative and prescriptive, while propositional knowledge gives way to practical knowledge. Medicine pertains causally to the world and engages us in reshaping the life of sentient beings. Medical statements and decisions

[53] Kendell and Jablensky (2003).

[54] Sturmberg (2018).

[55] Danto (1968, 232–240).

become a question of correct vs. incorrect use, rather than that of truth or falsity. Clinical manifestations turn into intentional and meaningful signs and symptoms.

Prevention and Health Maintenance

One of the central purposes of medicine and public health is to decrease the occurrence of diseases and on that account, to promote a healthy lifestyle in a healthy environment. Effective prevention and preventive activities are prudential interventions. They intend to reduce the incidence or the prevalence of a disease in the population, its fatality rate, to improve its course and/or eradicate it.

If treatment focuses on sufficient causes, prevention, namely tertiary care, tends to focus on avoidance or protection against necessary or quasi-necessary causes that are amenable to manipulation, are acceptable, economic, and effective and whose removal carries negligible unwanted effects.

Yet, there are no sharp boundaries between prevention and treatment.

1. Prevention is based on the identification of risk factors.
2. Preventive measures are based on the intention to reduce the incidence of diseases or other departures from health, to reduce their prevalence, namely improve their course and shorten their duration, as well as decrease the risk of unmanageable disability, comorbidity, and mortality. They cover avoidance through immunization, changing lifestyle, and screening. *Tolle causam* is an old medical maxim, but it was bacteriology that gave it a new meaning.

   Thus, the therapy for a condition detected by early diagnosis must not merely advance the time when diagnosis is made, but it is also intended, when feasible, to favourably modify the natural history of the disease in terms of functional ability, duration, survival, and quality of life.
3. Actually, prevention is best defined in the framework of levels usually called primary, secondary, and tertiary prevention. *Primary prevention* is the protection of health by individual and community interventions, such as preserving good physical fitness, nutritional status, and emotional well-being, as well as immunization programmes and making the physical, biological social environment safe. *Secondary prevention* includes measures available to individuals and the community for the early detection and control of departures from good health, such as screening colonoscopy[56] and early detection. *Tertiary prevention* seeks to reduce the impact of complications and the number of long-term impairments and disabilities. It covers clinical medicine, namely treatment and measures striving to reduce or eliminate the risk of comorbidity and long-term impairments, and to promote adjustment to disease chronicity.[57] Sometimes *Quaternary*

[56] Doubeni et al. (2016).

[57] Last (1995, 130).

prevention is added, which identifies patients who are at risk of over-medications and protects them from excessive interventions.[58]

4. Preventive measures are based on the belief in their effectiveness, and preferably the attributable rather than their relative effectiveness. One holds the belief on the ground of both evidence and theoretical reasons—it must be biologically sensible i.e., consistent with existing knowledge.

We are thus dealing with a scale of preventiveness: eradication of an infectious disease is better than immunization, and immunization, if effective, is better than therapeutic healing, which is itself better than palliative care for disability resulting from a chronic turn of the disease. The extremely high cardiovascular mortality in an eastern province North Karelia in Finland caused great concern among the local population. Action to reduce the problem was demanded in a petition to the Finnish government signed by local representatives of the population. In response, the North Karelia project was launched in 1972 to carry out a comprehensive community-based prevention program. After the first 5 years, prevention activities were also started nationally. The main aim was to reduce the extremely high serum cholesterol, blood pressure and smoking levels with lifestyle changes and improved drug treatment, especially for hypertension. Major declines were seen in serum cholesterol, blood pressure and smoking levels. Coronary mortality reduced in middle age population by 84% from 1972 to 2014. About 2/3 of the mortality decline was explained by risk factor changes and 1/3 by improvement of new treatments developed since 1980s. Population-based prevention through changes in lifestyle and environment is the most cost effective and sustainable way of controlling cardiovascular and other major non-communicable diseases. In the current global situation, it is a powerful lesson.[59]

Screening

The idea of screening or early detection is to advance in time the moment of diagnosis, relying on the assumption that the people diagnosed early will be those destined to develop problems. However, the reality is different since when we look hard for early forms of disease, we find that more people have them, so that all of them will not develop problems. And if we are treating those who would not have problems, the overdiagnosed cannot be helped but might be harmed. *Screening* methods need to reach the largest possible proportion of truly diseased, persons at the risk of identifying non-diseased individuals. In a contrary manner, *diagnosis* should avoid identifying non-diseased persons. So, screening may include false negatives while diagnosis must avoid false negatives.

[58] Ali and Katz (2015).

[59] Puska and Jaini (2020).

There are two considerations of immediate interest: overdiagnosis and the notion of lead time.

Overdiagnosis is the notion that some tissues histologically malignant are not a threat and do not need to be treated, so that cancers detected at screening would not have otherwise become clinically apparent in the woman's lifetime: it is not an incorrect diagnosis but a failure to prognosticate. A lot of harm from mammography could be reduced if the threshold would be looking for things 1 cm or bigger. A study from Denmark estimated that 15% to 39% of cancers diagnosed in women aged 50–69 years represented overdiagnosis.[60] Overdiagnosis of cancer means that the diagnosed cancer would have never caused any clinical problems during the patient's lifetime, since in overdiagnosis, *the tumour is not supposed to progress.* But overdiagnosis may also give the wrong impression of breast cancer is being more common than it really is.

To sum it up, how many people we screen, how aggressively we screen them, and how often we do so, can potentially bring to light fallacious self-fulfilling risk factors for a particular cancer, and mislead us regarding how prevalent that cancer is.

Lead time is the length of time between detection of a disease, for instance by screening, and its usual clinical presentation and usual diagnosis. It is defined as the time gained in treating a disease when detection is earlier than usual, namely in the presymptomatic stage. Yet, with screening, the clock starts earlier, so that earlier detection by screening overestimates survival duration. This is lead time bias. *In lead time bias, the tumour progresses.* The question is then whether overdiagnosis is merely lead time bias, suggesting a spurious survival benefit. Indeed, if the excess of small tumours never progresses, then the overdiagnosis will be higher; but if the excess merely reflects lead time bias, the rate of overdiagnosis will be lower.

An Independent UK Panel on Breast Cancer Screening concluded in 2012 that screening reduces breast cancer mortality but that some overdiagnosis occurs. The Panel added that "*Since the estimates provided are from studies with many limitations and whose relevance to present-day screening programmes can be questioned, they have substantial uncertainty and should be regarded only as an approximate guide.*"[61]

New guidelines from the European Commission Initiative for Breast Cancer Screening and Diagnosis recommend against annual mammography and screening in women aged 40–44 years, and instead recommend screening every 2–3 years for those aged 45–74 years.

Breast cancer screening practices in European countries differ sharply from those in the United States, Joann G. Elmore, a professor of medicine at the David Geffen School of Medicine at UCLA, and Christoph I. Lee, a professor of radiology at the University of Washington School of Medicine, wrote in an editorial accompanying the study. One reason for these differences it that European countries have a single-payer health system, while women in the United States need to navigate the

[60] Jorgensen et al. (2017) and Brawley and Jemal (2017).

[61] Independent UK Panel of Breast Cancer Screening (2012).

"wild-type" health system in the Unites States. In addition, most European nations have two radiologists interpret mammograms, which has been shown to be more accurate than the single reading typically seen in the U.S. "*Indeed, the false-positive rate is 2 to 3 times higher in the United States than in some European countries,*" Elmore and Lee wrote.

Scrutiny—Dependent Incidence and Risk Factors of Cancer

The act of clinical observation can affect the incidence and the apparent risk factors for cancer. The various forms of scrutiny can result in what H.G. Welch called "misleading feedback loops". They are sort of interactive kinds. The concept of cancer risk factors in epidemiology needs to change in order to make it more meaningful: risk factors for cancer death are different from risk factors from cancer diagnosis.

Aggressive screening for 'early detection' in the breast, prostate, melanoma or even lung cancer, may lead to find more scrutiny-dependent cancers, namely cancers with a substantial disease pool of well-differentiated, indolent subclinical forms, which in turn give the false impression of an *increasing incidence* of such cancers. The problem is especially clear with prostate, breast, lung, and thyroid cancers as well as for melanomas. Pushing for alleged early detection leads to more scrutiny-dependent cancers being found, which suggests an increased incidence of some cancers.

H.G. Welsh and O.W. Brawley[62] have been arguing, in an important publication, that the number of some cancers we diagnose is sensitive to the degree of scrutiny. Since breast cancers detected by imagery are much more likely to be harmless, as we screen for breast cancers with more advanced technologies able to *overdiagnose* and find smaller and smaller tumours, and more cases that are being indolent and that will not develop cancer, a process that affects the apparent incidence of the disease.

If so, then, screening affects the incidence of certain subtypes of cancers that are slow-growing, subclinical or even pre-cancerous, but that may progress to cause health problems or even death, and which are sometimes termed *scrutiny-dependent* cancers.[63]

The degree of scrutiny can also *affect the assessment of risk factors*, including genetic risk factors. Breast cancer, like prostate cancer or thyroid cancer, is a scrutiny-dependent cancer.[64] It follows that family history is a risk factor for breast cancer since it is sensitive to the degree of scrutiny. Wealthier, better educated women living in high socioeconomic neighbourhoods get more mammography, MRIs, and ultrasonography.

[62] Welch and Bradley (2018).

[63] Palmer et al. (2012).

[64] Whittemore (2003).

Scrutiny may also apparently *weaken a genuine risk factor*. A study made in Japan found after spiral computed tomography 10 times more lung cancer than previously found in the same population using chest X-rays. At the end of a 3-year screening lung detection programme, the investigators found virtually similar detected prevalence in smokers as that in non-smokers, with a relative risk near 1. Since the risk of smokers dying from lung cancer is at least 15 times higher than that of never-smokers, this investigation indicates that overdiagnosis may be a substantial problem with spiral computed tomography scanning.[65] This publication highlights the importance of the impact of the act of observation and how it can affect the phenomenon being observed, a feature that physicists have long acknowledged. On this score, the issues of incidence, prevalence and risk of some diseases and the concept itself of disease are being called into question.

Herd Immunity

Herd immunity the indirect protection from an infectious disease that happens when a population is immune either through vaccination either through immunity developed through previous infection. It is a kind of immunity that occurs when a sufficient portion of a population is immunized, so that it provides an immunological protection to those who either have not developed immunity or have not been immunized. The greater the immune fraction of the population in a community, the smaller the probability that those vulnerable not immune will come in contact with an infectious individual. When the prevalence of immunization falls, herd immunity may break, giving rise to an increase in the number of new cases: this explains why measles outbreaks in Europe and pertussis outbreaks in the United States occurred, due to declining herd immunity.

In case of herd immunity, the risk to the unvaccinated is nil while the risk remains. On this score, the prisoner dilemma when prisoner A defects so that he goes free, illustrates a case when someone *not immunized* is being protected by those immunized. Contrariwise, vaccine refusal–though unjustified–leads to a resurgence of the disease. All in all, as in several aspects of life, only a sense of responsibility and a belief in truth and reliability allows us to achieve the collectively optimum outcome for the collectivity.

[65] Sone et al. (1998) and Sone et al. (2001).

The Paradox of Health Education

Primary and secondary prevention prioritizes the welfare of the general population and the interests of the sick people in general. Even when patient-centred, such as cancer screening, it is usually part of some public health programme which relies on the belief that it saves human lives and money. And even when directed at an individual or groups of individuals, it rests on ordinary moral obligations we all have with one another. Prevention is antecedent to the doctor-patient relationship since the latter embodies supererogatory moral obligations to their patients, which come over and above ordinary moral commitments. When one goes to the veterinary for immunizing one's dog for Carré's disease, or when a woman has a mammography, these procedures belong to preventive programmes which should be applied not to some, but to all dogs, or to all concerned women.

Geoffrey Rose has christened '*the preventive paradox*' an intervention which brings large benefit to the community, but which may offer little to each participating individual. The result, Rose shows, is that one should not expect too much from health education because there is little to expect for individuals. Thus, the response to health education by which individuals learn to behave in ways conducive to the promotion, maintenance or restoration of health cannot be justified by the prudential prospect of protecting their own personal health. For instance, the odds that wearing car seat belts will ever be beneficial to a motorist are of the order of several hundred to one, even though it halves the risk of death of the driver. This shows that the common justification of health promotion is fallacious. The claim that health education safeguards our autonomy and helps us to take control of our own life is misguided. Those attempts at justification offer no reason and merely poor legitimate motivation for individuals. Hence health education as well as health-preserving behaviour are scientifically grounded, and warranted not by act utilitarianism, but by some rule-utilitarian argument, namely by the extent to which they provide valued effects to the greatest number of people. Yet, health education acts upon an individual's habits, lifestyle, and health practises on the tacit and unwarranted assumption of influencing his future risks of becoming ill or of dying from selected causes. The antinomy springs from the bifocal ploy of health education which hinges, on the one hand, on a rule-guided approach which evaluates the worth of actions people do for the protection and promotion of the health of the total population; and on the other hand, on a prudential, agent-centred view based on how an action will affect the agent's long-term own well-being. It follows that health education attempts to reach its rule-utilitarian goal by enticing individuals through the agency of a persuasive but poorly warranted prudential argument.

Conclusion

There are three levels of prevention improving the overall health of the population:

- preventing disease and injury before it occurs (primary prevention)
- altering unhealthy or unsafe behavior that can lead to disease or injury (secondary prevention)
- treatment and recovery (tertiary prevention).

1. The goal of medical interventions is to prevent, control, alleviate or cure medical ailments, diseases or injuries and their consequences, to reduce or eliminate suffering and pathological processes and to avoid premature death.
2. By definition, treatment is a matter of need, not of demand.

   It is supposed to do what does better than a placebo in improving the natural history of a disease: this surplus is what we call efficacy.
3. Philosophically, the therapeutic attitude is based alternatively on scientific realism or on instrumentalism.
4. Concerning the assessment of medical interventions, mention is made of the difference between effectiveness and efficiency, and the use of Randomized Controlled Trials (RCT) that aim to reduce some sorts of bias.
5. Medical interventions, preventative or therapeutic, proceed from practical prudential reasoning and they are part of the definition of disease.
6. There is no magic bullet: there is no absolute safety in therapeutic interventions and the residual harm after reasonable precautions has to be weighed against their expected benefit. Yet biomedical research and medical care are increasingly engaged in targeted treatments of cancers or of congenital conditions such as hemophilia, retinal dystrophy, or thalassemia.
7. Placebos and nocebos are at the centre of what we mean by treatment.

   First, a placebo is a non-efficacious intervention that is used as a control in RCT's. Second, it is a non-efficacious intervention that may be used deliberately and as a stand-in for treatment, or unwillingly be either self-administered or administered by the health team.
8. *Evidence based medicine* (EBM) is the conscientious, explicit, judicious, and reasonable use of modern, best evidence in making decisions about the care of individual patients. EBM integrates clinical experience and patient values with the best available research information.
9. Cancer *screening* helps to detect cancer before symptoms appear. However, there are two problems: one thing is that PSA tests for prostate cancer and mammograms for breast cancers often flag cancers that pose no risk. Quite another is that widespread screening for "scrutiny-dependent cancers" may increase the apparent incidence of some cancer.
10. '*Herd immunity* is a form of immunity that occurs when the vaccination of a significant portion of a population (or herd) provides a measure of protection for individuals who have not developed immunity.'
11. The *Paradox of health education* arises because many interventions that aim to improve health have relatively small influence on the health of most people. Thus, health education is of no interest for the individual person.

Chapter 15
The Clinical Relationship. The Tale of Two Stories

The treatment of a disease may be entirely impersonal; the care of a patient must be completely personal.

Francis Peabody (1927, 877)

Contents

A naturalistic view construes the world as if there were some objective way to describe it and as if there were no observer, no scientists in it who could affect this description. According to this view the fact of observing does not need to be considered in our final description of the world. This is the basic standpoint of natural sciences. Scientific language, as it were, is *external* to the world. Sherrington has argued that the world had only been constructed at the price of taking the self, that is mind, out of it.[1]

But medicine requires something more: the mutual dyadic interaction which binds the doctor and his patient and the process by which the latter is being known and acted upon is an intrinsic value-laden part of medical insight and therapy. So, the dialogue of the clinical relationship is *internal* to its exercise. A pure biological science, with its externalist worldview would leave suffering and diseases medically unintelligible.

[1] Sherrington (1940).

L. Karhausen, *Analytical Philosophy of Medicine*, Philosophy and Medicine 152,
https://doi.org/10.1007/978-3-031-81257-6_15

Narrative Medicine

If evidence-based medicine is a necessary condition for good medical care, it is not a sufficient condition. Raymond Tallis writes: "*The recent emphasis on narrative-based medicine, taking account of individual practice is a healthy corrective to the notion that medical practice can be reduced to a series of algorithms.*"[2] Evidence-based medicine provides the general background knowledge and narrative medicine does the work of customizing this knowledge to particular interest.[3]

Narrative medicine's central claim is that attention to narrative, vs in the form of both the patient's story and the physician's story, is essential for medical care. Narrative medicine,[4] on this score, is the "art", while "evidence-based medicine" is cast as the science. Telling a narrative is telling a sequence of connected events. For physicians, the medical gaze gives way to chunks of particular clinical facts to which they try to give some medical meaning. For the patients, the narrative is an apparent staccato of discrete events, with a private, inner phrasing that ligates them together. The tension created by the dialogue between the patient and the physician is made of deep existential considerations that are tacitly implied.

Narratives in medicine includes telling a sequence of connected events, i.e., telling and analysing stories of illnesses, such as those told be patients and physicians, although it often goes further to include any discourse or activity that has narrative features. It is an exhortation to listen, which requires appropriate communicating skills and attention. Narrative competence is an ability to listen, which supposes an ability that is called empathy, to experience vicariously what the patient is experiencing emotionally. Empathising with a patient, i.e., addressing the suffering of the patient makes sense of the suffering: personal identity is being identified with personal narrative, with the view that selves and stories are unique, singular. This narrative repair is how making meaning is part of patient care.

However, narratives are not always singular since narratives may often have structural similarities with one another, so that they can serve several purposes. Each story is the member of a class of stories. There are rites of passage for instance with standard breast cancer narratives, those illness narratives with the hope of restitution, or those that descent into chaos or despair; there are military metaphors of 'fighting' disease as well as those of initial despair finding purpose, usually in others, family, or friends, to sustain the decision to go through an excruciating treatment.

The ubiquity of similar breast cancer stories and the negative reactions to stories that depart from the norms indicates that some kinds of narratives need to be discouraged such as those that describe personal struggle that do not end in personal victory. The standard story of personal successful struggle is appealing since gives women the capacity to act independently and to make free choices and it reminds women of their responsibility to get regular mammograms.

---

[2] Tallis (2004b, 35).

[3] Meza and Passerman (2011).

[4] Charon and Wyer (2008).

Human beings are usually receptive to some narratives that attempt to weave experience into a coherent story by omitting some facts resulting in an intentional or nonintentional distortion or falsification. Patients who get cancer often have stories about why they got cancer that blame themselves or something in the environment (such as pesticides or occupational exposures).[5]

Narrative therapy may be a kind of psychotherapy when it aims at empowering the patient and encouraging positive change.[6]

The Therapeutic Alliance

Three ideas shape one another and are united in the following scenario:

1. The clinician armed with the book of medicine intends to identify what is the case with his patient and how to handle his condition.
2. He then faces unique, multifaceted, and uncertain plights that are not in the books since every patient is different and each and every one of them is a moving target.
3. Furthermore, the book tacitly assumes that the patient will assent to the diagnosis and participate readily in the treatment procedures, but patient's adherence is far from being the rule: the patient has the right of self-determination based on accurate information and his own values and preferences as well as social context, and he should be encouraged to take as much responsibility as possible in making informed choices.

Medical care then operates at the boundaries between radically different territories, between knowing that and knowing how, translating research outcomes into clinical sense, or between what Virchow called scientific medicine and the art of medicine.[7]

The difficulty here turns on the necessity to bridge the gap of incommensurability that separates the physicians from the patient's narratives: doctors and patients do not speak the same language.[8] The patient tells a subjective, inalienable story which expresses his experience of illness and embodies what it means to him, his fear, anger, loss of dignity and desire for control over the process and its consequences. On these feelings and other cognate themes that are deeply meaningful and vitally important to the patient's stories, medical science itself is mostly silent.

The clinical story, that one which is written in the medical record or presented at the hospital's weekly medical conference, is objective, formulaic, seemingly cold and concerned with mere facts. A clinician must first map disease categories and generalizations of medical science onto unstable sets of signs and symptoms

[5] Solomon (2015).

[6] White and Epston (1990).

[7] Virchow (2018) and Peabody (1927, 875).

[8] Hunter (1991, 123–147).

presented by a particular patient. The medical account translates the patient's narrative, rearranges the clinical facts it supervenes, or reframes them in its own terms, namely into a kind of meta-narrative, a medical representation of the patient's story. This cognitively *externalist* standpoint identifies meaning with descriptive meaning namely with *what is sayable*: it is made of assertions that must, if not to be meaningless, be either true or false, albeit preferably true. The patient becomes a case. His story, by now medically retold, becomes irrelevant: so is his suffering, bewilderment, apprehension, and helplessness, fear of death, infirmity, or loss of control.

Were we able to subtract the information the doctor's story shares with the patient's story from the patient's narrative, signs, and symptoms, we would be left, *pace Wittgenstein,* with residual subjective features *that cannot be said* but that *can be shown,*[9] and that makes the relevant disease a single individual's illness: *individuum est ineffabile*.

This process by which the patient's story is neutralized when translated in medical discourse, resulting in distancing these two narratives from one another, is a *sine qua non*-condition for the diagnosis and care the patient has sought. Becoming a patient is then the first step to the depersonalization that anyone undergoes in the medical representation of illness and that is part of what Peter Strawson called the *objective attitude* to another human being seeing him as an object of social policy or as a subject of treatment. Strawson contrasts reactive attitudes (or range of reactive attitudes) of involvement or participation in a human relationship with objective scientific attitudes (or range of scientific attitudes) to another human being; this sets off interpersonal attitudes towards persons *qua* subjects, from scientific attitudes towards individuals *qua* objects.

He insists these two attitudes are not exclusive of each other but are profoundly opposed to each other. The patient, Strawson might contend, is seen as excluded within the medical discourse from the range of reactive feelings and attitudes, which belong to involvement or participation with others in interpersonal human relationships.[10]

On the other hand, from the sick person's vantage point, things look quite different. The patient seeks to understand the medical facts and in addition he expects the doctor to help him to make sense of them and to translate medical knowledge into the terms of his life. For him, medical factual statements are expected not only to inform him, as to console him, alert him, warn him or whatever; for him, their meaning is not so much externally determined—as in the medical narrative—but rather internally conferred; it bestows *internalist* perspective to his illness narrative as a piece of his larger life story.

Yet, for the therapeutic alliance to succeed one must be able to cross the yawning treacherous rift which divides two non-congruent narratives; this daunting task will remain an irreducible hurdle of patient-physician communication, obtruding perfect understanding of one another. Plainly, if clinical medicine is subject to the ordinary

---

[9] Wittgenstein (1961, 4.1212).

[10] Strawson (1974, 9).

approaches of natural sciences with their standards of coherence, it requires, when put into practice with its commitment to action, tacit inferences, and an 'intuitive' address of a different order. Practitioners must then be able to reach across what divides Dilthey's *Erklären* from *Verstehen,* explanation from understanding, or even objective from empathic understanding. They must know how to interpret and convert theoretical, that is to say convergent medical knowledge into medical practice with its unique, divergent, chancy, and unpredictable features, involving medical theory merged with techniques as well as skills.[11] Practitioners are then shaping up the clinical situation into a sophisticated bond[12] and make themselves part of it, inside the clinical transaction.

The Doctor's Authority

"*Scientific medicine* writes Raymond Tallis,[13] *is an appropriate response to the needs of human beings caught up in the inhuman situation of being betrayed by the infirmities of their bodies.*" The mutual relationship between doctor and patient—and more broadly between health care services and population—is strongly asymmetrical inasmuch as they are not equal partners: doctors think in terms of needs, namely of optimizing medical care, patients in terms of demand and expectations.[14] The doctor provides a service to the patient at the latter's request, but the patient does not know as much about his disease as his doctor and may feel anguished, vulnerable and isolated. On that account, patients find themselves in a dependent position with respect to their physician. Plato writes: "*For as we agreed, the business of the physician, in the strict sense, is not to make money for himself, but to exercise his power over the patient's body.*"[15]

Be that as it may, within a proper therapeutic alliance—that is to say for the interaction pattern to proceed correctly and successfully—the doctor, or the care provider, is someone people or patients regard as being in authority, that is to say that he occupies a certain favourable condition with respect to the facts. He is someone they would defer to on matters related to their health and whose relevant advice they would be inclined to accept; what the doctor says is or should be a sufficient reason for the patient to believe it or at least to suspend his disbelief in it. He confers the sick role on individuals. So, a physician has expertise, competence and Aesculapian authority,[16] i.e., sapiential, moral, and charismatic authority, but he also

[11] Schön (1983, 45).

[12] Barondness (1983).

[13] Tallis (1997, 134).

[14] Nussbaum (1994, 27, 74).

[15] Plato (1945, 23–24).

[16] Siegler and Osmond (1974, 93).

has a fiducial responsibility to help his patient in a balanced mixture such that it produces the greatest possible amount of benefice to him.

Yet, 15% of clinical interactions with patients are perceived as "difficult" by doctors.[17] Physicians and patients may hold differing ideas on issues such as diagnosis, investigations, and management options. Difficulties arise when there appears to be no common ground, which is often the result of unrealistic expectations or when patients push for a diagnosis or treatment that the doctor is not comfortable providing.

To be sure, medical authority, the ability to get the therapeutic proposal accepted may vary in degrees since it depends on deference to the doctor's role as well as on personal characteristics such as professional qualifications, charisma, sympathetic understanding, or persuasive ability. For all that, medicine is fallible and sometimes discredited in favour of alternative methods. A physician is often aware of the uncertainty that pervades his diagnoses and decisions, but this should not prevent him from conveying to his patient the assurance that he is treating him with the most effective means available. To achieve that aim he must give advice to his patient and obtain his assent.

Truth and Truthfulness in the Clinical Transaction

The patient-physician mutual relationship shares some common traits with many human activities such as various co-operative schemes and tacit contracts, and one of the central one is the importance of truthfulness. If the medical worker does not speak the truth, communication breaks down: this is one of the reasons why truthfulness is a necessary condition for the patient to listen to and participate in an intelligible dialogue. Informed consent is impossible without truthfulness. Truthfulness involves two virtues of truth: *accuracy*, namely, to make sure that we have the best evidence available and *sincerity*, namely speaking the truth and avoiding deception.[18] The importance of truth telling in the patient-physician mutual relationship has been amply discussed in the literature. Most discussions concern truthfulness understood as the moral side of truth telling, namely not intending to mislead or to deceive. Patients, it is argued, have a right to be informed, which rests on the general principle that knowledge of one's condition is likely to help keeping control of one's affairs and to improve compliance.

But the return of the descriptive content of the patient's narrative once reshaped into the medical account is often quite unrecognizable to the person that is concerned. To be sure, knowledge itself is transitive but knowledge cannot be transferred without understanding: truth does not imply understanding. This represents a haunting difficulty that follows medical care as a shadow.

---

[17] Hahn et al. (1996) and Jackson and Kroenke (1999).

[18] Williams (2002) and Trilling (1971).

Then, beyond this, the exchange of information is not limited to the narrow assertive content of what is communicated. Within the doctor-patient mutual relationship, the doctor starts off with his medical account as a *res cogitans* but intends to end up as a *res agens*.

Be that as it may, the doctor-patient mutual relationship being essentially a therapeutic one will quite often be of a more complex nature. In a way, truthfulness simply dissolves into the reality of the clinical transaction. What it leaves is a different linguistic interaction altogether. The clinical set-up opens up a vast array of possibilities for communication where linguistic exchanges, which are a constitutive part of the therapeutic process, inhere in a mixture of truth telling or descriptive, persuasive, hortatory and imperative statements. The doctor, then, is not discussing a clinical case with his patient the way he might do in a clinical conference. He must now divest himself from his formulaic medical narrative. His words are acts, verbal acts. He does believe what he is saying, and he wants his patient to concur and share his own beliefs. Persuasion is then central to the medical dialogue, calling on a hypothetical prudential 'ought' instead of a categorical one.

Actually, there is a familiar divergence between what a sentence strictly and literally means and what a speaker intends on some occasion to mean when using it: sentences may have secondary meaning, i.e. what HP. Grice (1813–1988), a British philosopher of language, called *conversational implicatures*, which are implicit statements, that are not said but implied within some conversational context.[19] When a doctor tells his patient: *"I shall give you a course of antibiotics"*, he is doing something beyond what someone would be describing, were he to say of him that "*he will give him a course of antibiotics*". The doctor is conveying to his patient a cognitive factual content that carries an additional non-cognitive meaning with a persuasive mode along, namely the core component of a placebo effect.

Both aspects are part of the therapeutic role and cannot be disentangled from one another. To tell one's patient that his clinical condition and his laboratory tests are improving is descriptive, but it is also persuasive. It is *as if* what the doctor says was good for his patient just because he is saying it and beyond what it strictly says: it is, as it were, a reason in itself to endorse or pursue what he says. The representative reconstruction of the patient's account and the interactive, transactional doctor-patient communication have a healing function provided they succeed bringing a meaningful story along, i.e., providing a meaning (descriptive and emotional) to the patient's illness experience.

Proceeding further, the principle of doing good and not doing harm ought in some cases, to take precedence over the need to tell the truth. To tell a patient he is suffering from a cancer may be a very traumatic experience because the connotation of some words for a patient differs from what a medical scientist means by it. Some diseases are metaphors for the patients[20] and the distance between what words like 'cancer', 'leprosy', 'schizophrenia' or 'multiple sclerosis' refer to, and what they

[19] Grice (1975).

[20] Siegler and Osmond (1974, 58–65) and Sontag (1983).

mean for patients, measure the gap between some designation and its emotive meaning for those it does concern. In some circumstances lying to a patient may be a virtue as long as it is not in the doctor's interest.

So, compromises may, occasionally be necessary between competing professional values with the resulting manipulation of information such as delaying, framing, or omitting information.[21] The case of chronic diseases is particularly important because the role of listening and of speech acts becomes crucial in care even more than in cure. One of the problems of contemporary medicine is precisely the decreased part given to speech acts and listening in the therapeutic alliance.

But the greatest hindrance to communication between doctor and patient stem from their distinct cognitive systems. The medical model, that is its way of theorizing about causes, mechanisms and treatment is diverging more and more from the popular ways of understanding and explaining illnesses. This increasing gap tends to be incommensurable. *"Just as the transition from Newtonian mechanics to Einsteinian relativity—or from astrology to astronomy—requires a different frame of reference, so too more is required than the explication of a few terms or a simple translation from one idiom to another for patients and doctors to comprehend each other's views about illness"*.[22]

Expectancy

We do not treat patients in order to satisfy their preferences but their needs. *Prima facie,* the patient's benefit should take priority over his satisfaction, granted that those outcomes are not always coextensive. There is actually little evidence suggesting that patients' satisfaction pertains to quality of care even though it may be valuable *per se* to the individual patient.[23] Besides, patients' dissatisfaction, may be more important in health care than their satisfaction.

On this score, can we utterly dismiss the significance of people's preferences and desires in regard to health interests and medical needs? It might be appropriate at this point to bring in the critical role in medical care of patients' expectations tied in with desires for relief in medical treatment. Jerome D. Frank writes: *"Physicians have always known that their ability to inspire expectations and trust in a patient partially determines the success of treatment."*[24]

Expectations are beliefs that anticipate the occurrence of specific responses or states of affairs but that do not exercise or possess volitions. They differ from intentions in that the latter are anticipations of voluntary states of affairs. Since expectations have the directedness or the aboutness of conscious states, they are intentional

---

[21] Rosenbaum et al. (2004, 404).

[22] Gillick (1985).

[23] LaCombe (1995), Weingarten et al. (1995), and Pickering (1996).

[24] Frank (1974).

states: the object of expectations towards which they are directed exists only in a patient's thinking and may never be realized. Furthermore, they are usually associated with emotional states that reflect or underlie the bearer's feelings about the relevant expectation.

A patient's expectation, when favourably oriented, is a forward-looking belief that some relief will be obtained or that things will somehow make sense and allow him to achieve some mastery over the course of his illness. Research has shown that various categories of patient expectations distinct from demand are pervasive features of medical care and that awareness of patients' concerns should be a key feature, often neglected, of the therapeutic alliance. A patient's expectation of what a drug can do may influence his response to a placebo, and thus, the drug's effectiveness. The capacity to spur expectations influences the success of a treatment and to prompt symptoms relief is related to the strength of expectations. In a contrary manner, pessimistic, anxious, or apprehensive patients may have negative expectations leading to symptoms or sickness. By way of illustration, the experience of depression is the most depressing feature of the life of depressed individuals. The expectancy of depression breeds depression as feedback takes place.

Whether expectations, satisfied or unmet, may significantly alter the natural course of a disease is a matter of debate although the arousal of hope may partly explain relief of anxiety by psychotherapy and alleviation of symptoms. Placebos are then the only form of treatment whose effectiveness depends exclusively on their capacity to marshal the patient's expectancy of help. Small wonder then that it has such a central role in medical care and particularly in unstructured functional disorders and symptom clusters.

The Fault Lines of the Caring Relationship

Doctors are primarily technical problem solvers. But clinical practice by necessity spills over the boundaries of medical knowledge.

Until fairly recently, nearly all medications were placebos, though often unknowingly since medicine had little effective remedies to offer. Medical care was hardly coming out of a history of prescientific doctrines characterized by ineffective medical interventions when, in the middle of the twentieth century the concept of placebo came to be recognized. First used for the evaluation of therapies, placebos initially tightened up what we mean by 'treatment'. Now real medicines are available, but this has not diminished the power of the placebo and of the ubiquitous placebo effect.

Placebos thus split medical care between disease-oriented, technologically intensive care, and the circumstantial effectiveness of patient-oriented therapies unsupported by scientific evidence. Medicine's technical rationality is about diseases, but with the coming of placebos and nocebos, medical care turns towards sick individuals while validity of diagnosis and effectiveness of therapies shifts from causal capacities towards causal roles.

The boundary, which separates therapeutic from placebo effect, was traditionally supposed to be at the perimeter of medical care but it now appears to be at its core. It thus threw medical thinking in disarray, when it appeared that a large fraction—if not the major part—of what happens in medical care falls out, as it were, of medical insight. Placebo effect is weaved in the whole gamut of medical care, in the small but growing core of effective therapies as well as in the growing mass of dummy medical interventions that may have unexpected real effects.

To make matters worse, it next became apparent that placebo and nocebo effects were accompanying active therapies as well as diagnostic gestures as a shadow, blurring even further the difference between being a healer and masquerading as one, or treating a patient and faking it. The centre of gravity of clinical care has thus been shifting away from a coherent intelligible world to a realm of uncertainty, instability, uniqueness, and conflict that bears little relation to the prior perspective of medical science and medical research.

Primary-care physicians often have great difficulties to dispel the idea that deliberately administering placebos is either unethical or deeply worrying. They have been making a legitimate time-honoured stand against quackery and alternative medicines in the name of intellectual integrity but the divide between physicians and quacks now is becoming thinner.

Each of the components invoked to define the contours of therapies or of placebos progressively became somewhat overlapping, intricate and predictably controversial. Medicine, I should conjecture, finds itself in an epistemological quandary as the picture of physicians as scientists, namely applied scientists, or rather as rational problem solvers has gone to pieces. Small wonder, then, that, considering the central function of therapeutic interventions in medicine, physicians may now feel uneasy about a growing unintelligibility precisely at these junctures where it is so much a matter of consequence. The further away doctors find themselves from the realm of technical rationality and its clear and unambiguous ends, the greater the risk they will lose their bearings and find themselves out of keeping.

The recent movement fostering evidence-based medicine endeavours to reinstate some modicum of coherence and map out the realm of medical intelligibility. This move represents a legitimate effort to draw some demarcation line within the fuzzy no-man's land that lies between full-blown diseases and sets of pure symptoms or functional disorders, or between the domain of application of efficacious, i.e., non-placebo remedies, and placebos.

Autonomy and Paternalism

First of all, disease-oriented medical care, which is fondly called paternalism, assumes that a clinician should act so as to fill his patients' health needs and do the best for them: we're doing the best for them means doing what is strictly medically best, even if to achieve this, implies denying them the information they need to make an informed choice. It sacrifices an interest of liberty of a patient to promote his overall best health interests. In clinical, mainly hospital care, as well as in public health decisions, disease-oriented health care assumes that people are rational, and that wisdom prevails over autonomy. This first stance is usually the principal and preponderant approach in secondary and tertiary care.

Dworkin defines paternalism as follows: "*the interference with a person's liberty of action justified by reasons referring exclusively to the welfare, good, happiness, needs, interests or values of the person coerced.*"[25]

On the other hand, patient-oriented care is also an approach, which stresses the patient's autonomy or critical competence, by fostering his independence as a self-directed person. Priority should give way to respecting the patient's health demands in deciding for himself what he thinks would be best for him, as well as his right to direct his life ahead of his interests over considerations of promoting his good or welfare. An autonomous patient assesses alternatives and decides which of the options for dealing with his health-care problem will be best for him, given his particular values, concerns, and goals.[26] Quality of life becomes here a primary concern.

The growing movement of patient assertiveness—giving the power to the people—has led to a greater self-determination exercised by patients. Without some sense of autonomy there is no responsibility. Autonomy establishes the doctor's client as a person, namely a bearer of rights and duties. Isaiah Berlin conveys the argument: "*I wish my life and decisions to depend on myself, not on external forces of whatever kind. I wish to be the instrument of my own, not of other men's acts of will. I wish to be a subject, not an object; to be moved by reasons, by conscious purposes, which are my own, not by causes which affect me, as it were, from outside ... I wish, above all, to be conscious of myself as a thinking, willing, active being, bearing responsibility for my choices and able to explain them by references to my own ideas and purposes.*"[27]

But then, room is opening for some of the difficulties peculiar to the respect of autonomy itself. Autonomy has a privileged status: it is an ideal character trait, a baseline that has intrinsic value. But autonomy and paternalism represent the opposite asymptotic ends of a spectrum. The very concept of non-compliance evinces the doctor's concern about patients' autonomy. Someone may possess the capacities to live autonomously, but, even so, he may not exercise them in all respects. And, even though each patient may decide as he wishes, not all his decisions are necessarily correct. Autonomy supposes a modicum of rationality and self-awareness but if autonomy is a cornerstone of medical ethics, it is not a premise for correctness or goodness; it may occasionally lead to choices, which are seriously unreasonable: autonomous choices may be neither prudent, nor rational. And should patients be always compelled to make each decision about their health care? It may be an autonomous choice to waive part of one's autonomous choice. Should trust and

---

[25] Dworkin (1971).

[26] Glover (1977, ch 5) and Downie and Telfer (1980, 50–66).

[27] Berlin (1969, 127).

good riddance be sometimes substituted to autonomy? Furthermore, what a patient wants now could come in conflict with what he will autonomously want later.

Actually, physicians retain exclusive control, for example, of patients' use of prescription drugs. People listen to their lawyer's advice, but they tend to comply with their doctor's orders. It is also clear that informed consent is based on a myth, i.e., the claim that patients understand the diagnosis and the nature of the proposed modes of treatment, which usually results in empty rather than meaningful consent. This principle grows in an adversarial atmosphere, which is more common in the US than in Europe, where the doctor-patient relationship relies more on trust.

"The simple phrase informed consent contains both manipulative and coercive aspects." Consent" of the patient is supposed to prove that he was not physically or psychologically forced into a procedure. *"We then insist that this consent be "informed", recognizing that if a patient readily agrees to something about which she understands little or about which she has false understanding, we have somehow or other abrogated or sidestepped her autonomous decision-making rights."* [28]

If so, then, in the reality of clinical care, the exercise of autonomy is much more chaotic than ethical theory would have us believe, inasmuch as medical decisions arise in confusing situations and in the quandary and uncertain realities of the bedside.

Plainly, the autonomy debate seems more and more to be obsolete.

The Three Situations

The silent contract between doctor and patient is a partnership where doctor and patient grow together, and which may present itself in various modes. Autonomy comes by degrees, and we may attempt to identify three theoretical positions which are mere paradigmatic sanctified sighting marks along the spectrum of medical care, namely a complex sequence of actual clinical and public health plights; each of them being epistemically prior to the following one.[29]

First Situation. The Model of Activity-Passivity

This is the traditional model of medical care that reflects credit on medical science. It also construes medicine as a science-based instrumental practice that is supposed to unseat superstitions, craft, and quackery. There is a radical inequality between the passive patient and the active benevolent physician. The doctor has knowledge or claims to knowledge as well as skills and acts irrespective of the patient's

[28] Gaylin and Jennings (1996, 159).

[29] Szasz (1956).

contribution. He is defining the patient's needs and voices his advice in a categorical form. The patient is construed as a phenomenon to be managed.

This is a type of fiduciary partnership that is well suited for general anaesthesia and emergencies, when the patient is unconscious, incompetent or when he in the grip of delirium and is not expected to respond to reason in a certain way. It may be suitable for the management of suicidal patients, severely mentally disabled persons, some patients in intensive care, or in cases of veterinary medicine, namely in patients that are unable to anticipate the consequences of medical decisions.[30] This objective attitude does not consist in taking another human or sentient being as a mere 'object', but in understanding him objectively through observation, fact gathering and inference leading to technical decisions.

The realm of the first situation is that of technical rationality, when converting knowledge into action or knowing that into knowing how is quite straightforward. It is the compelling interest of a young comatose diabetic showing up with a blood sugar of 350 mg% to be treated urgently with insulin shots if he wants to avoid serious complications.

The task of medicine within this first situation is purely prudential inasmuch as health care needs and interests are basically inescapable primary values. On that account, ethical issues are never raised, and they are never raised because they are housed right from the beginning in the discourse of medical science. Even then, the nature of such clinical problems with their multiple constraints may be ambiguous and uncertain and their resolution may require continuing input and open deliberation to make the best of a bad plight. The boundaries of the means best suited to achieve the doctor's ends are those of facts. And the border of the first situation, which thus coincides, with the confines of medical science fences them in.

Second Situation. The Model of Guidance-Cooperation

In an ideal world, the doctor would rely on the underlying rationale for this contract, which is that if his patient were to foresee the consequences of the treatment, he would accept it. The doctor asserts his recommendations on purely prudential grounds expressed either in the second (in clinical medicine) or in the third person (in public health). Much of medical practice is subsumed under this model of asymmetric relationship.

Meanwhile, the therapeutic alliance, here, becomes a consensual transaction, a tacit contract in which the patient gives up certain freedoms in order to gain some prudential benefits. This model is suitable for acute conditions in which a conscious patient is ill with pain, distress and anxiety and calls for help. It often applies to the healing and elimination of an illness. It may also be employed for the management

[30] DeGrazia (2002, 19, 84; 95).

of patients handled by senior consultants, or admitted in secondary or tertiary care hospitals, which implies some sort of tacit consent.

The patient seeks help from his physician. The doctor is in authority and expects co-operation, while the patient occupies a dependent role and surrenders part of his autonomy. Both actors are active, but the authority willingly conferred by the patient to the physician brings the latter in a leading role on this partnership. Paternalism, i.e., making someone do what is good for him—or rather what is intended, whether it succeeds or not, to promote or preserve his own good—even if he turns it down, inserts itself in the space that divides need from preferences. The major drawback of this model is that it is limiting the patient's liberty or autonomy for what is taken to be his own good. To be sure, paternalism in medical care is not an all-or-nothing feature, but it comes in degrees. The doctor's interference may be soft, weak, or imperceptible. It may be brief or limited instead of constant and fully adequate. What's more, medical interference may be permissible in children or in some mentally ill whose decisions lack autonomy.

Yet, we cannot always practice what we teach. Not surprisingly, if medical care in the first situation relates with diseases, facts, and phenomena, when it takes over the second situation, it moves into a minefield of misgivings and uncertainty. In striving after clinical sense, and trying to melt knowing that into knowing how, physicians now construe diseases in their particularity: a case of disease is seen as something unique, built into an individual, instead of something that is being absorbed in a general class of patients. As the language of medicine yields to persuasive instrumental use, clinicians relate to patients through reciprocal transaction. Treatment decisions become more of a joint undertaking between physician and patient and increasingly tailored to the latter. Medical care may now edge across the boundaries of the world of fallible facts into ethical issues. As the day-to-day exercise of clinical medicine inheres in trade-offs between needs and demands, the physician may part company with unalloyed medical doctrine.

Third Situation. The Model of Mutual Participation

In this type of covenant, the doctor should be a consultant to the patient helping to support his autonomy: doctor and patients are dependent on one another and ought to be dependable. They tend to be equal partners. The physician still has scientific authority, and the patient allows him to make some important decisions, but the physician also strives to identify with his patient and to conceive him in terms of himself.

General practice and management of chronic diseases serve as a paradigm for this model. As the day-to-day exercise of clinical medicine inheres in trade-offs between needs and demands, the physician parts company with pure, bookish, untainted medical doctrine. Demand may gradually take precedence over needs as the clinical alliance between doctor and his patient implies decisions, transactions, agreements, and deals. It presupposes that the doctor will make concessions on the patient's demands and meet him halfway if the latter will acknowledge his medical needs. When there are differing interventions available, the patient should be given proper information about competing options and their outcomes, attending the patient's preferences, and trying to adjust the medical evidence for the particular patient. The process of 'informed consent' encapsulates this position. The weakness of this model of mandatory autonomy stems from the difficulties involved in insuring consent and compliance and the potential consequences of its failure. Patients' satisfaction is not always synonymous with good medical care.[31]

Conclusion

The clinical tête-à-tête of the clinical relationship is *intrinsic* to its exercise.

Yet, the cognitive relationship is not symmetrical. Medical care harbours a tension that is revealed by the medical precedence of *needs over wants* alongside the incommensurability between the patient's and the doctor's narratives. But the patient *expectancy* rather than his satisfaction may reach across the gap that divides these two narratives.

Three models of partnership represent three positions, which dethextend from *paternalism to autonomy*: doctor's authority, guidance-cooperation, and mutual partnership. First, medicine gives a third-person account of what are first-person account issues, bestowing descriptive status to the patient's narrative. Next, it tunes in a *prescriptive attitude mingled with empathy*, (even though empathizing with some patients may be extremely upsetting for the physician, e.g., in cases of severely psychotics). Except possibly for the treatment of brief acute conditions, medical care should be *comprehensive*, that is to say that understanding of the patient's disease implies understanding of the whole person. Integrative medicine is a healing-oriented medicine that take account of the whole person including all aspects of his lifestyle.

[31] Kassirer (1994), Pickering (1996), Covinsky et al. (1999), and Cassell (2005).

Chapter 16
Context and Limits of Medicine

> *The task of science, therefore, is not to attack the objects of faith, but to establish the limits beyond which knowledge cannot go and to find a unified self-consciousness within these limits.*
>
> Rudolph Virchow

Contents

A doctor's knowledge encompasses the knowledge of his limits. The trajectory of life from birth to death and the functional range of on-going life set the *prima facie* limits of medicine. Medicine is not an end in itself, neither does it end when there is no more prospect of adjusting our biological processes or with death, but it allows people to work towards their own ends. And limits are also reached when demand for health care exceeds needs, or when acceptable demand exceeds accepted needs.

Whenever physicians leave this cornerstone of medical rationality, namely whenever they get involved in arguments or interventions which do not arise from the prudential need resulting from the original negativities, they are gazing across the limits of medicine.

Physicians, as it were, are standing at the centre of their compass card. When they look at the outside world and get involved, *qua physicians,* in the prevention of motor accidents, in civil violence, criminal deviance or war, they are still talking from the narrow vantage point revolving around the centre of their dial. The more considerations they introduce which are not derived from the initial convention—be they social, economic, moral, or political—and the further away they find themselves from the medical system of reference, the closer they get to the borderlands of their discipline, merely speaking and acting *qua citizens*.

L. Karhausen, *Analytical Philosophy of Medicine*, Philosophy and Medicine 152,
https://doi.org/10.1007/978-3-031-81257-6_16

If the limits of medicine were within the field of medicine, they could not possibly constitute their limits. The limits of medical experience cannot be within the range of medical experience. So, medicine, as it were, has no limits.

II we want to avoid ill-suited broadening of the alleged perimeter of medicine, we need to draw our boundaries in order to know what is and what is not relevant to health care. We can set limits to medicine by working inwards through what is not medically relevant.

Epistemic Limits of Clinical Care

For one thing, a practitioner gathers his scientific know-how and its doubts from his training as well as from medical texts and medical publications. Medical knowledge is construed, elaborated, criticized, and proclaimed through the editorials of the leading medical magazines. What gives a patient's sickness episodes, their clinical identity and unity is their shoreline. In applying their skills to persons and their individual illnesses, physicians meet with the issue of vagueness and borderline cases.

For another, a physician assumes that his patient's medical history, signs and symptoms are facts of the matter and that they are true (or false) prior to, and independently of the clinical encounter. Plainly, scientific knowledge is not sufficient and needs to be supplemented by the skills necessary for diagnosis and management of the patient. Although scientific findings are the core of medicine, the discipline acquires its consummation and autonomy through the clinical narratives: the process of clinical assessment, transaction and therapeutic management that constitutes the patient-doctor relationship arises within this sequential narrative account.

It follows that the physician, *nolens volens*, is a nearly *participant observer* much more than could ever be the case in the natural sciences. We can only observe biological processes and describe some of their causal uniformities. But a physician gains knowledge of his patient's condition inasmuch as both of them are participants in the epistemic diagnostic process. There is no such a thing as the truth of a statement being fixed in advance of its actual employment. A doctor needs a coherent and objective account, but he has to mesh with the version, intentions, and desires of his patients.

Furthermore, the symptoms described by the patient are voiced from his own perspective. That is to say that they occur only within quotation marks. Problems arise with the *disquotation process,* when the physician understands the content and structure of his patient's narrative from outside. The grammar of the first-person account and that of the third-person account are such that what is valid in the first case may not be so in the second case. The often-unavoidable plight to mend this gap to a satisfactory extent may cause uncertainty about the diagnostic conclusion to which divergent interpretations may be assigned.

What does seem to follow is that the further away from the scientific core and the more involved physicians are in specific concrete issues of particular cases, the closer they find themselves to the limits of medicine.

Professional Limits

Medical errors are also common. Art Papier, a dermatologist specialising in medical informatics and a professor at the University of Rochester, stated that "*We don't know how many diagnostic errors we make*". Diagnostic errors have become and are accepted as part of the medical process. Many of them concern common diseases such as cancer. Patients usually suffer from fairly trivial conditions, but a clinician is often fooled by variations on common diagnoses. Paul Epner founded the SIDM Society to Improve Diagnosis in Medicine in the US in September 2018, which brings together 40 organisations, including the major medical societies and academic medical centres. Marc Graber, president, and co-founder of SIDM, says that the first and most important step is for "everyone to understand that we have a problem". The measures to be taken concern doctors, nurses, care organisations and patients, who must also understand that the diagnostic process must "help co-produce the diagnosis".

The American College of Physicians (ACP) believes that internists make errors in 10–15% of cases. The ACP is now introducing continuing education courses to encourage thinking about diagnosis and decision-making, how problems can occur, and how errors are likely to affect patients and clinicians.

Fortunately, new tools are available to streamline the diagnostic process and help clinicians more reliably find the right answer. Some of these tools use artificial intelligence that can mimic human behaviour, or they use machine learning that uses artificial intelligence and statistical methods that allow devices to improve by solving new problems.

Ethical Limits

Furthermore, medical thinking is harnessed to human valuing right from its foundations. A large part of what we call medical ethics is actually *a posteriori* necessary, internal to medicine and constitutive of medical care. Implicit moral issues are usually converted into the medical language of clinical judgement. Medical obligations thus lie outside every moral system postulated by morality *qua* morality.

Yet, beyond this initial tacit mating of knowledge with value, there is a further range of unprecedented ethical questions that penetrates medicine from outside and which deals with issues of paternalism, autonomy, means and ends, moral and non-moral goods, right and wrong, quality of life, cost vs. benefit.

The straightforward commitment to medical intervention that is embedded in the concept of disease must now relinquish to considerations of consequences and values other than medical ones, as well as to the patient's aspirations and demands. What's more, with the irruption of new moral issues and community medical concern making their way in the borderland of medicine, clinicians may face dilemmas and values, which are in conflict with traditional Hippocratic and Maimonidean precepts.

By way of illustration, autonomy can either be a socially expected aspiration, vested with great moral legitimacy, but, if overblown, it can also undermine social order and foster extreme forms of deviance and disrespect for one's peers and kin. If so then, autonomy appears to be health promoting but it can also potentiate health risks such as smoking, drug and alcohol abuse, and other high-risk behaviours. Autonomy is one of the cornerstones of the ethics of medical care, but it is confined to the edges of medicine. That is to say that the limits of a patient's autonomy are of a piece with the limits of medicine, i.e., the limits of a doctor's obligations.[1] What is valuable in medical theory may not be so in specific social and clinical circumstances. Patients' preferences, wants, demands, and desires, especially nonessential ones, barge in the sphere of needs and lead the physician to step out of his strict role *qua physician*. It is perfectly legitimate for a patient who is well aware of the dangers he faces, to refuse a lifesaving procedure. This is why comprehensive medicine is cogently concerned with patients' general well-being and their non-medical needs, but therethrough, it crosses the confines of medicine.

"Doctor's anguish, writes Renée Claire Fox, *seems to come from violating every day what they know they ought to be doing". "The pain is from the degree to which they still espouse values but can't live up to them."* Professional dissatisfaction grows out of bureaucracy, loss of autonomy, diminished prestige, patients' noncompliance, that is to say factors which keep doctors from carrying out the task of medicine which is to treat, alleviate and prevent diseases. The milder the patient's condition, the more the balance between the physician's responsibility and the patient's autonomy shifts in the direction of the patient, and conversely.

But the closer one gets to the core of medicine, the greater the doctor's responsibility. The clinical perspective comes to the fore in a patient suffering from a curable cancer or a psychotic episode when it is important that he should receive prompt advice: in such a case, the physician needs more control over his patient's behaviour because it will be his own fault if things go wrong. Yet this is often a moot point: the problem here is that the public lacks the professional expertise of the physician and may in such a case underestimate or downgrade the doctor's advice and think in mere moral terms. The balance sheet may drive credits or debits one way or another but the more remote the decision wanders from medical considerations, the more perplexed the physician might find himself.

[1] Brett and McCullough (1986).

Ontological Limits

Biomedical research assumes that medical conditions and the patho-physiologic mechanisms that explain them are basically simple and regular and reflect a linear vertical or pyramidal hierarchy based one layer of organization upon the next, each level being explained by derivation from some more fundamental theory. Thus, to discover the cause of a disease is to find out what has happened that caused the patient to be different from a healthy control, what is it that we could have wished to be different in the causal antecedents, and what is it that needs to be altered. Medicine construes pathological events as lying at the end of causal chains. Going back along such causal chains allegedly reaches *some or several initial abnormalities* (bacteriological, biochemical, or genetic) that constitutes an ultimate causal system. The time-honoured empiricist programme of disease-specific categories and their pathophysiological explanation is not a medical credo but rather some sort of working hypothesis in search of magic bullets. It gained some ascendancy with the germ theory in the end of the nineteenth century. The Central Dogma of molecular biology is called into question by the now known one gene, one protein, and one disease model.

So, the ontological model assumes the existence of some sort of initial instance or set of necessary causes, such as the internalization of a bacillus. The disease can thus be traced back to these original causal factors. This root causes initiate the ensuing upstream sequence of pathological and pathogenic bodily changes that constitute a unifying basis for the relevant disease.

Meanwhile, the value-loaded predicate 'being abnormal' is discernible at the clinical level of observation since it belongs to any disease process. From a logical vantage point, judgements of value (e.g., being harmful or abnormal) cannot be deduced from statements of pure fact, unless there is at least one statement of value in the premises. If so, values are deductively inherited with the result that the consequent of any negatively valued step of the deductive explanatory model is negatively valued. Each step of the upstream explanatory ladder is being value-flagged until it reaches the clinical level. It follows that if we make our way up in the explanatory model, the negativity that started with the initial abnormalities reaches the surface of clinical or public health language, having been transmitted transitively through the successive hierarchical layers of the explanatory chain.

Beyond this initial abnormality, the events and processes are non-medical, they are value-free biochemical of physiological types of occurrences that may normally occur. Within the ontological paradigm, this conjectural initial statement is the grail of biomedical research; it defines the confines of medicine on one side of which (that is downstream of which) dwells physiology and on the other side (that is upstream of which), pathology.

The Media and Misinformation

Medical misinformation has several dimensions. Most doctors and healthcare professionals are unaware of the ubiquity of dubious or poor-quality data and evidence that contribute to over- and under-consumption, avoidable adverse effects, missed opportunities and wasted healthcare resources. First and foremost, we need high-quality randomised controlled trials of the effectiveness of certain interventions. But we also need to be warned that any evidence-based practice can be circumvented by, for example, clinical trials sponsored by the pharmaceutical industry. Journalists must have the requisite training and qualifications: journalism schools should include the principles of epidemiology and statistics in their courses for future medical and public health writers.

What's more, instead of arguing over words like 'an important breakthrough', 'a major discovery', 'alarming', 'epidemic', etc., scientists and journalists often differ over the use of anecdotes of public interest to illustrate research: for the scientist, such a story is simply a particular piece of information in a large database. But for the journalist, the story is likely to leave an indelible mark that will rule out any discussion or qualification. If the story is representative, it may be useful, but if it is not, it could be misleading and potentially misleading.

As Mark Twain wrote: "*Beware of reading health books, you might die from a misprint.*"

Medicine and Culture

Medicine is a science—theoretical and applied—in the same way as physics, chemistry, or astronomy. But how many sciences make up medical science? We're not talking about molecular biology or genetics, but about practising medicine, the day-to-day routine of diagnosis and treatment. Is the relationship between doctor and patient comparable to that of an experimental laboratory, or is it more a question of a commitment that is both rational and subjective, binding doctor and patient together?

If medicine is to be scientific, it must meet two conditions. The effectiveness of medical practices and procedures must be clearly and rigorously established, and it must not differ from one region or country to another. Just as a physicist will reach the same conclusion if he measures the speed of light in France or Russia, a patient with the same signs and symptoms should receive more or less the same diagnosis and treatment wherever he is. However, the expression and verbalisation of symptoms depend on the cultural context, particularly in the case of psychiatric illnesses.

All this is true, however, and generally applies to academic medicine, but the reality of practising medicine shows that differences sometimes outweigh similarities. Some countries have "rubbish" diagnoses, which contribute to their national character, such as heart failure in Germany or liver failure in France, or intestinal problems in England. In Germany, the diagnosis of *Herzinsuffizienz* (heart failure) does not have the same meaning as in other countries: many patients suffer from, and are treated for, *Herzinsuffizienz* even though they have no shortness of breath, no oedema of the lower limbs and a normal electrocardiogram. According to sociologists, this diagnosis is also a mark of standing, a sign of social success. It has its origins in many of the great literary figures of German Romanticism, who were very concerned with their sensibilities, their "hearts" and the German soul.

Similarly, in France, the liver crisis is a national pathology that corresponds to gastrointestinal disorders, dyspepsia, but also migraines, paleness, generalised fatigue, painful menstruation, acne, skin rashes or herpes even though the liver function tests are normal. The diagnosis of mild liver failure dates back to the 1880s and became popular in France in the first half of the twentieth century. "Between the 1920s and 1950s," said Henri Péquignot (1914–2003), Chair of Medicine at the Hôpital Cochin, radiologists and gastroenterologists invented "a great many details and spurious scientific explanations in response to patient demand, rather as a biscuit manufacturer diversifies its production in response to the demand created by a good advertising campaign".

Louis Pasteur, the father of microbiology, insisted on the importance of the terrain, just as much as that of bacteria. The importance attached to the terrain shapes and distinguishes French medical practice from that of the English, for whom disease is more like an invasion. French medicine seeks to favourably modify the terrain, using tonics, vitamins, minerals, and spa treatments, but it also favours alternative medicines (such as homeopathy or aromatherapy, cellulotherapy, micromineral therapy, opotherapy, lithotherapy, isotherapy), which are much rarer in other countries, and non-existent in the United States. It tends to be less "aggressive". It also has a much more relaxed attitude to germs and uncleanliness than is found in other countries. "Name me one disease transmitted by toilet lids", asked the French allergist Georges Halpern (1935-).

France is one of the countries that consumes the most drugs per capita, while England consumes three times less. In England, on the other hand, there is a growing number of doctors—often from Eastern Europe and encouraged by the Royal Family—who practise homeopathy, osteopathy, acupuncture, laying on of hands and other bonesetter techniques, in addition to the medicines, potions and manipulations that are the hallmark of their discipline.

With the exception of tranquillisers, English doctors prescribe fewer drugs and no comfort drugs than the French or Germans. They also carry out fewer screening tests and less surgery, and when it is necessary, it is less extensive; the diagnostic threshold is also higher, so that high blood pressure, urinary tract infections and obesity are much less frequent. Generally speaking, the English are more sceptical about medicine than other European citizens.

The English's respect for factual information, combined with the primacy of society over the individual, explains why England has been the main and first promoter of social medicine, epidemiology and clinical trials, an area where France has lagged behind, because it does not place as much value on data collection, in the name of protecting privacy.

Finally, psychiatry is the branch of medicine where concepts are the most blurred. One of the most widespread illnesses in Germany, under the influence of Romanticism, is Erschöpfung or Leistung, which is linked to categories such as melancholia, neurasthenia, and today, depression, chronic fatigue syndrome and burnout. Otto von Bismarck, Friedrich Nietzsche, Cosima Wagner, Franz Kafka, Thomas Mann, Rainer Maria Rilke, and others suffered from it. The nineteenth century not only saw the emergence of this form of exhaustion, but also numerous therapeutic strategies to counter its effects, such as nudism, vegetarianism, gardening, expressive dance, and macrobiotic diets. It is a variety of behaviour, acedia, already described by Thomas Aquinas who, in his *Summa Theologica*, identified it with the pain of the world, *Weltschmerz*, which was also the subject of a drawing by Breughel the Elder. Walter Benjamin, in The Origin of German Tragedy, wrote that Prince Hamlet's indecision "*is nothing other than saturnine acedia*".

The French prefer the term *bouffées délirantes* to schizophrenia. The Russians use the concept of schizophrenia in a non-psychotic form to lock up political dissidents. What's more, the term "*neurosis*" is still commonly used in France, even though it was removed from medical vocabulary in 1994 and abandoned by the World Health Organisation's Tenth Classification of Diseases: This bastard term is therefore empty, and the disorders formerly described as neurotic cover a wide range of clinically and causally distinct entities, such as anxiety disorders, phobias, obsessive disorders, post-traumatic stress disorders, dissociative disorders, somatoform disorders and depersonalisation syndromes.

Let's end with the example of autism. In the 1950s, psychiatry, under Bruno Bettelheim, considered that autism was the result of a disorder in the mother-child relationship. However, this psychoanalytical view has since been largely overturned by medical research, and psychiatry now recommends treating these children using educational, behavioural, and developmental techniques whose effectiveness is well established. The problem of autism treatment was the subject of a decision by the Council of Europe in 2004, which criticised France for not fulfilling its duty to educate children suffering from autism. In 2012, the French High Authority for Health (HAS) declared that psychoanalytical approaches "have not been proven to be either effective or ineffective". Finally, the 3rd autism plan 2013–2017 presented in May 2014 was accepted by the HAS: Marie-Arlette Carlotti, the minister in charge of disabled people concluded, "With this 3rd autism plan we are finally taking the direction of respectful support for autistic people and their families." However, the war is still raging in France between the advocates of psychoanalytic and behaviourist approaches because academic circles, like the media, are still largely steeped in psychoanalysis and its rearguard battles. What's more, we now know that autism is a neurodevelopmental anomaly (30% of autistic children suffer from epilepsy), with a strong genetic component and genome anomalies that they share with schizophrenia and bipolar syndrome.

It is true that until the end of the Second World War, psychoanalysis was the only psychotherapy practised in Europe. The almost imperialist control of psychoanalysis over psychiatry and its lobbies is in decline throughout Europe, except in France and, to a lesser extent, Italy, and it is highly controversial in England. It has never managed to gain a foothold in Spain or the Slavic countries. Peter Medawar, an English biologist, and winner of the Nobel Prize for Medicine in 1960, added that Freud gets away with "*a concept of veracity that belongs essentially to the literature of the imagination, where the opposite of truth is not error, but another truth*".

In conclusion, humanity has always sought to resolve difficulties rather than avoid them. Given the inherent limitations of any therapeutic intervention, the twenty-first century will be one of preventive medicine. What's more, medical procedures are becoming increasingly targeted, aiming, in the case of cancers for example, at their genomic specificity rather than their anatomical location.

Contextual Limits: Medicalization

The concept of 'treatment', writes Thomas Szasz, *is the grand legitimiser of our age. Call whatever you want to do 'treatment', and, pronto, you are hailed as a great humanitarian and scientist."*

Ivan Illitch (1926–2002) denounced the obsession with perfect health. In his pamphlet *Medical Nemesis,*[2] he suggested that medicine tends to expropriate many aspects of life, such as pain, ageing, death, patients' expectations, and prevention techniques. Turning life events into pathological processes and converting risks into diseases means extending medical concepts beyond the remit and mission of medicine. It means describing everything that is abnormal as pathological. Thus, alcohol abuse, homosexuality, the menopause, masturbation, sadomasochism, social deviance, delinquency and even crime, political opposition and revolutionary beliefs have been described as diseases. Illitch declared, in the tradition of Rousseau, that society is unhealthy to the point of consciously threatening the survival of humanity and of causing, in all clarity of mind, serious and fatal illnesses.

There is a whole current of ideas in medical literature which sees war as the expression of the psychopathology of power and an abnormal contempt for human life, as in the case of Napoleon or Hitler. Are juvenile delinquency, paedophilia, alcohol abuse, tobacco consumption and addiction to heroin or cocaine medicine? If we extend our idea of health beyond biological or psychological abnormalities to include social and cultural conditions of personal or community interest, this has the effect of fuelling expectations on the part of the public, and of forming misleading hopes about what medicine should be able to achieve or prevent.

However, if these threats are real, they are external to medicine. That humanity is in the process of destroying its ecological niche is undoubtedly true, but this issue lies outside the field of medicine except in terms of its consequences for public health. What Illitch was highlighting was rather the illegitimate invasion of social, but non-medical, concerns into what falls within the remit of medicine, as well as the effort to drag medicine away from its objectives. Medicalisation extends

---

[2] Illitch (1978).

medical thinking to public order or public interest and pushes back the limits of its prudential interests.

The difficulty lies in the fact that the term 'health' is not medical: it is vague, shifting and more political than scientific in nature. Doctors, as doctors, are not responsible for society's state of health.

The concept of health is at the confines of medicine too, inasmuch as it is not a medical but a vernacular notion. The term quality of life pertains to a cluster of overlapping concepts uniting personal health and functional status with social well-being, life satisfaction and restoring lost functional ability. When used counterfactually in medical care, for instance in cases of depression or physical handicap, it refers to the impact of an illness and its treatment on the individual's physical, psychological, and social functioning and lifestyle, as compared to his prior health status. But when we detach quality of life from medical conditions and talk of individual well-being, which may depend on the social environment, social participation, family relationship, professional fulfilment, and social support, we are leaving the realm of medicine.

Ivan Illitch has argued within a 'Rousseauist' tradition, that society is unhealthy to the point of knowingly threatening the survival of humankind and of knowingly producing much of severe and lethal diseases in individuals. But those threats to the survival of the species are external to the scope of medical care. That humankind is destroying its 'ecological niche' is true but lies beyond the scope of medicine except to the extent that it portends public health consequences, for instance the risk of life-threatening world-wide epidemics or viral diseases. Any condition leading to limited incapacitation is of medical concern, but where do we begin? Perhaps all Illitch actually meant to stress was the illegitimate invasion of non-medical social concerns in the field of medicine and the effort to drag medicine away from its central remit.

The limits of medicine depend on how broad our views of prudential reasons are: if the latter are coherent with the overall interests and experiences of human beings, the medical umbrella will be larger than if minimally limited to our welfare. In the case of the broad view, medicine encompasses notions of full enjoyment of one's physical and mental abilities, full blooming of one's personality and happiness. This is tantamount to WHO's illusory conception of positive health which covers every human state. Renée Claire Fox contends that a diffuse definition of health has increasingly come to be advocated: *"This conception of health extends beyond biological and psychological phenomena relevant to the functioning, equilibrium and fulfilment of individuals, to include social and cultural conditions of communal as well as personal import."* This view *"is connected with higher expectations on the part of the public about what medicine ideally ought to be able to accomplish and to prevent."*[3]

Likewise, preventive procedures, which slow down ageing, actually delay the onset of cancer, age-related diseases and disabilities and extend the duration of

[3] Fox (1979a, b).

healthy life. Improving living conditions, better nutrition, physical exercise, housing, medical care, public health facilities and accident prevention decreased the risk of death towards limiting values. Thus, ageing is a major risk factor, a threat to health, and thus is of medical concern. More precisely, what is pathological is not ageing in itself, but whatever adds or removes years to a person's biological age.

In like manner, medicalization of normal processes such as menopause or pregnancy tends to equate therapeutic with non-therapeutic interventions. There is no denying that pregnancy and childbirth are instances of biologically 'normal' processes. Yet, both of them increase the risk of morbidity and mortality, but this does not prevent women from getting pregnant because, on the balance, pregnancy for most women is valuable. This is why, although pregnancy is not a disease, a pregnant woman assumes the sick role.

Risk factors such as tobacco smoking, alcohol and drug abuse, environmental exposures, air pollution, occupational hazards, the contamination of the food chain by various chemicals or waste products, the consequences of overpopulation and climate changes and wars, have resulted in new or shifting health-related prudential concerns, inasmuch as medical and public health practises have broadened their scope and penetrated into the territory of public policy. In this way, prudential needs ought to define the limits of medicine. And the absence of present or potential suffering and of health care needs, seems to stand as a paradigm case of the *non*-medical health.

Is Medical Enhancement a Faustian Deal?

What we often call enhancement technologies can be described as treatment. Viagra treats erectile malfunction and plastic surgery can enhance the body, treat disfigurements, or treat "inferiority complex", which Elizabeth Haiken called 'psychiatry with a scalpel'.[4]

But then, abruptly, the concept of enhancement becomes unwieldy. It carries its connotation *from a second order standpoint,* and far beyond present medical practice and its caring ends, that is to say *beyond the limits of medicine*. A momentous shift inheres in this move, from a stance germane to medicine to a biotechnological programme for helping healthy human beings to perfect themselves. Enhancement then becomes a primary intention. Alvin Weinberg christened *'technological fix'* a cultural fascination with technology for the solution of social problems, i.e. with remedies that address only part of a problem and usually neglect related difficulties and create problems of their own right. The situation is compounded by people's blind faith and unbridled enthusiasm for prescription drugs. Drug spending represents 11.9% of health-care spending in the United States and 16.4% in France. We

[4] Haiken (1997).

must indeed remember Osler's witty definition of man as *"the medicine-taking animal"*.

The point at issue is that we start here with normal, healthy people who have no disability, no cause for complaint, and no reason to be dissatisfied with their life but who wish to boost a state they think is a normal level. Enhancement supposes that human beings are perfectible. But to assert that man is perfectible is to assert either that he has the capacity to realize his potentialities, Aristotle's natural ends, or else, that, through the promises of biotechnology, he is able to improve his aptitudes and to redesign his nature and hence to enhance some of human species' traits. The latter form of enhancement forays in an open-ended process, with a demand for ever more fulfilment, more happiness in the absence of the restraint imposed by medical care. Furthermore, if perfectibility means that the capacity to be improved is boundless, man's ability to perfect himself must be understood with the performance of a task, i.e., the ability to perfect himself technically in some tasks. Improvement of athletic performance with its deleterious side effects is of the most conspicuous example of this mode of enhancement. Or the administration drugs such as modanifil to those whose tiredness might endanger others, such as airline pilots or truck drivers.

Spinoza has introduced a useful distinction here between an inability and a disability. Human babies are unable to walk but this is no disability: they merely lack the ability to walk just like adults are unable to fly. On the other hand, the incidence of disability arising from musculoskeletal diseases rises with age. Since this represents a natural trend why do we not call it an inability? The grounds for this distinction should be clear: no one would complain to the paediatrician that the newborn cannot walk; but hobbling and shambling along is a handicap for elderlies and this is why they may need help to learn methods of ambulation, safety measures, the use of a cane, crutches, or a wheelchair. The need for help distinguishes inability from disability and normal from pathological. Medical care pertains to need and disability, and enhancement to demand and inability.

Briefly, enhancement embodies a growing number of physical or mental interventions, which may be relevant to the concerns of practicing physicians, the goals of preventive care and the social demand of the public. On this score, enhancement is thereby accepted as kind of 'treatment' although it pertains to demand, which is intentional, not to medical need.

Enhancement may have his side effects. Hair colouring that moved from the hair salon to the home is a risk factor of bladder cancer. Moderate exposure to sunlight contributes to the production of melanin and vitamin D by the body, but excessive exposure to ultraviolet rays has negative health effects, including sunburn and increased risk of melanoma, as well as depressed immune system function and accelerated aging of the skin.

No doubt, medical enhancement though *prima facie* not contrary to the goals of medicine, spreads *in the vicinity of surrounding medical boundaries*. The discussion of the last section, paragraphs shows that enhancement is part of actual medical practice when people feel there is something wrong in their life, physically, emotionally, or socially in the absence of any clinical abnormality. Enhancement hence deals with people's demand about conditions or traits that lie at the edges of medicine.[5]

"*Man is something that is to be surpassed*", writes Nietzsche. Enhancement is a concept that defines improvements to the human organism through biomedical intervention in genetics, neurosciences, pharmacology, and physiology, for conditions, which are beyond the limits of medicine. It refers to various non-therapeutic or perfective procedures ranging from increase or optimization of the functioning of human abilities to cosmetic alterations or to the medicalization of unhappiness.

Enhancement intends to correct or improve some alleged social disadvantage related to some bodily or mental trait. It does not pertain to features from which people are being deprived or dispossessed through biological lack or loss, but rather something nonessential they have been prevented from having but they can dispense with. This has led to a renewed interest in the possible transcendence of human nature, experienced as a challenge and a promise.[6]

"*Du mußt dein Leben ändern*" ("You must change your life") writes Rilke in the *Archaic Torso of Apollo*. But if we are able to change things, we may elect to do so or not to do so. If medicine tries to make the human body less inhuman, medical enhancement tries to make it more human. Those "*to whom the past no longer belongs, and not yet the future*" may choose between what is being given and what should be done. In either case one take responsibility for what is going to happen. Granted, then, that Jonathan Glover[7] has provided the most extensive and the most effective discussion of the ethical and social policy dimensions of enhancement and of transcending human boundaries, I shall forfeit these aspects here.

Enhancement in medicine means either an effort to regain health or to go beyond what is required to gain health. Yet the divide is not always clear. Do we not constantly apply biological knowledge to improve animals and plants? And surely, diseases are natural events; therapeutic medical interventions attempt to correct them, and thus tamper with nature. Treating baldness pertains to enhancement, but taking care of pregnancy, pertains to medicine. Sabin and Daniel argue that it is not necessary and probably fruitless to establish an exact line between legitimate "treatment" and nonmedical "enhancement" by attempting to limit the former to curing diseases and the latter to improve positive health.[8]

Furthermore, the boundary between preventive medicine and enhancement is also rather fuzzy. The classic book of Jonathan Glover on *Choosing Children*[9] approaches the issue of enhancement and places the topic at the border of preventive health care. He states that "… *What we care about is our children's flourishing …*". Correcting body disabilities such as club foot, or congenital heart defects is not essentially different from correcting genomic defects, such as Huntington's disease.

---

[5] Harris (2007) and Elliott (2003).

[6] Savulescu (2007, 2009).

[7] Glover (1984) and Harris (2007).

[8] Sabin and Daniels (1994).

[9] Glover (2007).

And repairing the genes of spina bifida is not so different from treating it with prenatal surgery.

Sime philosophical objections are essentially ideological since they are tainted with the assumption that some normative use of human nature is guiding decisions in medicine: human enhancement with biotechnology and genetic engineering beyond what would be considered as a therapeutic intervention are allegedly alterations of human nature, a vague multivalent concept. But surely, enhancement appears like a contentious issue for many medical scientists because it forays into the heuristics of medicine and allows it to spill beyond the boundaries inside which it has legitimate application: *if it ain't broke, don't fix it.* Gregor Wolbring from the University of Calgary's Cummings School of Alberta observed that the impact of such interventions on health care and ethics is rarely discussed in the literature. He sees risks in seeking too rapidly change patient's cognitive abilities.[10]

Be that as it may, outright rejection of enhancement would meet strong legitimate resistance. Failure to satisfy some human desire, be it a treatment for unhappiness or depression, or some demand for fulfilment or mental tranquillity or for a higher quality of life might appear as a breach of the doctor-patient relationship. Genomics will, no doubt revolutionize medicine: the gene editing of CRISPR is being used to develop treatment for a range of diseases. It could also eradicate some genetic conditions. But it could also be used to perfect *Homo sapiens.*

However, if humans have an essence, it is constantly transforming the world, and consequently reinventing itself. Even then, human beings have the right to control their own bodies and minds, and this is desirable, since humanity has good and bad characteristics.

To begin with, we might try to save the situation and ask why we should call in doubt the drive to improve ourselves if it crosses the blurry line between needs and demand? What are the snares human beings are stepping into when they develop various biomedical technologies for purposes other than treating diseases? The questions raised by enhancement are not whether or not we are trying to play God,[11] but rather that it responds to someone's demand in the absence of need.

N. Daniels claims that "*...it is our norms and values that define what counts as disease, not merely biologically based characteristics of persons.... Pointing to the line between treatment and enhancement is not, then, pointing to a biologically drawn line but it is an indirect way of pointing to the valuations we make*".[12]

Then, beyond this, there is a momentous difference between the meliorist position and the medical one. The need for medical interventions is constitutive of and is *internally* related with what we mean by "disease". But the case of enhancement is very different. Enhancing interventions have the human bodies, or human beings as their object, being *externally* related to them. Nothing ligates medical or biological knowledge or technology with medical enhancement except for people desires

[10] Wolbring et al. (2013).

[11] Taylor (1969).

[12] Daniels (2000, 313).

or intentions. Enhancement uses externally conferred medical technologies and remains outside the medical realm as it responds to demand not to needs.

Meanwhile, for one thing, medical care is thoroughly involved in some of the most basic aspects of the human condition: birth, growth, survival, hope and despair, physical and emotional abilities and their breakdown Physicians are confronted with the disempowerments of illness, suffering, accidents, handicap, old age, anxiety and the agony of nearing death and those distinct tensions which are integral parts of health care. "*A paradoxical and potentially disturbing mix of the sacred and the profane is part of their work*", writes Renée Claire Fox.

For another thing, enhancement ignores the tragedy of transience and is deprived of empathic understanding of disease and of what I might call the medical sacralization of the body stemming from the built-in relationship between disease and needs. The upshot is that, at least in the typical case, being stripped off the existential and soul-searching dimension of medicine, it meets with the bare inhuman nature or impersonal element of our bodies "*that, in relation to certain things, we are pieces of matter, subject to the laws of physics and chemistry* …".[13]

What clearly emerges from this is that the medical stance differs, as the case may be, from the tranquillity and the open-ended self-confidence of technological enhancement, since the latter lacks the seal on inevitability and the *Angst* that characterize medical care.

Sleep is often considered as a waste of time, and it could become an issue pertaining to economic competition for individuals as well as for society. There is a true war on sleep at all levels of society in this world of accelerating change. Jonathan Crary wrote a book about *Late Capitalism and the Ends of Sleep.*[14] Furthermore, the *Defence Advanced Research Projects Agency* of the United States is financing a project aiming at the creation of a super-soldier who has conquered the need for sleep, at least for a few days. Incidentally, such research project could possibly help to treat sleeping disorders.

Is ageing a disease? Ageing is a process of accumulation of changes in the cells and tissues that increase the risk of disease and death.[15] Ageing is not a disease. It is a biological process or a set of conjoined and merging biological processes, which depend on some inborn predisposition, the environment, some morbid conditions, or genetic defects if any. But the prevention of death is a basic medical concern. If so, then, ageing is a contributing and a risk factor of any disease or condition whose incidence, manifestations and recovery are affected by age. Although they are part of the ageing process, loss of muscle bulk, diminishing cardiac and respiratory function, decreased resistance to infection, increased risk of depression and delirium, urinary incontinence, and impairment of memory are considered in medicine to be 'normal' ageing processes, yet remedial procedures are proposed or requisite for them.

---

[13] Tallis (1997, 134).

[14] Crary (2013).

[15] Harman (1996).

The same applies to osteoarthritis, osteoporosis or dementia that are manifestations of ageing and are constitutive of what we mean by ageing; still, we actually call them 'diseases' because of the set of conventions, which constitute medical care. Furthermore, bisphosphonates are slowing bone breakdown, effectively maintain bone density and have been shown to decrease the risk of a fracture.

Carl Elliott has tackled some of the basic existential questions in his absorbing book *Better Then Well.* With this move, a different vision of enhancement offers the possibility of a positive standpoint. Small wonder then, that people should see themselves as having needs that can be fulfilled by drug companies, all the more so since those who manufacture and market those medications often claim that they are merely responding to needs originating in human nature. A case in point is oestrogen replacement therapy: it was at first promoted as a risk-free way to help women to preserve their youthfulness. Then, by blurring the dividing line between enhancement and preventive care and thereby presenting menopause as a deficiency disease, it was marketed as a hormone replacement therapy and was recast as a treatment of postmenopausal osteoporosis. But in 1974, when it was shown to increase risk of endometrial cancer, the manufacturers proposed a combination therapy grouping together oestrogen and a progestin. But risks still outweigh benefits as reflected by an increased incidence of breast cancer, ischemic stroke, coronary artery disease, pulmonary embolism, and dementia.

Descartes wrote that "*... the mind is so intimately dependent upon the condition and relation of the organs of the body, that if any means can ever be found to render men wiser and more ingenious than hitherto, I believe that it is in medicine they must be sought for*".[16] The goal of this work is the creation of *nootropics,* a generic term for chemical compounds that may increase the positive aspects of mental functioning with negligible or no negative side effects. There are a growing number of randomized control trials describing real efficacy in healthy subjects. Caffeine associated with the amino acid L-theanine, which is found in tea, has increased performance and attention tasks.[17] In 2009, the American Academy of Neurology proposed ethical and legal guidance to neurologists consulted by healthy patients asking for medications with the purpose of improving their memory, cognitive focus, or attention span.

Psychotropic drugs are among the big blockbusters. It follows that one of the hottest topics in the promised capabilities of biotechnology is neuroenhancement, that is to say, any pharmacological agent, neuromodular technique or genetic technology used to meliorate mentally healthy individual's mood or cognition, psychostimulants such as Ritalin and Adderall to enhance memory, Provigil for attention or concentration or Aricept used as a treatment of Alzheimer's disease. Psycho stimulants are widely abused by students swotting for exams, artists, or writers (Jean-Paul Sartre) to improve intellectual ability, the way Honoré de Balzac used coffee. A readership poll published in the journal *Nature* indicated that one in five

[16] Descartes (1978, 49).

[17] Dodd et al. (2015).

respondents has used medications to improve memory or concentration. On this score, the question now before us is whether we can improve cognitive skills, levels of cognition, mood states, learning abilities, or memory? To put it in another way, can we feel better than well?[18]

What are the benefits, risks, and utility of the new technologies? What are the unexpected collateral effects, considering the available evidence those mood stabilizers threaten the health and lives of those taking them? Were such cosmetic drugs proved to be efficacious, what would be the risks of addiction or of abuse, of suffering from side effects like those of Prozac, i.e., withdrawal syndrome, constant agitation, violent behaviour, horrible nightmares, anxiety, dizziness, nausea, dependency as well as a growing tolerance and the consequent risk of higher dangerous doses? Could cognitive enhancement lead to changes in the direction of antisocial behaviour? Memory is a delicate and selective process: might a memory drug improve our retention of all memories, including traumatic or trivial memories that the brain tends to repress?[19] Are we going to bring some greater good to someone, at the cost of some damaging health effects? And even if we bow to this suggestion, would a cost-benefit analysis such as those used in medical care be legitimate? What level of acceptable harm, if any, will pay for what level of betterment? All in all, the word "betterment" is treacherous. If the goodness of the goal should be ethically legitimate, either the goals or its goodness could be illusory.

"*Science,* writes René Dubos,[20] *provides methods of control for the problems inherited from the past generations, but it cannot prepare solutions for the specific problems of tomorrow because it does not know what these problems will be. Physicians, like soldiers, are always equipped to fight the last war. What may be worth asking is whether medical science can help the individual and society to develop a greater ability to successfully meet the unpredictable problems of tomorrow.*"

Just like Faust promising never to pause striving for more progress and perfection, are we striking some sort of Faustian bargain in exchange for nothing with this biologically driven human betterment? In a situation of scarce health-care resources, what might be the larger social effects of embracing those technologies too enthusiastically? Are we going to foster individual well-being at the expense of some larger medical or health care good? Is there something deeply wrong in the fashionable promotion of some faultless biotechnological enhancement, a strategy that is tainted with legitimate suspicion? This is not a matter of denial but of moderate scepticism. Besides, and contrary to medical treatments, enhancement cannot be forfeited through evidential overthrow because there is no evidence yet. Nevertheless, what undergirds the promise of human enhancement is the utopian longing, with the underlying age-old question: is man perfectible?[21] And surely, it is very hard to get

---

[18] Maher (2008).

[19] Whitehouse et al. (2004, 264).

[20] Dubos (1961, 85).

[21] Passmore (1971).

rid of the feeling that man has the ability to become much superior to what he now is. Utopias are actions, kind of pipe dreams, which look promising, as well as tacit injunctions to be fulfilled in some desirable future. History suggests that in case of conflict between new technological developments and ethical principles, the new technologies have a strong tendency to win. In saying this, we must recognize that, despite their potential dangers, biomedical revolutions are inevitable.

"*Become who you are*" wrote Nietzsche.

Even so, to quote Dryden in his drama *The Indian Emperor:*

MONTEZUMA …*To mankind one equal way to bliss is not designed for though some more may know, and some know less, yet all must know enough for happiness.*

CHRISTIAN PRIEST *If in this middle way you still pretend to stay, your journey never will have end.*

MONTEZUMA *Howe'ver, 'tis better in the midst to stay, than wander farther in uncertain way.*

Could it be that the enlighteners were right in the view that human beings must remain human beings, a worried, emotional, passionate, and dissatisfied being, which is at his best when he renounces the quest for some fantasy world of utopian perfection?

Summary

The further away we move from the scientific and prudential core of medical thinking, the closer we get to the confines of medicine. Yet, medicine needs boundaries, which must rely on the least disputable grounds that is to say on cases in which an individual becomes a casualty of some condition of his own body or mind. These boundaries may be epistemic, ethical, or ontological, and going further beyond these limits is a matter of medicalization or medical enhancement.

Just as the limits of medicine become nearer, its objectives seem to be adulterated or equivocal at the edges, by the medicalization of society and the foray of legal, political, existential, or new ethical concerns that intrude from outside into the central core of medical thinking. Clinical knowledge is knowledge by participation much more than could ever be the case in pure biological science or in medical research. Clinical understanding is thus more likely to be influenced by the physician's interest and emotions than is the case in medical or biological research.

If we surrender our *first move* and our disease concept, it becomes difficult to articulate some explorative theories because the very idiosyncratic propositions and ordering concepts of medicine cannot be allowed to spill beyond the boundaries within which they have been defined and have found their meaning. Beyond these boundaries what is being treated is no more a disease and the treatment is no more a treatment

Chapter 17
Tragedy

Contents

It is often said that medicine is an art, a view vigorously rejected by Claude Bernard.[1] But a scientific-naturalistic view must remain forever incapable of describing the ends to which we aspire and to identify what counts as a life worth living, since medical care is confronted with situations that are a threat to a meaningful life, inasmuch as diseases and suffering are cries in the night, and death or the approaching death remove all meaning from life. "*Medicine,* writes Renée Claire Fox, *deals with life and death and suffering, the human body and the human psyche, in physical, emotional, and symbolic ways that are inherently disturbing.*"

Contrasting Two Types of Illness-Narratives

There is, as it were, a radical distinction between apprehending natural phenomena subject to the ordinary approach of the natural sciences, and the understanding of human beings' states of mind such as purposes, motives, or desires. There are two sorts of illness narratives, the medical record and the patient's story.

[1] Bernard C, 1920, 4323–325.

Sunt lachrimae rerum et mentem mortalia tangunt. Virgil

L. Karhausen, *Analytical Philosophy of Medicine*, Philosophy and Medicine 152,
https://doi.org/10.1007/978-3-031-81257-6_17

For one thing, the 'detached' medical perspective requires a scientific replicable mould. The physician thinks in terms of diagnosis, prognosis, and management. He has spent years studying in depth the ins and out of biology and biochemistry. He takes his patient's sickness and complaints as objective phenomena inasmuch as they exist independently and are publicly observable.[2]

Even then, if the patient bestows narrative status upon his lived illness experience, he apprehends his condition from his subjective standpoint, from within: his situated' narrative is part of what Jonathan Glover calls "our inner story". "*We think of it as the truth from which other stories may deviate a bit.*"[3] It is embedded in feelings, emotions and activity experienced within a nexus of interpersonal feelings that cannot be captured by the medical narration. So, if the patient's subjective account is no piece of the medical and scientific explanatory enterprise, it still has an authority that clinical medicine cannot dismiss.[4]

The patient is no spectator, but he is involved in his narrative as agent and patient. Whenever he talks about it, he is living through his illness. Small wonder, then, that his illness narrative describes the objective features (the symptoms and characteristics) of his sickness, at the same time as it tacitly brings forward the mode of his subjective commitment to the claims of life. Hence this two-dimensional personal account is constitutive of his illness experience. In so far as he is in it, the patient is thus *internally* related to his illness experience.

But only its descriptive features are thoroughly passed on by the patient and incorporated in the medical record. The declining sense of direction and the dwindling meaningfulness of the patient's life are only incompletely translated into medical language. And this trimmed account of the patient's narrative embodied in the medical record is externally related to his illness process. The psychological meaning (or whatever is left of it) that was internal to the patient's narrative must now be given an external, referential, and medical meaning in the doctor's narrative. Whether the patient tells the doctor, he is suffering from pain in the elbow or butterflies in the stomach, the story the patient relates is occurring within quotation marks in the doctor's narrative. The patient's narrative is nonetheless not something that can be detached (by the 'observer status') from the course of his illness but it is embedded in its process instead, so that the diagnostic interpretation in search for objective evidence must here get on with a delicate process of disquotation.

The physician himself in his attempt to elicit an anamnesis sorts out the patient's complaints and plots, reconstructs, and analyses the symptoms into a more or less integrated story line. There may actually be many facts in the patient's story but from amongst all of these only a certain few are being selected for the medical history, sometimes mistakenly missing out or omitting some important details. Completely describing a sickness event is something, which cannot be done. The imposition of a medical narrative organisation should be minimally involved with

[2] Cassell EJ, 2004, 156–163.

[3] Glover J, 1988, 139.

[4] Baron RJ, 1965.

subjective considerations beyond relevant symptoms. The patient's story, namely the narrative he furnishes, as well as the replotted and reshaped account provided by the physician are thus the object of cognitive asymmetry. The unbridgeable gap between medical science and the patient's experience, between description and injunction, between the ethics of truthful discourse and the moral core of healing, are being buttressed by the incommensurability of the two narratives which requires a cognitive address of a quite distinct order.

We thus have two kinds of descriptive narratives, an internal first-personal and an external objective standpoint. Both the patient and the physician are seeking some sort of narrative unity. The patient's narrative is not a mere staccato of autonomous events but rather a legato of connected time-separated events, so that his way of phrasing them might not be medically meaningful. The patient tries to broach the history of his ailment and accommodate the loss of his quality of living to the inner story and conflicting layers of his personal experiences. The doctor, for his part, does not explain the disease as such but rather under some intended description. The patient's sickness needs to be described and fitted into tenseless sentences in order to be intelligible and to lead to the conclusion of some diagnostic argument.

On this score, the weekly clinical-pathological conferences of the Massachusetts General Hospital published in *The New England Journal of Medicine* exhibit a retrospective narrativization and a completed account of a patient's medical history from the initial signs or symptoms to his death. The case presentation lacks a first-person tense and is innocent of the inner perspective. The audience of the clinical-pathological conference does not know, any more than the patient or his GP did, how the sickness episode is going to come out. But the audience knows that it will soon know, namely by the end of the conference. The case presentation proceeds through successive stylized steps in order to provide some reflective diagnostic expectation. The narration achieves a thorough description as the available evidence is disclosed to the public so as to complete the story from beginning to end.

What then differentiates the perspective afforded by the clinical-pathological conference from the properly clinical one (or from the patient's account), is that the first one is true but the second one is still unsealed. The patient and to a lesser extent the doctor lack pieces of knowledge as long as the disease has not come to an end which might have been cure, chronicity or death. A firm diagnosis, explanation and reconstruction of the sickness can only be fashioned retrospectively after the condition has run its full course. *Hic est locus ubi mors gaudet succurrere vitae.*[5] Neither the patient nor the doctor could spell out the content of the future clinical-pathological conference because earlier events could not have been described and explained in terms of yet unknown facts that would later be made available.

Ignoring how things are going to end is the mark of an incomplete, open-ended narration. *Prognostic* knowledge may be available beforehand, albeit medical care might improve the course of the sickness or even heal a disease initially diagnosed as incurable. But if *predictive* knowledge was available, little could have been done.

[5] Here is where death rejoices to come to the aid of life.

It would have been on the cards for quite a long time. And this would abolish the causal and logical ligatures between present and future, between the illness's course and medical interventions that warrant any sort of narration. If true, the integral sum of a patient's suffering could never be less than what it has been. As it happens, the clinical-pathological conference is like 'fate' as it can only be conjugated in the past tense. "*Fate, in a tragedy,* writes Northrop Frye, *normally becomes external to the hero only after the tragic process has been set going.*"[6]

Aristotle on the Tragic[7]

Medicine is tragic, physics is not. "*The awareness of death and of the tragic aspect of life,* writes Erich Fromm, *whether dim or clear, is one of the basic characteristics of man.*"[8] The Greeks from tragedy derived the word tragic, although it stems from the Greek word for goat (τραγος). The tragic, according to Aristotle, is made up of παθος (some destructive or painful event) and περιπετεια (reversal). The tragic παθος involves bad fortune to a person: it is "*an event of a destructive or painful nature*"[9] such as "*death in its various forms, bodily injuries and afflictions, old age, diseases, starvation.*"[10] It evokes fear and pity, the emotional move towards an object or away from it. And "*the people for whom we feel pity are those we know, if only they are not very closely related to us, since in that case we feel about them as if we were in peril ourselves.*"[11] Peripeteia is the means by which the tragic change takes place. It introduces the role of unexpected suffering, of "fearful and pitiable things". "*Such things come to be most of all when they occur unexpectedly and at the same time when they follow as cause and effect*",[12] that is to say, "*according to the principle of probability and necessity.*"[13]

A tragic situation thus pertains to some "destructive and painful events" which arouse "pity and fear" because they occur unexpected and bring about one another. The movement of the story line between good and bad fortune is an inexorable sequence of events in which contingency and necessity take a leading if not devastating part. Tragedy arises from inevitability. It reflects the process by which some disease moves forward inexorably with its own nature as well as the fact that, by the time it is clear what is going on, it is too late, if it ever was, to do anything about it.

---

[6] Frye N, 1957, 210.

[7] Belfiore ES, 1992.

[8] Fromm E, 1960, 211.

[9] Aristotle. Poetics 11.1452b. 11–13.

[10] Aristotle. Rhetoric. 2.8 1386a. 6–9.

[11] Aristotle. Rhetoric. 2.8 1386a 18–19.

[12] Aristotle. Poetics 9.1452a 1–4.

[13] Aristotle. Poetics 9.1451a 36–38.

Spinoza differentiated sad from tragic: 'sad' designates any loss of values while 'tragic' designates any loss of values which is not necessary, which could have been avoided and which occurs, nevertheless. "*It is tragic what could have been avoided*", writes Alfred Stern.[14]

The clinical condition is that of a random inarticulate victim, isolated by his sufferings from the social group to which he attempts to belong, victim of a process that does not depend on his moral status. Tragic suffering brings home the perception that one's life is meaningless, and that the world appears to be empty. Medicine's quest then seeks to make life possible and tolerable and to dodge the unreturningness that lies in the nature of things. Tragedy celebrates suffering but medical care attempts to solve it. The victim of tragedy is morally innocent as the tragedy comes from without. Conditions, which are easily healed, are usually not tragic but those that are difficult to heal are more tragic than those that are less so. Death seen as a deprivation of a good thing, namely life, is quite irreversible and is thus preeminently tragic in this respect.[15]

Medicine brings out the tragic dimension of the human predicament. What characterizes sentient beings is that each of us, humans or dogs or swallows or cheetahs, is unique and irreplaceable. No other being, no one else is going to stand in for us when we die. That life is a unique experience poignantly brings home the transient nature of things. This once-for-all quality echoes Rilke's answer to the question:

> "*Why have to be human?*"
> *Once everything, only once. Once and no more. And we, too, once. Never again. But having been this once, even though only once; having been on earth does not seem revocable.*

By contrast, Nietzsche's eternal recurrence is the idea that whatever has happened before will happen again; he was seeking, *nolens volens*, to dodge the unreturningness that lies in the nature of things.

The Value of Life

"*When the prospect of dying is near at hand*, writes Plato, *a man begins to feel some alarm about things that never troubled him before.*" The experience of disease and suffering herewith conveys existential anguish since it seems to capture the intuition as to why life is worth living. According to Karl Jaspers, we come to know ourselves most vividly in what he called limiting situations such as diseases or mortal threats.

Science has forcefully taught us that 'why questions' are unanswerable and perhaps senseless, "Why does the heart beats?" "Why shall I die?", "Why am I sick?" and so forth. "How or through what process does the heart beats?" is the proper question. But then, what sense are we to make of all diseases, of suffering, sorrow,

---

[14] Stern A, 1971, 94.

[15] Gowans CW, 1994.

and death? "*Why should we live? If life has no purpose*, asks Tolstoy, *if it has been given to us for its own sake, we have no reason for living.*"[16]

So, is a meaningful life, perhaps, a function of the degree to which a person's major purposes are being at least reasonably and successfully pursued? Answers Spinoza: "*Each individual thing in so far as it is in itself, endeavours to persist in its own being. The effort wherewith a thing endeavours to persist in its own being is nothing else than the actual essence of that thing.*"[17] Each thing (and each human being), Spinoza insists, is internally determined by a power wherewith it strives to remain in existence. In so far as it is itself, it will act in such a way as to persevere in existing. A person who commits suicide is a case in point since, Spinoza avers, he is not himself but is driven to do what he does by external factors. So that, at least in the typical case, a patient strives to live as long as possible and medical care tends to persuade him by word and deed that life will continue. Not surprisingly, medicine is predicated on the prima facie assumption that death is pathological and must be fought and postponed as long as possible.

It might well be that the question of what constitutes a meaningful life might not be intelligible in so far as the issue tends to fade away as soon as it becomes an object of inquiry. Jonathan Glover has introduced the "*concept of a life worth living*", and although he is not prepared to say, "*what sorts of things do make life worth living*"; he believes that "*being conscious is intrinsically valuable*". He does not claim that life is always or everlastingly good but that particular lives are worthwhile living. He adds that he has no argument to convince those who disagree with this matter.[18] Likewise, Hobbes' axiom states that each man is afraid of death and that it makes no sense to ask why. We may thus hold that a sentient being's life is valuable not merely because no individual is quite like another but because being alive and being conscious is intrinsically valuable.

Against this background, a patient faced with the disarray of his predicament, wonders whether his tragic condition has meaning or whether it is tragic because it is meaningless.

The Induced Disconnection of Life Events

A human life flows from birth to death. It is experienced as a single span and as a schema of personal identity. Seen from outside, a person's life is a loose aggregation of parts and pieces of the world encountered by chance as well as a finite series of things, people, and events: it looks like a mere *staccato* of separate and disparate moments. But from inside, we value our ability to construct a life plan and to strive to act in accordance with it. The ascription of biographical significance ties together

[16] Tolstoy L, The Kreutzer Sonata.

[17] Spinoza B, 1982, part 3, propositions VII & VIII.

[18] Glover J, 1977, 51–53.

or rather phrases the person's actions and experiences in meaningful sequences and "projects of self-creation".[19] It relates past and future to present into several narratives which are then woven into an uninterrupted nearly necessitated flow which must be appreciated in the light of some inner occurrences, such as purposes, motives, intentions, emotions and desires, achievements, and failures.

Be that as it may, people begin to value life when the valuing that is placed on it appears to be draining away. There is no room for harping on the meaning of one's existence in ordinary discourse until one's life runs the risk of sounding unstructured and worthless: "*The question of the meaning of life arises only at those times when such a meaning had become doubtful.*"[20] It is a default notion. Physical and mental suffering and diseases hence tend to disestablish those meaningful connections from the line path of a person's life whilst taking its toll on the point, purpose, or end of human existence. Raymond Tallis writes: "*I have a prejudice against pain, believing that, once it has done its job of warning us of danger, it is meaningless; or worse than meaningless, since it drains meaning from everything else and commands rapt and exclusive attention.*"

The vicissitudes of an individual's sickness and the increased attention focused on his illness erase those slurs that indicate the phrasing of his life. The disease-experience awareness results in loss of wilfulness and autonomy, threatened or disrupted personal identity, and depressive preoccupations with painful or harmful bodily processes.

A meaningless or a pointless life is a life into which one sees nothing outside the sheer pursuit of life, a life that seems devoid of any non-instrumental interests in the world considered as lasting longer than one's own life course. The medical sociologist, Renée Claire Fox writes: "*I have learned, too, that in the face of such ultimate experiences, Job-like questions are wrenched out of most people, and that it is the quality of their Angst, rather than of their education, that determines how lucidly and lyrically these questions are answered.*"[21]

The patient's attitude toward his disease is influenced by hope as much as by fear. He hopes his miseries will end and that eventually life will be worth living again. He fears the prospect of being slowly tortured to death or of becoming increasingly helpless and a burden to others. He then finds himself striving, with more or less success, to reappropriate some of the lost connectiveness, which was canvassing the pattern of his past experiences. For some patients, the why of the problem of suffering begs for a purpose.

Whatever our religious or non-religious beliefs, and even though some people claim to have a concept of a life after death, most of us—when the scope of our life is being narrowed by illness—pretend that suffering of any kind and death are not part of reality as we get round them and escape into distractions of various kinds:

---

[19] Glover J, 1988, 131.

[20] Reininger R, 1947, 65.

[21] Fox RC, 1979a, b, 11.

while we try, at that point, to up value and extend our lives, disease represents the darker side of life.

In the interface of the caring relationship and the existential *Angst* that pervades it, patients and physicians ineffably share a preoccupation with dauntlessness and dread, hope, and despair. That is to say that they search a compromise between an ontological and an attitudinal account, between a complete account and an open-ended one and between the detachment of the view from nowhere and the fear of the unexpected. But reconciliation between the physician's and the patient's narrative, that is, between the needs of the third-person account and the demands of the first-person account may sometimes not be possible anymore. This irreconcilable divorce between the two narratives then opens a gap which may bring about a "*conspicuous discrepancy between pretension or aspiration and reality*"[22] resulting in a feeling of loneliness and isolation.

Human and Non-human Sentient Beings

The question before us now is raised by the Ecclesiastes: "*He that increaseth knowledge increaseth sorrow*." At least in human beings, that of suffering matches the degree of awareness. To be sure, animals can suffer in much of the same ways as human beings and their suffering bears no less moral weight than human suffering. Sentient beings such as a dog know and remember who someone is by taking account of his facial appearance, height, speech, and odour. But they are presumably dimly aware of their own identity through time, just like infants. In a contrary manner, human beings are able to compare their consciousness of their self at any two moments.

"*Being a person may sometimes be a matter of degrees*", writes Jonathan Glover but he adds that human beings seem to differ from other sentient beings in that they tend to form pictures and to be aware of the sort of persons they want to be.[23] A person wrote Leibniz, conserves *"the consciousness, or the reflective inward feeling of what it is"*. Able-bodied human beings have the ability to contemplate themselves from outside and hence to dissociate their personal perspective from the agent-neutral clinical view. Schopenhauer observed that the modality of awareness renders human chronic suffering more cumulative than that of animals, since sorrows, he suggests, might not accumulate in like manner in higher animals (or in infants) as they do in adult humans through the weaker role played by anticipation and memory in their conscious awareness and cognitive architecture. Suffering is personally bad as it refers to how individuals feel, but it is also objectively bad because it is intrinsically bad for the individual who experiences it.

[22] Nagel T, 1979a, b.

[23] Glover J, 1988, 61, 131.

Non-observable knowledge, the sort of knowledge we have of our bodies is such that to be suffering or in distress and to present symptoms implies to be aware of them. But it seems plausible to assert that human beings are likely to be abler to consider their suffering from outside or from a neutral standpoint than other sentient beings, which does not mean that the latter cannot have a meaningful life. It may well be that, contrary to other cognitive beings, human beings when aware of being afflicted by a chronic ailment, are actually conscious of the fact that they are conscious of it, or at least more so than non-human beings. Casting one's own story as a tragedy is a way of making sense of it; and we tend to understand our own live as if it was acting out a meaningful plot. This feeling is specifically human: a dog may be deeply unhappy, just like a human being, but he is probably unable to envision his life as a connected narrative.

The Physician's Predicament

Physicians have to meet various conflicting perspectives, which may be of medical or of ethical nature. They have to deal with the problem of uncertainty. It is always difficult for a clinician to tell between his personal ignorance and the limitation of available knowledge. Does this patient suffer from psoriasis? Or is it mycosis fungoides? The clinical picture might be the same and the pathological one ambivalent, but they differ dramatically in prognostic terms.

For one thing, besides learning to cope with uncertainty, the physician may every so often be confronted with deaths and suffering that one views with horror, or with suffering and illness. In Susan Sontag's view: "*Illness is the night-side of life, a more onerous citizenship. Everyone who is born holds dual citizenship, in the kingdom of the well and in the kingdom of the sick. Although we all prefer to use only the good passport, sooner or later each of us is obliged, at least for a spell, to identify ourselves as citizens of that other place.*"[24] Diseases such as tuberculosis or AIDS or cancer or Huntington's disease may then be lived through as metaphors, as mysterious and capricious invasions, and as something to hide.

But for another, what about the physician himself, the medical student, or the nurse? Arthur Kleinman writes "*we know more about the patient than the healer … What the doctor feels is most at stake—what is most relevant to practise—slips through our crude analytical grids.*"[25]

How do physicians react to their patients' illness and the tragic outcome that awaits many of them? And to the inefficacy or unwanted side effects of their interventions? How do they face issues of suffering, uncertainty, life and death, triage and decisions involving other people's life as well as the negative consequences of the state of the art?

[24] Sontag S, 1983, 7.

[25] Kleinman AZ, 1988, 210.

"*Medical work* writes RC. Fox *is morally and existentially serious*." Physicians have a double allegiance, to the constant inrush of science as well as to clinical empathic understanding. Clinical care, then, oscillates between the sacred and the profane. The pendulum is constantly swinging between the two somewhat incommensurable sides of the boundary which separates what Hegel called nature in contrast to spirit.

Admittedly, a physician knows more about breast cancer than his patient does and what constitutes knowledge will be embodied in the medical record. But would it seem abusive to claim that a physician although well versed in the science of medicine, cannot account for his patient's condition until he has experienced it in his own person? If he understands perfectly well that his patient suffers from an incurable breast cancer, does he understand what it means to her? Does he have the special feelings, emotions or attitudes that are aroused in her by having undergone certain experiences or participated in certain kinds of events and being faced with certain sort of anticipation? And all in all, what is the sense of this question?

Rita Charon, a physician, here writes "*we get emotionally and existentially wounded by patients who arouse anxieties or who, through transference and counter transference, force us to reexperience painful episodes in our own lives or to face unresolved ongoing conflicts*." "*Much of medicine,* she concludes, *if practised effectively, strikes home.*"[26] In 1643 Thomas Brown wrote: "*Let me be sick myself, if sometimes the malady of my patient be not a disease unto me; I desire rather to cure his infirmities than my own necessities: where I do him no good, methinks it is scarce honest gain; though I confess 'tis but the worthy salary of our well-intended endeavours.*"[27] Further, Renée Claire Fox writes that physicians' anguish comes "*from the degree to which they still espouse values but can't live up to them*".

The loss of a person in the family or the death of a friend is a major event in whoever's life. But when a physician turns professional and comes face to face with his emotions, it may become pretty tough for him to accept the loss of a patient or his suffering beyond the fact that death represents a failure: he has to get rid of some of his feelings for his life to remain tolerable. He feels obliged, every so often, to block out the fact that his patient is a living soul that resembles his own and has internal sensations, and to see him as a mere embodied expectation of consciousness describable in the public language of science with reference to some set of observations. When confronting terminally ill patients, he is as much at a loss as any one of us and struggles to find some midway between inappropriate attachment and cold indifference.[28]

So, does the style of internal medicine come sometimes in for criticism: it is claimed that it lacks a human dimension, that it is inescapably depersonalized, that it strips medical care of its humanity and that it ends up searching knowledge for its

[26] Charon R, 1993, 158.

[27] Browne T, 1845, 185–186.

[28] Chen PW, 2006.

own sake rather than for the patient's benefit.[29] *"The physician in our culture, then, is often a man with a carapace"* writes Alex Comfort. *"The carapace serves most of us well, unless we are disturbed or very thin-skinned, and it has served medicine for many years."*[30] A clinician needs what Renée Claire Fox calls *'detached concern'*, a kind of emotional and moral distancing, namely the ability to combine the counter attitudes of detachment and concern to attain the balance between objectivity and empathy expected of mature physicians. Doctors are once removed from their patients. In their pursuit of objectivity, they attempt to maintain *'the view from nowhere'*, namely an outer perspective in which "*nothing seems to have value of the kind it appears from inside*".

It is in the context of the autopsy that medical students are first called upon to meet death, *"and to carry out the dissection of a human body with the relatively impersonal attitude of a scientist"*. The following comment made by a medical student and quoted by Fox, seems impersonal and stylized: "*We pretty much talk about the autopsy the way we would if you were presenting a case.*"[31] Pathology and necropsy, life, and death, are both part of clinical medicine. Doctors, as it were, are like the angels[32] mentioned by Rilke, who do not distinguish the dead from the livings:

> *"Angels, it is said, are often unable to tell*
> *Whether they move among the living or dead"*

An embarrassing gap thus opens between physician and patient precisely when the latter feels estranged, and is gaining a sense of distance, and experiences a gulf between himself and his own life. It follows that the physician finds himself officiating on one side of the boundary, within the epistemic frontiers and the resultant inherently restricted scope of his discipline, not to mention the uncertainty and the finite capability of correcting disease processes. And what makes a disease tragic is that their involvement is perceived and understood by physicians as a loss: "*Tragedy is in the eye of the observer, and not in the heart of the sufferer*" writes Emerson.[33]

Bertrand Russell concluded: "*The doctor occupies a seat in the front row of the stalls of the human drama, and is constantly watching, and even intervening, in the tragedies, comedies and tragicomedies which form the raw material of the literary art.*"[34]

---

[29] Shorter E, 1985, 194–201.

[30] Comfort A, 1980, 12.

[31] Fox RC, 1979a, b, 56, 66.

[32] Rilke RM, 1929. Rilke's angels have nothing in common with Christian angels. They are poetic images and belong to a realm to which we can aspire.

[33] Emerson RW, 1946, 219.

[34] Russell W, 1968, 413.

Summary

1. Medicine deals with death and suffering in physical, emotional, and symbolic ways that are inherently disturbing.
2. Terre are two types of Illness-Narratives, an internal first-personal, the patient's subjective account, and an external objective standpoint, the medical perspective.
3. In the case of disease, the first type of narrative is influenced by hope and fear and existential Angst; but in the external standpoint, people's life is experienced from birth to death as a loose aggregation of things, people and events which look like a *staccato* of separate and diverse moments.
4. Non-Human-Sentient Beings can suffer in much of the same way as human beings and their suffering bears no less moral weight than human suffering. But they are presumably dimly aware of their own identity through time.
5. An embarrassing gap opens between physician and patient. It follows that the physician finds himself officiating on one side of the boundary, even though what makes disease tragic is that their involvement is perceived and understood by physicians as a loss.

Correction to: Analytical Philosophy of Medicine

Correction to: L. Karhausen, *Analytical Philosophy of Medicine*, Philosophy and Medicine 152, https://doi.org/10.1007/978-3-031-81257-6

The book was inadvertently published with incorrect volume number "151," which should be "152". The volume number has now been updated in this book.

The updated version of the book can be found at
https://doi.org/10.1007/978-3-031-81257-6

L. Karhausen, *Analytical Philosophy of Medicine*, Philosophy and Medicine 152, https://doi.org/10.1007/978-3-031-81257-6_18

Bibliography

Aaron LA. and Buchwald D. A review of the evidence for overlap among unexplained clinical conditions. *Annals of Internal Medicine*. 2001; **134**: 868–880

Achinstein P. The Concept of Evidence. Oxford University Press. 1983

Agich GJ. Autonomy and long-term care. Oxford. Oxford University Press. 1993

Agich GJ. Evaluative judgement and personality disorder, in: Philosophical Perspectives on Psychiatric Diagnostic Classification, ***JZ. Sadler, OP. Wiggins and MA. Schwartz*** (eds.) Baltimore. The Johns Hopkins University Press. 1994

Agich GJ. Toward pragmatic theory of disease, in: ***JM. Humber and RF. Almeder,*** (eds.) What is a Disease? Totawa. New Jersey. Humana Press. 1997

Aggleton P. Health. Psychology Press. 1990

Aggleton P. Health. London. Routledge. 1991

Akil H., Brenner S., Kandel E. et al. The future of psychiatric research: genomes and neural circuits. *Science*. 2010; **327**: 1589–1581

Albert DA, Munson R and Resnik M. Reasoning in Medicine. An Introduction to Clinical Inference. Baltimore. The Johns Hopkins University Press. 1988

Alexander F. Psychosomatic Medicine. New York. Norton. 1950

Alexandrovna A. Well-Being, pp. 9–30, in: ***N. Cartwright & E. Montuschi*** (eds.), Philosophy of Social Science. Oxford. Oxford University Press. 2014

Allan FN. and Kaufman M. Nervous factors in general practice. *JAMA*. 1948; **138**: 1135–1138

Ali A., Katz DL. Disease Prevention and Health Promotion. *American Journal of Preventive Medicine*. Supplement, **49**: S230-S240, 2015

Al-Lamee R., Thompson D., Dheby H-M., et al. Percutaneous coronary intervention instable angina (ORBITA): a double-blind, randomised controlled trial. *Lancet*. 2018; **391**: 31–40

Allison W. Radiation and Reason. London. YPD Books. 2009

Allison DB., Downey M., Atkinson RL., et al. Obesity as a disease: a white paper on evidence and arguments. *Obesity*. 2008; **16:** 1161–1177

Alloy LB. and Abramson LY. Depressive realism: four theoretical perspectives, in: ***Alloy LB*** (ed.) Cognitive Processes in Depression. New York. Guilford. 1988

Althubaiti A. Information bas in health research: definition, pitfalls, and adjustment methods. *Journal of Multidisciplinary Healthcare*. 2016; 211–217

Alvergne A., Jenkinson C., Faurie C. (eds.) Evolutionary Thinking in Medicine. Springer. 2016

American Diabetes Association. **Diagnosis and Classification of Diabetes.** *Diabetes Care*. 2013; **36:** 67–74

American Association Task Force on Clinical Practice Guidelines. *Journal of the American College of Cardiology*. 2018; 71: e127–e248

L. Karhausen, *Analytical Philosophy of Medicine*, Philosophy and Medicine 152, https://doi.org/10.1007/978-3-031-81257-6

American Medical Association. AMA Adopts New Policies on Second Voting at Annual Meeting. 2013 [Internet, cited 2014 April 7]

American Psychiatric Association. Diagnostic and Statistical Manual of Mental Disorders. DSM III. Third Edition, Revised. Washington, D.C. American Psychiatric Association. 1987

American Psychiatric Association. Diagnostic and Statistical Manual of Mental Disorders. DSM IV. Fourth Edition. International Version. Washington, D.C. American Psychiatric Association. 1995

American Psychiatric Association. Diagnostic and Statistical Manual of Mental Disorders. DSM-5. Fifth Edition. International Version. Washington, D.C. American Psychiatric Association. 2013

Ananthaswamy A. When cystitis might be good news. *New Scientist*, 16 November, 2002

Anderson DM., Keith J., Novak PD. and Elliott MA. Dorland's Illustrated Medical Dictionary. Philadelphia. Saunders. 1985

Andreasen NC. Creativity and mental illness; prevalence rates in writers and their first-degree relatives. *American Journal of Psychiatry*. 1987; **144:**1288–1292.

Angell M. Disease as a reflection of the psyche. *The New England Journal of Medicine*. 1985a; **312:** 1570–1572

Angell M. Letter. *The New England Journal of Medicine*. 1985b. **313**: 1358–1359

***Angell, M* 2000,** "Is academic medicine for sale?", *New England Journal of Medicine,* 2000. **342:.1516–1518.**

Annals of the New York Academy of Sciences. ***Volume 1036.*** Youth Violence. Specific Approach to Prevention. The New York Academy of Sciences. 2004

Anscombe GEM. The intentionality of sensations: a grammatical feature, in: ***Butler RJ.* (ed.)** Analytical Philosophy. Second series. Oxford. Blackwell. 1965

Annual Report to the Nation on the Status of Cancer (1973 through 1998), Featuring Cancers with Recent Increasing Trends. Washington. US Government Printing Office. 2001

Anthony JC., Folstein M., Romanoski AJ., et al. **Comparison of the lay Diagnostic Interview Schedule and a standardized psychiatric diagnosis**. Experience in Eastern Baltimore. *American General Psychiatry*. 1985; **42:** 667–675

Aquinas T. Summa Theologica

Ariew A., Cummins R. and Perlman M. (ed.). Functions. New essays in the philosophy of psychology and biology. Oxford. Oxford University Press. 2002

Aristotle. Categories. Oxford. JL. Ackrill (ed.). Oxford. Clarendon Press. 1969

Aristotle. The Ethics of Aristotle. The Nichomachean Ethics, *JAK. Thomson, H. Tredennick and J. Barnes* (eds.) London. Penguin Books. 1976

Aristotle. Metaphysics Books G, D, E. Oxford. Clarendon Press. 1971

Aristotle. Physics I, II. *W. Charlton* (ed.). Oxford. Oxford University Press. 1970

Aristotle. The Politics of Aristotle. *E. Barker* (ed.). Oxford. Oxford University Press. 1952

Aristotle. De motu animalium. Princeton University Press. 1986

Armitage CJ et al. An exploratory randomized trial of a simple, brief psychological intervention to reduce subsequent suicidal ideation and behaviour in patients admitted to hospital for self-harm. *British Journal of Psychology*; **208**: 470–476, 2016

Armstrong D. Commentary: The discovery of hidden morbidity. *International Journal of Epidemiology*. 2013; **42:** 1617–1619

Arnold M. Culture and Anarchy: With Friendship's Garland and Some Literary Essays. Ann Arbor. University of Michigan Press. 1960

Aronowitz RA. From myalgic encephalitis to yuppie flu: a history of chronic fatigue syndrome, 155–181, in: ***CE. Rosenberg and J. Golden,* (eds.).** Framing Disease. New Brunswick. Rutgers University Press. 1992

Aronowitz RA. Making Sense of Illness. Science, Society and Disease. Cambridge. Cambridge University Press. 1998

Aronowitz RA. When do symptoms become a disease? *Annals of Internal Medicine*. 2001; **134:** 803–808

Aucouturier V, Demazeux S. The concept of mental disorder, in: ***Carel H.*** (ed.), Cooper R. Health, Illness and Disease. Acumen Publishing. 2013

Austin JL. Truth. *Proceedings of the Aristotelian Society*. 1950; **24:** 111–128

Austin JL. Philosophical Papers. London. MacMillan. 1961

Austin JL. How to Do Things with Words. Oxford. Clarendon Press. 1962

Austin JL. A plea for excuses. Presidential dress. *Proceedings of the Aristotelian Society. New Series*. **57**: 1–30, 1956–1967, p. 6

Austin JL. Philosophical Papers. Clarendon Press. 1970

Baines J. Hermetic Philosophical Morals for the 21st Century. Baines Institute Incorporated. 1991

Baines DL. and Parry DJ. Analysis of the ability of the new needs adjustment formula to improve the setting of weighted capitation prescribing budgets in English general practice. *British Medical Journal*. 2000**; 320:** 288–290

Balint M. The Doctor, His Patient and the Illness. London. Pitman Medical. 1971

Bambrough R. Universals and family resemblances, in: ***MJ. Loux*** (ed) Universals and Particulars. New York. 1970

Barbour A. Caring for Patients. A Critique of the Medical Model. Stanford. Stanford University Press. 1995

Bar-Hillel Y. Aspects of Language. Amsterdam. North-Holland; 1970

Barker DJP. and Rose G. Epidemiology in Medical Practice. Edinburgh. Churchill Livingstone. 1979

Baron R. An introduction to medical phenomenology: I can't hear you while I'm listening. *Annals of Internal Medicine*. 1965; **103:** 606–611

Barondess JA. Disease and illness—crucial distinction. *American Journal of Medicine*. 1979; **66:** 375–376

Barondess JA. The clinical transaction: theme and descants. *Perspectives in Biology and Medicine*. 1983; **27:** 25–38

Barraclough B. et al. A hundred Cases of Suicide: Clinical Aspects. *British Journal of Psychiatry*. 1974; **125:** 355–373

Barron AJ., Zaman N., Cole GD. Systematic review of genuine versus spurious side-effects of beta-blockers in heart failure using placebo control.: Recommendations for patient information. *International Journal of Cardiology*. 2013; **168:** 3572–3579

Barsky AJ. The paradox of health. *The New England Journal of Medicine*. 1988; **318:** 414–418

Barsky AJ. Palpitations, arrhythmias, and awareness of cardiac activity. *Annals of Internal Medicine*. 2001a; **134:** 832–837

Barsky AJ. The patient with hypochondriasis. *The New England Journal of medicine*. 2001b **345:**1395–1399

Barsky AJ. and Borus JF. Functional somatic syndromes. *Annals of Internal Medicine*. 1999; **130:** 910–921

Barsky AJ., Saintfort R., Rogers, MP. and Borus JF. Nonspecific medication side effects and the nocebo phenomenon. *Journal of the American Medical Association*. 2002; **287:** 622–627

Bartlett E. An Essay on the Philosophy of Medical Science. 1844. Medical heritage library. US National Library of Medicine.

Baybrooke D. Meeting Needs. Princeton. Princeton University Press. 1987

Beauchamp TL. and Childress JF. Principles of Biomedical Ethics. Oxford. Oxford University Press. 2001

Beckner M. The Biological Way of Thought. Berkeley. University of California Press. 1968

Beecher H. Medical Research and the Individual, in: ***E. Shils*** (ed) Life or Death Ethics and Options. Reed College. Portland. Ohio.1968

Behr MA., Divangahi M. & Lalande JD. What's in a name? The (mis)labelling of Crohn's as an autoimmune disease. *Lancet*. 2010. **376:** 202–203

Behrmann M, Avidan G. Congenital prosopagnosia: face-blind from birth. *Trends Cognitive Science*. **9**: 180–7, 2005

Belfiore ES. Tragic Pleasures. Aristotle on Plot and Emotion. Princeton. Princeton University Press. 1992

Benedetti F. Placebo Effects: Understanding the Mechanisms in Health and Disease. Oxford University Press. 2009

Benedetti F., Carlino E., Pollo A. How Placebos change the Patient's Brain. *Neuropsychopharmacology.* 2011; **36:** 339–354

Benedict R. Anthropology of the Abnormal. *The Journal of General Psychology;* **10**: 59–82, 1934 (2002)

Bennett J. Linguistic Behaviour. Cambridge. Cambridge University Press. 1979

Bennett MR. and Hacker PMS. Philosophical Foundations of Neuroscience. Oxford. Blackwell. 2003

Bennett MR, Hacker PMS. Philosophical Foundations of Neuroscience. Second Edition. Wiley Blackwell. 2022

Bennett MR, Dennett D, Hacker PM. Neuroscience and Philosophy: Brain, Mind and Language. Columbia University Press. 2007

Bensing JM. and Verhaak PFM. Somatisation: a joint responsibility of doctor and patient. *New England Journal of Medicine.* 2006; **367:** 452–453

Bentham J. An Introduction to the Principles of Morals and Legislation, JH. Burns and HLA. Hart, (eds.). Oxford. Clarendon Press. 1996

Bergström J. Hemodialysis Induced Hypertension. *The International Journal of Artificial Organs*: **1**: 202–202, 1978

Berkeley G. Introduction to the Principles of Human Knowledge. London. Everyman's Library. Dent. 1983

Berlin I. Four Essays on Liberty. London. Oxford University Press.1969

Bermudez JL., Marcel A. & Eilan N. (eds.). The Body and the Self. Cambridge, Mass. The MIT Press. 1995

Bernard C. Leçons de pathologie expérimentale. Paris. Baillère. 1872

Bernard C. Introduction à l'Étude de la Médecine Expérimentale. Paris. Delagrave. 1920

Bernard C. Principes de médecine expérimentale. Les Presses Universitaires de France. 1947

Bernard C. An Introduction to the Study of Experimental Medicine. New York. Dover. 1961

Berrios GE. The History of Mental Symptoms. Cambridge. Cambridge University Press. 1996

Bhopal RS. Concepts of epidemiology: an integrated introduction to the ideas, theories, principles and methods of epidemiology. Oxford University Press. 2002

Bhopal RS. Concepts of Epidemiology. Integrating the Ideas, Theories, Principles and Methods of Epidemiology. Oxford University Press. 2016

Bicchieri C. The Grammar of Society. Cambridge University Press. 2006

Bicchieri C. Norms, Conventions, and the Power of Expectations, p. 208–229, in: ***N. Cartwright & E. Montuschi,*** (eds.). Philosophy of Social Science. A New Introduction. Oxford. Oxford University Press. 2014

Biering-Sorensen. Physical Measurements as risk indicators for low-back trouble over a one-year period. *Scandinavian Journal of Rehabilitation Medicine.* 1929

Biering-Sørensen F and Friis Bendix A. Working off low back pain. *The Lancet.* 2000; **355**: 1929–1930

Bigelow J and Pargetter R. Functions. *The Journal of Philosophy.* 1987; **84:** 181–196

Bittner R. What the Reason demands. Cambridge. Cambridge University Press. 1983

Black M. Vagueness: an exercise in logical analysis, 25–58, in: ***M. Black.*** Language and Philosophy. Ithaca, N.Y. Cornell University Press. 1949

Black, M. The Gap Between "Is" and "Should". *The Philosophical Review.* 1964; **73:** 165–181

Blackburn S. Think. Oxford. Oxford University Press. 1999

Blanshard B. The Nature of Thought. Vol. II London. Allen & Unwin. 1939

Blood Pressure Lowering Treatment Trialists' Collaboration. Blood pressure-lowering treatment based on cardiovascular risk: a metaanalysis of individual patient data. *Lancet.* 2014; **384:** 591–598

Blois SM. Medicine and the nature of medical reasoning. *New England Journal of Medicine.* 1988; **318:** 847–851

Bluhm R., Borgerson K. Evidence Based Medicine, in: ***F. Gifford*** (eds.) Handbook of Philosophy of Science, Vol. 16 Philosophy of Medicine. Amsterdam. Elsevier. 2011

Bok S. Lying. Moral Choice in Public and Private Life. Hassocks. Sussex. The Harvester Press. 1978

Bolton D. What is psychiatric disease. A commentary on Dr Ghaemi's paper, in: ***Kendler S and Parnas*** **J** (eds.) Philosophical issues in psychiatry II. Nosology. Oxford University Press. 2012

Bolton D. Clinical significance, disability: Shifts in thinking between DSM-IV and DSM-5, 8–16, in: ***KS. Kendler, J. Parnas,*** (eds.) Philosophical issues in psychiatry IV classification of psychiatric illness. Oxford University Press. 2017

Bolton D. and Hill J. Mind, Meaning and Mental Disorder. The Nature of Causal Explanation in Psychology and Psychiatry. Oxford. Oxford University Press.1996

Bonjour L. The Structure of Empirical Knowledge. Harvard University Presses. 1985

Boorse C. On the distinction between disease and illness. *Philosophy and Public Affairs*. 1975; **5:** 49–68

Boorse C. Wright on function. *Philosophical Review*. 1976; **85:** 70–86

Boorse C. Health as a theoretical concept. *Philosophy of Science*. 1977. **44:** 542–573

Boorse C. A rebuttal on health, in: ***JM Humber and RF Almeder*** (eds). Totawa. New Jersey, Humana Press. 1997

Boorse C. Concepts of Health and Disease, in ***F. Gifford*** (ed) Philosophy of Medicine. Elsevier. 2011

Boorse C. A Second View of Health. Copyright by Christopher Boorse. 2014

Borsboom D. Mental disorders, network models, and dynamical systems, pp. 80–97, in: ***KS. Kendler and J. Parnas*** (eds.) Philosophical issues in psychiatry IV. Oxford University Press. 2017

Bortolotti L. Rationality and sanity: the role of rationality judgments in understanding psychiatric disorders, pp. 480–496, in: ***Fullford KWM., Davies M., Graham G.*** (eds.), 2013

Bowen JL. Educational strategies to promote clinical diagnostic reasoning. *New England Journal of Medicine*. 2006; **355:** 2217–2225

Bowling A. Measuring Health. A review of quality-of-life measurement scales. Buckingham. Open University Press. 1997a

Bowling A. Research Methods in Health. Investigating Health and Health Services. Open University Press. 1997b

Bradley R and Swartz N. Possible Worlds. An Introduction to Logic and Philosophy. Oxford. Basil Blackwell. 1979

Bradley C. Importance of differentiating health status from quality of life. *Lancet*. 2001; **357:** 7–8

Bradley CK., Wang TY., Li S., et al. Patient-Reported Reasons for Declining or Discontinuing Statin Therapy: Insights from the PALM Registry. *Journal of the American Heart Association*. 2019; 8: e011765

Bradwell AR., Carmalt MHB. and Whitehead TP. Explaining the unexpected abnormal results of biochemical profile investigations. *The Lancet*. 1974; **304:**1071–1074

Braithwaite RB. Scientific Explanations. Cambridge. Cambridge University Press. 1953

Brandom RB. Between Saying and Doing. Towards and Analytic Pragmatism. Oxford. Oxford University Press. 2008

Brawley O., Jemal A. Cancer Statistics. The impact of ethnicity, socioeconomic and racial disparities. https://doi.org/10.1016/bs.ircmb. 2017

Braybrooke D. Meeting Needs. Princeton University Press. Princeton. 1987

Brennan TA., Gawande A., Thomas E. and Studdert D. Accidental deaths, saved lives and improved quality. *New England Journal of Medicine*. 2005; **353:**1405–1409

Brent DA. & Elhem N. Familial transmission of suicidal behavior. *Psychiatric Clinics of North America*. 2008; **31:** 157–177

Breslow NE. & Day NE. Statistical methods in cancer research. Vol I. The analysis of case-control studies. Lyon. International Agency for Research on Cancer. 1980

Brett AS. and McCullough LB. When patients request specific interventions. *New England Journal of Medicine*. 1986; **315:**1347–1351

Brewster D. More Worlds than One. The Creed of the Philosophers and the Hope of the Christians. London. Murray. 1854.

Broad CD. Scientific Thought. London. Kegan Paul, Trench, Trubner & Co. 1923

Broadbent A. Philosophy of Epidemiology. Palgrave MacMillan. 2013

Broadbent A. Causation and Prediction in Epidemiology: A Guide to the Methodological Revolution. *Studies in History and Philosophy of Biological and Biomedical Sciences.* 2015; **54:** 72–80

Broadbent A. Philosophy of Epidemiology. London. MacMillan. 2017

Broadbent A. Philosophy of Medicine. Oxford University Press. 2019

Brock D. Quality of life measures in health care and medical ethics. Pp. 95–132, in: ***MC. Nussbaum & A. Sen*** (eds.). The Quality of Life. Oxford. Clarendon Press. 1993

Brody H. Placebos and the Philosophy of Medicine. Chicago. University of Chicago Press. 1977

Brody H. The doctor as therapeutic agent: a placebo effect research agenda, 77–92, in: ***Harrington H.*** (ed.) The Placebo Effect. Cambridge, Mass. Harvard University Press. 1997

Brown PJ and Konner M. An anthropological perspective on obesity, in: Human Obesity, *RJ. Wurtman and JJ. Wurtman* (eds.). *Annals of the New York Academy of Sciences* 1987; **499:** 29–46

Browne T. Religio Medici. William Pickering. London. 1845

Browne KD & Hamilton-Giachritsis C. The influence of Violent Media on Children and Adolescents: A Public Health Approach. *Lancet*; **365**:702–720, 2005

Brunström M., **Carlberg B.** Association of blood pressure lowering with mortality and cardio-vascular disease across blood pressure levels: a systematic review and meta-analysis. *JAMA Internal Medicine*. 2018**. 178:** 28–36

Buber M. I and Thou. Edinburgh. T. & T. Clark. 1958

Buchanan AE. and Brock DW. Deciding for Others. The Ethics of Surrogate Decision Making. Cambridge. Cambridge University Press. 1990

Buckholz JW and Meyer-Lindenberg A. Psychopathology and the Human Connectome: Toward a Transdiagnostic Model of Risk for Mental Illness. *Neuron*; **74**: 990–1004, 2012

Bunge M. How does it work? The search for explanatory mechanisms. *Philosophy of the Social Sciences*. 2004; 182–210

Bunge M. Medical Philosophy. Conceptual Issues in Medicine. World Scientific Publishing. 2014

Burton R. The Anatomy of Melancholy. New York. *New York Review of Books*. 2001

Butler S. *Erewhon*. London. Penguin Books. 1970

Butler CC., Rollnick S., Pill R. et al. Understanding the culture of prescribing: qualitative study of general practitioners' and patients' perceptions of antibiotics for severe throats. *British Medical Journal*. 1998; **317:** 637–642

Butterworth G. An ecological perspective on the origins of the self, pp. 87–105, in**: *JL. Bermudez A. Marcel and N. Eilan,*** (eds.). The Body and the Self. Cambridge, Mass. The MIT Press.1995

Bynum B. Discarded diagnoses. DaCosta's Syndrome. *The Lancet*. 2001a; **358:** 1736

Bynum B. Discarded diagnoses. Nostalgia. *Lancet*. 2001b; **358:** 2176

Bytzer P., and Talley NJ. Dyspepsia. *Annals of Internal Medicine*. 2001; **134:** 815–823

Callahan D. & Nulan SB. The Quagmire. *The New Republic*. 2011; July

Camiret M.S. Picasso, un génie de la déconstruction. Du reniement d'une esthétique au Wokisme. Paris. L'Harmattan. 2023

Campaner R. Philosophy of Medicine: Causality, Evidence and Explanation. Bologna. Archetipo Libri. 2012

Campbell EJM., Scadding JG. and Roberts RS. The concept of disease. *British Medical Journal*. 1979**; 2:** 757–762

Canguilhelm G. Le normal et le pathologique. Paris. Presses Universitaires de France. 1966

Cannon WB. The Wisdom of the Body New York. Norton & Company. 1963

Caplan A.L, Engelhardt HT. and McCartney JJ. Concepts of health and disease. Interdisciplinary Perspectives. Reading, Massachusetts, Addison-Wesley Publishing Company. 1981

Caplan AL. Does the Philosophy of Medicine Exist? *Theoretical Medicine*. 1992; **13:** 67–77

Caplan AL. The concepts of health, illness, and disease, 233–248, in: ***Bynum WF. and Porter R*** (eds.), Companion Encyclopedia of the History of Medicine. London. Routledge. 1993

Caplan AL., MacCartney JJ. & DA. Sisti, (eds.). Health, Disease and Illness. Concepts in Medicine. Washington. Georgetown University Press. 2004

Carel H. & Cooper RV. (eds.). Health, Illness and Disease: Philosophical Essays. London. Routledge. 2014

Carmalt MHB. in: Fifth Symposium on Advanced Medicine held at the Royal College if Physicians, *R. Williams* (ed.), pp. 268–278. London. Pitman Medical. 1969

Carnap R. Meaning and Necessity. A Study in Semantics and Modal Logic. Chicago. University of Chicago Press. 1956

Cartwright A. Patients and Their Doctors: a Study of General Practice. London. Routledge. 1967

Cartwright N. How the Laws of Physics lie. Oxford. Oxford University Press. 1983

Cartwright N. Nature's Capacities and their Measurement. Oxford. Oxford University Press. 1989

Cartwright N. Fundamentals vs the patchwork of laws. *Proceedings of the Aristotelian Society.* 1994; **2:** 279–292

Cartwright N. What is wrong with Bayes Nets? *Monist.* 2001, April 01

Cartwright N. Are RCTs the gold standard? *Biosocieties.* 2007a; **2:** 11–20

Cartwright N. Hunting Causes and Using Them. Cambridge University Press. 2007b

Cartwright N. The Dappled World: A Study of the Boundaries of Science. Cambridge University Press. 2008

Cartwright N. Causal Inference, pp. 308–326, in: ***N. Cartwright. & E. Montuschi,*** (eds.). 2014

Cartwright N. & Montuschi., (eds.) Philosophy of Social Science. A New Introduction. Oxford. Oxford University Press. 2014

Casarotto S., Comanducci A., Rosanova M., Sarasso S., Fecchio M., et al. Stratification of unresponsive patients by an independently validated index of brain complexity. *Annals of Neurology.* 2016; **80:**718–729

Casey J. Pagan Virtue. An Essay on Ethics. Oxford. Clarendon Press. 1990

Cassell EJ. The Nature of Suffering and the Goals of Medicine. *New England Journal of Medicine.* 1982; **306:** 639–645

Cassell EJ. The Nature of Suffering and the Goals of medicine. Oxford. Oxford University Press. 2004

Cassell EJ. Consent or Obedience? Power and Authority in Medicine. *New England Journal of Medicine*; **352**: 328–330, 2005

Cates C. An evidence-based approach to reducing antibiotics use in children with acute otitis media: controlled before and after study. *British Medical Journal.* 1999; **318:** 715–716

Centre for Addiction and Mental Health. **(CAMH)** Association of translocator protein total distribution volume with duration of untreated major depressive disorder: a cross-sectional study. *Lancet* Psychiatry. 2018, 26 February. DOI: https://doi.org/10.1016/S2215-0366(18)30048-8

Chadwick R. Normality as Convention and as Scientific fact, 17–28, in: ***T. Schramme & S. Edwards.*** Handbook of the Philosophy of Medicine. Volume 1. Dordrecht. Springer. 2017

Chalmers DJ. The Conscious Mind, in: Search of a Fundamental Theory. Oxford. Oxford University Press. 1996

Changeux JP. L'homme neuronal. Paris. 1983

Charland LC. Why psychiatry should fear medicalization, 159–175, in: ***Fullford KWM., Davies M., Graham G.*** (eds.). 2013

Charlton BG. Attribution of causation in epidemiology: chain or mosaic? *Journal of Epidemiology.* 1996a; **49:** 105–107

Charlton BG. The scope and nature of epidemiology. *Journal of Epidemiology.* 1996b; **49:** 623–626

Chapman RH., Benner JS., Petrilla AA., et al. Predictors of adherence with antihypertensive and lipid-lowering therapy. *Archives of Internal Medicine.* 2005; **163:** 1147–1152

Charon R. The narrative road to empathy, pp. 147–159, in: ***H. Spiro, MGM Curnon, E. Peschel and D. St. James,*** (eds.). Empathy and the Practice of Medicine. New Haven. Yale University Press. 1993

Charon R and Wyer P. Narrative evidence based medicine. *The Lancet*; **371**: 296-297, 2008

Chen PW. Final Exam. A Surgeon's Reflections on Mortality. New York. Alfred Knopf. 2006

Childress JF. Who Should Decide? Paternalism in Health Care. New York. Oxford University Press. 1982

Chisholm MR. Theory of knowledge. Englewood Cliffs. Prentice Hall. 1989

Chitty RN. Why clinicians are natural Bayesians? *British Medical Journal.* 2005; **330**: 1390.

Churchland P. Neurophilosophy: Toward a Unified Science of the Mind-Brain. MIT Press. 1989

Churchland PM. Eliminative materialism and the propositional attitudes, 382–400, in*: **J. Heil,*** (ed.). The Nature of Mind, Oxford: Oxford University Press. 2004

Cicero MT. Brutus. Sammlung Tusculum. De Gruyter Auflage. 2014

Cipriani A., Furukawa T., Salanti G., et al. Comparative efficacy and acceptability of 21 antidepressant drugs for the acute treatment of adults with major depressive disorder: a systematic review and network meta-analysis. *Lancet.* February 21, 2018; published on line.

Claridge G. Schizophrenia and human individuality, pp. 29–41, in: ***C. Blakemore and S. Greenfield*** (eds.). Mindwaves. Thoughts on Intelligence, Identity and Consciousness. Oxford. Basil Blackwell. 1987

Clark, A. Psychological Models and Neural Mechanisms Oxford. Clarendon Press. 1980

Clarke Jr DS. Practical Inferences. London. Routledge Kegan and Paul. 1985

Clarke B. Causation in medicine, pp. 297–322, in*: **JA. Marcum*** (ed.). The Bloomsbury Companion to Contemporary Philosophy of Medicine. London. Bloomsbury. 2017

Clarke DM., Piterman L., Byrne CJ., Austin DW. Somatic symptoms, hypochondriasis and psychological distress: a study of somatisation in Australian general practice. *Medical Journal of Australia*. 2008. **189:** 560–564

Clausen AJ. and Wyle CJ. Paths to the mental hospital. *Journal of Social Issues*. 1955; **11:** 25–32

Cleophas TJM. The importance of placebo effects. *JAMA.* 1995. **273:** 283–284

Cloninger CR. Diagnosis of somatoform disorders. A critique of DSMIII, 243–259, **in: *GL. Tischle*r (ed.)** Diagnosis and classification in psychiatry. A critical appraisal of DSM-III. Cambridge. Cambridge University Press. 1987

Clouser KD., Culver CM. and Gert B. Malady, in: ***JM. Humber and RF. Almeder*** (eds.) **Totawa, New Jersey, Humana Press**. 1997

Cochrane AL. The history of the measurement of ill health. International. *Journal of Epidemiology*. 1972**; 1:** 89–92

Cochrane AL. Effectiveness and Efficiency. Random Reflections on Health Services. London. British Medical Journal. The Nuffield Provincial Hospital Trust. 1989

Cohen MR. and Nagel E. Logic, and Scientific Method. 1934

Cohen LJ. Bayesianism versus Baconianism in the Evaluation of Medical Diagnosis. *British Journal of the Philosophy of Science*, 1980; **31:**45–62

Cohen LJ. An Essay on Belief and Acceptance. Oxford. Oxford University Press. 1992

Coldman A., Philips N., Wilson C. et al. Pan-Canadian study of mammography screening and mortality from breast cancer. *Journal of the National Cancer Institute.* 2014; doi: 10

Collingwood RG. An Essay on Metaphysics. Oxford. Clarendon Press. 1940. 1998

Collins H. Tacit & Explicit Knowledge. Chicago. Chicago University Press. 2010

Collins R., Armitage J. et al. *Lancet.* Authors Reply. 2017. **389:** 1099–1100

Comfort A. What is a Doctor? Philadelphia. Georges Stickley. 1980

Conan Doyle A. Sherlock Holmes; The Sign of Four. London. Fox Publishing. 2014

Conn JH. The Decline of Psychoanalysis. The End of an Era. Here we Go Again. *JAMA.* 1974; **228:** 711–712

Conrad P. The Medicalization of Society. Baltimore. John Hopkins. 2007

Cooper JE. The diagnosis and psychopathology of schizophrenia in New York and London. *Schizophrenia Bulletin;* **1**: 80–102, 1972

Cooper R. Disease. *Studies in History and Philosophy of Biological and Biomedical Sciences*. 2002; **33:** 263–282

Cooper R. Psychiatry and Philosophy of Science. Stockfield. Acumen. 2007

Cooper R. Natural kinds, 950–965, in: ***Fullford KWM., Davies M., RGT. Gipps, Graham G. et al.*** (eds.). 2013

Cooper JE. and Sartorius N. A Companion to the Classification of Mental Disorders. Oxford University Press. 2013

Copeland DD. Concepts of disease and diagnosis. *Perspectives in Biology and Medicine.* 1977; **20:** 528–538

Copeland J. What is a 'case'? A case for what? **in:** ***Wing JK., Bebbington P. and Robins LN.,*** (ed.). What is a Case? Problem of Definition in Psychiatric Community Surveys. London. Grant McIntyre. 1981

Copp D. Reason and needs, 112–137, in: ***Frey RG. and Morris CW.*** (eds.). Value, Welfare and Morality. Cambridge. Cambridge University Press. 1993

Copp, D. Morality, Normativity & Society. Oxford. Oxford University Press. 1995

Corbellini G. Storia e teorie della salute e della malatia. Roma. Carocci editore. 2014

Corradi Fiumara G. The Symbolic Function. Oxford. Blackwell. 1992

Coronary Drug Project Research Group. Influence of adherence treatment and response of cholesterol on mortality in the Coronary Drug Project. *New England Journal of Medicine.* 1980; **303:** 1038–1041

Council of Scientific Affairs, American Medical Association. Clinical Ecology. *JAMA.* 1992; **268:** 3465–3467

Covinsky KE., Wu AW., Landefeld CS., Connors AE., Philips RS. et al. Health status versus quality of life in older patients: does the distinction matter*? American Journal of Medicine.* 1999; **106:** 435–440

Cowan JL. Pleasure and Pain. London. MacMillan. 1968

Crary J. Late Capitalism and the Ends of Sleep. London. Verso. 2013

Craver CF. Role functions, mechanisms, and hierarchy. *Philosophy of Science.* 2001; **68:** 53–7

Creed FH. & Barsky A. A systematic review of somatisations disorder and hypochondriasis. *Journal of Psychosomatic Research.* 2004; **56:** 391–408

Creed FH., Davies I., Jackson J., Littlewood A., et al. The epidemiology of multiple somatic symptoms. *Journal of Psychosomatic Research.* 2012: **72:** 311–317

Crews F. Talking back to Prozac. *New York Review of Books*; December 6, 2007

Crick F. The Astonishing hypothesis: the scientific search for the soul. New York. Scribner. 1994

Critchley M. Congenital indifference to pain. *Annals of Internal Medicine.* 1956; **45:** 737–747

Crossley NA., Scott J., Ellison-Wright I., Mecheli A. Neuroimaging distinction between neurological illnesses. *British Journal of Psychiatry.* 2015; **207:** 429–434

Crookshank FGT. First Principles of Epidemiology. *Proceedings of the Royal Society of Medicine.* 1920; **13:** 159–184

Crookshank FGT. The Theory of Diagnosis. *Lancet.* 1926. **208:** 995–999

Crow TJ. A Darwinian Approach to the origins of psychosis. *British Journal of Psychiatry.* 1995; **167;** 12–25

Culver CM. and Gert B. Philosophy in Medicine. Conceptual and Ethical Issues in Medicine and Psychiatry. Oxford. Oxford University Press. 1982

Cummins R. Functional analysis. *The Journal of Philosophy.* 1975. **72:** 741–765

Damasio AR. Descartes' Error. Emotion, Reason and the Human Brain. Chatham, Kent. Picador. 1994, 112–117

Daniels N. Normal functioning and the treatment/enhancement distinction. Cambridge *Quarterly of Health Care Ethics.* 2000; 309–322

Danielyan A. & Nasrallah HA. Neurological disorders in schizophrenia. *Psychiatric Clinics of North America.* 2009; **32:** 719–757

Dante. Paradisio, in: **A Mandelbaum,** (ed.), The Divine Comedy. New York. Bantam Books. 1986

Danto A. Analytical Philosophy of Knowledge. Cambridge. Cambridge University Press. 1968

Danto A. Analytical Philosophy of Action. Cambridge. Cambridge University Press. 1973

Danto A. Comment on Gewirth. Constructing an epistemology of human rights: a pseudo-problem? *Social Philosophy & Policy.* 1984; **1:** 25–30

Danto A. Narration and knowledge. Columbia University Press. 1985

Danto A. Connections to the World. The Basic Concepts of Philosophy. New York. Harper & Row, Publishers. 1989

Danto A. The Body/Body Problem. Berkeley. University of California Press. 1999

Darby RR., Horn A., Cushman F. and Fox MD. Lesion network localization of criminal behavior. *Proceedings of the National Academy of the United States*. 2018; **115:** 601–606

Darden L. Mechanisms and Models, 139–159, in**:** ***D. Hull & M. Ruse,*** (eds.). The Cambridge Companion to the Philosophy of Biology. Cambridge. Cambridge University Press. 2007

Darrah LW. The difficulty of being normal. *Journal of Nervous and Mental Diseases*. 1939; **90:** 730–737

Darwin C. On the Origin of Species by Means of Natural Selection. London. John Murray. 1859. Harvard University Press.

David, M. Correspondence and Disquotation: An Essay on the Nature of Truth, New York: Oxford University Press. 1996

David AS., Nicholson T. Are neurological and psychiatric disorders different? *British Journal of Psychiatry*. 2015; **207:** 373–374.

Davidson D. Causal relations. *Journal of Philosophy*. 1967; **64:** 691–703

Davidson D. Essays on Action and Events, Oxford. Oxford University Press. 1980.

Davies PS. Norms of Nature. Naturalism and the nature of Functions. Cambridge, Massachusetts. The MIT Press. 2003

Davis, J. (November 2006). "Face Blind". Wired. *Retrieved 31 December 2014.*

Deahl M. Smoke, mirrors, and Gulf War illness. *Lancet*. 2005. **365:** 635–638

Deary IJ. Editorial. A taxonomy of medically unexplained symptoms. *Journal of Psychosomatic Research*. 1999; **47:** 51–59

De Caro M. & MacArthur D. Naturalism ad Normativity. Columbia University Press. 2010, pp. 214–218

DeGrazia D. Animal Rights. A Very Short Introduction. Oxford. Oxford University Press. 2002

De Haan L, Sutterland AL, Schotborgh JV. Association of Toxoplasma gondii Seropositivity with Cognitive Function in Healthy People. A Systematic Review and Meta-Analysis. *JAMA Psychiatry*. Published online July 14, 2021. Doi:https://doi.org/10.1001/jamapsychiatry.2021.1590

Delaporte F. Le Savoir de la maladie: Essai sur le cholera de 1832 à Paris. Presses Universitaires de France. 1990

Del Mar C. Sore throats and antibiotics. Applying evidence on small effects is hard; variations are probably inevitable. *British Medical Journal*. 2000**; 320:** 130–131

Dennett DC. Consciousness explained. London. Penguin Books; 1991

Descartes R. The Essential Descartes**,** ***Margaret D. Wilson and Robert Paul Wolff*** (ed.) New York, A Mentor Book. 1969

Descartes R. A Discourse on Method. Meditation and Principles. London. Dent & Sons. 1978

Devereaux MW. Neck and low back pain. *Medical Clinics of North America*. 2003; **87:** 643–662

Devine J., Giligan J., Miszek KA. et al. (eds.) Youth Violence. Scientific Approaches to Prevention. 2004. New York. The New York Academy of Sciences**.**

Diehl AR. Epidemiology of gallbladder cancer: a synthesis of recent data. *Journal of the Royal Cancer Institute*. 1980; **65:** 1209–1214

Dilts SL. Models of the Mind. A Framework for Biopsychosocial Psychiatry. Philadelphia. Brunner-Routledge. 2001

Dixon B. Beyond the Magic Bullet. London. George Allen & Unwin. 1978

Dodd FL., Kennedy DO., Rilby LM., Haskell-Ramsay CF. A double-blind, placebo-controlled study evaluating the effects of caffeine and L-threanine both alone and in combination on cerebral blood-flow, cognition and mood. *Psychopharmcology*. 2015**; 232:** 2563–2576

Doll R., Peto R., Hall E., Wheatley K. and Gray R. Mortality in relation to consumption of alcohol: 13 years' observations on male British doctors. *British Medical Journal.* 1994; **309:** 911–918

Dorland's Illustrated Medical Dictionary**.** Philadelphia. W.B. Saunders Company. 1988

Doubeni CA., Corley DA., Quinn VP. et al. Effectiveness of screening colonoscopy in reducing the risk of death from right and left colon cancer: a large community-based study. 2016; *Gut* **67:** 291–298

Douros A., Dell'Aniello S., Deghan G., et al. Degree of serotonin reuptake inhibition of antidepressants and ischemic risk. A cohort study. *Neurology* 2019; **93**: DOI: https://doi.org

Downen MK., Stampfer MJ., Matthew R., Cooperberg MR. Declining Incidence Rates of Prostate Cancer in the United States. Is This Good News or Not? *JAMA Oncology.* 2017; **3:**1623–1624

Downie RS. and Telfer E. Caring and Curing. London. Methuen. 1980

Downie RS. and Calman KC. Healthy Respect. London. Faber and Faber. 1987

Downie RS. and Macnaughton J. Clinical Judgment. Evidence in Practice. Oxford. Oxford University Press. 2000.

Dowty DR. Word Meaning and Montague Grammar. Dordrecht. Reidel. 1979

Doyal L. et al. A Theory of Human Need. MacMillan. 1991

Doyal L and Gough I. A Theory of Human Need. London. MacMillan. 1991

Dragulinscu S. Inference to the Best Explanation and Mechanisms in Medicine. *Theoretical Medicine and Bioethics.* 2016; **37:** 211- 232

Dretske FI. Contrastive statements. *Philosophical Review*. 1972**. 81:** 411–437

Dretske FI. Explaining Behavior. Reasons in a World of Causes. Cambridge. MIT Press. 1995

DSM-III American Psychiatric Association. Diagnostic and Statistical Manual of Mental Disorders (3d. ed.) Washington, D.C. Psychiatric Press. 1980

DSM-III-R. *Diagnostic and Statistical Manual of Mental Disorders.* Washington. American Psychiatric Association. 1987

DSM-IV. *American Psychiatric Association.* Diagnostic and Statistical Manual of Mental Disorders. Fourth Edition. International Version. Washington, D.C. American Psychiatric Association. 1995

Dubos R. and Dubos J. Consumption and the romantic age, 65–66, in: The White Plague. Boston. Little Brown and Co. 1952

Dubos R. The Mirage of Health. New York. Harper and Row. 1959

Dubos R. The Dreams of Reason. Science and Utopias. New York. Columbia University Press. 1961

Ducassé CJ. Explanation, mechanism and teleology. *Journal of Philosophy.* 1925; **22:** 150–155

Ducassé CJ. Truth, Knowledge and Causation. London Routledge & Kegan Paul. 1968

Dunlap D.M., Henderson T.L. & Inch R.S. A survey of 17,301 prescriptions on from E.C. 10. *British Medical Journal.***1:** 1952; pp. 292–295

Dunnell K and Cartwright A. Medicine Takers, Prescribers and Hoarders. London. Routledge and Kegan Paul. 1972

Dupré J. Natural kinds and biological taxa. *The Philosophical Review.* 1981; **90:** 66–90

Dupré J. The Disorder of Things. Cambridge, MA. Harvard University Press. 1993

Dupré J. Natural Kinds, 311–319, in**: Newton-Smith W** (ed.), A Companion to the Philosophy of Science. Oxford. Blackwell. 2000

Dupré J. Human Nature and the Limits of Science. Oxford. Clarendon Press. 2001

Dupré J. How to be Naturalistic Without Being Simplistic in the Study of Human Nature, pp. 289–303, in: ***M De Caro & D Macarthur*** (eds.). Naturalism and Normativity. New York. Columbia University Press. 2010

Dupré J. Natural Kinds, Chapter 46, in**: *Newton-Smith W* (ed.),** A Companion to the Philosophy of Science. Oxford. Blackwell. 2017

Dworkin G. Paternalism, in: ***R Wasserstrom*** (ed.) Morality and the Law, Belmont. Wadsworth. 1971

Dworkin G. The Theory and Practice of Autonomy. Cambridge. Cambridge University Press. 1988

Dyson F. The Scientist as a Rebel. New York. *New York Review of Books;* 2006: 269–283

Edelman GM. Bright Air, Brilliant Fire – On the Matter of the Mind. London. Penguin. 1994

Edelman GM. & Gally JA. Degeneracy and complexity in biological systems. *Proceedings of the National Academy of Sciences of the USA*. 2001; **98:** 13763–13768

Eastwood M. The Relation between Physical and Mental Illness. Toronto. University of Toronto Press. 1975

Editorial. Insensitivity to pain. *British Medical Journal*. 1973; **2:**187–188

Editorial. Therapeutic orphans. *Lancet*. 1985; **326:** 702–703

Editorial. The concept of disease. *British Medical Journal.* 1979**; 2:** 752–752

Editorial. Treatment of syndromes or of symptoms in psychiatry. *Lancet.* 1988; **332:** 373–374

Editorial. Focus on psychiatric disorders. *Nature Neuroscience*. 2016; **19:**1381–1382

Editorial. Is exposure to media violence a public-health risk? *The Lancet*. 2008; **371:** 1137–1214

Edwards RB. Pleasures and Pain. A Theory of Qualitative Hedonism. Ithaca. Cornell University Press. 1979

Ehrenreich H. & Nave KA. Phenotype-Based Genetic Association Studies (PGAS)-Towards Understanding the Contribution of Common Genetic Variants to Schizophrenia Subphenotypes. *Genes* (Basel). 2014**; 5:** 97–105

Eilan N. & Roessler J. Agency and Self-Awareness. Oxford. Oxford University Press. 2004

Eisenberg L. Disease and Illness. Distinction between professional and popular ideas of sickness. *Cultural Medical Psychiatry*; **1**: 9–23, 1977

Eisenberg L. Mindlessness and brainlessness in psychiatry. *British Journal of Psychiatry*. 1986; **148:** 497–508

Eldredge N. Reinventing Darwin? The Great Debate at the High Table of Evolutionary Theory. New York. John Wiley & Sons. 1995

***Elliott AM., McAteer A.* & Hannaford PC.** Revisiting the symptom iceberg in today's primary care: results from a UK population survey. *Family Practice*. 2011; **12:** 16–27

Elliott C. Better than well. American medicine meets the American dream. New York. W.W. Norton. 2003

Ellis BJ. & Boyce WT. Biological Sensitivity to Context. *Current Directions in Psychological Science*. 2008**; 17:** 183–186

Elveback IR., Giullies CI. and Keating FR. Jr. Health, normality and the ghost of Gauss. *JAMA*. 1970: **211:** 69–75 ***Elwood PC. and Wood MM.*** Effect of oral iron therapy on the symptoms of anaemia. *British Journal of Preventive and Social Medicine*. 1966; **20:** 172–175

Emmet D. The Passage of Nature. London. Macmillan. 1992

Emmet D. Functions, Purpose and Powers London. Macmillan. 1972

Emerson RW. The portable Emerson, M. van Doren, (ed.) New York. The Viking Press. 1946

Enç B. Nonreducible supervenient causation, pp. 169–186, in*: **EE. Savellos and UD. Valçin*** (eds.) Supervenience. Cambridge. Cambridge University Press. 1995

Engel GL. The need of a new medical model: a challenge for biomedicine, 589–607, in: ***Caplan AL. Engelhardt HT. and McCartney J.J,*** (eds.). Concepts of Health and Disease. Reading, Mass. Addison Wesley. 1981

Engel GL. and Salzman LF. A double standard for psychosomatic papers? *The New England Journal of Medicine.* 1973; **288:** 44–46

Engelhardt HT. The concepts of health and disease, in*: **HT. Engelhardt and SF. Spicker*** (eds.). Evaluation and Explanation in the Biomedical Sciences. Dordrecht. Reidel Publishing Company. 1975

Engle RL. and BJ. Davis. Medical diagnosis: past, present and future. 1. Present concepts on the meaning and limitations of medical diagnoses. *Archives of Internal Medicine*. 1963; **112:** 512–523

Epicurus. Letters, Principal Doctrines, and Vatican Sayings. New York. Bobbs-Merrill. 1964

Epstein R., McKinney P., Fox S., Garcia C. Support for a fluid continuum model of sexual orientation; A large-scale internet study. *Journal of Homosexual Studies*. 2012; 1356–1381

Ereshefsky M. Defining 'health' and 'disease'. *Studies in History as Philosophy of Biological and Biomedical Sciences*. 2009; **40:** 221–227

Escobar JI., Burnam A., Karno M., Forsyte A. and Golding J. Somatization in the community. *Archives of General Psychiatry*. 1987**. 44:** 713–718

Evans AS. Causation and disease: the Henle-Koch postulates revisited. *Yale Journal of Biology and Medicine*. 1976; **49:** 175–195

Evans RW. Headaches case studies for the primary care physician. *Medical Clinics of North America*. 2003; **87:** 589–607

Evans SE et al. Exposure to domestic violence: a meta-analysis of child and adolescent outcomes. *Aggression and Violent behaviour*; **13**: 131–140, 2008

Ey H. Outline of an Organo-dynamic Conception of the Structure, Nosography, and Pathogenesis of Mental Diseases, 110–163, in: ***Natanson M.*** (ed.), Psychiatry and Philosophy. New York. Springer-Verla. 1969

Eysenck HJ. The Measurement of creativity, 199–242, in: MA. Boden (ed.). Dimensions of Creativity. Cambridge, Massachusetts. MIT Press. 1994

Engelhardt and JJ. McCartney (eds.). Concepts of Health and Disease. Reading, Massachussetts, Addison-Wesley. 1981

Fabrega H. Concepts of Disease: Logical Features and Social Implications. 493–522, in: ***AJ. Caplan, HT. Engelhardt and JJ. McCartney*** (eds.). Concepts of Health and Disease. Reading, Massachusetts, Addison-Wesley. 1981

Fabrega H. Psychiatric stigma in non-Western societies. *Comparative Psychiatry*; **32**: 534–551, 1991

Fabrice G., Guillem F., Tiberghien G., Stip E. Mismatched expressions decrease face recognition and corresponding ERP old/new effects in schizophrenia. *Neuropsychology*. 2012; **26** 568–77

Fagot-Largeault A. Les causes de la mort. Histoire naturelle et facteurs de risque. Paris. Vrin. 1989

Faires JS. and McCarthy DJ. Acute arthritis in man and dog after intrasynovial injection of sodium urate crystals. *The Lancet*. 1962: **280**: 682–685

Falagas ME., Vardakas KZ., Vergidis P. Under-diagnosis of common chronic diseases: prevalence and impact on human health. *International Journal of Clinical Practice*. 2007; **61:**1569–1579

Falk WD. **Morality, Self, and Others, in:** ***H-N. Castañeda and G. Nakhnikian*** (eds.). Morality and the Language of Conduct. 25–47. The Wayne State University Press. 1963

Farmer A. and McGuffin P. The classification of depressions. Contemporary confusion revisited. *British Journal of Psychiatry*. 1989. **155:** 437–443

Farrison PJH. Dopamine and schizophrenia—proof at last? *Lancet*. 2000; **356**: 958–959

Fazel S. and Danesh J. Serious mental disorder in 23000 prisoners: a systematic review of 62 surveys. *Lancet*. 2002; **359:** 545–550

Feder K.F*, in: *M. Solomon, JR. Simon, & H. Kincaid (eds.), 2017

Feinberg J. Doing and Deserving. Essays in the Theory of Responsibility. Princeton. Princeton University Press. 1970

Feinberg J. Social Philosophy. Englewood Cliffs. Prentice-Hall Inc. 1973

Feinberg J. Harm to others. The Moral Limits of the Criminal Law. Oxford University Press. Oxford. 1984

Feingold M., Johnson TJ. Television violence. Reactions from physicians, advertisers and the networks. *The New England Journal of Medicine*. 1977; **294:** 423–427

Feinstein A. Clinical Judgment. Baltimore. Williams & Wilkins. 1967

Feinstein A. The inadequacy of binary models for the clinical reality of three-zone diagnostic decisions. *Journal of Clinical Epidemiology*. 1991. **43:** 109–113

Feldman MD. and Ford CV. Patient or Pretender. Inside the Strange World of Factitious Disorders. New York. John Wiley & Sons. 1994

Fenton JJ., Weyrich MS., Durbin S. Prostate-specific antigen-based screening for prostate cancer. A systematic evidence review for the U.S. Preventive Task Force. Agency for Healthcare Research and Quality. Evidence Synthesis. N° 154. AHRQ Publication N° 17-05229.1498

Fenton JJ., et al. Prostate-specific Antigen-Based Screening for Prostate Cancer. Evidence Report and Systematic Review for the US Preventive Services Task Force. *JAMA*; **319**: 1914–1931, 2018

Fergusson D., Glass KC., Waring D. and Shapiro S. Turning a blind eye: the success of blinding reported in a random sample of randomised, placebo-controlled trials. *British Medical Journal*. 2004; **328:** 432

Field D. The social definition of illness, in: *Tuckett D.* (ed.), An Introduction to Medical Sociology. London. Tavistock; 1976

Figdor C. Neuroscience and the multiple realization of cognitive functions. *Philosophy of Science.* 2010; **77:** 419–456

Fine K. Essence and modality: The Second Philosophical Perspectives Lecture. *Philosophical Perspectives.* 1994; **8:** 1–16

Fingarette H. Heavy Drinking. The Myth of Alcoholism as a Disease. Berkeley. University of California Press. 1988

***Fink M., Akimova E., Spindelegger* C.** Social anxiety disorder. Epidemiology, Biology and Treatment. *Psychiatria Danubina.* 2009. **21:** 533–542

Finnis DG., Kaptchuk TJ., Miller F. & Benedetti F. Biological, clinical and ethical advances of placebo effects. *Lancet.* 2010; **375:** 686–694

First MB. Factors in the development of psychiatric epidemics, 130–142, in**: KS. Kendler and J. Parnas** (eds.) Philosophical issues in psychiatry IV. Oxford University Press. 2017

Fisher DH. Historians' Fallacies. London. Routledge & Kegan Paul. 1971

Fitzgerald FT. The tyranny of health. *The New England Journal of Medicine.* 1994; **331:**196–199

Fitzgerald M. Do psychiatry and neurology need a close partnership or a merger? *British Psychiatric Bulletin.* 2015; **39:** 105–107

Fleck L. Genesis and Development of a Scientific Fact. University of Chicago Press. 1935. 1979

Foddy B, Kahane G, Savulescu J. Practical Neuropsychiatric Ethics, pp. 1185–1201, in*: **KWM. Fulford, T. Thornton & G. Graham*** (eds.) Oxford Textbook of Philosophy and Psychiatry. Oxford. Oxford University Press. 2006

Fodor J. In Critical Condition. Cambridge. MIT Press. 1998

Fodor J., Piattelli-Palmerini M. What Darwin got wrong. Farrar, Straus and Giroux. 2021

Foot P (ed.). Moral Beliefs. Proceedings of the Aristotelian Society. 1958–9

Foot P (ed.). Theories of Ethics. Oxford University Press. 1967

Foster JA. & McVey K-A. Gut-brain axis: how the microbiome influences anxiety and depression. *Trends in neurosciences.* 2013; **36:** 305–312

Foucault M. Maladie mentale et psychologie. Paris. Presses Universitaires de France. 1954

Fox RC. Training for Uncertainty. in*: **RK. Merton, G. Reader and P. Kendall*** (eds.) The Student-Physician: Introductory Studies in the Sociology of Medical Education. Cambridge. Harvard University Press. 2008. pp. 207–241.

Fox RC. Essays in Medical Sociology. John Wiley and Sons. 1979a

Fox RC. The Human Condition of health professionals. School of Health Studies. University of New Hampshire. 1979b

Fox E. Predominance of the curative model of medical care. *Journal of the American Medical Association.* 1997; **278:** 761–763

Fox RC. and Swazey JP. The Courage to Fail. Chicago. The University of Chicago Press. 1974

Frangou S. A Systems Neuroscience Perspective of Schizophrenia and Bipolar Disorders. Schizophrenia Bulletin. 2014**; 40:** 522–531

***Frank JD.* Persuasion and Healing. New York. Schocken Books.** 1974

Frankfurt HG. Necessity and desires. *Philosophy and Phenomenological Research.* 1984; **45:** 1–13

Frankfurt JG. The Importance of what we Care about: Philosophical Essays. Cambridge University Press. 1988

***Freedman LZ.* and Hollingshead AB.** Neuroses and social class. I. Social interaction. *American Journal of Psychiatry.* 1971; **113:** 769–775

Frege G. Über Sinn und Bedeutung. *Zeitschrift für Philosophie und philosophische Kriti*k. 1892, NF **100:** 25–50

Freid TR., Bradley EH., Towle VR. & Allore H. Understanding the treatment preferences of seriously ill patients. *The New England Journal of Medicine.* 2002; **346:** 1061–1066

Freidson E. Profession of Medicine. A Study of Sociology of Applied Knowledge. New York. Dodd, Mead & Company. 1970

Frieden TR. Evidence for Health Decision Making – Beyond Randomized, Controlled Trials. *The New England Journal of Medicine*. 2017; 465–475

Friedman RA. Violence and mental illness–how strong is the link? *New England Journal of Medicine*. 2006; **355:** 2064–2066

Fromm E. The Fear of Freedom. London. Routledge & Kegan Paul. 1960

Fromm E. Beyond the Chains of Illusion. New York. Simon & Schuster, 1962

Frye N. Anatomy of Criticism. Princeton. Princeton University Press. 1957

Fuchs T. Delusion, Reality and Intersubjectivity. *Philosophy, Psychiatry and Psychology,* **27:** 61–79, 2020

Fulford KWM. Moral Theory and Medical Practice. Cambridge. Cambridge University Press. 1989

Fulford KWM., Thornton T. & Graham G. (eds.) Oxford Textbook of Philosophy and Psychiatry. Oxford. Oxford University Press. 2006. 2013

Furmark T., Appel L., Henningson S. *A* Link between Serotonin Related Gene Polymorphisms, Amygdala Activity, and Placebo Induced Relief from Social Anxiety. *The Journal of Neuroscience*. 2008. **28:** 13066–13074

Fuso S. Naturale = buono? Roma. Carocci Editore. 2016

Gabbani C. Epistemologia e clinica. Pisa. Edizioni ETS. 2013

Gaebel W., Wölver W. Facial expression and emotional face recognition in schizophrenia and depression. *European Archives of Psychiatry and Clinical Neuroscience.* 1992; **242**: 46–52

Galen. On the Therapeutic Method. Book I and II. Oxford. Oxford University Press. 1991

Galen and his patients. *The Lancet*; **378**: 478–479, 2011

Galen R., Robert S. and Gambino SR. Beyond Normality: The Predictive Value and Efficiency of medical Diagnoses. New York. John Wiley and Sons. 1975.

Galison P, Stump DJ. The Disunity of Science. Boundaries, Contexts, and Power. Stanford University Press. 1996

Gambarota F, Sessa P. Visual Working Memory for Faces and Facial Expressions as a Useful 'Tool' for Understanding Social and Affective Cognition. *Frontiers of Psychology,* **10:** 1–7, 2019

Gambill JD. The relevance of sociobiology for mental illness. *Perspectives in Biology and Medicine.* 1981; **25:** 155–165

Gandal MJ., Haney JR., Parikshak NN., Leppa V. et al. Shared molecular neuropathology across major psychiatric disorders parallels polygenic overlap. *Science.* 2018; **359:** 693–697

Gardner S. Irrationality and the Philosophy of Psychoanalysis. Cambridge. Cambridge University Press. 1993

Garfinkel A. Forms of Explanation. New Haven. Yale University Press. 1981.

Garson J. A Critical Review of Biological Functions. Springer. 2016; pp. 17–32

Gasking D. Causation and recipes. *Mind.* 1955; **64:** 479–487

Gaylin W. and Jennings B. The Perversion of Autonomy. New York. The Free Press. 1996

Geach P. Mental Acts. London. Routledge and Kegan Paul. 1957

Geach P. **Teleological Explanation, in:** ***S Körner*** (ed.). Explanation. New Haven. Yale University Press. 1975

Gerber D.J and Tonegawa S. Psychomimetic effects of drugs–a common pathway to schizophrenia? *New England Journal of Medicine.* 2004; **350:** 1047–1048

Gert B., Culver CM., Clouser KD. Bioethics. A Return to Fundamentals. Oxford University Press. 1997

Ghaemi SN. The Concepts of Psychiatry. Baltimore. The John Hopkins University Press. 2007

Gibbs WW. All in the Mind. Fact or Artifact? The Placebo Effect may be a Little of Both. *Scientific American*. **285:** 26, 2001

Giere RN. Statistical hypotheses, 251–271, in: ***LJ Cohen and M Hesse*** (eds.) Applications of Inductive Logic. Oxford. Clarendon Press. 1980

Giere RN. Understanding Scientific Reasoning. Fort Worth. Harcourt Brace and Jovanovich. 1991

Gifford F. Philosophy of Medicine. Handbook of Philosophy of Science. Volume 16. Amsterdam. Elsevier. 2011

Gill C., Sabin L. & Schmid C. Why clinicians are natural Bayesians. *British Medical Journal*.2005; **330:** 1080–1083
Gillett G. The Mind and its Discontents. Oxford. Oxford University Press. 1999
Gillies D. The Russo-Williamson thesis and the question of whether smoking causes heart disease 110–125, in*:* ***PM. Illari, F. Russo and J. Williamson,*** (eds.). Causality in the Sciences. Oxford. Oxford University Press. 2011
Gilligan J. Violence in public health and preventive medicine. *Lancet*.2000; **355:** 1802–1804
Gillick MR. Common-sense models of health and disease. *New England Journal of Medicine*; **313**: 700–703, 1985
Gillon R. Philosophical Medical Ethics. New York. Wiley & Sons. 1985
Giroux E. Définir objectivement la santé: une évaluation du concept biostatistique de Boorse à partir de l'épidémiologie moderne. *Revue philosophique de la France et de l'étranger*. **134:** 35–38, 2009
Glas G. Anxiety and Phobias: Phenomenologies Concepts, Explanations, 551–573, in: ***FWM. Fulford, T. Thornton & G. Graham*** (eds.) Oxford Textbook of Philosophy and Psychiatry. Oxford. Oxford University Press. 2013
Glenn AL. & Raine A. The neurobiology of psychopathy. *Psychiatric Clinics of North America*. **31:** 463–475, 2008
Glover J. Responsibility. London. Routledge & Kegan Paul. 1970
Glover J. Causing Death and Saving Lives. London. Penguin Books. 1977
Glover J. What Sort of Persons should there be? London. Penguin Books. 1984
Glover J. I: The Philosophy and Psychology of Personal Identity. Penguin Books. 1988
Glover J. *Choosing Children: Genes, Disability, and Design*. Clarendon Press: Oxford, 2007.
Glymour C. Theory and Evidence. Princeton University Press. 1980
Glymour B. Contrastive, non-probabilistic statistical explanations. *Philosophy of Science*. **65:** 448–471, 2003
Glymour G, Spiegelman D. Evaluating public health interview. Causal inference in public health research. *American Journal of Public Health*; **107**: 81–85, 2017
Goethe W. The Sorrows of Young Werther. 1774
Goethe JW. Sprüche in Prosa. Frankfurt. Insel; 2005
Goetzsche PC., Joergensen KJ. Screening for breast cancer with mammography. *Cochrane Database System Review*. 2013; **6:** CD001877
Goldberger E. How Physicians Think. Springfield. Charles C. Thomas.1965
Goldenberg MJ. Iconoclast or Creed? Objectivism, Pragmatism, and the Hierarchy of Evidence. *Perspectives in Biology and Medicine*. 2009**. 52:** 168–187
Goldman L and Shafer AI. Cecil Textbook of Medicine, 2019
Golub ES. The Limits of Medicine. How science shapes our Hope for the Cure. Chicago. Chicago University Press. 1997
Gonzalez-Moreno M., Saborido C., Teira D. Disease-Mongering through clinical trials. *Studies in History and Pathology of Biological & Biomedical Sciences*. 2015**; 51:** 11–18
Good I. Good Thinking. The Foundations of Probability and its Applications. Minneapolis. University of Minnesota. 1983
Good I. Causal tendency: a review, 23–50, in: ***B. Skyrms and WL. Harper,*** (eds.) Causation, Chance and Credence. Vol.1. Dordrecht. Kluwer Academic Publishers. 1988
Goody W. Syndromes. *Lancet*. 1961; **277:** 1–3
Goodman N. Fact, Fiction and Forecast. Boston. Harvard University Press. 1955
Goodyear-Smith FA., van Driel ML., Arroll B., Del Mar C. Analysis of decisions made in meta-analyses of depression screening and the risk of confirmation bias: a case study. *BMC Medical Research Methodology*. 2012; **12:** 76–93
Goodwin DW., Guze SB. Psychiatric Diagnosis. Oxford University Press. 1979
Goosens WK. Values, health and medicine. *Philosophy of Science*. 1980; **47**: 100–115
Gorowitz S. Doctors' Dilemmas. New York. Oxford University Press. 1982
Goshen CE. Documentary History of Psychiatry. London. Vision. 1967

Gould MS. Suicide and the Media, 200–221, in: ***H. Hendin and JJ. Mann*** (eds.). *Annals of the New York Academy of Sciences*, 2001; vol. **932:** 200–224
Gould SJ. The Mismeasure of Man. New York. W.W. Norton and Company. 1981
Gould SJ. The exaptive excellence of spandrels as a term and prototype. *Proceedings of the National Academy of Sciences*. 1991; **94:** 10750–10755
Gould SJ. La mesure de l'homme. Paris. Odile Jacob. 1997
Gould SJ. & Lewontin RC. The spandrels of San Marco and the panglossian paradigm: a critique of the adaptationist programme. *Proceedings of the Royal Society of London*. 1979; **205:** 581–598
Gowans CW. Innocence Lost. New York. Oxford University Press. 1994
Graham G. The Disordered Mind. An Introduction to Philosophy of Mind and Mental Illness. London. Routledge. 2010.
Grant M. The Raymond Tallis Reader. London. Palgrave. 2000
Gräsbeck R. Health and disease from the point of view of the laboratory, in: ***L. Nordenfelt and B. Lindahl*** (eds.). Health, Disease, and Causal Explanations in Medicine. Dordrecht, Reidel Publishing Company. 1984
Greco M. Medically unexplained symptoms. Social Science & Medicine. 2012
Green R. Homosexuality as a mental illness, 333–351, in: ***AJ. Caplan HT. Engelhardt and JJ. McCartney.*** (eds.). Concepts of health and Disease. Reading, MA. Addison-Wesley Publishing Company. 1981
Green JA. Prescribing by Numbers. Baltimore. John Hopkins. 2007
Green LA., Fryer GE., Yawn BP., Lanier D. and Dovey SM. The ecology of medical case revisited. *The New England Journal of Medicine.* 2001; **344:** 2021–2025
Green R. & Palpant N. (eds.) Bioethics and suffering. Oxford. Oxford University Press. 2014
Greenberg BG. Discussion, in: ***Zubin J.*** (ed.) Field Studies in Mental Disorders. Grüne & Stratton. New York. 1961
Greenspan G. The galloping ghost of Gauss and the "normal" radioiodine uptake. *California Medicine* 1970; **112:** 57–59
Greenwood Major FRS. Epidemics and Crowd-Diseases. London. Williams and Norgate. 1935
Grene M. The Understanding of Nature. Dordrecht. D. Reidel Publishing Company. 1974
***Grene* M.** Philosophy of Medicine; prolegomena to a philosophy of science. *PSA*. 1976, **2:** 77–83
Grice HP. Logical Conversation, in: ***P. Cole and J. Morgan*** (ed) Syntax and Semantics. New York. Academic Press. 1975
Grice GR. Motive and reason, 168–171, in: ***J. Raz*** (ed.). Practical Reasoning. Oxford. Oxford University Press. 1978
Grice HP. Logic and conversation. 1967; 22–40, in: **Studies in the Way of Words, *HP. Grice*** (ed.). Cambridge, MA: Harvard University Press. 1991
Grice GR. Studies in the Way of Words. Cambridge. Harvard University Press. 1989
Griffin J. Well-Being. Its meaning, measurement, and moral importance. Oxford. Oxford University Press. 1986
Griffin J. Quality of life measures in health care and medical ethics, 133–139, in: ***MC. Nussbaum & A. Sen*** (eds.). The Quality of Life. Oxford. Clarendon Press. 1993
Grinker RR. Psychosomatic Concepts. New York. Jason Aronson; 1973
Groopman J. How Doctors Think. Boston. Houghton Mifflin Company. 2007
Grüter T, Grüter M, CC Carbon. "Neural and genetic foundations of face recognition and prosopagnosia". *Journal of Neuropsychology. 2* : 79–97, 2008 797.
Guillain G. *J-M.* Charcot, 1825–1893. His Life-His Work. New York. Paul B. Hoeber. 1959
Guillaume B. L'esprit de la précaution. *Revue de Métaphysique et de Morale;* **4**: 491–509, 2012
Gupta A., Thompson D., Whitehouse A. et al. Adverse events associated with unblinded, but not with blinded, statin therapy in the Anglo-Scandinavian Cardiac Outcomes Trial. *The Lancet.* 2017; **389:** 2473–2481
Gur RE. & Gur RC. Functional magnetic resonance imaging in schizophrenia. *Dialogues in Clinical Neurosciences*. 2010; **12:** 333–343

Gur-Ozmen S., Nirmalanantha N., van Oetzen TJ. Change of pitch due to carbamazepine anox-carbazepine independently. *Seizure.* 2013; **22:** 162–163

Guyatt G., Owman AD., Akl EA., et al. GRADE Guidelines: 1. Introduction–GRADE evidence profiles and summary of findings tables. *Journal of Clinical Epidemiology*. 2011; **64:** 383–394

Haack S. Warrant, Causation, and the Atomism of Evidence Law. *Episteme.* 2008; **5:** 253–266

Hacking I. Representing and Intervening. Cambridge. Cambridge University Press. 1983

Hacking I. The sociology of knowledge of child abuse. *Nous.* 1988. **22:** 53–63

Hacking I. The Taming of Chance. Cambridge. Cambridge University Press. 1990

Hacking I. The social construction of what? Cambridge. Harvard University Press. 1999

Hacking I. An Introduction to Probability and Inductive Logic. Cambridge. Cambridge University Press. 2001

Hacking I. *Kinds* of People: Moving Targets. Proceedings of the British Academy. 2007; **151:** 258–318

Hacking I. Plasmare le persone. Corso al College de France (2004–2005). Urbino. Quattrocento. 2008

Hafer A. The Not-So-Intelligent Designer. Eugene. Oregon. Cascade Books. 2015

Hahn RA. The nocebo phenomenon: scope and foundations, 56–76, in: **_Harrington H._** (ed.). The Placebo Effect. Cambridge, Mass. Harvard University Press. 1997

Hahn SR. Physical symptoms and physician-experienced difficulty in the physician-patient relationship. *Annals of Internal Medicine*. 2001 **134:** 807–904

Hahn SR., Kroenke K., Spitzer RL., Brody D., Williams JB., Linzer M., et al. The difficult patient: prevalence, psychopathology, and functional impairment. *Journal of General Internal Medicine.* 1996; **11:** 1–8

Haiken E. Venus Envy a History of Cosmetic Surgery. John Hopkins University Press. 1997

Hall TS. Ideas of Life and Matter. The University of Chicago Press. 1969

Hall H. Too Many Medical Tests. *Skeptical Inquirer.* 2019; **43:** 25–27

Hall N. & Paul LA. Causation and Pre-emption, 100–130, in: **P. _Clark & K. Hawley_** (eds.). Philosophy of Science Today. Oxford. Oxford University Press. 2003

Hamdy FC., Donovan JL., Lane JA. et al. 10-Year Outcomes after Monitoring, Surgery, or Radiotherapy for Localized Prostate Cancer. *The New England Journal of Medicine.* 2016; **375:** 1415–1424

Hamilton WF. The Lewis A. Connor Memorial Lecture. The physiology of the cardiac output. *Circulation*. 1953; **8:** 527–543

Hampshire S. Thought and Action. London. Chatto and Windus. 1965

Hankey GJ. New drugs, or new trials of current drugs, for the treatment of acute ischaemic stroke? *The Lancet* 2001; **358:** 683–684

Hannay DR. The Symptoms Iceberg. A study of community health. London. Routledge & Kegan Paul. 1979

Hare RM. Pain and evil. *Proceedings of the Aristotelian Society, Suppl.* 1964**: 38:** 91–106

Hare RM. The Language of Morals. Oxford. Oxford University Press.1970

Harman G. The Inference to the Best Explanation. *The Philosophical Review.* 1965; **74:** 88–95.

Harrington A. The Cure Within. W.W. Norton & Company. 2008

Harris J. Enhancing Evolution. The Ethical Case for Making Better People. Princeton University Press. 2007

Haslam N. Reliability, validity, and the mixed blessing of operationalism, 987–1002, in: **_Fullford KWM., Davies M., RGT. Gipps, Graham G. et al._** (eds.). 2013

Heath I. Treating violence as a public health problem. *British Medical Journal*. 2002; **325:** 727–728

Heinz A., Friedel E., Krüger H-P., Wackerhagen C. Philosophical Implications of Changes in the Classification of Mental Disorders in DSM-5, 1025–1039, in: **_T. Schramme & S. Edwards_** (eds.). Handbook of the Philosophy of Medicine, vol. 2. Dordrecht. Springer. 2017

Harman G. Change in View. Cambridge. The MIT Press. 1986

Harman D. Aging and disease: extending the life span, 321–336, in: ***K. Kitani, A. Aoba & S. Gotto*** (eds.). Pharmacological intervention in aging and age-associated disorders, *Annals of the New York Academy of Sciences*; 1996; **786:** 1–460

Harrington A. (ed.) The Placebo Effect. An Interdisciplinary Exploration. Cambridge. Harvard University Press. 1997

Hart HLA. and Honoré T. Causation in the Law. Oxford. Oxford University Press. 2d Edition. 1985

Harvey W. Exercitationes duae Anatomicae de Circulatione Sanguinis ad J. Riolanum. Cambridge 1649. in: ***Harvey W.*** Opera Omnia; Leyden; 1737

Hatfield G. The Natural and the Normative. Cambridge, Massachusetts, MIT Press. 1990

Havens LL. Approaches to the Mind. Boston. Little Brown and Company. 1973

Haynes RB., Sackett DL., Taylor DW. et al. Increased absenteeism from work after detection and labelling of hypertensive patients. *New England Journal of Medicine.* 1978; **299:** 741–744

Healy D. Let Them Eat Prozac: The Unhealthy Relationship Between the Pharmaceutical Industry and Depression. New York. New York University Press. 2006

Heart Protection Study Collaborative Group. ***MRC/BHF*** Heart Protection Study of cholesterol lowering with sinvastatin in 20536 high-risk individuals: a randomised placebo-controlled trial. *Lancet.* 2002. **360:** 7–22

Heinz A., Friedel E., Krüger H-P, Wackerhagen C. Philosophical Implications of Changes in the Classification of Mental Disorders in DSM-5, 1025–1039, in: ***T. Schramme & S. Edwards*** (eds.) Handbook of the Philosophy of Medicine, vol 2. Dordrecht. Springer. 2008

Hellman CG. Culture, Health and Illness. Oxford. Butterworth. Heinemann. 1994

Hempel C. Aspects of Scientific Explanation and Other Essays. New York. Free Press; 1965

Henderson L. The Problem of Induction, in: ***EN. Zalta*** (ed.) The Stanford Encyclopedia of Philosophy. Metaphysics Research Lab. Stanford University. 2010

Henn F., Sartorius N., Helmchen H., Lauter H. Contemporary Psychiatry. Vol 1. Foundations of Psychiatry. New York. Springer. 2001

Hennekens CH. and Buring JE. Epidemiology in Medicine. Lippincott Williams and Wilkins. 1987

Hernan MA. A definition of causal effect for epidemiologic research. *Journal of Community Health.* 2004; **58:** 265–271

Herzlich C. Health and Illness. London. Academic Press. 1973

Hesse M. The Structure of Scientific Inference. Berkeley. University of California Press. 1974

Hesslow G. What is A Genetic Disease? 183–193, in: ***L. Nordenfelt and BIB. Lindahl*** (eds.). Health, Disease and Causal Explanation in Medicine. Dordrecht. Reidel. 1984

Hesslow G. Do we need a concept of disease? *Theoretical Medicine.*1993; **34**: 1–14

Heywang SH., Schreer I., Hacker A. et al. Conclusions for mammography screening after 25-year follow-up of the Canadian Screening Study (CNBSS). *European Radiology.* 2016; **26:** 342–350

Hieronymus F., Lisinski A., Nilsson S. and Eriksson E. Efficacy of selective serotonin reuptake inhibitors in the absence of side effects. *Molecular Psychiatry.* On line publication, 25 July, 2017

Hill AB. Principles of Medical Statistics. London. The Lancet Ltd. 1967

Hill JA. Medical Misinformation: Vet the Message. *European Heart Journal.* 2019; **40:** 404–405

Hinkle LE., Redmont R., Plummer N. and Wolff HG. An examination of the relation between symptoms, disability and serious illness in two homogeneous groups of men and women. *American Journal of Public Health.* 1960; **50:** 1327–1336

Hippocrates. The Genuine Works of Hippocrates. F. Adams, (ed.). Baltimore. The Williams Wilkins Company. 1939

Hobbes T. Leviathan or the Matter, Form & Power of a Commonwealth Ecclesiastical and Civil. 1904

Hodges W. Logic. London. Penguin Books. 1977

Hoehndorf R., Axel-Cyrille Ngonga Ngomo A-C N, Kelso J. Applying the functional abnormality ontology pattern to anatomical functions. *Journal of Biomedical Semantics.* 2010; **1:** 4–12

Hoey J. The one and only one Mrs. Jones. *Canadian Medical Journal Association.* 1998; **159**: 241–242

Höffler M. Causal inference based on counterfactuals. *BMC Medical Research Methodology.* 2005; doi: https://doi.org/10.1186/1471-2288-5-28.

Hofstadter D. I am a strange loop. New York. Basic Books. 2007

Hofmann B. Suffering: Harm to Bodies, Minds, and Persons, 129–145, in: ***T. Schramme & S. Edwards,*** (eds.) Volume 1, Dordrecht, Springer. 2017

Hofman B. Do health professionals have a prototype concept of disease? The answer is no. *Philosophy, Ethics, and Humanities in Medicine.* 12: 1, article 6, 2017

Hogarthy GE & Goldberg GE. Drug and sociotherapy in the aftercare of schizophrenic patients. *Archives of General Psychiatry.* 1973; **28:** 54–64

Holland WW. (ed.), Evaluation of Health Care. Oxford. Oxford University Press. 1983

Holland WW. and Karhausen L (eds.). Health Care and Epidemiology. London. Henry Kimpton. 1978

Holland PW. Comment: Causal Mechanisms or Causal Effect: Which is Best for Statistical Science? *Statistical Science.*1988; **3:** 186–188

Holland S. Death as a biological category, 189–206, in: ***T. Schramme, S. Edwards,*** (eds.). 2017

Hollingshead AB. and Redlich FC. Social Class and Mental Illness. New York. 1958

Hórbjartsson A. and Gotzsche PC. Is the placebo powerless? An analysis of clinical trials comparing placebo with no treatment. *The New England Journal of Medicine* 2001; **344:**1594–1602

Horrobin D. The Madness of Adam and Eve. How Schizophrenia shaped up Humanity. Bantam Press. 2001

Horwitz RI., Viscoli CM., Berkman L. et al. Treatment adherence and risk of death after myocardial infarction. *Lancet.* 1990; **336:** 542–545

***Hospers* J.** An Introduction to Philosophical Analysis. London Routledge & Kegan Paul. 1967

Hospers J. An Introduction to Philosophical Analysis. London Routledge & Kegan Paul. 1990

Howick J. Justification of Evidence-Based Medicine, 113–146, in: ***JA. Marcum*** (ed.). The Bloomsbury Companion to Contemporary Philosophy of Medicine. London. Bloomsbury. 2017

Howick J. The Philosophy of Evidence-Based Medicine. BMJ Books. Wiley-Blackwell. 2011

Howie JGR. Clinical judgment and antibiotic use in general practice. *British Medical Journal.* 1976; **30:**1061–1064

Hruby A, Hu FB. The epidemiology of obesity. *Pharmacoeconomics.* 2013; 673–689

Huang Z., Liu X., Mashour GA. & Hudetz G. Timescales of intrinsic BOLD signal dynamics and functional connectivity in pharmacology and neuropathologic states of consciousness. *Journal of Neurosciences.* 2018**; 38:** 2304–2317

Hubley S, Uebelacker L, Eaton C. Managing medically unexplained symptoms in primary care. *American Journal of Lifestyle Medicine.* 2014; **10:** 109–119

Hull DL. The Effect of Essentialism on Taxonomy: Two Thousand Years of Stasis. *British Journal of the Philosophy of Science.* 1965; **15:** 314–326

Hull D. Textbook of Biological Science. Prentice-Hall, 1974

Hull DL. Are species really individuals? *Systematic Zoology.* 1976; **25:**174–191

Humber JM. and Almeder RF. (eds.). What is a Disease? Totowa, New Jersey, Humana Press. 1997

Hume D. Enquiries Concerning the Human Understanding and Concerning the Principles of Morals. Oxford, in: ***LA. Selby-Bigge*** (ed.) Oxford. Clarendon Press. 1975

Hume D. A Treatise of Human Nature, ***LA. Selby-Bigge*** (ed.). Oxford. Clarendon Press. 1973

Hume D. Dialogues Concerning Natural Religion. London. Nelson. 1947

Humphreys P. The Chances of Explanation. Princeton, N.J. Princeton University Press. 1989

Hurley SL. Natural Reasons. Personality and Polity. Oxford. Oxford University Press. 1989

Hunter KM. Doctors Stories. The narrative structure of medical knowledge. Princeton. Princeton University Press. 1991

Husserl E. The Crisis of European Sciences. Northwestern University Press. 1970

Hutchison GB. Evaluation of preventive services *Journal of Chronic Diseases.* 1960; **11:** 497–508

Huyse FJ. and Stiefel FC. Integrated Care for the Complex Medically Ill. *Medical Clinics of North America.* 2006; **90:** 756–767

Hyler SE. and Spitzer RL. Hysteria split asunder. *American Journal of Psychiatry.* 1978; **135:** 1500–1504

Hyman SE. The diagnosis of mental disorders: the problem of reification. *Annual Review of Clinical Psychology.* 2010; **6:** 155–179

IASP Subcommittee on Taxonomy Pain Terms**.** Pain terms: a list with definitions and notes on usage. *Pain*. 1979; **6:** 249–252

Illary PM., Williamson J. Mechanisms are real and local, 818–844, in*: **PM. Illari, F. Russo and J. Williamson,*** (eds.). Causality in the Sciences. Oxford. Oxford University Press. 2011

Illari PM. & Russo F. Causality. Philosophical theory meets scientific practice. Oxford. Oxford University Press. 2014

Illic D., Neuberger MM., Djulbegovic M. et al. Screening for prostate cancer. *Cochrane Database of Systematic Reviews,* Issue1. Art. NO.: CD004720. 2013. doi: https://doi.org/10.1002/14651858. CD004720/pub3.

Illitch I. Limits to Medicine. Medical Nemesis. London. Penguin Books. 1978

Independent UK Panel of Breast Cancer Screening. The benefits and harms of breast cancer screening: an independent view. *The Lancet.* 2012; **380:** 1778–1786

Insel TS. and Quirion R. Psychiatry as a clinical neuroscience. *JAMA*. 2005; **294:** 2221–2224

Ingelfinger FJ. Violence on TV: an unchecked environmental hazard. *The New England Journal of Medicine*. 1977; 837–838

Institute of Medicine. Crossing the quality chiasm: a new health system for the 21st century. Washington, D.C. National Academy Press. 2001

International Classification of Impairments, Disabilities and Handicaps. World Health Organisation. 1980

International classification of diseases and related health problems. ICD 10. 10th revision. Geneva. World Health Organisation. 1992

Jablensky A. Classification of nonschizophrenic disorders. National Institutes of Health (NIH); **3**:326–331, 2001

Jablensky A. Research Methods in Psychiatric Epidemiology: An Overview. *Australian & New Zealand Journal of Psychiatry;* **36:** 298–310, 2002

Jablensky A. The syndrome-an antidote to spurious comorbidity? *World Psychiatry.* 2004; **3:** 24–25

Jablensky A. A Commentary on 'A Proposal for a meta-structure for DSM-5 and ICD-11'. *Psychological Medicine*. 2009; **39:** 2099–2103

Jablensky A. The Nosological Entity in Psychiatry, 76–94, in: ***KS. Kendler, J. Parnas.*** Philosophical Issues in Psychiatry. Oxford University Press. 2012

Jablensky A. Psychiatric classifications: validity and utility. *World Psychiatry.* 2016; **15:** 26–31

Jablensky A and Kendell RE. Criteria for assessing a classification in psychiatry, 1–24, in: ***Maj M., Gaebel W., Lopez-Ibor JJ. and Sartorius N.*** (eds.). Psychiatric Diagnosis and Classification. Chichester. Wiley & Sons. 2002

Jablensky A and Sartorius N. What did the WHO studies really find? *Schizophrenia Bulletin*; **34**: 253–255, 2008

Jablensky A., Sartorius N., Ernberg G., Anker M., et al. Schizophrenia: manifestations, incidence and course in different cultures. A World Health Organization Ten-Country Study. Geneva. World Health Organization. 1992

Jackson F. Epiphenomenal Qualia. *Philosophical Quarterly*. 1982; **32:** 127–136

Jackson JL., Kroenke K. Difficult patient encounters in the ambulatory clinic: clinical predictors and outcomes. *Archives of Internal Medicine*. 1999; **159:** 1069–75.

Jackson JL., and Kroenke K. The effect of unmet expectations among adults presenting physical symptoms. *Annals of Internal Medicine*. 2001; **134:** 889–897

James W. Pragmatism. Harvard University Press. 1907. 1979

Jameson R. Personal view. *British Medical Journal*. 1985; **291:** 541

Jameson L. et al.. Harrison Principles of Internal Medicine. MacGraw-Hill. 2018

Jamison KR. Touched with Fire. Manic-Depressive Illness and the Artistic Temperament. New York. The Free Press. 1993

Janovic J. Tourette's syndrome. *The New England Journal of Medicine* 2002; **345:**1184–1192
Januzzi JL., Stern TA., Pasternak RC., et al. The influence of anxiety and depression on outcomes of patients with coronary artery disease. *Archives of Internal Medicine*. 2000; **160:** 1913–1921
Jaspers K. General Psychopathology. Manchester. Manchester University Press; 1962
Jaspers K. General Psychopathology. 1923. Manchester University Press. 1963
Jaynes J. The Origins of Consciousness in the Breakdown of the Bicameral Mind. London. Allen & Unwin. 1976
Jaspers K. General Psychopathology. 2013. Manchester University Press. 1982
JNC VII. The seventh report of the Joint National Committee on Prevention, Detection, Evaluation and Treatment of High Blood Pressure. *JAMA*. **289:** 2560–2572. 2003
Johansson I. & Lynoe N. Medicine & Philosophy. London. Walter de Gruyter. 2008
Johnson HA. Diminishing returns on the road to certainty. *JAMA*. 1991; **265:** 2229–2231
Johnson WD. **Tuberculosis, in:** ***Kiple KF.*** (ed.). The Cambridge World History of Human Disease. Cambridge. Cambridge University Press. 1993
Joplin DA. Placebos Effects in Psychiatry and Psychotherapy. 1202–1226, in: ***KW. Fulford, M. Davies, RGT. Gipps, et al.*** (eds.). The Oxford Handbook of Philosophy of Psychiatry. 2013
Jorgensen KJ., Gotzche PC., Kalager M. et al. **Breast** Cancer Screening in Denmark. *Annals of Internal Medicine*. 2017; **166:** 313–323
Judd LL., Schettler PJ. and Akiskal HS. The prevalence, clinical relevance and public health significance of subthreshold depressions. *Psychiatric Clinics of North America*. 2002; **25:** 685–698
Kahneman D. & Tversky A. Prospect Theory: An Analysis of Decision and Risk. *Econometrica*. 1979**, 47:** 263–291
Kandel ER. Principles of Neural Science. Columbia University. 1972
Kandel ER. A new intellectual framework for psychiatry. *American Journal of Psychiatry* 1998; **155:** 457–469
Kandel ER. et al. Essentials of neural science and behavior. Appleton and Lange. 1995
Kant I. Versuch über die Krankheiten des Kopfes, 259–271, in: Kant's Werke, Bd II–Gesammelte Schriften, Vorkritische Schr. II, 17571777, Berlin. Reimer. 1905
Kant I. Critique of Teleological Judgment, Oxford. Clarendon Press 1952
Kant I. Critique of Pure Reason. translated by M Müller, New York. Dolphin Books. Doubleday. 1961
Kant I. The Metaphysical Principles of Virtue. Indianapolis. Bobbs Merrill Company. 1964
Kant I. Critique of Pure Reason. New York. St. Martin's Press. 1965
Kant I. The Moral Law. Kant's Groundwork of the Metaphysics of Morals. London. Hutchinson University Library. 1972
Kant I. Lectures of Ethics. Gloucester. Peter Smith. 1978
Kant I. "Essay on the Maladies of the Mind" 1764 and Observations on the Feeling of the Beautiful and the Sublime. Cambridge University Press. 2000
Kant I. Premium Collection: Complete Critiques, Philosophical Works and Essays. 2001
Kant I. Critique de la faculté de juger. GF Philosophie. 2015
Kantrowitz JT., Epstein ML., et al. Neurophysiological mechanisms of cortical plasticity impairments in schizophrenia and modulation by the NMDA receptor agonist D-serine. *Brain*. 2016; **139:** 3281–3295
Kaptchuk TJ., Kelley JM., Conboy LA. et al. Components of placebo effect: randomised controlled trial in patients with irritable bowel syndrome. *British Medical Journal*. 2008; **336:** 999–1003
Karhausen L. Anatomical Vales for Reference Man, 8–272, in: ***WS. Snyder, MJ. Cook., ES. Nasser, L. Karhausen, G. Parry Howells, IH Tipton.*** Report of the Task Group on Reference Man. Oxford. Pergamon Press. 1975
Karhausen L. Re: On the logic of causal inference. *American Journal of Epidemiology*. 1987a; **126:** 556–557
Karhausen L. From ethics to medical ethics, in S Doxiadis (ed.). Ethical Dilemmas in Health Promotion. New York. Wiley and Sons. 1987b

Karhausen L. The Logic of Causation in Epidemiology. *Scandinavian Journal of Social Medicine.* 1995a; **23:**1–5

Karhausen L. The poverty of Popperian epidemiology. *International Journal of Epidemiology.* 1995b; **24:** 869–874

Karhausen L. Theoretical versus 'Atheoretical' Epidemiology. *International Journal Journal of Epidemiology*. 1997; **26:** 1

Karhausen L. Causation: the elusive grail of epidemiology. *Medicine, Health Care and Philosophy*. 2000; **3:** 59–67.

Karhausen L. Causation in epidemiology: a Socratic dialogue. *International Journal of Epidemiology*. 2001a; **30:** 704–711

Karhausen L. Commentary: Coda – a Socratic dialogue: Plato. *International Journal of Epidemiology*. 2001b; **30:** 710–711, 2001

Karhausen L. Exposures, mutations and the history of causality. *Journal of Epidemiology & Community Health*. 2001; **55:** 607–608

Karhausen L The Bleeding of Mozart. Xlibris. 2011

Karhausen L. Mythologies médicales. Paris. 2014. Paris. L'Harmattan

Karhausen L. Dr Georges Canguilhem Médecin anomal. Paris. L'Harmattan. 2017

Karhausen L Analytic Philosophy of Clinical and Community Medicine. Xlibris. 2022

Karhausen L. The Fragility of Philosophy of Medicine. Springer. 2023

Karpman B. Criminality, insanity, and the law. *Journal of Criminal Law and Criminal Police Science*. 1949; **39:** 584–604

Kassirer JP. Incorporating patients' preferences into medical decisions. *New England Journal of Medicine*.1994; **330:** 1895–1896

Kassirer JP. On the Take: How Medicine's Complicity with Big Business Can Endanger Your Health. Oxford. Oxford University Press. 2005

Kasl S. **and Kobb S.** Health behavior, illness behavior and sick role behavior. *Archives of Environmental Health*. 1966; **12:** 246–266

Katon W., Sullivan M. and Walker E. Medical Symptoms without identified pathology: relationship to psychiatric disorders, childhood and adult trauma, and personality traits. *Annals of Internal Medicine*. 2001. **134:** 917–925

Kaufmann W. The Faith of a heretic. New York. Doubleday & Company. 1963

Kaufman, F. Disease: definition and objectivity, in: ***JM. Humber and RF. Almeder*** (eds.). What is Disease? Totowa, New Jersey. 1997

Kaunitz AM., Pinkerton JV., Ghate SV. et al. Screening Mammography for Average-Risk Women. *Menopause*. 2018; **25:** 343–345

Keating P. & Cambrosio A. Biomedical Platforms. Realigning the Norma and the Pathological in Late-Twentieth Century Medicine. Cambridge. The MIT Press. 2003

Keller MB et al. A comparison of Nefazodone, the cognitive behavioral-analysis system of psychotherapy, and their combination for the treatment of chronic depression. *New England Journal of Medicine;* **342**: 1462–1467, 2000

Kendell RE. The Concept of Disease and its Implications for Psychiatry. *British Journal of Psychiatry.* **127:**305–315; **1975**

Kendell RE. The concept of disease and its implications for psychiatry, 443–458, in: ***A. Caplan HT., Engelhardt and JJ. McCartney*** (eds.). Concepts of Health and Disease. London. Addison-Wesley. 1981

Kendell RE., Jablensky A. Distinguishing between the validity and utility of psychiatric diagnoses. *American Journal of Psychiatry*. 2003; **160:** 4–12

Kendler KS., Parnas J. (eds.) Philosophical issues in psychiatry. Explanation, Phenomenology, and Nosology. Johns Hopkins University Press. 2008

Kendler KS., Zachar P., Craver C. What kinds of things are psychiatric disorders? *Psychological Medicine*. 2011; **41:** 1143–1150

Kendler KS. The nature of psychiatric disorders. *World Psychiatry*. 2016; 5–12

Kendler KS., Parnas J. (eds.) Philosophical issues in psychiatry IV classification of psychiatric illness. Oxford University Press. 2017

Kessler RC., McGonacle KA., Swartz M., Blazer DG. and Nelson CB. Sex and depression in the National Comorbidity Survey, I. Lifetime prevalence, chronicity and recurrence. *Journal of Affective Disorders*. 1993; **29:** 85–96

Kessler RC. Need for attention to mental health in young adults. *Lancet*. 2002; **359:** 1956–1957

Kessler RC., Bromet EJ. The epidemiology of depression across cultures. *Annual Review of Public Health*. 2013; **34:** 119–138

Khalidi MA. Interactive kinds. *British Journal of the Philosophy of Science*. 2009; **61:** 335–360

Khan AA., Khan A., Harezlak J. et al. Somatic symptoms in primary care: etiology and outcome. *Psychosomatics*. 2003; **44:** 471–478

Khosla P., Bajaj V., Sharma G. and Mishra K. Background noise in healthy volunteers—a consideration in adverse drug reaction studies. *Journal of Physiological Pharmacology*. 1992; **36:** 259–262

Kim J. Causes and counterfactuals. *Journal of Philosophy*.1973; **70:** 570–572

Kim J. Supervenience, emergence, realization, reduction, 556–584, in: ***MJ. Loux and DW. Zimmerman*** (eds.). The Oxford Handbook of Metaphysics. Oxford. Oxford University Press. 2003

Kim J. The Layered Model: Metaphysical Considerations. *Philosophical Explorations*. 2002; **5:** 2–20

Kim J. Supervenience, or something near enough. Princeton. Princeton University Press. 2005

Kim H., Hudetz G., Lee J., Mashour GA. et al. Estimating the Integrated Information Measure Phi from High-Density Electroencephalography during States of Consciousness in Humans. *Frontiers in Human Neurosciences*. 2018; 20 February

King LS. What is a disease? in: ***Caplan AL., Engelhardt T., McCartney JJ.*** (eds.). Concepts of Health and Disease, London. Addison-Wesley. 1981

King LS. Medical Thinking. A Historical Perspective. Princeton. Princeton University Press. 1982

Kingma E. Disease as Scientific and as Value-Laden Concept, 45–63, in: ***T. Schramme & S. Edwards*** (eds.). Handbook of the Philosophy of Medicine, Volume 1. Dordrecht. Springer. 2004

Kingma E. What Is It to Be Heathy? *Analysis*. 2007; **67:** 128–133

Kingma E. Naturalist Account of Mental Disorder, 263–384, in: ***KWM. Fulford, M. Davies, RGT. Gipps, et al.*** (eds.) The Oxford Handbook of the Philosophy Psychiatry. Oxford University Press. 2013

Kinkaid H. Philosophical Foundations of the Social Sciences. Analyzing Controversies in Social Research. Cambridge. Cambridge University Press. 1993

Kiple KF. (ed.). The Cambridge World History of Human Diseases. Cambridge. Cambridge University Press. 1993

Kirkbride JB., Fearon P., Morgan C., Dazzan P. et al. Heterogeneity in Incidence Rates of Schizophrenia and Other Psychotic Syndromes. *Archives of General Psychiatry*. 2006; **63:** 250–258

Kirmayer LJ., Robbins JM. Three forms of somatization in primary care. Prevalence, co-occurrence, and sociodemographic characteristics. *The Journal of Nervous and Mental Disease*. 1991; **179:** 647–655

Kirsch I. Specifying nonspecific psychological mechanisms of placebo effects, 166–186, in: ***Harrington H.*** (ed.) The Placebo Effect. Cambridge, Mass. Harvard University Press. 1997

Kitcher P. The hegemony of molecular biology. *Biology and Philosophy*. 1999; **14:** 195–210

Kitcher P. Science, Truth, and Democracy. Oxford University Press. 2001

Klaus K., Rief W., Brähler E. et al. The distinction between "medically unexplained" and "medically explained" in the context of somatoform disorders. *International Journal of Behavioral Medicine*. 2013; **20:**161–71

Kleinman A. The Illness Narratives. New York. Basic Books. 1988

Klempner MS. and Shapiro DS. Crossing the species barrier–One small step to man, one giant leap to mankind. *New England Journal of Medicine*. 2004**; 350:** 1171–1172
Knowles JH. *(ed.)*. Doing Better and Feeling Worse, New York. Norton & Company. 1977
Knox J. Can the self survive the death of its mind*? Religious Studies*. 1969; **5:** 85–97
Knutson D. Modern Medicine. Public Health Policy and Infant Health Evidence From a Preventive Health Program in Sweden. Stockholm University. 2018
Koberstein JN., Poplawski SG., Wimmer ME. Learning-dependent chromatin remodeling highlights noncoding regulatory regions linked to autism. *Science Signal.* 2018; 11: 16 Jan
Koerner BI. Disorders Made to Order. *Mother Jones*.2002**; 27:** 58–63
Komaroff AL. 'Minor' illness symptoms. The magnitude of their burden and of our ignorance. *Archives of Internal Medicine*. 1990; **150:** 1586–1587
Konstam MA. Systolic and diastolic dysfunction in heart failure? Time for a new paradigm. *Journal of Cardiac Failure*. 2003; **9:** 1–3
Koroukian S., Dong W., Berger NA., Changes in Age Distribution of Obesity-Associated Cancers. *JAMA Network Open*. DOI:https://doi.org/10.1001/jamanetworkopen.2019.9261
Kraepelin E. Lectures on Clinical Psychiatry. New York. William Wood and Company. 1913
Kramer PO. Listening to Prozac. New York. Penguin. 1993
Kraut, R. The rationality of prudence. *Philosophical Review*. 1972; 351–359
Krieger N, Davey Smith G. The tale wagged by the DAG: broadening the scope of causal inference and explanation for epidemiology. *International Journal of Epidemiology*; **46**: 1787–1808, 2016
Kriegbaum M., Liisberg KB., Wallach-Kildemoes H. Pattern of statin use changes following media coverage of its side effects. *Patient Preference Adherence*. 2017**; 11:** 1151–1157
Kripke, SA. Naming and Necessity. Cambridge. Harvard University Press. 1980
Kretschmer E. Physique and Character. London. Kegan Paul, Trench, Trubner. 1925
Kroenke K. Studying symptoms: sampling and measurement issues. *Annals of Internal Medicine*. 2001; **134:** 844–853
Kroenke K. and Mangelsorff AD. Common symptoms in ambulatory care: incidence, evaluation, therapy, and outcome. *American Journal of Medicine*. 1989**; 86:** 262–266
Kroenke K. and Rosmalen JGM. Symptoms, syndromes, and the value of psychiatric diagnostics in patients who have functional somatic disorders. *Medical Clinics of North America*. 2006; **90:** 603–626
Krohs U. & Kroes P. (eds.). Functions in Biological and Artificial Worlds. London. MIT Press. 2009
Kubzansky LD., Huffman JC., Boehm JK, et al. Positive psychological well-being and cardiovascular disease. *Journal of the American College of Cardiology*. 2018; **72:** 1382–1396
Kurz P. Decision and the Condition of Man. New York. Dell Publishing Company. 1965
Lach G., Schellekens H. et al. Anxiety, Depression, and the Microbiome: A Role for Gut Peptides. *Neurotherapeutics*. 2017; Nov 23: 1–24
LaCombe MA. What is it patients want? Editorial. *The American Journal of Medicine*. 1995; **99:** 588–589
Lader M. and Marks I. Clinical Anxiety. London. William Heinemann Medical Books. 1971
Laing RD. The Divided Self. London. Penguin Books. 1965
Laing RD. The Politics of Experience and the Bird of Paradise. New York. Ballantine Books Inc. 1967
Lakoff G. Women, Fire and Dangerous Things: What Categories Reveal about the Mind/ University of Chicago Press. 1987
Lalumera E. Understanding Schizophrenia Through Wittgenstein: Empathy, Explanation and Philosophical Clarification, in: ***Hippolito I, Gonçalves J, Pereira J*** (eds.) Schizophrenia and Common Sense. Springer Nature. 2017
Langley LL. Homeostasis. New York. Van Nostrand Reinhold Company. 1965
Largier N. In Praise of the Whip. A Cultural History of Arousal. Brooklyn. Zone Books. 2007
Last JM. The Iceberg: 'Completing the clinical picture' in general practice. *Lancet*. 1963**: 282:** 28–31
Last JM. A Dictionary of Epidemiology. Third Edition. Oxford. Oxford University Press. 1995

Last JM. Commentary: The iceberg revisited. *International Journal of Epidemiology*. 2013; **42:** 1613–1617

Leaders. Sex and Science. *The Economist*, February 18th, 2017; 7

Lee TH. Ecology in evolution. *The New England Journal of Medicine*. 2001; **344:** 2018 -2020

Lee K., The Philosophical Foundations of Modern Medicine. London. Palgrave MacMillan. 2012

Leff J. Psychiatry around the Globe. A Transcultural View. New York. Marcel Dekker Inc. 1981

Leff J, Sartorius N., Jablensky A., Korten A. and Ernberg G. The International Pilot Study of Schizophrenia: five-year follow-up findings. *Psychological Medicine*. 1992; **22:** 131–145

Legrenzi P., Umiltà C. Neuromania. On the Limits of Brain Science. Oxford. Oxford University Press. 2011

Leigh H. and Reiser MF. Major trends in psychosomatic medicine. The psychiatrist's evolving role in medicine. *Annals of Internal Medicine*. 1977; **87:** 233–239

Lemoine M. Introduction à la philosophie des sciences médicales. Paris. Hermann. 2017

Lende D. The Research Domain Criteria of the NIMH and the RDoC Vision for Mental Health Research and Diagnosis. *American Journal of Psychiatry;* **171**: 385–397, 2014

Leor J., Poole WK., Kloner RA. Sudden cardiac death triggered by an earthquake. *New England Journal of Medicine*. 1996; **334:** 413–419

Leplège A. and Hunt S. The problem of quality of life in Medicine. *JAMA*. 1997; **276:** 47–50

Levine GN. The Mind-Heart-Body Connection. *Circulation* 2019; **140:** 1363–1365

Levitin DJ. and Tirovolas AK. Current advances in the cognitive neuroscience of music. *Annals of the New York Academy of Sciences*. 2009; **1156:** 211–231

Lewis. A. Prognosis in the manic-depressive psychosis. *Lancet*. 1936; **55:** 997–999

Lewis DK. Convention. A Philosophical Study. Cambridge. Massachusetts. Harvard University Press. 1969

Lewis DK. Counterfactuals. Blackwell. 1973

Lewis, DK. On the Plurality of Worlds, Blackwell, Oxford, 1986a

Lewis DK. Philosophical Papers. Vol. II. Oxford. Oxford University Press. 1986b

Lewontin RC. Adaptation, in*: **E. Sober*** (ed.). Conceptual Issue in Evolutionary Biology. An Anthology, 235–251, Cambridge, MIT Press. 1984

Lewontin RC. Not so natural selection. *New York Review of Books*. 2021, 25/07/21

Lewy G.* G.E. Moore on the naturalistic fallacy, in: *PF. Strawson (ed.). Studies in the Philosophy of Thought and Action, 134–146, Oxford. Oxford University Press. 1968

Lilienfeld SO, Marino L. Mental Disorder as a Roschian Concept: A Critique of Wakefield's "Harmful Dysfunction" Analysis. *Journal of Abnormal Psychology*. **104:** 411–420, 1995

Lindqvist PG., Epstein E., Nielsen K., et al. Avoidance of sun exposure as a risk factor for major causes of death: a competing risk analysis of the Melanoma in Southern Sweden cohort. *Journal of Internal Medicine*. 2016**; 280:** 375–387

Lipsitt DR. Consultation-Liaison Psychiatry and Psychosomatic Medicine: The Company They Keep. *Psychosomatic Medicine*. 2001; **63:** 896–909

Lipton P. Inference to the Best Explanation. London. Routledge; 1991. 2nd ed. 2004

Lizza JP. Brain Death, 453–487, in: ***F Gifford.*** 2011

Lloyd GG. Hysteria: a case for conservation? *British Medical Journal*. 1986; **293:** 1255–1256

Locke J. First letter to Stillingfleet. 1697

Locke J. An Essay Concerning Human Understanding**, AC Fraser** (ed.). New York. Dover Publications, Inc. 1959

Lockwood CJ. Prediction of pregnancy loss. *The Lancet*. 2000; **355:** 1292–1293

Loeser JD. and Melzack R. Pain: an overview. *Lancet*. 1999; **353:** 1607–1615

Lombardi VC., Ruscetti FW., Das Gupta J. et al. Detection of an infectious retrovirus, XMRV, in blood cells of patients with chronic fatigue syndrome. *Science*. 2009; **326:** 585–589

Long CW. An account of the First Use of Sulfuric Ether by Inhalation as anaesthetic in Surgical Operations. *The American Journal of Medical Sciences*; **37**: 275–279, 1991

Lorenz KA., Shapiro MF., Asch SM., Bozzette SA. and Hays RD. Association of symptoms and health-related quality of life: findings from a national study of persons with HIV infection. *Annals of Internal Medicine*. 2001; **134:** 854–860

Louhiala P. Placebo Effects: The Meaning of Care in Medicine. Springer. 2020

Louw A., Diener I., Ferandez-de-la-Perias C., et al. Sham Surgery in Orthopedics: A Systematic Review of the Literature. *Pain Medicine*. 2017; **18:** 736–750

Lovelock J. Novacene. London. Allen Lane. 2019

Lucretius. De Rerum Natura. CreateSpace Independent Publishing Platform. 2014

Luper S. Death, 115–123, in: ***M. Solomon et al.*** 2017

Lutz T. American Nervousness 1903. An Anecdotal History. Ithaca. Cornell University Press. 1991

Lycan W. Logical Form in Natural Language. Cambridge, Massachusetts. The MIT Press. 1985

Macartney FJ. Diagnostic Logic, 59–99, in: ***CI. Philips*** (ed.) Logic in Medicine. London. BMJ Publishing Group. 1995

Macarthur D. Taking the Human Sciences Seriously, 125–141, in: ***M. de Caro & D. Macarthur,*** (eds.) Naturalism and Normativity. New York. Columbia University Press. 2010

MacIntyre A. After Virtue. Notre Dame. Notre Dame Press. 1984

Mackenbach JP. Editorial. Carl von Linné, Thomas McKeown, and the inadequacy of disease classifications. *European Journal of Public Health*. 2004; **14:** 225

Mackie JL. The Cement of the Universe. A Study of Causation. 2d Ed. Oxford. Clarendon Press. 1986

MacCorquodale K. & Meehl PE. On a Distinction between Hypothetical Constructs and Intervening Variables. *Psychological Review*. 1948; **55:** 95–107

MacMahon B. Concepts of Multiple Factors, in: ***HH Lee and P Kotin*** (eds). Multiple Factors in the Causation of Environmentally Induced Disease. New York. Academic Press. 1972

MacMahon B. Causes and Entities of Disease, 17–24, in: ***DW. Clark and B. MacMahon*** (eds.). Preventive and Community Medicine. Boston. Little, Brown and Company. 1981

MacMahon B. and Pugh TF. Epidemiology: principles and methods. Boston. Little Brown. 1970

MacMahon B. and Trichopoulos D. Epidemiology. Principles and Methods. Second Edition. Boston. Little Brown and Company. 1999

MacMahon B., Pugh TF. and Ipsen **J**. Epidemiologic Methods. Boston. Little Brown. 1960

MacSorley K. An investigation into the fertility rates of mentally ill patients. *Annals of Human Genetics*. 1964; **27:** 247–256.

Magni G., Marchetti M., Moreschi C. et al. Chronic musculoskeletal pain and depressive symptoms in the National Health and Nutrition Examination. *Pain*. 1993; **53:** 163–168

Maher B. Poll results: look who's doping. *Nature*. 2008; **452:** 674–675

Mainx F. Foundations of Biology, in: ***O. Neurath, R. Carnap*** and ***C. Morris*** (eds.). Foundations in the Unity of Science. University of Chicago Press. 1950

Maj M. and Sartorius N. Schizophrenia. Chichester. John Wiley and Sons. 1999

Maj M., Gaebel W., Lopez-Ibor JJ. and Sartorius N. Psychiatric Diagnosis and Classification. Chichester. John Wiley and Sons. 2002

Malafouris L. Mind into Matter. *New Scientist*. 2013; September: 28–29

Malcolm N. Thought and Knowledge. Ithaca. Cornell University Press. 1977

Maldonado G., Greenland S. Estimating Causal Effects. *International Journal of Epidemiology*. **31:** 422–429. 2002

Mandelblatt JS., Cronin KA., Bailey S., et al. Effects of mammography screening under different screening schedules: model estimates of potential benefits and harms. *Annals of Internal Medicine*. 2009; **151:** 738–747

Mannix LK. Epidemiology and impact of primary headache disorders. *The Medical Clinics of North America*. 2001; **85:** 887–695

Marcum JA. (ed.). The Bloomsbury Companion to Contemporary Philosophy of Medicine. London. Bloomsbury. 2017

Margolis J. Psychotherapy & Morality. A study of two Concepts. New York. Random House. 1966

Margolis J. Values and Conduct. Clarendon Press. 1971

Margolis J. Negativities. The Limits of Life. Colombus. Ohio. Charles E. Merrill Publishing Company. 1975

Margolis J. The concept of disease. *Journal of Medicine and Philosophy*. 1976**;** **1:** 238–255

Margolis, J. Persons and Mind. The Prospects of Nonreductive Materialism. Dordrecht. D. Reidel Publishing Company. 1978

Margolis, J. The Concept of Disease, 561–577, in: ***AA. Caplan, HT. Engelhatdt and JJ. McCartney*** (ed.). Concepts of Health and Disease. Interdisciplinary Perspectives. Reading, Massachusetts. Addison-Wesley Publishing Company. 1981

Margolis J. Taxonomic puzzles, in**: *JZ. Sadler, OP. Wiggins and MA. Schwartz*** (ed.). Philosophical Perspectives on Psychiatric Diagnostic Classification. Baltimore. The Johns Hopkins University Press. 1994

Margolis J. Puzzles about the Causal Explanation of Human Action, 125–141, in**: *L. Laudan*** (ed.) Mind and Medicine. Problems of Explanation and Evaluation in Psychiatry and the Biomedical Sciences. Berkeley. University of California Press. 1983

Maris RW. Suicide. *Lancet*. 2002; **360:** 319–326

Marks EM., Hunter MS. Medically Unexplained Symptoms: an acceptable term? *British Journal of Pain* 2015; **9:** 109–114**.**

Martin M. Defining irrational action in medical and psychiatric contexts. *The Journal of Medicine and Philosophy*. 1986; **11:** 179–184

Mary BE. Science and Health. London. 1875

Mashour GA. Neural Correlates of Unconsciousness in Large-Scale Brain Networks. *Trends in Neurosciences*. 2018; **41:** 150–160

Matthews JR. Quantification and the Quest for Medical Certainty. Princeton. Princeton University Press. 1995

Mattthews E. Mental disorder: can Merleau-Ponty take us beyond the "Mind-Brain" problem? in: ***KWM. Fulford, M. Davies, RFT. Gipps et al.*** (eds.). 2013

Maugham S. The Moon and Sixpence. New York. Harcourt, Brace and Company. 1919

Mayr E. The Growth of Biological Thought. Cambridge. Harvard University Press. 1982

Mayr E. The multiple meanings of teleological, in: ***Mayr E*** (ed.) Towards a New Philosophy of Biology. Harvard University Press. 1988; 38–86

Mayerfield J. Suffering and Moral Responsibility. Oxford University Press. 1999

McAteer A., Elliott AM., Hannaford PC. Ascertaining the size of the symptom iceberg in a UK-wide community-based survey. *British Journal of General Practice*. 2001; **10**: e1-e10

McCartney J. Concepts of Health and Disease. Interdisciplinary Perspectives. Reading, Massachusetts. Addison-Wesley Publishing Company. 1981

McCawley JD. Everything that Linguists have always Wanted to Know about Logic but were afraid to ask. Oxford. Blackwell. 1981

McCawley JD. Everything linguists have always wanted to know about logic but were unaware to ask. University of Chicago Press. 1993

McDowell J. Mind and the World. Cambridge. Harvard University Press. 1996

McFarland JD. Kant's Concept of Teleology. Edinburgh. University of Edinburgh Press. 1970

McGilchrist I. The Master and His Emissary. The Divided Brain and the Making of the Western World. Yale University Press. 2010

McGinn C. The Subjective View. Secondary Qualities and Indexical Thoughts. Oxford. Clarendon Press. 1983

McGinn C. Mental Content. Oxford. Blackwell. 1989

McHugh PR and Slavney P. The Perspectives of Psychiatry. Wolters Kluwer Health, 1983

McHugh PR., Slavney PR. The Perspectives of Psychiatry. John Hopkins University Press. 1998

McKay JR., Hiller-Sturmhöfel S. Treating Alcoholism as a Chronic Disease. *Alcohol Resarch & Health*. 2011**; 33:** 356–370

McKeown T. The origins of human disease. Blackwell. 1988

McLellan F. Do violent movies make violent children? *Lancet*. 2001; **359:** 502–503

McWhinney IR., Epstein RM. and Freeman TR. Rethinking Somatization. *Annals of Internal Medicine.* 1997; **126:** 747–750

Mead, GH. Mind, Self and Society. Chicago. The University of Chicago Press. 1934

Meador CK. The art and science of nondisease. *The New England Journal of Medicine.* 1965; **272:** 92–95

Mechanic D. Medical Sociology. New York. The Free Press. 1978

Mechanic D. Response factors in illness: the study of illness behavior. *Social Psychiatry.* 1966; **1:** 11–20

Mechanic D. The concept of illness behavior, in: ***AA. Caplan, HT. Engelhatdt and J. McCartney*** (eds.). Concepts of Health and Disease. Interdisciplinary Perspectives. Reading, Massachusetts. Addison-Wesley Publishing Company. 1981

Medawar P. The Art of the Soluble. London. Methuen. 1967

Medawar P. The Hope of Progress. London. Methuen. 1972

Medical Research Council. Streptomycin Treatment of Pulmonary Tuberculosis. *A Medical Research Council Investigation.* 1948; **2:** 769–782

Meehan K.B., De Panfilis C., Cain N.M., et al. Facial emotion recognition and borderline personality pathology. *Psychiatric Research;* **255:** 347–354, 2017

Melzack R. and Wall PD. The Challenge of Pain. London. Penguin Books. 1996

Merskey H. and Spear FG. Pain. Psychological and Psychiatric Aspects. London. Baillère, Tindall & Cassell. 1967

Merskey H., Lindblom U., Mumford JM. et al. Pain terms and notes on usage. *Pain.* 1986; Supplement 3: 215–221

Merton RK. Social Theory and Social Structure. New York. The Free Press. 1949

Meza J.P, Passerman DS. Integrating narrative medicine and evidence-based medicine: the everyday social practice of healing. London. Radcliffe Publishing. 2011

Midgley M. Are you an illusion? London. Acumen. 2014

Mill JS. On Liberty Oxford. Oxford University Press. 1859/1971

Mill JS. Philosophy of Scientific Method, **E. Nagel (ed.)** New York. Hafner Publishing Co. 1950

Mill JS. Nature and Utility of Religion. Indianapolis. Bobbs-Merrill. 1958

Mill JS. Utilitarism. Penguin. 1970

Miller WR. Haunted by the Zeitgeist: Reflections on contrasting treatment goals and concepts of alcoholism in Europe and the United States. *Annals of the New York Academy of Sciences.* 1986; **472:** 110–129

Miller G. Beyond DSM: seeking a brain-based classification of mental illness. *Science.* 2010; **327:** 1437

Miller AH., Haroon E., Felger JC. Therapeutic implications of brain immune interactions. *Neuropsychopharmacology.* 2017; **42:** 334–359

Millikan RG. In Defense of Proper Functions. *Philosophy of Science.* 1989; **56:** 288–302

Millikan R. Biofunctions: Two Paradigms, 113–156, in: ***A. Ariev, R. Cummins & M. Perlman,*** (eds.) Functions New Essays in the Philosophy of Psychology and Biology. Oxford University Press. 2002

Millstein R & Skipper R. Population Genetics. 22–43; in: ***D. Hull & M. Ruse,*** (eds.). The Cambridge Companion to the Philosophy of Biology. Cambridge. Cambridge University Press. 2007

Mirsky AF. & Duncan CC. Neuropsychological and Psychophysiological Contributions to the Diagnosis of schizophrenia, in: ***M.J Maj & N. Sartorius*** **(eds.).** Schizophrenia. New York. John Wiley & Sons. 1999

Mitchell JR. Hearts and minds. *British medical Journal.* 1984; **289:** 1557–1558

Moerman D. Meaning, Medicine and the 'Placebo Effect'. Cambridge University Press. 2002

Monson RA. and Smith GR. Current Concepts in Psychiatry. Somatization disorder in primary care. *New England Journal of Medicine.* 1983; **308:** 1464–1465

Moore GE. Philosophical Studies. London. Routledge & Kegan Paul. 1960

Morley A. Quantifying leukaemia. *New England Journal of medicine.* 1998; **339:** 627–629

Morris JN. Uses of Epidemiology. Edinburgh. E. & S. Livingstone. 1970

Moser RP., Hesse BW., Shaikh AR. et al. Grid-Enabled Measures. *American Journal of Preventive Medicine*. 2011; **40:** 134–143

Moyal-Sharrock D. Understanding Wittgenstein' s On Certainty. New York. Palgrave. 2007

Moynihan R., Brodersen J., Heath I., et al. Reforming disease definitions: a new primary care led, people-centred approach. 2019; *BMJ Evidence-Based Medicine*. https://doi.org/10.1136/bmjebm-2018-111148

Morris DB. The Culture of Pain. Berkeley. University of California Press. 1991

Mourelatos APD. Events, Processes, and States. *Syntax and Semantics*. 1978; **14:** 191–210

Moustgaard H, Clayton GM, Jones et al., in: ***Karhausen L.*** Reconsidering Medicine. 2019

Mumford L. The Myth of the Machine: Techniques and Human Development. New York. Harcourt & Brace. 1967

Mundt C. & Spitzer M,* 13–19**, in*: **E. Henn, N. Sartorius, H Helmchen, H Lauter (eds.). 2001

Munro E. Evidence-based policy, 48–67, in*: **N. Cartwright & E. Montuschi**,* (eds.). 2014

Munson B. Relations of Sex and Dialect to Reduction. *Speech Communication*: **15**: 39–54, 1994

Munson R. and Roth P. Testing normative naturalism: the problem of scientific medicine. *British Journal of the Philosophy of Science*. 1994; **45:** 571–584

Murphy D. Psychiatry in the Scientific Age. Cambridge. The MIT Press. 2006

Murphy EA. The Logic of Medicine. 2d Edition. Baltimore. The Johns Hopkins Press. 1974. The Johns Hopkins University Press. 1997**.**

Murphy CJL. and Lopez, *A.* Global mortality, disability, and the contribution of risk factors: Global Burden of Disease Study. *Lancet*. 1997; **349:** 1436–1442

Murphy E. The Logic of Medicine. Baltimore.

Murray S. and James MA. What is Hypertension? *Lancet*. 1999; **354:** 593–594

Muscari PG. The structure of mental disorder. *Philosophy of Science*. 1981; **48:** 553–572

***Nagel E.* (ed.).** John Stuart Mill's Philosophy of Scientific Method. New York. Hafner Publishing Company. 1950

Nagel T. Observation and Theory. John Hopkins Press. 1970-1971

Nagel E. Teleology Revisited and Other Essays in the Philosophy of Science. New York. Columbia University Press. 1979a

Nagel T. Mortal Questions. Cambridge. Cambridge University Press. 1979b

Nagel T. The View from Nowhere. Oxford. Oxford University Press. 1986

Nagel T. The Last Word. Oxford. Oxford University Press. 1997

Narod SA, Iqbal J, Miller AB. Why have breast cancer mortality rates declined? *Journal of Cancer Policy*. 2016**; 5:** 8–17

Narod SA., Sun P., Wall G. et al. Impact of screening mammography on mortality from breast cancer before age 60 in women 40 to 49 years of age. *Current Oncology*. 2014; **21:** 217–221

Neander K. The teleolgical notion of function. *Australasian Journal of Philosophy*. 1991; 454–468

Nesse RM. What good if feeling bad? *The Sciences*. 1991; **31:** 30–37

Nesse RM. Testing evolutionary hypotheses about mental disorders, 260–266, in*: **SC. Stearns*** **(ed.).** Evolution in Health and Disease. Oxford. Oxford University Press. 1999

Nesse RM. Is stress an adaptation? *Arch. Gen. Psychiatry*. 2000; **57:** 14–20

Nesse RM. Good Reasons for Bad Feelings. Penguin. Dutton. 2019 Nettleton S. 'I just want permission to be ill': towards a sociology of medically unexplained symptoms. *Social Science & Medicine*. 2006**; 62:** 1167–1178

Nesse RN and Williams G. Why we get sick: the new science of Darwinian medicine. Doubleday. 1995

Neurath O. Protocol statements. 91–99, in**: *RS. Cohen and M. Neurath,*** (eds.) Philosophical Papers 1913–1946. Dordrecht. Reidel. 1983

Neuwirth ZE. Physician empathy–should we care? *Lancet*. 1997; 350: 606

Newton-Smith WH. A Companion to the Philosophy of Science. 2000. 2017. Blackwell

Nicolle LE. Asymptomatic bacteriuria—important or not? *The New England Journal of Medicine*. 2000**; 343:** 1037–1039

Nietzsche F. The Portable Nietzsche, W Kaufmann (ed.) New York. The Viking Press. 1967

Nietzsche F. The Gay Science. Random House. 1974

Nilsen, A.S et al. Are we really unconscious in « unconscious » states ? Front Hum Neurosci. 2022; 16: 987051. Published online 2022 Oct 5. doi: https://doi.org/10.3389/fnhum.2022.987051

Njolstad I., Arnesen E., Lund-Larsen PG. Smoking, serum lipids, blood pressure, and sex differences in myocardial infarction. *Circulation.* 1996; **93:** 450–456

Noë A. Out of our minds. Why you are not your brain and other lessons from the biology of consciousness. London. Hill and Wang. 2009

Nolan JL. The Therapeutic State. Justifying Government at Century's End. New York. New York University Press. 1998

Nordenfelt L. On the Nature of Health: an action-theoretic approach. Dordrecht. Kluwer Academic Publishers. 1995

Nordenfelt L. The opposition between naturalistic and holistic theories of health and disease, **in: H. Carel and R. Cooper** (eds.), Health, Illness and Disease. Philosophical Essays. London. Routledge. 2013

Nussbaum M. The Therapy of Desire. Theory and Practice in Hellenistic Ethics. Princeton. Princeton University Press. 1994

Nussbaum M. Free to Choose. *The Philosophers' Magazine.* 2000**; 11:** 37–40

O'Connor PG. and Schottenfeld RS. Patients with alcohol problems. *New England Journal of Medicine.* 1998**; 338:** 592–601

Odden MC., Peralta CA., Haan MN., et al. Rethinking the association of high blood pressure with mortality in elderly adults; the impact of frailty. *Archives of Internal Medicine.* 2012; **172:** 1162–1168

Odgers CL. Group based trajectory modelling in clinical research. *Archives of General Psychiatry.* 2010; **64:** 476–484

Offer D. and Sabshin M. Normality. Theoretical Concept of Mental Health. New York. Basic Books. 1974

Ohayon MM., O'Hara R., Vitiello MV. Epidemiology of Restless Legs Syndrome: A Synthesis of the Literature. *Sleep Medical Review.* 2012; **16:** 283–295

Ohlsson R. **The Moral Import of Evil.** On counterbalancing death, suffering and degradation. Stockholm. Filosofiska studier 1. Akademilitteratur. 1979

O'Malley PG., Jones D., Feuerstein IM. and Taylor AJ. Lack of correlation between psychological factors and subclinical coronary artery disease. *The New England Journal of Medicine.* 2000; **343:** 1298–1303

Osler W. A Way of Life. New York. Dover Publications. 1951

O'Sullivan S. It's All in Your Head. Vintage. Penguin. 2015

Owen AM., Coleman MR., Boly M., Davis MH., Laureys S., Pickard JD. "Detecting awareness in the vegetative state" *Science.* 2006; **313:**1402

Owen GS., Cutting J., David AS. Are people with schizophrenia more logical than healthy volunteers? *British Journal of Psychiatry.* 2007**; 191:** 453–454.

Oxford Dictionary**.** The Compact Edition of the Oxford Dictionary. London. Book Club Associates. 1971

Paci E. **EUROSCREEN WORKING GROUP.** Summary of the evidence of breast cancer service screening outcomes in Europe and first estimate of the benefit and harm balance sheet. *Journal of Medical Screening.*2012; **19** (Suppl 1): 5–13

Pagano JS. Viruses and lymphoma. *New England Journal of Medicine.* 2002; **347:** 78–79

Pain terms**:** a list with definitions and notes on usage. *Pain* 1979; **6:** 249–252

Palmer JR., Boggs DA., Wise LA., et al. Individual and neighbourhood socioeconomic status in relation to breast cancer incidence in African-American Women. *American Journal of Epidemiology.* 2012; **176:** 1141–1146

Papineau D. & Dennett DC. Neuronal spike trains all the way. *Times Literary Supplement.* 2017, August 4

Parascandola M. Epistemic risk: empirical evidence and the fear of being wrong. *Probability and Risk.* 2010: **9:** 201–214

Parfit D. Personal identity. *Philosophical Review.* 1971**; 80:** 3–27

Parfit D. Prudence, Morality and the Prisoner's Dilemma. Oxford University Press. 1981

Parfit D. Experiences, Subjects and Conceptual Schemes. *Philosophical Topic*s. 1999; **26:** 217–270

Parfit D. Is Personal Identity What Matters? The Ammonius Foundation. 2007; ammonius.org/assets/pdfs./ammoniusfinal.pdf

Paris J. Is Psychoanalysis Still Relevant to Psychiatry? *Canadian Journal of Psychiatry.* 2017**; 62:** 308–312

Parnas J. Diagnostic epidemics and diagnostic disarray: The issue of differential diagnosis, 143–145, in**: *KS. Kendler and J. Parnas*** (eds.) Philosophical issues in psychiatry IV. Oxford University Press. 2017

Parnas J., Urfer-Parnas A. The ontology and epistemology of symptoms: The case of auditory verbal hallucinations in schizophrenia, 201–216, in**: *KS. Kendler and J. Parnas*** (eds.) Philosophical issues in psychiatry IV. Oxford University Press. 2017

Parsons T. The Social System. London. Routledge and Kegan Paul. 1951

Parsons T. Social Structure and Personality. London. The Free Press. Collier-MacMillan. 1964

Parsons T. Definitions of health and illness in the light of American values and social structure, 57–82, in**: *AJ. Caplan, HT. Engelhardt* and *JJ. McCartney*** (eds.). Concepts of Health and Disease. Reading, Massachusetts, Addison-Wesley. 1981

Passmore J. The Perfectibility of Man. New York. Charles Scribner's Sons. 1971

Payer L. Medicine and Culture. London. Viking Penguin. 1990

Paynter N., Chasman, DI., Paré G. et al. Association between a literature-based genetic risk score and cardiovascular events in women. *JAMA.* 2010**; 303:** 631–637

Peabody FW. The care of the patient. *JAMA.* 1927**; 88:** 877–882

Pearce N. & Lawlor DA. Causal Inference—so much more than statistics. *International Journal of Epidemiology.* 2016; **45:** 1895–1903

Pearl J. Causality. Cambridge University Press. 2000

Pearl J. Causality. Models, Reasoning, and Inference. Cambridge. Cambridge University Press. 2009

Pearl J. An Introduction to Causal Inference. *International Journal of Biostatistics.* 2010; 6: Issue 2. Article 7

Pearl J. The structural theory of causation, in: ***H McKay et al*** (ed) Causality in the Sciences. Oxford University Press. 2011

Pearl J, Glymour M, Jewell NP. Causal Inference in Statistics. A Primer. Chichester. Wiley. 2016

Pedro-Botet J. & Rubiés-Prat J. Statin-associated muscle symptoms: Beware of the nocebo effect. *Lancet.* 2017; **389:** 2445–2446

Pellegrino ED. and Thomasma DC. The Virtues in Medical Practice. New York. Oxford University Press. 1993

Peschel RE. and Peschel E. Selective empathy, 110–120, in*: **H. Spiro, MG. McCrea Curnen and D. St James*** (eds.). Empathy and the Practice of Medicine. New Haven. Yale University Press. 1993

Peters RS. The Concept of Motivation. London. Routledge & Kegan Paul. 1960

Peters K. Exceptional Matters. *Lancet.* 2004; **364**: 2142–2151

Philips CI. (ed.). Logic in Medicine. London. BMJ Publishing Group. 1995

Pickar D. Pharmacogenomics of psychiatric drug treatment. *Psychiatric Clinics of North America.* 2003; **26:** 303–321

Pickering GW. The genetic factor in essential hypertension. *Annals of Internal Medicine.* 1955**; 43:** 457–64

Pickering GW. Does medical treatment mean patient benefit? *Lancet.* 1996; **347:** 379–380

Pickering TG. What is Hypertension? *Lancet.* 1999**; 354:** 593

Pigliucci M. Design, yes, intelligent, no. *Philosophy Now.* 2001; **32:** 26–29.

Pitcher G. The Philosophy of Wittgenstein. Englewood Cliffs. Prentice-Hall. 1964

Pitcher G. The awfulness of pain. *Journal of Philosophy*. 1970; **14:** 481–493
Pittendrigh CS. Adaptation, natural selection, and behavior, 390–416, in: ***A. Roe, GG. Simpson*** (eds.). Behavior and Evolution**.** New Haven. 1958
Pizzo PA., Robichaud KJ., Edwards BK., Schumaker C., Kramer BS. and Johnson A. Oral antibiotic prophylaxis in patients with cancer: a double-blind randomized placebo-controlled trial. *Journal of Pediatrics*. 1983**; 102**: 125–133
Plato. The Great Hippias, in: Platon. Œuvres complètes, I, Gallimard. Pléiade, 1950
Plato. Cratylus. **B. Jowett (ed.).** New York. C. Scribner Son. 1971a
Plato. Gorgias. T. Irwin (ed.). Oxford. Oxford University Press. 1979
Plato. Phaedo, D. Gallop (ed.). Oxford. Clarendon Press. 1975
Plato. Phædrus. Indianapolis. Hackett Publishing Company. 1995
Plato. The Republic of Plato, ***FM. Cornford*** (ed.). Oxford. Oxford University Press. 1945
Plato. Timaeus. Plato's Cosmology, ***FM. Cornford*** (ed.) London. Routledge and Kegan Paul. 1971b
Platts M. Ways of Meaning. An Introduction to a Philosophy of Language. Cambridge. Massachusetts. The MIT Press. 1997
Popper K. Logik des Forschung. Tübingen. Mohr Siebeck. 1934
Popper K. The Open Society and its Enemies. London. Routledge; 1966 London. Hutchison. 1982
Popper K. The logic of the social sciences, 87–104, in**: *G. Adey and D. Frisby*** (eds). The Positivist Dispute in German Sociology. London. Heinemann. 1976
Popper K. Conjectures and Refutations. Routlege and Kegan Paul. 1982
Popper K. The Open Universe. An Argument for Indeterminism. 1988. Routledge**.**
Porter R. What is Disease? in**: *Porter R.*** (ed.) The Cambridge History of Medicine. Cambridge University Press. 2006, 83
Porter I et al. Framework and guidance for implementing patient-reported outcomes in clinical practice: evidence, challenges and opportunities. *Journal of Comparative Efficient Research*; **5:** 507–519, 2018
Potochnik A. Levels of explanation reconceived. *Philosophy of Science*. 2010**; 77:** 59–72
Prasad V., Lenzer J, Newman DH. Why cancer screening has never been shown to « save lives »– and what we can do about it. *British Medical Journal*. 2016; **352**: h6080
Prins JB., van der Meer JW. & Bleijenberg G. Chronic fatigue syndrome. *New England Journal of Medicine*. 2006; **467:** 346–355
Proust M. Sur la Lecture, préface du traducteur, in**: J Ruskin,** Sésame et les Lys. Paris. Payot. 2011
Ptacek R., Kuzelova H., Stefano GB. Dopamine D4 receptor gene DRD4 and its association with psychiatric disorder. *Medical Science Monitor*. 2011; **17**: 215–220
Puliga D. La depressione è una dea. I Romani e il male oscuro. Bologna. Il Mulino. 2017
Pull CB., Cloos JM. & Pull-Erpelding MC. Clinical assessment instruments in psychiatry, in: ***Maj M., Gaebel W., Lopez-Ibor J. & Sartorius N.*** (eds.). Psychiatric Diagnosis. Chichester: Wiley. 2002; 177–218.
Puska P., Jaini P. The North Karelia Project: Prevention of Cardiovascular Disease in Finland Through Population-Based Lifestyle Interventions. *American Journal of Lifestyle Medicine*. 2020 **14:** 495–499
Putnam H. Brains and Behavior, 1–19, in*: **J. Butler*** (ed.). Analytical Philosophy. Second Series. Oxford. Basil Blackwell. 1968.
Putnam H. Meaning and the Moral Sciences. London. Routledge & Kegan Paul. 1974
Putnam H. Mind Language and Reality. Philosophical papers. Volume 2. Cambridge. Cambridge University Press. 1975
Putnam H. What is Realism? *Proceedings of the Aristotelian Society;* **76**: 177–194, 1976
Putnam H. The nature of mental states, 223–231, in**: *N. Block* (ed**.). Readings in Philosophy of Mind, volume one. Cambridge, Mass. Harvard University Press. 1980a
Putnam H. Philosophy and our mental life, 134–143, in**: *N Block*** (ed.). Readings in Philosophy of Mind, volume one. Cambridge, Mass. Harvard University Press. 1980b
Putnam H. Reason, Truth and History. Cambridge. Cambridge University Press. 1981
Putnam H. (ed) Realism and Reason. Cambridge University Press. 1983

Putnam H. The Many Faces of Realism. LaSalle, Ill. Open Court; 1987
Putnam H. The Nature of mental states, 47–56, in: ***Lycan WG*** (ed.). Mind and Cognition. Oxford. Basil Blackwell. 1990
Putnam H. Renewing Philosophy Cambridge. Harvard University Press. 1992
Putnam H. The Collapse of the Fact-Value Dichotomy. Cambridge, Mass. Harvard University Press. 2002
Putnam H. Naturalism, Realism and Normativity. Harvard University Press. 2016
Quine WV. Mathematical Logic. New York. Harper & Row. 1951
Quine WV. Word and Object. Cambridge, Mass. MIT Press. 1960
Quine WV. Two dogmas of empiricism, in: From a Logic Point of View. New York. Harper Torchbooks. 1961a
Quine WV. On what there is, in: From a Logic Point of View. New York. Harper Torchbooks. 1961b
Quine WV. Methods of Logic London. Routledge & Kegan Paul; 1974
Quine WV. Philosophy of Logic. Englewood-Cliffs, N.J. Prentice Hall. 1970
Quine WV. States of mind. *Journal of Philosophy*. 1985; **82**: 5–8
Quine WV. From Stimulus to Science. Cambridge. Mass. Harvard University Press. 1995
Quine WW. and Ullian JS. The Web of Belief. New York. Random House. 1970
Raballo A. & Laroi F. Clinical staging: a new scenario for the treatment of psychosis. *New England Journal of Medicine*. 2009; **374:** 365–367
Radcliff-Brown A. The Concept of Function in Social Science. *American Anthropologist*. 1935; **37:** 394–402
Radden J. Madness and Reason. Harper Collins Publishers Ltd. 1985
Radden, J. Madness and Reason. London. Allen & Unwin. 1988
Ralston S. Davidson's Principles of Medicine. Educa Books, 2018
Randall F. & Downie RS. The Philosophy of Palliative Care. Critique and Reconstruction. Oxford University Press. 2006
Rathod S., Phiri P. & Kingdon D. Cognitive Behavioral Therapy for schizophrenia. *Psychiatric Clinics of North America*. 2010; **33:** 527–536
Ravindrarajah R., Hamada S., et al. Systolic blood pressure trajectory, frailty and all-cause mortality >80 years of age: cohort study using electronic health records. *Circulation*. 2017: **135:** 2357–2368
Rawls, J. A Theory of Justice. Boston. Harvard University Press. 1971
***Raz J.* (ed.).** Practical Reasoning. Oxford. Oxford University Press. 1978
Read R. On approaching schizophrenia through Wittgenstein. *Philosophical Psychology*. **14:** 449–475, 2001
Read R. Wittgenstein among the Sciences. Routledge. 2012
Read R. On Delusions of Sense: A Response to Coetzee and Sass. *Philosophy, Psychology and Psychiatry*. **10:** 135–141, 2003
Redlich FC. The Concept of Normality. *American Journal of Psychotherapy*. 1952; **6:** 551–76
Reger DA., Narrow WE., Rae DS. et al. The de facto US mental and addictive disorders system: epidemiologic catchment area prospective 1-year prevalence rates of disorders and services. *Archives of General Psychiatry*. 1993: **50:** 85–94
Reid, T. Essays on the Intellectual Powers of Man. *AD Woozley* (ed.). 1785/1978. London. MacMillan.
Reidenberg MM. and Lowenthal DT. Adverse non-drug reactions. *New England Journal of Medicine*. 1968; **279:** 678–679
Reininger R. Wertphilosophie und Ethik. Wien. 1947
Rescher N. Welfare. The Social Issue in Philosophical Perspective. Pittsburgh. University of Pittsburgh Press. 1972
Reznek L. The Nature of Disease. London. Routledge. 1987
Reznek L. The Philosophical Defence of Psychiatry. London. Routledge. 1991
Reznek L. Delusions and the Madness of the Masses. Rowland & Littlefield Publishers. New York. 2010

Rice N., Dixon P., Lloyd DCEF. and Roberts D. Derivation of a needs based capitation formula for allocating prescribing budgets to health authorities and primary care groups in England: regression analysis. *British Medical Journal.* 2000; **320:** 284–287

Richards DW. Homeostasis versus hyperexis: or Saint George and the Dragon. *The Scientific Monthly.* 1953; **77:** 289–294

Richards DW. Homeostasis: its dislocations and perturbations. *Perspectives in Biology and Medicine.* 1960; **3:** 218–251

Richards AR. Species and Taxonomy, 161–187, in ***Ruse M*** (ed.) The Oxford Handbook of Philosophy of Biology. Oxford University Press. 2008

Richardson JTE. The premenstrual syndrome: a brief history, in: ***AL. Caplan, JJ. McCartney, DA. Sisti*** (eds.). Health, Disease, and Illness. Concepts of Medicine. Washington. Georgetown University Press. 2004

Riese W. The Conception of Disease. New York. The Philosophical Library. 1955

Rieckman N., Gerin W., Kronish IM., et al. Course of depressive symptoms and medication adherence after acute coronary syndromes: an electronic medication monitoring study. *Journal of the American College of Cardiology.* 2006; **48:** 2218–2222.

Rilke RM. Duino Elegien. Frankfurt-am-Main. Insel Verlag. 1929

Ripley HS. Long-term study of combat area schizophrenic reactions. *American Journal of Psychiatry.* 1951; **108:** 409–416

Rippere V. 'Hysteria', 'functional' or 'psychogenic'. *Journal of the Royal Society of Medicine.* 1992**; 85:** 59–60

Risch N., Herrell R., Lehner T. et al. Interaction between the serotonin transport gene (5-HTTLPR), stressful life events, and risk of depression. A metanalysis. *JAMA.* 2009; **301:** 2462–2471

Roberts CJ. Epidemiology for Clinicians. London. Pitman Medical. 1977

Robins E. The Final Months: A Study of the Lives of 134 Persons Who Committed Suicide. Oxford University Press. 1981

Robins LN., Helzer JE., et al. Lifetime prevalence of specific psychiatric disorders in three sites. *Archives of General Psychiatry.* 1984; 949–958

Rochester S. & Martin J. Crazy Talk. A Study of the Discourse of Schizophrenic Speakers. New York. Plenum Press. 1979

Roessler J. & Eilan N. Agency and Self-Awareness. Oxford. Oxford University Press. 2004

Röh J. and Jansen L. Why functions are not special dispositions. *Journal of Biomedical Semantics.* 2014**; 5**: 1–6

Rosch E. and Mervis CB. Family Ressemblances. Studies in the Internal Structure of Categories. *Cognitive Psychology*, **7:** 573–605**, 1975**

Rose G. Cardiovascular diseases, 133–144, in**: *WW. Holland, R. Detels and K. Knox.*** (eds.) Oxford textbook of public health, vol. 4 Specific applications. Oxford. Oxford University Press. 1985a

Rose G. Sick individuals and sick populations. *International Journal of Epidemiology.* 1985b; **14:** 32–38

Rose G. The mental health of populations, 77–85, in**: *P. Williams, G. Wilkinson and K. Rawnsley*** (eds.). The Scope of Epidemiological Psychiatry. London. Routledge. 1989.

Rose G. The Strategy of Preventive Medicine. Oxford. Oxford University Press. 1992

Rose G. and Barker DJP. What is a case? Dichotomy or continuum? *British Medical Journal.* 1978; **2:** 873–874

Rosenbaum JR., Bradley EH., Holmboe et al. Sources of conflict in medical housestaff training: a qualitative study. *American Journal of Medicine.* 2004; **116:** 402–407

Rosenberg A. The supervenience of Biological Concepts. 1978; *Philosophy of Science.* **45:** 368–386

Rosenberg A. The Structure of Biological Science. Cambridge. Cambridge University Press. 1985

Rosenberg JH. Epidemiologic and Methodologic Problems in Determining Nutritional States of Older Persons. *American Journal of Clinical Nutrition.* **50**: 1231–1233, 1989

Rosenberg A. Instrumental Biology or the Disunity of Science. Chicago. Chicago University Press. 1994

Rosenberg A. Reductionism (and Antireductionism) in Biology, in: ***D. Hull & M. Ruse.*** (eds.). The Cambridge Companion to the Philosophy of Biology. Cambridge. Cambridge University Press. 2007

Rosenberg C. The Tyranny of Diagnosis: Specific Entities and Individual Experience. *The Milkbank Quarterly.* 2002; **80:** 237–260

Rosenberg C, Golden J. Framing Diseases. Rutgers University Press. 1992

Rosenblueth A., Wiener N. and Bigelow J. Behavior, purpose and teleology. *Philosophy of Science.* 1943**; 10:** 18–24

Rosendal M., Hartman TCO., Aamland A., et al., "Medically unexplained" symptoms and symptom disorders in primary care: prognosis-based recognition and classification. *BMC Family Practice.* 2017**; 18:**18

Rosenhan DL. and Seligman MEP. Abnormal Psychology. New York. W.W. Norton & Company. 1995

Rozenzweig F. Understanding the Sick and the Healthy. Cambridge. Harvard University Press. 1999

Rosenzweig P., Brohier S., Zipfel A. The placebo effect in healthy volunteers: influence of experimental conditions on the adverse events profile during phase I studies. *Clinical Pharmacological & Therapeutic*s. 1993; **54:** 579–583

Ross M. and Olson JM. An expectancy attribution model of the effects of placebos. *Psychological Review.* 1981**; 88:** 408–437

Rosso BD et al. On the meaning of work: a theoretical integration and review. *Research in Organization Behaviour*; **30**: 91–127, 2010

Rothman K. Causes. *American Journal of Epidemiology.* 1976; **104:** 587–592

Rothmann K. Modern Epidemiology. Boston. Little Brown and Company. 1986

Rothmann K. (ed.). Causal inferences. Chestnut Hill. Epidemiologic Resources. 1988

Rothman K., Greenland S. & Lash TL. Modern Epidemiology. Philadelphia. Lippincott. 2008

Rothschild BM., Tanke D., Carpenter. Tyrannosaurs suffered from gout. 1997; *Nature***: 387:** 357

Rowbotham DJ. Endogenous opioids, placebo response, and pain. *Lancet*. 2001; **357:** 1901–1902

Rozanski A. Behavioral cardiology: current advances and future directions. *Journal of the American College of Cardiology*. 2014; **64:** 100–110

Ruben, DH. Explaining explanation. London. Routledge. 1993

Ruse M. Is Biology Different from Physics**, in: *RG Colodny*** (ed.). Logic, Laws, & Life. Some Philosophical Complications. Pittsburgh. University of Pittsburgh Press. 1977

Ruse M. The Oxford Handbook of Biology. Oxford University Press. 2008

Russell B. Mysticism and Logic and Other Essays. London. Allen & Unwin. 1959

Russell B. Human Knowledge: its Scope and Limits. Allen and Unwin. 1961

Russell W.* in: *MB. Strauss (ed.). Familiar Medical Quotations. Boston. Little, Brown and Company. 1968

Russell, B. On 'Insolubilia' and their Solution by Symbolic Logic, in: ***D. Lackey*** (ed.) Essays in Analysis. London: George Allen & Unwin, 190–214. 1973 [1906]

Russell PH. How Best to Determine the Mortality Benefit from Screening Mammography: Dueling Results and Methodologies from Canada. *Journal of the National Cancer Institute.* 2014; Volume 106, dju317

Ryan M., Slevin JT. Restless Legs Syndrome. *American Journal of Health System Pharmacy.* 2006; **63:** 1599–1612

Ryle JA. The Natural History of Disease. Oxford. Oxford University Press. 1936

Ryle JA. The meaning of normal. *Lancet*. 1947; **249:** 1–4

Ryle G. Internal relations. *Proceedings of the Aristotelian Society, Suppl.* 1935: **14:** 154–185

Ryle G. The Concept of Mind. London: Hutchinson. 1949

Ryle G. Dilemmas. Cambridge. Cambridge University Press. 1954

Ryle G. The Concept of Mind. Penguin. 1963

Ryle G. Collected Papers, Vol. II, Collected Essays 1929–1968, London. Hutchinson of London. 1971

Sabin J., Daniels N. Determining Medical Necessity, in: *Mental Health Practice*. Hastings Center Report. 1994; **24:** 5–13

Sackett DL. Bias in analytic research. *Journal of Chronic Diseases*. 1979; **32:** 51–63

Sackett DL., Haynes RB., Guyatt GH. and Tugwell P. Clinical Epidemiology. A Basic Science for Clinical Medicine. Boston. Little Brown and Company. 2d Edition. 1991

Sackett DL. Clinical Reality, binary models, babies and bathy water. *Journal of Clinical Epidemiology*. 1991; **44:** 217–218.

Sackett DL. Clinical Epidemiology: a Basic Science for Clinical Medicine. Little Brown and Company. 1996

Sacks O. Awakenings. London. Picador. Pan Books. 1982

Sadegh-Zadeh K. The Prototype Theory of Disease. *Journal of Medicine and Philosophy* 2008; **33:**106–139

Sadegh-Zadeh K. Handbook of Analytic Philosophy of Medicine. London. Springer. 2017

Sadegh-Sadeh K. The Logic of Diagnosis, 357–424, in: ***Gifford F.*** (ed.). Philosophy of Medicine. Amsterdam. Elsevier. 2017

Sadler JZ., Wiggins OP. & Schwartz MA. (ed.) Philosophic Perspectives on Psychiatric Diagnostic Classification. Baltimore. The Johns Hopkins University Press. 1994

Salmon WC. Scientific Explanation and the Causal Structure of the World. Princeton. Princeton University Press. 1984

Samuels RL. Delusions as a natural kind, in: ***Broome M & Bortolotti*** L. (eds). Psychiatry as Cognitive Neuroscience: Philosophical Perspectives. Oxford. Oxford University Press. 2009.

Sandison RA. Depression: illness, social disease, or natural state? *Lancet.*1972; **299:** 1227–1229

Saquib N., Saquib J., Ioannidis JPA. Does screening save lives in asymptomatic adults? Systematic review of meta-analyses and randomized trials. *International Journal of Epidemiology.* 2015; 264–277

Saracci R. Epidemiology. A Very Short Introduction. Oxford University Press. 2010

Sass LA. Madness and Modernism. Basic Books. 1992

Sass LA. The Paradoxes of Delusion. Cornell University Press. 1994

Sass LA. Incomprehensibility and Understanding: On the Interpretation of Severe Mental Illness. *Philosophy, Psychiatry & Psychology*. **10:** 125–132, 2003

Sass LA. Contradictions of Emotions in Schizophrenia. *Cognition & Emotion.* **21:** 351–390, 2007

Sartorius N., Holt RIG., Maj M. Comorbidity of Mental Disorders and Physical Disorders. Basel. Karger. 2015

Savulescu J. Genetic interventions and the ethics of enhancement of human beings, 516–535, **in:** ***B. Steinbock*** (ed.). The Oxford Handbook of Bioethics. Oxford. Oxford University Press. 2007

Savulescu J. The human prejudice and the moral status of enhanced beings: what do we owe the gods? 211–250, **in:** ***N. Bostrom & J. Savulescu*** (eds.). Human enhancement. Oxford. Oxford University Press. 2009

Scadding JG. Diagnosis: the clinician and the computer. *Lancet*. 1967; **290:** 877–882

Scadding JG. The semantic problem of psychiatry. Psychological Medicine. 1990; **20:** 243–248

Scadding JG. Essentialism and nominalism in medicine: logic of diagnosis in disease terminology. *Lancet*. 1996; **348:** 594–596

Scambler A., Scambler G. and Craig D. Kinship and friendship and women's demand for primary care. *Journal of the Royal College of General Practitioners* 1981; **26:** 746–750

Schaffner KF. Explanation and Causation in the Biomedical Sciences, in: ***L. Laudan*** (ed.). Mind and Medicine. Problems of Explanation and Evaluation in Psychiatry and the Biomedical Sciences. Berkeley. University of California Press; 1983: 79–124

Schaffner KF. Discovery and Explanation in Biology and Medicine. Chicago. University of Chicago Press. 1993

Schaffner KF. Reduction in Biology and Medicine, 137–157, in: ***F. Gifford*** (ed.). Philosophy of Medicine. Amsterdam. Elsevier. 2011

Schaffner KF. Reduction and reductionism in psychiatry, 1003–1022, **in:** ***KWM. Fulford, M. Davies, RGT. Gipps, et al*** (eds.), 2013

Scheff TJ. Being Mentally Ill: A Sociological Theory. Chicago. Aldine Publishing Company. 1966
Scheffler, I. The Anatomy of Inquiry. Indianapolis. Bobbs-Merrill Company. 1963
Scheffler, I. Human Morality. Oxford. Oxford University Press. 1972
Schizophrenia Working Group of the Psychiatric Genomics Consortium**.** Biological insights from 108 schizophrenia associated genetic loci. *Nature.* 2014**; 511:** 421–427
Schneider C. Die Psychologie der Schizophrenen und ihre Bedeutung für die Klinik der Schizophrenie. Leipzig. Thieme. 1930
Schoeman F. Alcohol Addiction and Responsibility Attributions, in: ***Graham G. and Stephens L*** (eds.) Philosophical Psychopathology. MIT Press. 1994
Schön D. The Reflective Practitioner. How Professionals Think in Action. New York. Basic Books. 1983.
Schopenhauer A. The World as Will and Representation. New York. Dover 1969
Schopenhauer A. Essays and Aphorisms in Life & Meaning. London. Penguin Classics. 1979
Schopenhauer A. Le monde comme volonté et représentation. Gallimard. 2009; pp. 251–252
Schramme T. & S. Edwards. Handbook of the Philosophy of Medicine. 2 vol. Springer. 2017
Schramme T. Philosophy of Medicine and Bioethics, 3–15, in*: **T. Schramme & S. Edwards,*** (eds.). 2017; vol. 2
Schroeder SA. Rethinking health: healthy or heathier than? *The British Journal of the Philosophy of Science.* 2013; **64:** 151–159
Schwarz P. Decision and Discovery in Defining 'Disease', p. 47–63, in: ***H. Kincaid and J. McKitrick*** (eds.) Establishing Medical Reality. Amsterdam. Springer. 2007
Schwartz P. Defining dysfunction: Natural selection, design, and drawing a line. *Philosophy of Science.* 2007; **74:** 364–385
Schwartz P. Progress in Defining Disease: Improved Approaches and Increased Impact. *The Journal of Medicine and Philosophy: A Forum for Bioethics and Philosophy of Medicine*, 2017; **42,** 485–502
Scott S. & Walter F. Studying Help-Seeking for Symptoms: The Challenge of Methods and Models. *Social and Personality Psychology Compass.* 2010**; 4:** 531–547
Scruton R. Emotion, practical knowledge and common culture, 519–536, in*:* ***A. Oksenberg & D. Rorty*** (ed.). Explaining Emotions. Berkeley. University of California Press. 1980
Scruton R. Sexual Desire. London. Weidenfeld and Nicholson. 1986
Scruton R. Modern Philosophy. London. Sinclair-Stevenson. 1994
Searle JR. Speech Acts. Cambridge. Cambridge University Press. 1969
Searle JR. Expression and Meaning. Cambridge. Cambridge University Press. 1979
Searle JR. Intentionality. Cambridge. Cambridge University Press**.**1983
Searle JR. The Rediscovery of Mind. London. The MIT Press. 1992
Searle JR. The Construction of Social Reality. London. Penguin Books. 1995
Searle JR. Mind. A Brief Introduction. Oxford. Oxford University Press; 2004
Searle JR. Making the Social World: The Structure of Human Civilization. Oxford University Press. 2010
Sechehaye M, (ed.) Autobiography of a schizophrenic girl. New American Library. 1970, p. 33
Seedhouse D. Health. The Foundations for Achievement. Chichester. John Wiley and Sons. 1986
Seedhouse D. and Cribb A. Changing Ideas in Health Care. Chichester, New York, Wiley, 1989
Seligman M. et al. Positive psychology progress: empirical validation of interventions. *American psychologist,* **60**: 410–421, 2005
Sellars W., Empiricism, and the Philosophy of Mind, 253–329, in**:** ***H. Feigl & M. Scriven*** (eds.). Minnesota Studies in the Philosophy of Science, vol. I. Minneapolis. University of Minnesota Press. 1956
Sellars W. Empiricism and the Philosophy of Mind, 127–196, in: Science, Perception and Reality. London. Routledge & Kegan Ridgeview. 1963
Selzer R. Letters to a Young Doctor. New York. Simon and Schuster. 1982
Seneca. Peace of Mind. De Tranquillitate Animi. CreateSpace Independent Publishing Platform. 2016

Serjeant R. The Spectrum of Pain. London. Rupert Hart-Davis. 1969

Seth V. An Equal Music. New York. Vintage International. 1999

Shapiro R., Sartorius N., Jablensky A., Kimura M. and Gulbinat W. (eds.). Schizophrenia. An International Follow-up Study. Chichester. John Wiley and Sons. 1979

Shapiro AK., Shapiro E. The Powerful Placebo. Baltimore. The Johns Hopkins University Press. 1997

Shapiro AK. The placebo: is it much ado about nothing? 12–36, in: ***H. Harrington*** (ed.). The Placebo Effect. Cambridge, Mass. Harvard University Press. 1997

Sharpe M., Carson A. "Unexplained" somatic symptoms, functional syndromes, and somatization: do we need a paradigm shift? *Annals of Internal Medicine*. 2001; **134:** 926–930

Shashy RG., Moore EJ. and Weaver A. Prevalence of the chronic sinusitis diagnosis in Olmsted County, Minnesota. *Archives of Otolaryngological Head and Neck Surgery*. 2004; **130:** 320–323

Shepherd M. Epidemiological perspective. Psychosomatic medicine. *International Journal of Epidemiology*. 1978; **7:** 201–205

Shepherd JP. Criminal deterrence as a public health strategy. *Lancet*. 2001; **358:**1717–1722

Sherrington CS. Man and his nature. London. Cambridge University Press. 1940

Shoemaker S. Physical Realization. Oxford University Press. 2007

Shorter E. Bedside Manners: The Troubled History of Doctors and Patients. New York. Simon & Schuster. 1985

Shorter E. From Paralysis to fatigue. A History of Psychosomatic Illnesses in the Modern Era. New York. The Free Press. 1992

Sidgwick H. The Methods of Ethics. Indianapolis. Hackett Publishing Company. 1981

Siegler M and Osmond H. Models of Madness, Models of Medicine. New York. MacMillan Publishing Co. 1974

Silverman J. Shamans and Acute Schizophrenia. *American Anthropologist*. 1967; **69:** 21–31

Simon M. Validity and Reliability in Qualitative Research. *Curationis*; **16**: 35–38, 2011

Simon GE., Gater R., Kisely S & Piccinelli M. Somatic symptoms of distress: an international primary care survey. *Psychosomatic Medicine*. 1996; **58:** 481–488

Simon GE., Von Korff M., Piccinelli M., Fullerton C. and Ormel J. An international study of the relation between somatic symptoms and depression. *The New England Journal of medicine*. 1999; **341:** 1329–1335

Simopoulos AP. Characteristics of obesity: an overview. in: ***RJ. Wurtman and JJ. Wurtma*n** (eds.). Human Obesity. *Annals of the New York Academy of Sciences*. 1987; **499**: 4–13

Sims ACP. Symptoms in the Mind. Philadelphia. Saunders. 2003

Sinnott-Armstrong W., Fogelin R. Understanding Arguments. An introduction to informal logic. Boston. Cengage Learning. 2015

Skodol AE. Longitudinal course and outcome of personality disorders. *Psychiatric Clinics of North America*. 2008; **31:** 495–503

Skrabanek P. The emptiness of the black box. *Epidemiology*, 1994; **5:** 553–555

Skyrms B. Conditional Chance, 161–178, in; ***Fetzer JH.*** (ed.). Probability and Causality. Dordrecht. Reidel Publishing Company. 1988

Slavney PR, McHugh PR. Psychiatric Polarities. Methodology & Practice. The Johns Hopkins University Press. 1987

Sloane PD., Coeytaux RR., Beck RS. and Dallara J. Dizziness: state of the science. *Annals of Internal Medicine*. 2001; **134:** 823–832

Smart JJC. Philosophy and Scientific Realism. London. Routledge & Kegan Paul. 1963

Smart B. Concepts and Causes in the Philosophy of Disease. London. Palgrave Macmillan. 2016

Smith, H. Plato and Clementine. *The Bulletin of the New York Academy of Medicine*. 1947; **23:** 352–377

Smith R. In search of "non-disease". *British Medical Journal*. 2002; **324:** 883–885

Smith PA. The tantalizing links between gut microbes and the brain. *Nature*. 2015; **201:** 312–314

Sneath H. Mathematics and Classification from Adamson to the prison in psychiatry to social mathematics. University of Chicago Press. 1975

Sober E. Philosophy of Biology, Oxford. Oxford University Press. 1993

Sober E. Evolution, Population Thinking and Essentialism. In: ***Sober*** (ed.) Conceptual Issues in Evolutionary Biology. MIT Press. 1984

Sober E. From a Biological Point of View: Essays in Evolutionary Philosophy. Cambridge University Press. 1994

Sokal RR. and Sneath PHA. Principles of Numerical Taxonomy. San Francisco. W.H. Freeman. 1963

Solomon RC. Emotions and choice, 251–282, in: ***A. Oksenberg Rorty*** (ed.). Explaining Emotions. Berkeley. University of California Press. 1980

Solomon M. Making Medical Knowledge. Oxford University Press. 2015

Solomon M., Simon JR. & Kincaid H. (eds.), The Routledge Companion to the Philosophy of Medicine. London. Routledge. 2017

Somers A. Violence, television and the health of American Youth. *The New England Journal of Medicine.* 1977; **294:** 811–817

Sone S., Takashima S., Li F. et al. Mass screening for lung cancer with mobile spiral computed tomography. *Lancet.* 1998; **351:** 1242–1245

Sone S., Li F., Yang Z. et al. Results of three-year mass screening programme for lung cancer using mobile low-dose spiral computed tomography scanner. *British Journal of Cancer.* 2001; **84:** 25–32

Sontag S. Illness as a Metaphor. Hammondsworth. Penguin Hooks.1983

Spencer WG. Discussion on Blood-Letting. Blood-Letting, its Past and Present Use. *Proceedings of the Royal Society of Medicine.* 1927; **20:** 1547–1574

Spiegel D. & Harrington A. What is the placebo worth? *British Medical Journal.* 2008; **336:** 967

Spinoza B. On the Improvement of the Understanding. The Ethics. Correspondence. New York. Dover Publications. 1955

Spinoza B. The Ethics and Selected Letters by Baruch Spinoza (1982-08-02) Paperback, Hackett Publishing. 1982

Spinoza B. Ethics. London. Penguin classics. 1998

Spiro HM. Visceral Viewpoints. Pain and perfectionism—the physician and the "pain patient" *The New England Journal of Medicine.* 1976**; 294:** 829–830

Spiro H. Doctors, Patients and Placebos. Yale University Press. 1986

Spitzer M. Symptôme et critère en médecine et psychiatrie. *Actes du 11e Symposium international de Wittgenstein.* 1987

Srinivasan S., Bettella F., Hassani S., et al. Probing the association between early evolutionary markers and schizophrenia. *PLoS One.* 2017 Jan 12: 12 (1): e0169227

Stanford Encyclopedia of Philosophy. *Natural Kinds.* 2018

Starcevic VS. Hypochondriasis and Health Anxiety. *The British Journal of Psychiatry.* 2013**; 202:** 7–8

Stearns SC. (ed.). Evolution in Health and Disease. Oxford. Oxford University Press. 1999

Stegenga J. Medical Nihlism. Oxford University Press. 2018a

Stegenga J. Care ad Cure. University of Chicago Press. 2018b

Stegenga J., Kennedy AG., **Tekin S., Jukola S., Bluhm R.** New Directions in Philosophy of Medicine, 343–367, in: ***JA. Marcum*** (ed.) Bloomsbury Companion to contemporary philosophy of medicine. London. Bloomsbury. 2014

Stehbens WE. The concept of cause in disease. *Journal of Chronic Diseases.* 1985; **11:** 947–950

Stehbens WE. Causality in medical science with paticular reference to heart disease and atherosclerosis. *Perspectives in Biology and Medicine.* 1992; **36:** 97–117

Steinbrecher N., Koerber S., Frieser D. and Hiller W. The Prevalence of Medically Unexplained Symptoms in Primary Care. *Psychosomatics.* 2011; **52:** 263–271

Steinbrook R. The cost of admission–tiered copayments for hospital use. *New England Journal of Medicine.* 2004**; 350:** 2539–2542

Stempsey WE. Disease and diagnosis. Value-dependent realism. Dordrecht. Kluwer Academic Publishers. 1999

Stempsey WE. Philosophy of Medicine is what Philosophers do. *Perspectives in Biology and Medicine.* **51:** 379–391, 2008

Stern A. The Search for Meaning. Memphis. Memphis state University Press. 1971

Stewart M., Brown JB., Weston WW. et al. **Patient-**Centered Medicine. Transforming the Clinical Method. Thousand Oaks, California. Sage Publications. 1995

Stoessl A and de la Fuente-Fernandez R. Willing oneself on placebo effective on its own right. *Lancet.* 2004**; 364:** 227–228

Stone JH. Incidentalomas–Clinical correlation and translational science required. *New England Journal of medicine.* 2006; **354:** 2748–2749

Stone MH. The Anatomy of Evil. Prometheus Books. 2009

Stone MH. Understanding the Emergence of Modern Violent Crime. Prometheus Books. 2019

Strawson PF. Persons, in: ***H. Feigl, M. Scriven and G. Maxwell*** (eds) Minnesota Studies in the Philosophy of Sciences, vol. II; 1958a.

Strawson PF. On justifying Induction. *Philosophical Studies.* 1958b; **9:** 20–21

Strawson PF. Individuals. An Essay in Descriptive Metaphysics. London. Methuen. 1959

Strawson PF. Truth, *Proceedings of the Aristotelian Society*, Supp. Vol. XXIV, 1950; 129–156, reprinted in: ***Strawson PF.*** Logico-Linguistic Papers. London. Methuen. 1971

Strawson PF. Subject and Predicate in Logic and Grammar. London. Methuen. 1973

Strawson PF. Freedom and Resentment and other Essays. London. Methuen. 1974

Strawson PF. "Causation and explanation", 115–135, in: ***B. Vermazen, and M. Hintikka*** (eds.). Essays on Davidson: Actions and Events. Oxford. Clarendon Press. 1985

Strik W. Psychiatric Neurophysiology, 143–178, in: ***E. Henn, N. Sartorius, H. Helmchen, H. Lauter*** (eds.). 2001

Stroll A. Why On Certainty Matters, in: ***D. Moyal-Sharrock and WH. Brenner*** (eds.). Readings of Wittgenstein's On Certainty. New York. Palgrave. 2007

Sturmberg JP. Knowledge Translation and Health Care—Towards Understanding its True Complexity. *International Journal of Health Policy Management.* 2018; **7:** 455–458

Sulmazy DP. Diseases are not natural kinds. *Theoretical Biology.* 2005; **26:** 487–513

Summerfield D. Depression: epidemic or pseudo-epidemic? *Journal of the Royal Society of Medicine.* 2006; **99:** 161–162

Suppes P. Probabilistic Theory of Causality. Amsterdam. North Holland. 1970

Susser M. Causal Thinking in the Health Sciences Oxford. Oxford University Press. 1973

Swales JD. (ed.) Platt versus Pickering. An episode in recent medical history. London. The Keynes Press. British Medical Association. 1985

Szasz T. Malingering: "Diagnosis" or social condemnation. *Archives of Neurology and Psychiatry*; **76**: 432–433, 1956

Szasz T. The Myth of Mental Illness. New York. Dell Publishing. 1961

Szasz T. Law, Liberty and Psychiatry. London. Routledge & Kegan Paul. 1974

Szasz T. Pain and Pleasure. A Study of Bodily Feelings. Syracuse University Press. 1975

Szasz T. The Untamed Tongue. A Dissenting Dictionary. La Salle, Ill. Open Court. 1990

Szasz T. and Hollender MH. Contribution to the philosophy of medicine. *A.M.A. Archives of Internal Medicine.* 1945; **97:** 585–592

Tabb K., Schaffner KF. Causal pathways, random walk, and tortuous paths: Moving from the descriptive to the etiological in psychiatry, 342–360, in: ***KS. Kendler and J. Parnas*** (eds.) Philosophical issues in psychiatry IV. Oxford University Press. 2017

Tallis R. Terrors of the body. Looking for the lost meaning of pain. *Times Literary Supplement*, May 1, 1992, 3–4

Tallis R. The Enemies of Hope. A Critique of Contemporary Pessimism. New York. St. Martin's Press. 1997

Tallis R. Why the mind is not a computer. Exeter. Imprint Academic. 2004a

Tallis R. Hippocratic Oaths. London. Atlantic Books. 2004b

Tallis R. The Knowing Animal. A Philosophical Inquiry into Knowledge and Truth. Edinburgh. Edinburgh University Press. 2005

Tallis R. Reflections on epilepsy. *Philosophy Now*. 2009; **75:** 46–47

Tallis R. Aping Mankind. Durham. Acumen. 2011

Tangen CM., Goodman PJ. et al. Biases in Recommendations for and Acceptance of Prostate Biopsy Significantly Affect Assessment of Prostate Cancer Risk Factors: Results from Large Randomized Clinical Trials. *Journal of Clinical Oncology*. 2016**; 34:** 4338–3447

Tas J. Psychical disorders among inmates of concentration camps and repatriates. *Psychiatric Quarterly*. 1951; **25:** 683–685

Taylor GR. The Biological Time Bomb. New York. New American Library. 1969

Taylor C. Philosophical Papers. 2 volumes. Cambridge. Cambridge University Press. 1985

Taylor JS. Dying and the End of Life, 529–538, in**:** ***T. Schramme, S. Edwards,*** (eds.). 2017

Tek C, Gold J, Blaxton T et al. Visual Perceptual and Working Memory Impairments in Schizophrenia. *Arch Gen Psychiatry*. 2002; **59**: 146–153

Temkin O. Galenism. Rise and Fall of a Medical Philosophy. Ithaca. Cornell University Press. 1973

Teresa d'Avila. The Complete Works of St. Teresa of Avila, 3 vol. Washington, D.C. Institute of Carmelite Studies. 1976–1985

Thagard P. How Scientists Explain Diseases. Princeton. Princeton University Press. 1999

Thagard P. Pathways to biomedical discovery. *Philosophy of Science*. 2003**; 70:** 235–254

Thakkar KN., Diwadcar VA., Rolfs M. Oculomotor Prediction: A Window into the Psychotic Mind. *Trends in Cognitive Science*.2017; **21:** 344–356

Thomas SL. Nuisance or natural and healthy: should monthly menstruation be optional for women? *Lancet*. 2000; **355:** 922–824

Thomson G. Needs. London. Routledge & Kegan Paul. 1987

Thomson RP. Causality, theories in medicine, 25–44, in: ***PM. Illari, F. Russo and J. Williamson,*** (eds.) Causality in the Sciences. Oxford. Oxford University Press. 2011

Thomson RP., Upshur R. Philosophy of Medicine: – An Introduction. London. Routledge. 2017

Thompson PR., Upshur R. Philosophy of Medicine. An Introduction. Routledge. 2018

Thoreau H.D. Walden, or Life in the Woods

Tierney R., Melfi CA., Signa W. et al. Antidepressant use and use in naturalistic settings. *Drug Benefit Trend*. 2000; 12: 7BH-12BH

Tiles M. The normal and pathological: the concept of a scientific medicine. *British Journal of the Philosophy of Science*. 1993; **44:** 729–742

Timmermann C. 'Just give me the best quality of life questionnaire': the Karnofsky scale and the history of quality of life measurements in cancer trials. Chronic Illn. 2013 Sep; 9(3):179–90. doi: https://doi.org/10.1177/1742395312466903. Epub 2012 Dec 13. PMID: 23239756; PMCID: PMC3837542.

Timmermans S. Seven warrants for qualitative health sociology. *Social Science Medicine*; **77**: 1–8, 2012

Tinetti ME. Potential pitfalls of disease-specific guidelines for patients with multiple conditions. *New England Journal of Medicine*. 2004; **351:** 2870–2874

Topiwala A., Valkanova V. et al. Moderate alcohol consumption as risk factor for adverse brain outcomes and cognitive decline: longitudinal cohort study. *British Medical Journal*. 2017; 357: j2353

Tormey A. The Concept of Expression. Princeton University Press. 1971

Toulmin S. Foresight and Understanding. An Inquiry into the Aims of Science. London. Hutchinson of London. 1961

Toulmin S. Human Understanding. Volume I, General Introduction and Part I. Oxford. Oxford University Press. 1972

Toulmin S. On the nature of the physician's understanding. *The Journal of Medicine and Philosophy*. 1976a; **1:** 32–50

Toulmin S. Concepts of function and mechanism in medicine and medical science, in**:** ***HT. Engelhardt and SF. Spiker*** (eds.). Evaluation and Explanation in the Biomedical Sciences, 51–66, Dordrecht: Reidel. 1976b

Toulmin S., Rieke R. and Janik A. An introduction to reasoning. New York. Macmillan Publishing Co. 1979

Toulmin S. Agent and patient is psychiatry. *International Journal of Law and Psychiatry*. 1980; **3:** 267–278

Toulmin S. The Return of Cosmology. Postmodern Science and the Theology of Nature. Berkeley. University of California Press. 1982

Tranoy KE. Asymmetries in ethics. *An Interdisciplinary Journal of Philosophy*; **10**: 351–372, 1967

Trigg R. Pain and Emotion. Oxford. Clarendon Press. 1970

Trilling L. Sincerity and Authenticity. Cambridge. Harvard University Press. 1971

Trimble MR. Functional diseases. *British Medical Journal*. 1982; **285:** 1768–1770

Tripoli G., Quattrone D., Ferraro L., et al. Facial Emotion Recognition in Psychosis and Associations with Polygenic Risk for Schizophrenia: Findings from the Multi-Center EU-GEI Case–Control Study. *Schizophrenia Bulletin*. 2022; **48:** 1104–1114

Trompet S et al. Apolipoprotein and genotype, plasma cholesterol, and cancer: Mendelian randomization study. *American Journal of Epidemiology*; **170**: 1415–1421, 2009

Turk D. and Okifuji A. Assessment of patients' reporting of pain: an integrated perspective. *Lancet*. 1999**; 353:** 1784–1788

Turner EH., Matthews AM., Linardatos E., et al. Selective Publication of Antidepressant Trials and Its Influence on Apparent Efficacy. 2008; *New England Journal of Medicine*. **358:** 252–260

Turner EH., Rosenthal R. Efficacy of antidepressants. 2008; *British Medical Journal*. **336:** 516–517

Tyrer SP. Learned pain behaviour. *British Medical Journal*. 1986**; 292:** 1–2

Ungar G. Inflammation and its control. A biochemical approach. *Lancet*. 1952**; 263:** 742–746

University of California Associates. The freedom of the will, 594–615, in**: *H. Feigl and W. Sellars*** (eds.). Readings in Philosophical Analysis. New York. Appleton-Century Crofts, Inc. 1949

Urmson JO. Aristotle on pleasure, in: Aristotle. A Collection of Critical Essays, 323–333, in: ***JME. Moravcsik*** (ed.). New York. Anchor Book. Doubleday & Company Inc. 1967

Üstün TB. and Sartorius N. Mental Illness in General Health Care. An International Study. Chichester. John Wiley and Sons. 1995

US Preventive Services Task Force Recommendation Statement**.** Screening for Ovarian Cancer. *JAMA*. 2018; **319:** 588–594

Vacha J. German constitutional doctrine in the 1920, and pitfalls of the contemporary conception of normality in biology and medicine. *The Journal of Medicine and Philosophy*. 1985**; 10:** 339–367

Van den Berg HM. Editorial. A Cure for Hemophilia within Reach. *The New England Journal of Medicine*. 2017**; 377:** 2592–2593

Vandenbroucke JP., Broadbent A. & Pearce N. Causality and causal inference in epidemiology: the need for a pluralistic approach. *International Journal of Epidemiology*. 2016; **45:** 1776–1786

Van der Feltz-Cornelis C., Elfeddali I., Wernecke U., et al. A European research agenda for somatic symptom disorders, bodily distress disorders, and functional disorders: results of an estimate talk-estimate delphi expert study. *Frontiers in Psychiatry*. 2018; **9:** 151

Van Duppen Z., Sips R. Understanding the Blind Spots of Psychosis: A Wittgensteinian and First-Person Approach. *Psychopathology*. 2018; **51:** 276–284,

Van Essen DC. & Glasser MF. The Human Connectome Project: Progress and Prospects. Cerebrum: The Dana Forum on Brain Science. 2016; cer-10–16

van Fraassen BC. The Scientific Image. Oxford. Clarendon Press; 1980

Vendler Z. Effects, results and consequences in**: *RJ. Butler*** (ed.). Analytic Philosophy. 1st Series. Oxford. Oxford University Press; 1962

Vendler Z. 'Verbs and Times', in**: *Z. Vendler*,** Linguistics in Philosophy. London. Routledge & Kegan Paul. 1967

Venning GR. Validiy of anecdotal reports of suspected adverse drug reactions. *British Medical Journal*, **284**: 249–251, 1982

Verbrugge LM. and Ascione FJ. Exploring and iceberg. Common symptoms and how people care for them. *Medical Care.* 1987; **25:** 539–563

Vesey, G. Perception. London. MacMillan. 1971

Vickers G. What sets the goals of public health? *The New England Journal of Medicine.* 1958; **258:** 589–596

Vikramseth. An Equal Music. New York. Vintage International. 1999

Virchow R. Concerning standpoints in scientific medicine. 1847, 188–190 in: ***AL Caplan, HT. Engelhardt and JJ. McCartney*** (eds.). Concepts of Health and Disease. London. Addison-Wesley. 1981

Virchow R. Archiv für Pathologische Anatomie und Physiologie und für Klinische Medizin. 1847. Forgotten Books. 2018

Von Domarus E. The specific laws of logic in schizophrenia, in: ***Kasanin*** (ed.). Language and Thought in Schizophrenia. Berkley. University of California Press. 1944

von Wright GH. The Varieties of Goodness. London. Routledge and Kegan Paul. 1963a

von Wright GH. Norm and Action. A Logical Inquiry. London. Routledge and Kegan Paul. 1963b

von Wright GH. Explanation and Understanding. Ithaca. Cornell University Press. 1971

von Wright GH. Causality and Determinism. New York. Columbia University Press. 1974

von Wright GH. Practical Reason. Philosophical Papers volume 1. London. Basil Blackwell. 1983

Wachbroit R. Normality as a biological concept. Philosophy of Science. 1994; **61:** 579–591

Waddington CH. Tools for Thought. London. Jonathan Cape. 1977

Wadsworth M., *Butterfield W. and Blaney R.* Health and Sickness: The Choice of Treatment. London. Tavistock. 1971

Waismann F. Verifiability, in: Logic and Language, 1st series, *A Flew* (ed.), Oxford. Oxford University Press. 1953

Wakefield JC. The concept of mental disorder: diagnostic implications of the harmful dysfunction analysis. *World Psychiatry*. 2007; **6:** 149–156

Walshe FMR. The nature and dimension of nosography in modern medicine. *Lancet.* 1956; **268:** 1059–1063

Walton DN. Defining Death. Montreal. McGill-Queen's University Press. 1979

Wartofsky MW. Organs, organisms and disease, in: ***HT. Engelhardt and SF. Spiker,*** (eds.) Evaluation and Explanation in the Biomedical Sciences, 67–83, Dordrecht: Reidel. 1976

Waterlow S. Passage and Possibility. Oxford. Clarendon Press. 1982a

Waterlow S. Nature, change, and agency in Aristotle's Physics. Oxford. Clarendon Press. 1982b

Watkins ES. Technophilia and the pharmaceutical fix. *Lancet.* 2010; **376:** 1638–1639

Weatherall D. The conflict between the science and art of clinical practice in the next millenium. *Annals of the New York Academy of Sciences.*1999; **882:** 240–246

Weber M. The Methodology of the Social Sciences. Glencoe, Ill.: Trade Cloth. 1949

Weiner H. and Fawzy IF. An Integrative Model of Health, Disease and Illness, 9–44, in: ***Cheren, S*** (ed.). Psychosomatic Medicine: Theory, Physiology and Practice, vol I. Madison, Conn. International Universities Press. 1989

Weingarten SR., Stone E., Green A. et al. A Study of Patient Satisfaction and Adherence to Preventive Care Practice Guidelines. *The American Journal of Medicine.* 1995; **99:** 590–596

Weinstein MC. and FHV. et al. Clinical Decision Analysis. Philadelphia. Saunders Company. 1980

Welch HG., Prorok PC., O'Malley AJ., Kramer BS. Breast-Cancer Tumor Size, Overdiagnosis, and Mammography Screening Effectiveness. *The New England Journal of Medicine.* 2016; **375:** 1438–1447

Welch HG., Bradley OW. Scrutiny-Dependent Cancer and Self-fulfilling Risk Factors. *Annals of Internal Medicine.* 2018; **168:** 143–144

Welch HG. Less Medicine Moore Health. Boston. Beacon Press. 2015

Welton D. Body and Flesh: a Philosophical Reader. Wiley-Blackwell. 2018

Wennberg JE. Tracking Medicine: A Researcher's Quest to Understand Health Care. Oxford. Oxford University Press. 2010

Wessely S. Responding to mass psychogenic illness. *New England Journal of Medicine*. 2000; **342:** 129–130

Wessely S. Chronic fatigue syndrome: symptom and syndrome. *Annals of Internal Medicine*. 2001; **134:** 838–843

Whelton PK., Carey RM. et al. Guidelines for the Prevention, Detection, Evaluation, and Management of High Blood Pressure in Adults: A Report of the American College of Cardiology.

Whelton PS., McEvoy JW., Shaw L. et al. Association of Normal Systolic Blood Pressure Level with Cardiovascular Disease in the Absence of Risk Factors. *JAMA Cardiology.* 2020: **5:** 1011–1018

Whitbeck C. Health Sciences. Oxford University Press. 1973

Whitbeck C. The relevance of philosophy of medicine for the philosophy of science. *PSA*. 1977a; **2:** 123–135

Whitbeck C. Causation in medicine: the disease entity model. *Philosophy of Science.* 1977b; **44:** 619–637

Whitbeck C. Four basic concepts of medical science. *PSA*. 1978; 210–221, 1978

Whitbeck C. What is a diagnosis. *Metamedicine,* **2:** 319–29, 1981a

Whitbeck C. A theory of health, 611–626, in: ***AL. Caplan, HT. Engelhardt and JJ. McCartney,*** (eds.). Concepts of Health and Disease. London. Addison-Wesley. 1981b

Whitbeck C. Ethics as design. Doing justice to moral problems. Hastings Center Report. May-June; 1996

White AR. Modal Thinking. Oxford. Basil Blackwell. 1975

White KL., Williams F. and Greenberg BG. The ecology of medical care. *New England Journal of Medicine*. 1961; **265:** 885–892

White M, Epston D. Narrative means to therapeutic ends. New York. Norton. 1990

White PD, Rickards H, Zeman AZJ. Time to end the distinction between mental and neurological disorders. *British Medical Journal*. 2012; **344:** e3454

Whitbeck C. A Theory of Health, in: AL Caplan, HT Engelhardt and JJ McCartney (ed) Concepts of Health and Disease. London. Addison Wesley, 1981

Whitehouse PJ., Juengst ET., Mehlman M. et al. From "enhancing cognition in the intellectually intact", in: ***AL. Caplan, JJ. McCartney, DA. Sisti*** (eds.). Health, Disease, and Illness. Concepts of Medicine. Washington. Georgetown University Press. 2004

Whittemore AS. Breast Cancer in Marin County. *Breast Cancer Research*. 2003; **5:** 232–234

Wiener N. The concept of homeostasis. *Transactions & Studies of the College of Physicians of Philadelphia*. 1953; **20:** 87–93

Wigal SB. Efficacy and safety limitations of attention-deficit hyperactivity disorder pharmacotherapy in children and adults. *CNS Drugs*. 2009; Supplement **1**: 21

Wiggins D. Needs, Values, Truth. Oxford. Basil Blackwell. 1987

Wiggins D. What is the force of the claim that one needs something? 33–55, in*:* ***G. Brock.*** (ed.). Necessary Goods. Our Responsibilities to Meet Other's Needs. Lanham, Md. Rowman & Littlefield Publishers Inc. 1998

Wiist H., Barker K., Arya N. The Role of Public Health in the Prevention of War: Rationale and Competencies. *American Journal of Public Health*. 2014**; 104:** 34–36

Wikipedia: Mental illness in Ancient Rome

Wilkerson TE Natural kinds. *Philosophical Books*. 1998; **39:** 225–233

Wilkes K. Real People. Personal Identity without Thought Experiments. Oxford. Oxford University Press. 1988

Williams RJ. Biochemical Individuality. John Wiley & Sons. 1956

Williams B. Ethics and the Limit of Philosophy. Harvard University Press. 1985

Williams B. Truth and Truthfulness. An Essay on Genealogy. Princeton. Princeton University Press. 2002

Williams GC. Adaptation and Natural Selection. A Critique of some Current Evolutionary Thought. Princeton. Princeton University Press. 1966

Williams RJ. Biochemical Individuality. New York. John Wiley & Sons. 1956

Williamson JD., Supiano MA., Applegate WB. et al. SPRINT Research Group: intensive vs standard blood pressure control and cardiovascular disease outcomes in adults aged >75 years: a randomized clinical trial. *JAMA*. 2016; **315:** 2673–2682

Wilson RN. The Sociology of Health: An Introduction. New York. Random House. 1970

Wilson RA. Boundaries of the Mind. The Individual and the Fragile Sciences. Cambridge. Cambridge University Press. 2004

Wing JK., Birley JTL., Cooper JE., Graham P., and Issacs AD. "A procedure for measuring and classifying Present Psychiatric State" *British Journal of Psychiatry*. 1967; **113:** 499–515.

Wing JK., Cooper JE., Sartorius N. Psychiatric diagnosis in New York and London. A comparative study of mental admissions. Maudsley Monograph. 20. 1974

Wing S et al. Whose Epidemiology. Whose Health? *Journal of Health Sciences;* **28**: 241–252, 1998

Wing JK., Sartorius N., Üstün TB. Diagnosis and clinical measurement in psychiatry. Cambridge. Cambridge University Press. 2017

Wirtz PH., von Känel R. Psychological stress, inflammation, and coronary heart disease. *Current Cardiology Reports*. 2017**; 19:** 111–115

Wise J. Mediterranean diet in UK shows positive effects in study. *British Medical Journal*. 2016**; 354**: i5286

Wittgenstein L. Philosophical investigations. Blackwell. 1953

Wittgenstein L. Tractatus Logico-Philosophicus London. Routledge & Kegan Paul. 1961

Wittgenstein L. Philosophical Investigations, by Pitcher G MacMillan. 1968

Wittgenstein L. Philosophical Investigations. Oxford. Wily-Blackwell. 2009

Wittgenstein L. Wittgenstein Notebooks 1914–1916, GH. Von Wright & GEM. Anscombe (eds.). Oxford University Press. 1961

Wittgenstein L. The Blue and Brown Book: Preliminary Studies for the Philosophical Investigations. 2nd revised edition. Blackwell. 1969a

Wittgenstein L. Lectures and Conversations on Aesthetics, Psychology and Religious Belief. Berkeley. University of California Press. 1969b

Wittgenstein L. Philosophische Grammatik. Suhrkamp Verlag. 2009 Wittgenstein L. Zettel. Oxford. Blackwell. 1981

Wittgenstein L. On Certainty. Oxford. Blackwell.1974

Wittgenstein L. Remarks on the Philosophy of Psychology, vol. 1. Oxford. Blackwell. 1980a, vol. 2 Blackwell, 1980

Wittgenstein L. Culture and Values. Oxford. Blackwell. 1980b Wittgenstein L. Philosophical Investigations. London. Wiley Blackwell. 2009

Wolbring G, Diep L, Yumakulov S, Ball N, Leopatra V; Yergens D. Emerging therapeutic enhancement enabling health technologies and their discourses: what is discussed within the health domain? *Healthcare* (Basel). 2013**; 1**: 20–52

Wolff HG. and Hardy JD. On the nature of Pain. *Physiological Review*. 1947; **27:** 167–199

Woloshin S., Schwartz LM. Giving Legs to Restless Legs: A Case Study of How the Media Helps Make People Sick. 2006; *PLoS Med* 3: e191

Wolpe PR. Treatment, enhancement, and the ethics of neurotherapeutics, in: ***Caplan A., McCartney JJ. and Sisti DA***. (eds.). Health Disease and Illness. Washington. Georgetown University Press. 2004

Wood MM. and Elwood PC. Symptoms of iron deficiency anaemia. A community survey. *British Journal of Preventive and Social Medicine*. 1966; **20:** 117–121

Woodfield A. Teleology. Cambridge. Cambridge University Press. 1976

Woodger JH. Biological Principles A Critical Study. London. Routledge & Kegan Paul. 1929

Woodger JH. Biological Principles: A Critical Study. Routledge and Kegan Paul. 1967

Woods M. Conditionals. Oxford. Clarendon Press. 1997

Wooton D. Bad Medicine. Doctors Doing Harm Since Hippocrates. Oxford. Oxford University Press. 2006

Wooton EA. Guide to the medical profession. A Comprehensive Manual Conveying the Means Of Entering The Medical Profession In The Chief Countries Of The World (1883). Kessinger Publishing. 2008

Wouters A. Four notions of biological function. *Studies in History of Biology and Biomedical Sciences*. 2002; **34:** 633–668

World Health Organisation. The international pilot study of schizophrenia. Geneva. World Health Organization. 1973

World Health Organisation. Constitution, 83–84, in: ***AJ. Caplan, HT. Engelhardt and JJ. McCartney*** (eds.). Concepts of Health and Disease. Reading, Massachusetts, Addison-Wesley. 1981

Worrall J. What evidence in evidence-based medicine? *Philosophy of Science.* 2002; **69:** S316-S330

Wright L. Functions. *Philosophical Review.* 1973; **82:**139–168

Wulff HR. and Lennard-Jones JE. Rational Diagnosis and Treatment. An Introduction to Clinical Decision-Making. Oxford. Blackwell. 1981

Wyss D. Depth Psychology. New York. Norton. 1966

Zheng P., Zeng B., Zhou C. et al. Gut microbiome remodeling induces depressive-like behaviors through pathways mediated by the host's metabolism. *Molecular Psychiatry.* 2016; 1–11

Zimmerman DW. Material People, in: ***MJ. Loux and DW. Zimmerman*** (eds.). The Oxford Handbook of Metaphysics. Oxford. Oxford University Press. 2003

Zola IK. Culture and symptoms—An analysis of patients' presenting complaints. American Sociological Review. 1966; **31**: 615–630

Zola IK. Pathways to the doctor: From person to patient. *Social Science Medicine;* **7**: 677–689, 1972

Zorumski C. & Rubin E. Psychiatry and Clinical Neuroscience: A Primer. Oxford University Press. 2014

Zuger A. Dissatisfaction with medical practice. *New England Journal of Medicine*. 2004; **350:** 69–75

The manufacturer's authorised representative in the EU is Springer
Nature Customer Service Centre GmbH, Europaplatz 3, 69115 Heidelberg,
Germany. If you have any concerns regarding our products, please
contact ProductSafety@springernature.com

Printed and bound by CPI Group (UK) Ltd, Croydon, CR0 4YY
21/09/2026
02219962-0017